AF899026

Preface

The farmer folks around the globe use N sources for crop fertilization; they either use synthetic chemical fertilizers or shift to organic farming, which is an awesomesauce accomplishment of the scientific fraternity. The nitrogen-fixing bacteria demonstrates to function as an alternative to meet the requirement of nutrients for the benefit of both plants and soil.

Biological nitrogen fixation (BNF) is synonymous with sustainability. Long-term sustainability relies on the contribution of naturally occurring beneficial microorganisms inhabiting beneath the soil. Among them, various types of associative interactions occur between diazotrophs and their host plants. In all the associations, the benefit of the interaction is the fixed nitrogen provided by the colonization of nitrogen-fixing bacteria.

Rhizobia are known for fixing nitrogen for more than a century. Their association with members of the family Fabaceae is well established. This family belongs to 750 genera and about 1800–1900 plant species, but their association with nitrogen-fixing bacteria observed only with 15% of the total genera. Taking cognizance of this group, bacterial interaction with non-legumes is very limited. Although, association of symbiotic, associative, and free-living N fixing bacteria including few archaeobacteria with limited herbs, shrubs, and trees have been discovered but yet to get a place on a suitable platform.

This book contains four parts:

1. Biological Nitrogen Fixation: Trends and Prospects
2. Plant Growth Promotion: Exploring Benefits
3. Application to Sustainable Agriculture
4. Future and Significance

The book contains 17 chapters including introduction and conclusions, authored by leading subject experts and researchers in their field of interest. The descriptive and illustrated literature in this book offers invaluable and updated resource material. Further, the whole content provides an authoritative overview for individuals interested in BNF in non-legume research. Thus, it will, therefore, be of immense interest to agronomists, microbiologists, ecologists, plant pathologists, molecular biologists, environmentalists, policymakers, conservationists, and NGOs who intend to use naturally beneficial microbes for sustainable crop production and protection.

We are deeply indebted to our internationally renowned contributors for their authoritative and cutting-edge scientific information to make this book a reality. The book presented under the series "Microorganisms for Sustainability" is entirely dedicated to various nitrogen-fixing microorganisms able to minimize nitrogen requirement in the soil judiciously and for boosting the sustainable growth of non-legumes.

We desire to pay our sincere gratitude to all the contributors, who lent their cooperation and patience in the completion of this book. Our research team members, who generously assisted in the compilation and completion of this project, are gratefully acknowledged. We extend our sincere thanks to Miss. Aakansha Tyagi and her colleagues for their mutual cooperation of scientific benefits.

Haridwar, Uttarakhand, India — Dinesh Kumar Maheshwari
Dehradun, Uttarakhand, India — Rajendra Dobhal
Palwal, Haryana, India — Shrivardhan Dheeman

Contents

About the Series Editor and About the Editors and Contributors

About the Series Editor

Naveen Kumar Arora received his PhD in Microbiology, is a Fellow of International Society of Environmental Botanists (FISEB), and is Professor and Head of the Department of Environmental Science at Babasaheb Bhimrao Ambedkar University (A Central University), Lucknow, Uttar Pradesh, India. He is a renowned researcher in the field of environmental microbiology and biotechnology. His specific area of research is plant–microbe interactions, particularly plant growth promoting rhizobacteria. He has more than 75 research articles published in premium international journals and several articles published in magazines and dailies. He is an editor of 25 books, published by Springer. He is a member of several national and international societies, Secretary-General of Society for Environmental Sustainability, serves on the editorial board of 4 journals, and a reviewer of several international journals. He is also the editor in chief of the journal *Environmental Sustainability* published by Springer Nature. He has delivered lectures at conferences and seminars around the globe. He has a long-standing interest in teaching at the PG level and is involved in taking courses in bacteriology, microbial physiology, environmental microbiology, agriculture microbiology, and industrial microbiology. He has been advisor to 134 postgraduate and 11 doctoral students. He has been awarded for excellence in research by several societies and national and international bodies/organizations. Although an academician and researcher by profession, he has a huge obsession for the wildlife and its conservation and has authored a book, *Splendid Wilds*. He is the President of Society for Conservation of Wildlife and has a dedicated website www.naveenarora.co.in for the cause of wildlife and environment conservation.

About the Editors

Dinesh Kumar Maheshwari, former Vice-Chancellor (Gurukul Kangri Vishwavidyalaya, Haridwar), a renowned Botanist and Microbiologist, holds MHRD-UGC-BSR Eminent Faculty Fellowship after his superannuation. Prof. Maheshwari worked in Biology Research Centre (Szeged), Hungry, in the year 1983–1984. He was Visiting Professor, Science University of Tokyo, Noda (Japan), University of Ulm (Germany), University of Gent (Belgium), University of Barcelona (Spain), University of Tokyo (Japan), and Daegu University (Korea). He has received several recognitions and has been awarded "Prof. Y.S. Murty Medal" and the most prestigious "Birbal Sahni Award" of Indian Botanical Society in the year 1992 and 2018, respectively, for his outstanding contribution and bestowed with the "Platinum Jubilee Lecture award" of the Indian Science Congress Association (2012) and unanimously nominated President—Plant Sciences, Indian Science Congress—2015–2016. Most recently, he was conferred with Lifetime Achievement Award on Feb. 11, 2020, at BBA University (A Central University), Lucknow, and was awarded "Vigyan Vibhuti Award" by the "Govt. of Uttarakhand" at UCOST, Dehradun, in 2020.

Rajendra Dobhal is currently working as Director General, Uttarakhand State Council for Science and Technology, Govt. of Uttarakhand. He had a distinct privilege to occupy positions like Chairman and Managing Director, National Research Development Corporation (NRDC), Govt. of India; Director, Uttarakhand Science Education and Research Centre (USERC); Director, Uttarakhand Space Application Centre (USAC); Senior Scientific Advisor/Project Director, Uttarakhand State Biotechnology Board; Scientist/Tech. Advisor to DG and Science Advisor, Govt. of Madhya Pradesh, Bhopal, in the past. He is also Chairman of the Sustainable Development Forum, Uttaranchal. Dr. Dobhal is a Fellow of the National Academy of Sciences (FNASc), India. Dr Dobhal is a known Intellectual Property (IP) professional trained from the Department of Science and Technology, GoI; Washington University, USA; International Law Development Institute (ILDI), Manilla, Philippines; and National Law University, Bangalore. To his credit, he has received major awards such as Royal Australian Chemical Institute award (2007), International Commonwealth Youth Silver Award (2008), and Pride of Uttarakhand (2016).

Shrivardhan Dheeman is working as an Associate Professor in the Department of Microbiology, School of Allied Health Sciences, MVN University, Palwal, Haryana, India. Dr. Dheeman has research interests in Plant-Microbe Interaction and Microbial Ecology. He is involved in teaching Microbiology and Biotechnology to different disciplines in undergraduate and postgraduate classes. He is an active member of several international reputed scientific bodies, including the Indian Science Congress Association, Asian PGPR Society for Sustainable Agriculture, and Indian Botanical Society. Dr. Dheeman was nominated for "Young Scientist" of Indian Science Congress in the year 2016 for his outstanding contribution to Microbial Diversity and Plant-Microbe Interactions. He has more than 30 publications in leading peer-reviewed & prestigious journals and books so far. His debut book entitled *Field Crops: Sustainable Management by PGPR* and following *Endophytes: Mineral Nutrient Management Volume 3* were published by Springer, Gewerbestrasse, Switzerland.

Contributors

Shreef Abdelfattah Abdelshafi Biotechnology and Its Application Program, Faculty of Science, Mansoura University, Mansoura, Egypt

Walaa R. Abou Zeid Cotton and Crops Acarology Department, Plant Protection Research Institute, Agriculture Research Center, Dokki, Giza, Egypt

Mona S. Agha Botany Department, Faculty of Science, Mansoura University, Mansoura, Egypt

Kannepalli Annapurna Division of Microbiology, ICAR-Indian Agricultural Research Institute, New Delhi, India

Naveen Kumar Arora Department of Environmental Science, School of Earth and Environmental Sciences, Babasaheb Bhimrao Ambedkar University, Lucknow, Uttar Pradesh, India

Asadullah The Peace Group of Schools and Colleges, Charsadda, KP, Pakistan

Asghari Bano Department of Biosciences, University of Wah, Wah Cantt, Pakistan

Aneta Buntić Institute of Soil Science, Belgrade, Serbia

Angelbert D. Cortes Department of Biological Sciences, College of Arts and Sciences, Cavite State University, Cavite, Philippines

Edson Cabral da Silva Goiano Federal Institute (IF Goiano), Rio Verde, Goiás, Brazil

Carlos Eduardo da Silva Oliveira São Paulo State University (UNESP), Ilha Solteira, São Paulo, Brazil

Nandita Das Department of Microbiology, Assam University, Silchar, India

Sampa Das Division of Plant Biology, Bose Institute, Kolkata, West Bengal, India

Sourav Debnath Department of Microbiology, Assam University, Silchar, India

Dušica Delić Institute of Soil Science, Belgrade, Serbia

Shrivardhan Dheeman Department of Microbiology, School of Allied Health Sciences, MVN University, Palwal, Haryana, India

Vagner do Nascimento São Paulo State University (UNESP), Dracena, São Paulo, Brazil

Rajendra Dobhal Uttrakhand Council for Science and Technology, Dehradun, Uttarakhand, India

Engy Atef Abou El-ftouh Biotechnology and Its Application Program, Faculty of Science, Mansoura University, Mansoura, Egypt

Hassan Etesami Soil Science Department, College of Agriculture and Natural Resources, University of Tehran, Tehran, Iran

Marcelo Carvalho Minhoto Teixeira Filho São Paulo State University (UNESP), Ilha Solteira, São Paulo, Brazil

Leandro Alves Freitas São Paulo State University (UNESP), Ilha Solteira, São Paulo, Brazil

Amir Hamzah Ahmad Ghazali School of Biological Sciences, Universiti Sains Malaysia, Gelugor, Pulau Pinang, Malaysia

Koj Haniya Division of Microbiology, ICAR-Indian Agricultural Research Institute, New Delhi, India

Arshad Jalal São Paulo State University (UNESP), Ilha Solteira, São Paulo, Brazil

Byoung Ryong Jeong Division of Applied Life Science (BK21 Four), Department of Horticulture, Graduate School, Gyeongsang National University, Jinju, Republic of Korea
Institute of Agriculture and Life Science, Gyeongsang National University, Jinju, Republic of Korea

Prashant Katiyar Department of Botany and Microbiology, Gurukula Kangri (Deemed to be University), Haridwar, Uttarakhand, India

Magdalena Knežević Institute of Soil Science, Belgrade, Serbia

Motohiko Kondo Graduate School of Bioagricultural Sciences, Nagoya University, Nagoya, Aichi, Japan

Sandeep Kumar Department of Botany and Microbiology, Gurukula Kangri (Deemed to be University), Haridwar, Uttarakhand, India

Upendra Kumar Crop Production Division, ICAR-National Rice Research Institute, Cuttack, Odisha, India

Dinesh Kumar Maheshwari Department of Botany and Microbiology, Gurukula Kangri (Deemed to be University), Haridwar, Uttarakhand, India

M. A. Baset Mia Department of Crop Botany, Bangabandhu Sheikh Mujibur Rahman Agricultural University, Gazipur, Bangladesh

Amr M. Mowafy Botany Department, Faculty of Science, Mansoura University, Mansoura, Egypt
Department of Biological Sciences, Faculty of Science, New Mansoura University, New Mansoura city, Egypt

Papri Nag Division of Plant Biology, Bose Institute, Kolkata, West Bengal, India

Shamdee Nahar-Cortes Graduate School, University of the Philippines Los Baños, College, Laguna, Philippines

Takanori Okamoto Graduate School of Bioagricultural Sciences, Nagoya University, Nagoya, Aichi, Japan

Piyush Pandey Department of Microbiology, Assam University, Silchar, India

Thilini A. Perera Department of Plant Sciences, Faculty of Science, University of Colombo, Colombo, Sri Lanka

Kedharnath Reddy Division of Microbiology, ICAR-Indian Agricultural Research Institute, New Delhi, India

Asmaa A. Sallam Botany Department, Faculty of Science, Mansoura University, Mansoura, Egypt

Mohammed Y. Sdiek Biotechnology and Its Application Program, Faculty of Science, Mansoura University, Mansoura, Egypt

Murugesan Senthilkumar Division of Basic Sciences, ICAR-Indian Institute of Pulse Research, Kanpur, Uttar Pradesh, India

Rina Shinjo Graduate School of Bioagricultural Sciences, Nagoya University, Nagoya, Aichi, Japan

Olivera Stajković-Srbinović Institute of Soil Science, Belgrade, Serbia

Karivaradharajan Swarnalakshmi Division of Microbiology, ICAR-Indian Agricultural Research Institute, New Delhi, India

Shamala Tirimanne Department of Plant Sciences, Faculty of Science, University of Colombo, Colombo, Sri Lanka

Swati Tyagi Division of Microbiology, ICAR-Indian Agricultural Research Institute, New Delhi, India

Ali Tan Kee Zuan Department of Land Management, Faculty of Agriculture, Universiti Putra Malaysia, Serdang, Selangor, Malaysia

Chapter 1
Biological Nitrogen Fixation in Nonlegumes: Introduction

Dinesh Kumar Maheshwari, Rajendra Dobhal, and Shrivardhan Dheeman

Abstract Sustainable agriculture (SA) is on prime importance in today's scenario. It is achievable via eco-safe application of nitrogen-fixing bacteria (NFB) biofertilizers, where these are applied on nonlegume crops and should not be limited to legume crop. NFB, including rhizobia or free-living rhizobacteria for the development of bioinoculants/biofertilizers/biopesticide, can be utilized for broad range of legume to nonlegumes crops, which contributing toward the sustainable development goal (SDG), "zero hunger". This introductory chapter provides an overview on the importance of its contents, and overall understand the role of nitrogen-fixing bacteria and their application in growth promotion of nonlegume crops, to achieve sustainable development.

Keywords Sustainable development · Biofertilizers · Rhizobia · Nitrogen fixation · Nonlegume

1.1 Introduction

The human race is on the edge of hunger due to the decline of the world's economy by the recent pandemic of COVID-19 that also caused food scarcity in many developing countries. It is hard to forecast a rise in hunger due to various bottlenecks, like climate change, an ever-growing population, a hike in food prices, etc. The excessive use of chemically produced fertilizers, pesticides, and herbicides is causing negative impacts on human health and agriculture. At this stage, using

---

D. K. Maheshwari (✉)
Department of Botany and Microbiology, Gurukula Kangri (Deemed to be University), Haridwar, Uttarakhand, India

R. Dobhal
Uttrakhand Council for Science and Technology, Dehradun, Uttarakhand, India

S. Dheeman
Department of Microbiology, School of Allied Health Sciences, MVN University, Palwal, Haryana, India

© The Author(s), under exclusive license to Springer Nature Singapore Pte Ltd. 2022
D. K. Maheshwari et al. (eds.), *Nitrogen Fixing Bacteria: Sustainable Growth of Non-legumes*, Microorganisms for Sustainability 36,
https://doi.org/10.1007/978-981-19-4906-7_1

biofertilizers and mitigation of food security as an eco-friendly alternative is an inevitable and precise way to attain scientific benefits for sustainable agriculture (SA).

Basically, in agriculture, plant life is cultivated in soil, which majorly demands nitrogen, phosphorus, and potassium as sole nutrients to drive their metabolic requirement. Other than these, soil organic matter (SOM) with some trace elements (e.g., Cu, Fe, Mn, Mg, etc.) is required. But, Nitrogen (N) requirement is often considered on the prime importance to the plants. Thereby, utilization of abundant fertilizer in soil, has snatched the soil fertility plus increased fertilizer dosage caused enormous financial burden in agriculture. Therefore, in the current scenario, eco-safe alternative biofertilizers or microbial inoculants are required to attain soil fertility back into course of sustainable agroecosystem. Nitrogen acts as one of the significant indicators for soil fertility, reflects the presence of good soil bacteria. As already known that nitrogen is present in gaseous form in the environment. Nitrogen-fixing bacteria (NFB) known for their ability of nitrogen fixation in legume crops (Soumare et al. 2020) and nonlegume crops (Behera et al. 2021) thereby used as biofertilizers to replace abundant input of chemical fertilizers, as an important approach for sustainable agriculture (Misra et al. 2020). Application of rhizobia to improve growth, yield, nutrient composition, and quality of nonlegume plants has been advocated due to direct and indirect plant growth-promoting activities (García-Fraile et al. 2012). The scope of using rhizobia (symbiont of legume) and free-living bacteria for nitrogen improvement in the soil has been increased with beneficial gears of plant growth promotion (Hayat et al. 2010). Biological nitrogen fixation in soil or rhizosphere has been reflected to increase crop productivity (Gaskins et al. 1985). In fact, rhizobia may associate with nonlegume via rhizosphere colonization and crack entry in xylem tissues exhibited ACC-deaminase activity, production of plant hormones, siderophore, HCN, and mineral (P and K) nutrient solubilization, which also supports growth and productivity of nonlegumes (Martínez-Viveros et al. 2010) other than forming true nodule in the roots and symbiotic fixation of nitrogen. The overall picture of rhizobia, free-living bacteria, and N-cycle can be understood with Fig. 1.1.

This sustains the quest of nonlegume crops with NFB, and judging their ecological roles (successful partnership among plant and bacteria, via production of plant root's exudates and molecular signals by bacteria necessary to engineer N-fixing association with nonlegume plants) in provisioning the benefits to nutrients transformation, soil organic matter mineralization, and carbon dynamics (Barrios 2007).

Application of biofertilizer is an alternative scheme to achieve environment-friendly sustainable crop production system (Seenivasagan and Babalola 2021). Cultivation and N-fertilization in rice is not limited to use rhizobia, because of several limitations, therefore, free-living rhizobacteria, as PGPR contribute significantly in order to achieve better productivity under field conditions (Yanni et al. 1997). Recent development of rhizobia-rice association in context to progress and challenges of developing suitable biofertilizers for rice cultivation has been addressed. Harnessing NFB for the development of bioinoculants/biofertilizers/biopesticide, applicable to broad range of legume to nonlegumes crops like cereals,

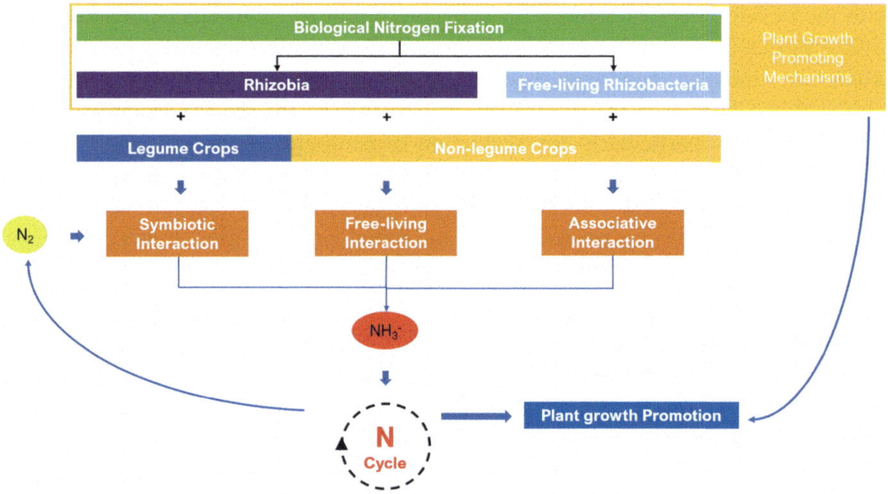

Fig. 1.1 Schematic relationship of rhizobia representing possible role of N cycling and plant growth promotion of nonlegumes. Differentiating this relationship with true symbiosis and free-living interaction of plants with other rhizobacteria

oil seed plants, vegetables, fruits, forages, and other important crops contribute in the sustainable development goal (SDG). On the other hand, a strategy to establish the C cycle coupled to N cycle in consideration to global warming is another route to achieve sustainable development goal.

1.2 Ecological Perspectives of Biological Nitrogen Fixation (BNF)

Ecology starts with its primary producers; and on Earth food chain starts with plants. An art or science of growing plant from the soil, requires nutrients, majorly nitrogen, that stands out as most important and more susceptible nutrient to plants as well as soil microbes. To build their proteins, and many components of life fixation of gaseous nitrogen by bacteria in specialized compartments, i.e., root nodules in legume plants have been studied vigorously. The benefits of these bacteria, in the form of nitrogen biofertilizers are inevitable. On the other hand, some free-living and associative bacteria, able to fix nitrogen biologically are in the concern to be utilized as biofertilizers in nonlegume crops. Behind BNF, biochemical genetics of symbiotic and asymbiotic nitrogen fixation has been reviewed in Chap. 2. It enunciates the potentials of symbiotic and free-living nitrogen-fixing bacteria in the transformation of green revolution to ever-green revolution. Application of rhizobia, as PGPR for nonlegume crops, and as a member of nitrogen fertilization has been advocated by many workers (Behera et al. 2021). As this association has received less attention but with the development of science on molecular (cellular communication) and ecological aspects has augmented its importance in nitrogen fertilization. Plant growth-promoting traits in beneficial bacteria felicitate application in nonlegume as reviewed in Chap. 3. It explores future application of rhizobia as biofertilizer for nonlegume crop, particularly for alleviation of ecological stress. Coupled to this, Chap. 4 is concerned to the biotechnological solutions in the form of certain bacteria and archaea for enhanced nitrogen fertilization and eco-safe crop production. This addresses challenges for production of microbial products and biotechnological approaches as solutions which may be implemented to improve N nutrition in nonlegume crops.

Nitrogen-fixing bacteria (NFB) as effective microorganisms (EM) have great contribution to the green revolution (Lynch 2007). Chapter 5 is opened with the past of NFB and their application for field crop, rice, and its productivity enhancement. Further, understanding of genetic engineering for transferring nitrogen-fixing genes in rice plants has been advocated; however, with due to some limitations of extremely complex process of BNF which regulated by absence of oxygen has been criticized, therefore, demanded to develop some newer technologies. It is an exclusive account to understand application of QTL regions for BNF in rice, as a result of advance molecular biology. Further, authors suggest more research to be carried out to re-discover rhizospheric colonization mechanisms in NFB at molecular level. The diverse genera of bacteria, archaea, etc. have also understood as potential agents for BNF (Raymond et al. 2004). Similarly, reviewed in following Chap. 6; BNF in nonlegume has been proved as an important approach for sustainable crop growth and productivity enhancement under agroecological practices.

1.3 Playbacks of Nitrogen-Fixing Bacteria (NFB)

Application of rhizobia to improve growth, yield, nutrient composition, and quality of nonlegume plants has been augmented (Santoyo et al. 2021). Direct and indirect PGP traits of bacteria have been documented by a majority of workers (Orozco-Mosqueda et al. 2021). Biocontrol mechanisms of NFB has increased their importance to promote growth and increase the yield of nonlegumes (Nosheen et al. 2021). All these benefits are summarized in Chap. 7 with current research advancement on rhizobia and nonlegume interaction with cereals, as a holistic approach has been covered in Chap. 8. The increase in productivity of cereals demonstrates as a central theme of this chapter, which explores the beneficial roles of diazotrophs in biological nitrogen fixation and plant growth promotion. With the advent of PGP mechanisms of rhizobia population, and research in support, have claimed the application of rhizobia for nonlegume crops (Behera et al. 2021).

Therefore, there is a scope of using rhizobia (symbiont of legume) free-living and associative bacteria for nitrogen elevation in the soil, well suit for plant growth promotion (Nosheen et al. 2021). Chapter 9 states, nitrogen fertilizers are essential for producing high crop yields and are used extensively by farmers, besides its abundant use has decreased the soil fertility (Rahman et al. 2021). Thus, pollution-free alternative to synthetic fertilizer in the form of diazotroph, those in play to fix atmospheric nitrogen and release in the form of NH_3 (Pankievicz et al. 2021). On the background of previous chapters, this chapter entrusts establishment of diazotroph as NFB, with a showcase of hurdles and success of creating microbial consortia as nitrogenous biofertilizers.

Exclusive benefits of nitrogen fixation in soil or rhizosphere often reflect in plant growth promotion and yield improvement (Zvinavashe et al. 2021). Analogous to this concept, production improvement of commercial crops such as Cacao and Coffee, using NFB has been reviewed in the Chap. 10. This highlights on NFB as eco-safe alternative over chemical fertilizers, as a successful outcome of diazotrophs with nonlegume plants. Recommendations to explore management of N-sources from the environment either via intercropping or inoculating diazotrophic rhizobacteria have been proposed.

Further, attaining answer to the quest of sustainability of nonlegume crop, and judging with the theme of the book, plant growth-promoting bacteria (PGPBs) and nitrogen-fixing bacteria (NFBs) have been studied for their ecological roles, imparting benefits to nutrients transformation, soil organic matter mineralization, and carbon dynamics (Prasad et al. 2021). In this context, Chap. 11 identifies avenue of development of microbial inoculants for various crops like cereals, oil seed plants, vegetables, fruits, forages, and other important crops. In continuation, Chap. 12 unravels the benefits of using rhizobia in cereal crops, imparts stress regulation via enzymatic ethylene regulation by ACC deaminase, production of plant hormones, siderophore for iron management, and mineral (P and K) nutrient management via solubilization, mobilization of other nutrients. Not forming true nodule in the roots, it exclusively enters in cereal's root through crack entry and colonizes in the intercellular spaces such as in the xylem tissues.

In an overview, rhizobia have been emerged not only a true symbiont of legume crops but also an associative bacterium for nonlegume crops where, in the nonhabitual niche, they perform like a contender and function as plant growth-promoting rhizobacteria (PGPR).

1.4 Biofertilizer: A Step Toward Sustainable Development Goals

As stated earlier, food production via sustainable agriculture (SA) is a fundamental concept for curbing food security. It directly connects to the sustainable development goals (SDG) as a "blueprint to achieve a better and more sustainable future for all." The outcome target of SDG, particularly sustainable food production systems and resilient agricultural practices, is achievable using biofertilizers, contributing to SA. Biofertilization toward SA is a driving force to counter Goal 2—"Zero Hunger" of the SDG. It involves promoting SA technologies in support to increase wheat, rice, and other nonlegume's crop productivity. To address the challenge of global food security, it is required to be headed toward sustainable approach of using biofertilizers.

In line with this aspect, Chap. 13 embodied research evidences of biological nitrogen fixation in nonlegumes and contribution to sustainable development goal (SDG). It attempts to understand the rational of using NFB, their ecological relationship with nonlegume, in-brief, besides major discussion focused on present challenges, future vision, and mission. Uninterruptedly, Chap. 14, explores the role of NFB, as potential producers of secondary metabolites modulate ecological behavior with nonlegume crops. It tries to identify mechanisms, involves in successful partnership among plant and bacteria, via production of plant root's exudates and molecular signals by bacteria necessary to engineer N-fixing association with nonlegume plants. A successful association between microbe and plants can increase nutritional ability of crop, as a result in improving the nutrient use efficiency, exclusively in the context to nitrogen use efficiency (Huang et al. 2022).

Application of biofertilizer is an alternative scheme to raise environment-friendly sustainable crop production system (Seenivasagan and Babalola 2021). Equally important as wheat, rice requires ample amount of nitrogen during cultivation and, therefore, application of NFB-based biofertilizer has gained prominence. Recent development of rhizobia-rice association in context to progress and challenges of developing suitable biofertilizers for rice cultivation has been addressed in Chap. 15.

N-fertilization in rice is not limited to use of rhizobia, because of several limitations; free-living rhizobacteria, as PGPR can contribute its production under field conditions, which is advocated in the Chap. 16. The past and present findings on the eco-physiological and agronomic aspects of free-living and endophytic N fixation in nonlegume crops with emphasis on rice have been reviewed. This is not limited to sustainable crop production of rice, besides embodied a clear commentary on the significant progress made on molecular-microbial aspects by development of meta-DNA/RNA analysis, indicating functioning N systems in the soil and plant. Uncovering metabolic aspects of NFB as microbial community, it identifies contributions to promote sustainable development. On the other hand, a strategy to establish the C cycle coupled to N cycle in consideration to global warming has been covered. Chapter 17 is concluding remarks on NFB and their role for sustainable growth of nonlegumes.

1.5 Conclusions

Future direction of development of biofertilizers to achieve sustainable agriculture and ever green revolutions is not limited to producing specialized biofertilizers for legume and nonlegume crops. With the recent science interventions of rhizobia-nonlegume interaction, free-living and associative rhizobacteria interaction with nonlegumes has created a notion to produce broad spectrum biological nitrogen-fixing biofertilizers for wide range of crops. This may emerge as an era of shaping future of human race as far as scarcity of food, safety, and security are concerned.

Acknowledgment DKM extends thanks to Uttarakhand Council for Science and Technology, Dehradun, India.

Conflict of Interest Author(s) declares no conflict of interest.

References

Barrios E (2007) Soil biota, ecosystem services and land productivity. Ecol Econ 64(2):269–285

Behera B, Das TK, Raj R, Ghosh S, Raza M, Sen S (2021) Microbial consortia for sustaining productivity of non-legume crops: prospects and challenges. Agric Res 10(1):1–4

García-Fraile P, Carro L, Robledo M, Ramírez-Bahena MH, Flores-Félix JD, Fernández MT, Mateos PF, Rivas R, Igual JM, Martínez-Molina E, Peix Á (2012) *Rhizobium* promotes non-legumes growth and quality in several production steps: towards a biofertilization of edible raw vegetables healthy for humans. PLoS One 7(5):e38122

Gaskins MH, Albrecht SL, Hubbell DH (1985) Rhizosphere bacteria and their use to increase plant productivity: a review. Agric Ecosys Environ 12(2):99–116

Hayat R, Ali S, Amara U, Khalid R, Ahmed I (2010) Soil beneficial bacteria and their role in plant growth promotion: a review. Ann Microbiol 60(4):579–598

Huang Z, Ruan S, Sun Y, Cheng X, Dai J, Gui P, Yu M, Zhong Z, Wu J (2022) Bacterial inoculants improved the growth and nitrogen use efficiency of *Pyrus betulifolia* under nitrogen-limited conditions by affecting the native soil bacterial communities. Appl Soil Ecol 170:104285

Lynch JP (2007) Roots of the second green revolution. Aus J Bot 55(5):493–512

Martínez-Viveros O, Jorquera MA, Crowley DE, Gajardo GM, Mora ML (2010) Mechanisms and practical considerations involved in plant growth promotion by rhizobacteria. J Soil Sci Plant Nutr 10(3):293–319

Misra M, Sachan A, Sachan SG (2020) Current aspects and applications of biofertilizers for sustainable agriculture. In: Yadav A et al (eds) Plant microbiomes for sustainable agriculture. Springer, Cham, pp 445–473

Nosheen S, Ajmal I, Song Y (2021) Microbes as biofertilizers, a potential approach for sustainable crop production. Sustainability 13(4):1868

Orozco-Mosqueda M, Flores A, Rojas-Sánchez B, Urtis-Flores CA, Morales-Cedeño LR, Valencia-Marin MF, Chávez-Avila S, Rojas-Solis D, Santoyo G (2021) Plant growth-promoting bacteria as bioinoculants: attributes and challenges for sustainable crop improvement. Agronomy 11(6): 1167

Pankievicz VC, do Amaral FP, Ané JM, Stacey G (2021) Diazotrophic bacteria and their mechanisms to interact and benefit cereals. Mol Plant-Microbe Interact 34(5):491–498

Prasad S, Malav LC, Choudhary J, Kannojiya S, Kundu M, Kumar S, Yadav AN (2021) Soil microbiomes for healthy nutrient recycling. In: Yadav AN et al (eds) Current trends in microbial biotechnology for sustainable agriculture. Springer, Singapore, pp 1–21

Rahman MH, Haque KS, Khan MZ (2021) A review on application of controlled released fertilizers influencing the sustainable agricultural production: a Cleaner production process. Environ Technol Innov 23:101697

Raymond J, Siefert JL, Staples CR, Blankenship RE (2004) The natural history of nitrogen fixation. Mol Biol Evol 21(3):541–554

Santoyo G, Guzmán-Guzmán P, Parra-Cota FI, Santos-Villalobos SD, Orozco-Mosqueda M, Glick BR (2021) Plant growth stimulation by microbial consortia. Agronomy 11(2):219

Seenivasagan R, Babalola OO (2021) Utilization of microbial consortia as biofertilizers and biopesticides for the production of feasible agricultural product. Biology 10(11):1111

Soumare A, Diedhiou AG, Thuita M, Hafidi M, Ouhdouch Y, Gopalakrishnan S, Kouisni L (2020) Exploiting biological nitrogen fixation: a route towards a sustainable agriculture. Plants 9(8): 1011

Yanni YG, Rizk RY, Corich V, Squartini A, Ninke K, Philip-Hollingsworth S, Orgambide G, Bruijn FD, Stoltzfus J, Buckley D, Schmidt TM (1997) Natural endophytic association between *Rhizobium leguminosarum* bv. trifolii and rice roots and assessment of its potential to promote rice growth. In: Ladha JK et al (eds) Opportunities for biological nitrogen fixation in rice and other non-legumes. Springer, Dordrecht, pp 99–114

Zvinavashe AT, Mardad I, Mhada M, Kouisni L, Marelli B (2021) Engineering the plant microenvironment to facilitate plant-growth-promoting microbe association. J Agric Food Chem 69(45): 13270–13285

Part I
Biological Nitrogen Fixation: Trends and Prospects

Chapter 2
Symbiotic and Asymbiotic Nitrogen Fixation: An Overview

Papri Nag, Shrivardhan Dheeman, and Dinesh Kumar Maheshwari

Abstract Biological nitrogen fixation (BNF) has evolved early during the evolution of life. Bacteria have nitrogenase enzyme as a central moiety for nitrogen fixation. However, the production, maturation, and function of nitrogenase are costly for the microbe. Molecular expression and regulation of nitrogenase are important for understanding BNF. Therefore, biochemical genetics of nitrogenase expression and regulation becomes important to be studied in the current scenario, to be utilized nitrogen fixation in variety of crops. Role of gene regulates enzymatic activity in symbiotic and nonsymbiotic system covered under this review. It is being predicted that symbiotic and asymbiotic nitrogen-fixing bacteria (NFB) both have the potential to bring ever-green revolution from the green revolution.

Keywords Nitrogen fixation · Rhizobacteria · Rhizosphere · Nonlegume · Crop production

2.1 Introduction

Nitrogen is one of the most important elements of life. Earth's atmosphere contains approximately 78% nitrogen. But this N_2 is not available for use in the biogeochemical cycle. Plants, animals, and most of the other microorganisms require combined form of nitrogen for incorporation into cellular biomass and growth. In an intensely cultivated agro-ecosystem, N_2 becomes limiting for crop growth as it is a highly stable molecule and requires huge amount of energy to break the triple bond which

P. Nag
Division of Plant Biology, Bose Institute, Kolkata, West Bengal, India

S. Dheeman (✉)
Department of Microbiology, School of Allied Health Sciences, MVN University, Palwal, Haryana, India

D. K. Maheshwari
Department of Botany and Microbiology, Gurukula Kangri (Deemed to be University), Haridwar, Uttarakhand, India

© The Author(s), under exclusive license to Springer Nature Singapore Pte Ltd. 2022

D. K. Maheshwari et al. (eds.), *Nitrogen Fixing Bacteria: Sustainable Growth of Non-legumes*, Microorganisms for Sustainability 36,
https://doi.org/10.1007/978-981-19-4906-7_2

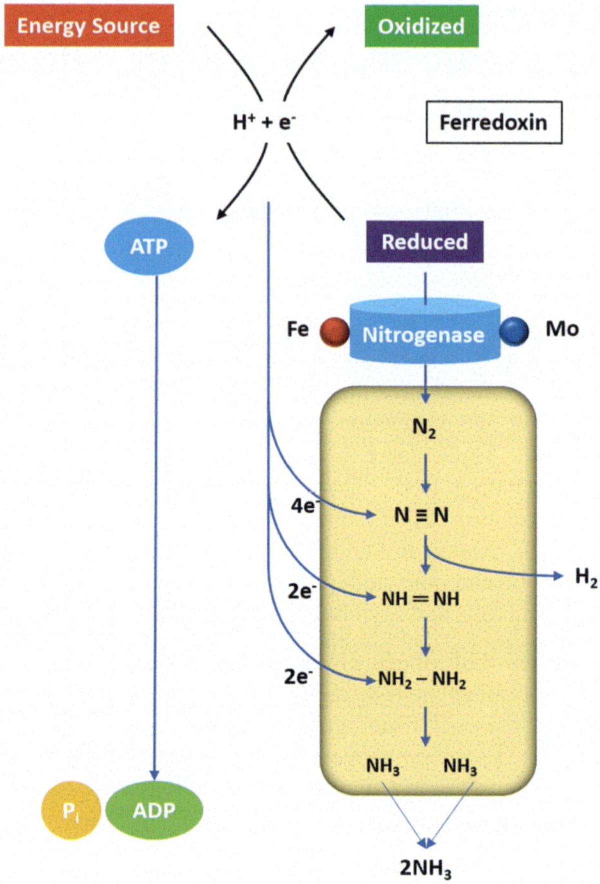

Fig. 2.1 Nitrogen fixation carried out by nitrogenase-producing bacteria fixes atmospheric nitrogen into ammonia (NH_3), at physiological pH, occurs as ammonium ion (NH_4^+) that can be used by other microorganisms, plants, and animals

limits reduction of nitrogen either chemically or biologically. In fact, the green revolution was accompanied by a massive increase in the crop production and use of synthetic fertilizers. Unfortunately, fertilizer use also became one of leading causes of pollution (John and Babu 2021). It is also evident that the use of nitrogenous fertilizers will increase further with growth in population (Bouwman et al. 2013). "Biological nitrogen fixation" (BNF) evolved in certain prokaryotes early in the history of evolution of life (Boyd et al. 2011; Raymond et al. 2004). Members of many bacterial phylum are known to fix nitrogen by breaking the triple bond of N_2 to form NH_3 (Fig. 2.1) which is utilized for growth, starting the N cycle. This process of nitrogen fixation is believed to be second most important biological process after photosynthesis and considered to be of tremendous importance to the environment and also contributes the sustainable inputs into the agriculture world. In this review, the involvement of Nod factor in bacteria-plant symbiosis, contribution to plant growth has been reviewed. Further, an overview on the biochemical genetics of symbiotic and asymbiotic bacteria for nitrogen fixation has been discussed.

2.2 Involvement of Nod Factor in Bacteria-Plant Symbiosis

Knight et al. (1986) studied the nodulation inhibition occurred due to *R. leguminosarum* multicopy *nodABC* genes, analyzed at early inhibition of infection in plants. On the other hand, Castillo et al. (1999) worked on an increase in nodulation in *Medicago sativa*, nitrogen fixation, and plant growth, inoculated with *Sinorhizobium meliloti*. Further the research outcome is that the common nodulation gene (*nodABC*) and *nif*N gene are essential for the process of BNF. The increased copy number into *S. meliloti* genome showed better symbiotic properties resulting in enhancement of plant growth and development. This suggests that increase in symbiotic activity is corresponded to the manipulation of structural and regulatory *nod* genes in rhizobia.

In the pioneer work of Spaink et al. (1991), the nod factor in the early stage of symbiosis played a vital role in formation of nodule primordium, resulting into enhancement of productivity. In fact, Souleimanov et al. (2002) reported a stimulatory effect similar to phytohormones obtained from purified *Bradyrhizobium japonicum* nod factor function in nonlegume (corn) as well as leguminous plant (soybean). This leads to identity of *nod* genes and nod factor cascade essential for endosymbiosis in angiospermic plants (Geurts et al. 2005). Also, earlier report by Prithiviraj et al. (2003) establishes that *B. japonicum* Nod factor induces seed germination of several diverse genera and Nod factor-induced genes are present in the genome of both legume and nonlegumes. Following this, Macchiavelli and Brelles-Marino (2004) reported a significant increase in nodule number after seed inoculation of *Medicago truncatula* with *S. meliloti*, and its lipo-chitooligosaccharides (LCOs). This also signifies by inducing root hair curling, re-initiation of cell proliferation, and sometimes elicitation of nodule-like structures (Gibson et al. 2008). The symbiotic activity of *Rhizobium leguminosarum* bv. trifolii was not improved due to competitive action in the presence of chemical signals in soil, suggesting the involvement of LCOs for enhanced nodulation in clover. Preincubation of *B. japonicum* with jasmonates accelerates the development of nodules and biological nitrogen fixation in *Glycine max*.

Nod factors have been reported to activate defense-related enzymes and induce synthesis of host nodulin proteins, which are important for the formation of infection thread (Fournier et al. 2015; Tsyganova et al. 2021). These factors induce a variety of physiological and biochemical reactions in plants such as cell division and embryo formation in temperature-sensitive carrot hybrid (Baier et al. 1999). Nonlegumes' root mass and length can be increased by applying a low concentration of lipo-chitooligosaccharides (LCO) (10^7–10^9 M) to the rhizosphere (Zhang et al. 2002). In the absence of auxins and cytokinins, Nod factors (LCO) can restore or restart cell division and embryogenesis in plants (Dyachok et al. 2000). Consequently, Nod factors induce seed germination and early seedling growth in nonlegumes like maize and cotton. Nod factors are now reported to play crucial role in establishment of plant-arbuscular mycorrhizal (AM) symbiosis. Application of very low concentrations of the bacterial metabolite stimulated AM fungal colonization in legume and nonlegume roots (Liang et al. 2013).

2.3 Biochemistry of Nitrogen Fixation

Bacterial genus capable of BNF can be categorized into three classes; prokaryotes which can perform BNF in free-living state (diazotrophs; asymbiotic nitrogen fixation), in association with other hosts (associative nitrogen fixation; also, a form of asymbiotic nitrogen fixation) or inside eukaryotic hosts as symbionts (symbiotic nitrogen fixation). All these categories have one feature in common, the enzyme "nitrogenase" capable of breaking the triple bonds present in dinitrogen at normal temperature and pressure to form ammonium. This enzyme complex consists of two components –the smaller dimeric component known as the iron (Fe) protein which is designated as dinitrogenase reductase and functions as an ATP-dependent electron donor to the larger heterotetrametric component known as the molybdenum-iron (MoFe) protein named as dinitrogenase (Fig. 2.2). Both of these components of

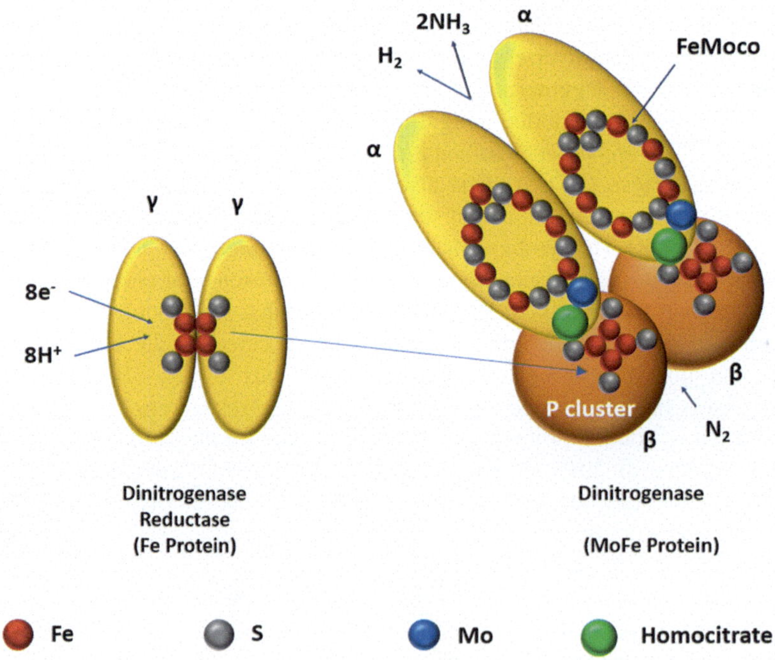

Fig. 2.2 Nitrogenase enzymes. Dinitrogenase reductase (Fe protein) and dinitrogenase (MoFe protein). Initially, electrons are transferred to dinitrogenase reductase enzyme (reaction center, Fe_4S_4). Followed by transferred to P cluster (Fe_4S_4) of dinitrogenase protein. Electrons are passed to FeMoco (iron-molybdenum cofactors) and Fe_7S_9Mo-homocitrate of the dinitrogenase. Ultimately, transferred to N_2, in which ammonia and hydrogen are evolved

nitrogenase are extremely sensitive to oxygen (Dixon and Kahn 2004) and inactivated under oxygen-rich environments. Inactivation of nitrogenase centrally depends on the availability of oxygen, and thus inactivation of nitrogenase can reverse at a lower concentration of oxygen. The fixation of nitrogen needs not only nitrogenase, but also ATP, reduced ferredoxin, and perhaps other cytochromes and co-enzymes. ATP for nitrogen fixation is provided by substrate phosphorylation. For example, *Rhodospirillum*, which has nitrogenase activity in the presence of light and no activity in dark, due to reversible modification of nitrogenase in different conditions. It is clear, that oxygen is not only responsible for the regulation of nitrogenase activity. This phenomenon has been reported in *Azotobacter, Klebsiella pneumoniae.* This nitrogen fixed by the BNF process is absorbed by plants and microbes to be incorporated into bio-geo-chemical cycle.

2.4 Expression and Maturation of FeMo Nitrogenase

Expression and maturation of FeMo nitrogenase have been studied in detail in *Azotobacter vinelandii* and are discussed in a brief simplified manner. The complex of genes that are responsible for the encoding of regulatory proteins associated with nitrogen fixation are known as *nif* genes. They are detected in many nitrogen-fixing bacteria. They act as an operon in free-living anaerobic nitrogen-fixing bacteria like *Rhodobacter capsulatus*, *Klebsiella pneumoniae,* etc. The expression of these Nif genes is induced in response to low concentrations of fixed nitrogen and oxygen. The structural genes, *Nif*H, *Nif*D, and *Nif*K, are arranged in one operon regulated by *Nif*A and *Nif*L (Martinez-Argudo et al. 2004). Although, some bacteria do not possess *Nif*L. The function of each *Nif* gene required maturation of the Fe- and MoFe-protein was determined by creating deletion mutants. The Fe protein is expressed from *Nif*H, the first gene in the operon. The Fe protein is simpler of the two subunits, it is a homodimer of two *Nif*H peptides and requires only the help of a peptidyl prolyl cis-trans isomerases (PPIase) containing *Nif*M. The 2[Fe_4S_4] clusters are first formed on *Nif*U and then transferred to apo-*Nif*H. *Nif*S, a cysteine desulfurase, supplies the sulfur for [Fe_4S_4] assembly (Rubio and Ludden 2005, 2008). For full maturation of MoFe protein, in addition to *Nif*H, several ancillary genes controlled by *Nif*A are required. The structural genes *Nif*D and *Nif*K products assemble to form a heterotetramer apo-FeMo protein. The metal clusters P and M are assembled separately and inserted into the apo-FeMo protein to form the holo-FeMo protein. The two P metal clusters containing the [Fe_8S_7] are assembled from 2 [Fe_4S_4] with the help of *Nif*U, *Nif*S, *Nif*Z, and *Nif*H. Two P-cluster precursors 2[Fe_4S_4] are inserted in each heterodimer as sequentially and converted into functional [Fe_8S_7] P clusters after the insertion of M-cluster in the apo-FeMo protein. The M-cluster assembly is much more complex and needs the help of *Nif*U, *Nif*S, *Nif*B, *Nif*E, *Nif*N, *Nif*W, *Nif*Y/*naf*Y, and *Nif*H. The core [Fe_4S_4] is assembled on *Nif*EN-B in two steps by formation of K-cluster supplied by *Nif*S and *Nif*U and formation of L-cluster on *Nif*EN in *S*-adenosyl-L-methionine (SAM)-dependent manner. *Nif*B is thought to be the SAM-dependent enzyme. The L-cluster on *Nif*EN is converted to M-cluster containing the [7Fe-9S-Mo-C-homocitrate] by insertion of Mo and homocitrate in an ATP-dependent manner with *Nif*H acting as the insertase. The M-cluster is transferred from *Nif*EN to apo-*Nif*DK by direct protein-protein interaction. Once the M-cluster is deposited on the apo-*Nif*DK, the P-cluster precursors also mature into P-clusters to form the holo-*Nif*DK. *Nif*Y is thought to protect α-Cys^{275} residue from rapid alkylation. FeMo-co is considered to be one of the most complex metallocluster and is the active site of N_2 reduction (Jimenez-Vicente et al. 2018). In *Klebsiella pneumoniae, Nif*Y aids in the insertion of FeMo-co into apodinitrogenase. *Nif*M is required for the maturation of *Nif*H. *Nif*J is involved in transporting electron to nitrogenase.

The nitrogen regulatory system activates some genes in response to nitrogen starvation and enables the organism to utilize the unusual nitrogen sources like histidine, proline, and N_2 itself. When enough fixed nitrogen is not available for the organism to use, *Ntr*C triggers *Nif*A expression which further activates the rest of the *Nif* genes and if sufficient amount of reduced nitrogen is available, *Nif*L is activated which inhibits the activity of *Nif*A which results in inhibition of nitrogenase formation. The genes for nitrogen fixation are organized into a regulon of 17 genes which consists of seven or eight operons each of which is transcribed into a single, usually polycistronic mRNA.

2.5 Nitrogenase and Its Regulation

Most of the information about genetics of nitrogenase has been obtained from the study of bacterium *Klebseilla pneumoniae* and this complex of nitrogenase is governed by several genes known as *Nif* clusters. This cluster consists of set of 20 genes called as *Nif* genes. Nitrogenase is an oxygen-labile heteromeric enzyme containing three metalloclusters which can function at less than 180 μM of dissolved O_2 (Oelze 2000). Nitrogenase also requires a high energy source as two molecules of NH_3 produced require 16 ATP molecules (Dixon and Kahn 2004). Further, rapid utilization/secretion of ammonium must be ensured as NH_3 can become toxic to the microbe at higher concentrations (Brewin et al. 1999). Thus, in addition to expressing the structural genes for nitrogenase, the microbe also has to express ancillary genes required for metallocluster organization and modulation of nitrogenase expression in concert to oxygen, nitrogen, and energy levels of the cellular environment (Dos Santos and Dean 2011). The complexity of BNF process increases from the free-living to symbiotic diazotrophs, so does the efficiency of incorporation of nitrogen into the living system. However, the basic biochemistry of nitrogenase remains the same in all diazotrophs.

The genes responsible for nitrogen fixation, including the *nif* genes for nitrogenase production, are under stringent genetic regulation (Evans et al. 1991; Fischer 1994) This is an extremely complex regulatory system that controls the expression of multiple *nif* genes required for the production of active nitrogenase. Genetic

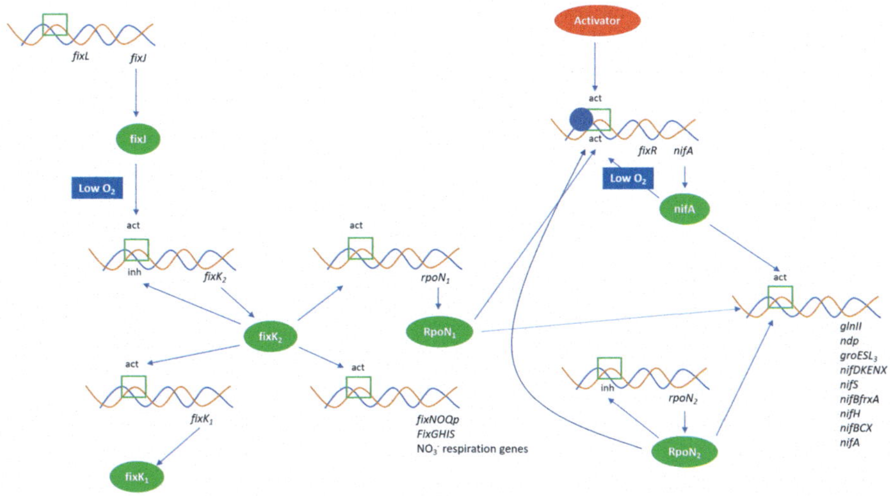

Fig. 2.3 Nitrogen fixation gene regulation in *Bradyrhizobium japonicum.* The regulation of the *nif* genes that control production of nitrogenase with other genes, such as *fix* genes under low oxygen conditions expressed nitrogenase. It appears functional, and especially important. Initially, fixJ and nifA are central systems of symbiotic regulation of N-fixation. Both *fixJ* and *nifA* are functional under low oxygen level. *act* activator, *inh* inhibitor, *open green box* nonsigma-dependent promotor, *blue circle* sigma-dependent promotor

regulation of nitrogen-fixation gene in *Bradyrhizobium japonicum* (in symbiotic system) has been illustrated in Fig. 2.3. The regulatory system for symbiotic nitrogen fixation is very complex. Sigma units facilitate switching of genes either turned on or off through the control of respective promotors (Fig. 2.3).

Three types of nitrogenases are known: MoFe nitrogenase, all ferrous nitrogenase, and vanadium nitrogenase depending on the metallocluster inserted in the enzyme. Among these, MoFe nitrogenase of *Azotobacter vinelandii* is the best studied system. The MoFe nitrogenase consists of two subunits: Fe protein (*Nif*H) is a homodimer with one ATP-binding site in each monomer and one shared [4Fe-4S] cluster. The second subunit is the MoFe protein (*Nif*DK), a heterotetramer ($\alpha_2\beta_2$) with one [Fe_8S_7] P-cluster and one [FeMo-co] ($MoFe_7S_9C$-homocitrate) M-cluster per heterodimer. The reduction of N_2 occurs when the Fe protein and MoFe protein associate to form a functional complex. The process of nitrogen reduction and production of NH_3 is an electrochemical process. Two models have been proposed to explain the kinetics. The Thorneley–Lowe Model (Rutledge and Tezcan 2002) states that the electron flow starts with the electron donation by electron donors like ferrodoxins and flavodoxins in an ATP-dependent manner to Fe protein. The electron is transferred from MgATP-bound Fe protein to the P-cluster of MoFe protein which in turn, passes it to the M-cluster, the active site. Eight rounds of ATP dephosphorylation occur and electrons accumulate before a one molecule of NH_3 is produced (Seefeldt et al. 2020). The second and recent model describes a transient electrostatic transduction state between Fe and MoFe protein. In this state, the transfer of electron from Fe protein to the P-cluster of MoFe protein occurs. This is followed by sequential hydrolysis of ATP and activation of the M-cluster by a thermodynamically induced change in the reducing potential and the subsequent reduction of N_2 (Howard and Rees 1996). The release of Pi forms of the nitrogenase complex is thought to be the rate limiting step (Yang et al. 2016). In addition to the reduction of N_2, nitrogenase can also reduce C_2H_2, C_2H_4, CO, NaCN, NaN_3, and H^+.

Several strategies are adopted to protect nitrogenase from oxygen damage and also for regulating its expression as excess O_2 can cause irreversible damage to the Fe protein. In *A. vinelandii,* three strategies are used for providing protection to nitrogenase from oxidative damage: conformational protection, respiratory protection, and transcriptional control. The FeSII or shethna protein is central to the conformational protection by binding to the nitrogenase and making the metallocluster of nitrogenase inaccessible for oxidation (Moshiri et al. 1994). During respiratory protection, the high respiration rate helps in scavenging oxygen from the cell (Oelze 2000). Central to respiratory control is the cytochrome bd oxidase (Kelly et al. 1990). The structural *Nif* genes encoding for nitrogenase are controlled at the transcriptional level by *Nif*A-*Nif*L (Martinez-Argudo et al. 2004). *Nif*A is the transcriptional enhancer of RNA polymerase σ^{54}, the NtrA gene product is a factor of RNA polymerase which recognizes the *Nif* and other Ntr-regulated genes. NtrA allows RNA polymerase to bind at *Nif* promoter and to initiate transcription, while *Nif*L binds *Nif*A functioning as its repressor. *Nif*L can inhibit the binding of *Nif*A to σ^{54} by hindering its ATPase activity. In proteobacteria lacking *Nif*L, the structure of *Nif*A contains an extra Cys–X_4–Cys motif. Several other genes are also expressed only when *Nif*A binds to σ^{54} (Dixon and Kahn 2004; Fischer 1994). These auxillary *Nif* genes are required for the maturation of nitrogenase and insertion of metal clusters. During transcriptional regulation of nitrogenase in *Klebsiella pneumoniae*, *Azoarcus* spp., and *A. vinelandii*, the oxidation of the FAD prosthetic group in *Nif*L induces binding to *Nif*A to inhibit transcription (Dixon and Kahn 2004). This inhibition is reversed when the flavin is reduced and *Nif*L can separate from *Nif*A. Regulation of *nif* genes in *Klebsiella pneumoniae* has two elements, *ntr* as external system and couple of *nifA* and *nifL* as internal system. The interrelationship of both systems is summarized in Fig. 2.4.

In *Rhodobacter capsulatus*, the two component RegB-RegA system controls the production of nitrogenase in relation to O_2 concentration. The RegB-RegA system was initially identified as an important system controlling the transition of the bacteria from aerobic to anaerobic growth conditions (Elsen et al. 2004; Torres et al. 2014). *R. capsulatus* has adapted a second level of control by using FdX during conformational change (as a counterpart of FeSII) and also by binding *Nif*A for inhibiting transcription of nitrogenase under semi-aerobic conditions (as a counterpart of *Nif*L) (Hoffmann et al. 2014). In symbiotic bacteria, three levels of controls are exerted to modulate the expression of nitrogenase gene: Direct inactivation of *Nif*A, through FixL-FixJ and through RegS-RegR. In symbiotic proteobacteria lacking *Nif*L, the *Nif*A protein is directly modulated by the oxidation state of the cell. This is done by controlling the metal ions present in the Cys–X_4–Cys motif

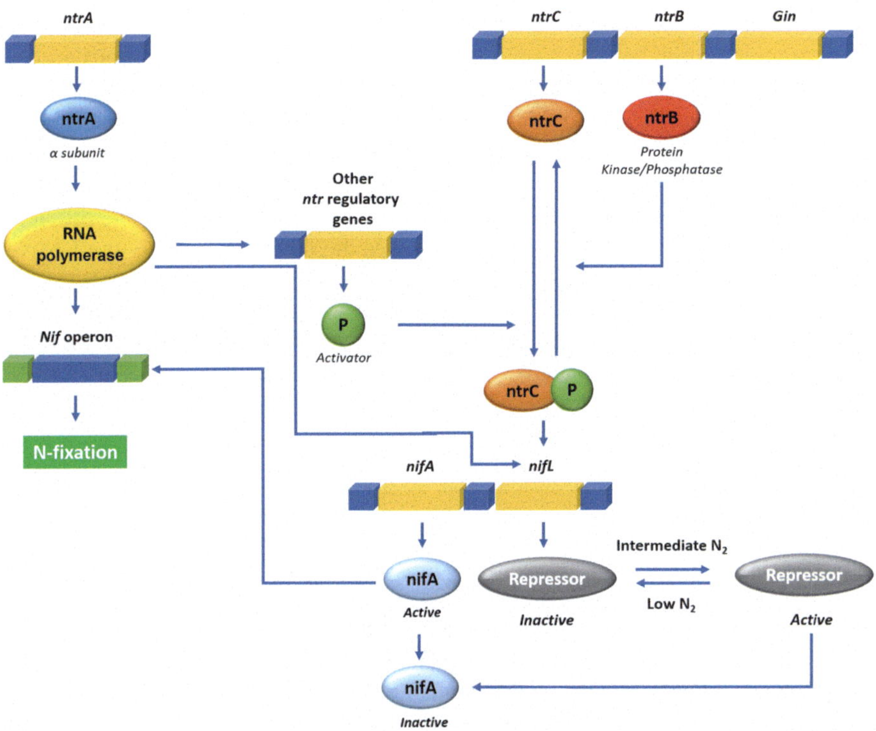

Fig. 2.4 Regulation of *nif* genes expression in *Klebsiella pneumoniae*: The ntrA protein from *ntrA* gene serves a factor for RNA polymerase, which recognizes *nif* and *ntr* genes. *ntrA* allows RNA polymerase to bind at *nif* promotors and initiate transcription. ntrB, a product of *ntrB* gene functions as kinase and phosphatase on the substrate ntrC, product of *ntrC* gene. Depending upon nitrogen concentration, particularly in starvation conditions, ntrC-P acts as inactivator of *nifL* and *nifA*. The nifA, activates *nif* transcription and nifL in presence of low nitrogen and oxygen, inactivates nifA, thereby, preventing transcription of *nif* genes. (Adapted from Tuli et al. 1982)

present in *Nif*A (Fischer 1994). The oxidation state of the histidine kinase, FixL, controls the phosphorylation state of FixJ which in turn controls the transcription of *Nif*A. In presence of high O_2, FixL is oxidized and is unable to phosphorylate FixJ resulting in FixJ dimerization which cannot bind the promoter to initiate the transcription of *Nif*A. Once oxygen stress is removed, FixL phosphorylates FixJ which in turn activates *Nif*A transcription. In addition, RegS-RegR is two component redox-sensing system which can control the production of nitrogenase by controlling the production of *NifA* (Emmerich et al. 2000).

In addition to oxygen, modulation of nitrogenase production in response to nitrogen status is very important for survival of the free-living diazotroph. In the symbiotic bacteroid, the cells keep on releasing ammonium for the host plant which diffuses out of the cell to be utilized by the host plant. BNF is a high energy-consuming process and is stringently controlled by the diazotroph. The regulatory cascades involved in controlling the production of nitrogenase under different nitrogen availability function at the global level with PII-like proteins (GlnK, GlnB, GlnZ), transcriptional level by NtrC and NtrB (He et al. 1997), and post-translational level by DraT and DraG (Masepohl et al. 2002). In some diazotroph like *A. vinelandii*, nonuridylylated GlnK prevents *Nif*L from repressing *Nif*A; while in others uridylylated GlnB binds *Nif*A to activate the *Nif* gene expression. The NtrC-NtrB global regulators function by controlling expression of *NifL* and *NifA* genes and the *glnK-amtB* operon, although regulation by the NtrC-NtrB varies among different bacteria (Dixon and Kahn 2004). GlnB activates the phosphatase activity of NtrC. Dephosphorylation of NtrB by NtrC prevents transcription of the *NifL-NifA* and *glnK-amtB* operons. The GlnB-GlnK system is utilized for modulation by the presence of energy sources like ATP and 2-oxogluterate. However, the nitrogen signal overrides the energy signal.

2.6 Conclusion

Biological nitrogen fixation is a costly process for the microbe as it is energy intensive. Hence, it is controlled at several levels from transcriptional control to post-translational control. Genetic regulations of nitrogenase are still complex and require to be understand in the current scenario. This review understood and provides a more improved picture of biochemical genetics of nitrogen fixation in symbiotic and free-living association level.

References

Baier R, Schiene K, Kohring B, Flaschel E, Niehaus K (1999) Alfalfa and tobacco cells react differently to chitin oligosaccharides and *Sinorhizobium meliloti* nodulation factors. Planta 210(1):157–164

Bouwman AF, Beusen AH, Griffioen J, Van Groenigen JW, Hefting MM, Oenema O, Van Puijenbroek PJ, Seitzinger S, Slomp CP, Stehfest E (2013) Global trends and uncertainties in terrestrial denitrification and N_2O emissions. Philos Trans R Soc Lond B: Biol Sci 368(1621): 20130112. https://doi.org/10.1098/rstb.2013.0112

Boyd ES, Hamilton TL, Peters JW (2011) An alternative path for the evolution of biological nitrogen fixation. Front Microbiol 2:205. https://doi.org/10.3389/fmicb.2011.00205

Brewin B, Woodley P, Drummond M (1999) The basis of ammonium release in nifL mutants of *Azotobacter vinelandii*. J Bacteriol 181:7356–7362

Castillo M, Flores M, Mavingui P, Martínez-Romero E, Palacios R, Hernández G (1999) Increase in alfalfa nodulation, nitrogen fixation, and plant growth by specific DNA amplification in *Sinorhizobium meliloti*. Appl Environ Microbiol 65(6):2716–2722

Dixon R, Kahn D (2004) Genetic regulation of biological nitrogen fixation. Nat Rev Microbiol 2(8): 621–631

Dos Santos P, Dean D (2011) Co-ordination and fine tuning of nitrogen fixation in *Azotobacter vinelandii*. Mol Microbiol 79(5):1132–1135

Dyachok JV, Tobin AE, Price NPJ, Von Arnold S (2000) Rhizobial Nod factors stimulate somatic embryo development in *Picea abies*. Plant Cell Rep 19(3):290–297

Elsen S, Swem LR, Swem DL, Bauer CE (2004) RegB/RegA, a highly conserved redox-responding global two-component regulatory system. Microbiol Mol Biol Rev 68(2):263–279. https://doi.org/10.1128/MMBR.68.2.263-279.2004

Emmerich R, Hennecke H, Fischer HM (2000) Evidence for a functional similarity between the two-component regulatory systems RegSR, ActSR, and RegBA (PrrBA) in alpha-Proteobacteria. Arch Microbiol 174(5):307–313. https://doi.org/10.1007/s002030000207

Evans H, Stacey G, Burris RH (1991) Biological nitrogen fixation. Chapman and Hall, New York

Fischer H-M (1994) Genetic regulation of nitrogen fixation in rhizobia. Microbiol Rev 58:352–386

Fournier J, Teillet A, Chabaud M, Ivanov S, Genre A, Limpens E, Barker DG (2015) Remodeling of the infection chamber before infection thread formation reveals a two-step mechanism for rhizobial entry into the host legume root hair. Plant Physiol 167(4):1233–1242

Geurts R, Fedorova E, Bisseling T (2005) Nod factor signaling genes and their function in the early stages of *Rhizobium* infection. Curr Opin Plant Biol 8(4):346–352

Gibson KE, Kobayashi H, Walker GC (2008) Molecular determinants of a symbiotic chronic infection. Annl Rev Genet 42:413–441

He L, Soupene E, Kustu S (1997) NtrC is required for control of *Klebsiella pneumoniae* NifL activity. J Bacteriol 179(23):7446–7455. https://doi.org/10.1128/jb.179.23.7446-7455.1997

Hoffmann MC, Müller A, Fehringer M, Pfänder Y, Narberhaus F, Masepohl B (2014) Coordinated expression of fdxD and molybdenum nitrogenase genes promotes nitrogen fixation by *Rhodobacter capsulatus* in the presence of oxygen. J Bacteriol 196(3):633–640. https://doi.org/10.1128/JB.01235-13

Howard JB, Rees DC (1996) Structural basis of biological nitrogen fixation. Chem Rev 96(7): 2965–2982. https://doi.org/10.1021/cr9500545

Jimenez-Vicente E, Yang ZY, Ray WK, Echavarri-Erasun C, Cash VL, Rubio LM, Seefeldt LC, Dean DR (2018) Sequential and differential interaction of assembly factors during nitrogenase MoFe protein maturation. J Biol Chem 293(25):9812–9823. https://doi.org/10.1074/jbc.RA118.002994

John DA, Babu GR (2021) Lessons from the aftermaths of green revolution on food system and health. Front Sustain Food Syst 5:644559. https://doi.org/10.3389/fsufs.2021.644559

Kelly MJ, Poole RK, Yates MG, Kennedy C (1990) Cloning and mutagenesis of genes encoding the cytochrome bd terminal oxidase complex in *Azotobacter vinelandii*: mutants deficient in the cytochrome d complex are unable to fix nitrogen in air. J Bacteriol 172(10):6010–6019. https://doi.org/10.1128/jb.172.10.6010-6019.1990

Knight CD, Rossen L, Robertson JG, Wells B, Downie JA (1986) Nodulation inhibition by *Rhizobium leguminosarum* multicopy nodABC genes and analysis of early stages of plant infection. J Bacteriol 166(2):552–558

Liang Y, Cao Y, Tanaka K, Thibivilliers S, Wan J, Choi J, Stacey G (2013) Non-legumes respond to rhizobial Nod factors by suppressing the innate immune response. Science 341(6152): 1384–1387

Macchiavelli RE, Brelles-Marino G (2004) Nod factor-treated *Medicago truncatula* roots and seeds show an increased number of nodules when inoculated with a limiting population of *Sinorhizobium meliloti*. J Exp Bot 55(408):2635–2640

Martinez-Argudo I, Little R, Shearer N, Johnson P, Dixon R (2004) The NifL-NifA system: a multidomain transcriptional regulatory complex that integrates environmental signals. J Bacteriol 186(3):601–610. https://doi.org/10.1128/JB.186.3.601-610.2004

Masepohl B, Drepper T, Paschen A, Gross S, Pawlowski A, Raabe K, Riedel KU, Klipp W (2002) Regulation of nitrogen fixation in the phototrophic purple bacterium *Rhodobacter capsulatus*. J Mol Microbiol Biotechnol 4(3):243–248

Moshiri F, Kim JW, Fu C, Maier RJ (1994) The FeSII protein of *Azotobacter vinelandii* is not essential for aerobic nitrogen fixation, but confers significant protection to oxygen-mediated inactivation of nitrogenase in vitro and in vivo. Mol Microbiol 14(1):101–114. https://doi.org/10.1111/j.1365-2958.1994.tb01270.x

Oelze J (2000) Respiratory protection of nitrogenase in *Azotobacter* species: is a widely held hypothesis unequivocally supported by experimental evidence? FEMS Microbiol Rev 24(4): 321–333. https://doi.org/10.1111/j.1574-6976.2000.tb00545.x

Prithiviraj B, Zhou X, Souleimanov A, Kahn W, Smith D (2003) A host-specific bacteria-to-plant signal molecule (Nod factor) enhances germination and early growth of diverse crop plants. Planta 216(3):437–445

Raymond J, Siefert JL, Staples CR, Blankenship RE (2004) The natural history of nitrogen fixation. Mol Biol Evol 21(3):541–554. https://doi.org/10.1093/molbev/msh047

Rubio LM, Ludden PW (2005) Maturation of nitrogenase: a biochemical puzzle. J Bacteriol 187(2): 405–414. https://doi.org/10.1128/JB.187.2.405-414.2005

Rubio LM, Ludden PW (2008) Biosynthesis of the iron-molybdenum cofactor of nitrogenase. Annu Rev Microbiol 62:93–111. https://doi.org/10.1146/annurev.micro.62.081307.162737

Rutledge HL, Tezcan FA (2002) Electron transfer in nitrogenase. Chem Rev 120(12):5158–5193. https://doi.org/10.1021/acs.chemrev.9b00663

Seefeldt LC, Yang ZY, Lukoyanov DA, Harris DF, Dean DR, Raugei S, Hoffman BM (2020) Reduction of substrates by nitrogenases. Chem Rev 120(12):5082–5106. https://doi.org/10.1021/acs.chemrev.9b00556

Souleimanov A, Prithiviraj B, Smith DL (2002) The major Nod factor of *Bradyrhizobium japonicum* promotes early growth of soybean and corn. J Exp Bot 53(376):1929–1934

Spaink HP, Sheeley DM, van Brussel AA, Glushka J, York WS, Tak T, Geiger O, Kennedy EP, Reinhold VN, Lugtenberg BJ (1991) A novel highly unsaturated fatty acid moiety of lipo-oligosaccharide signals determines host specificity of *Rhizobium*. Nature 354(6349):125–130

Torres MJ, Argandoña M, Vargas C, Bedmar EJ, Fischer HM, Mesa S, Delgado MJ (2014) The global response regulator RegR controls expression of denitrification genes in *Bradyrhizobium japonicum*. PLoS One 9(6):e99011. https://doi.org/10.1371/journal.pone.0099011

Tsyganova AV, Brewin NJ, Tsyganov VE (2021) Structure and development of the legume-rhizobial symbiotic interface in infection threads. Cells 10(5):1050

Tuli R, Fisher R, Haselkorn R (1982) The ntr genes of Escherichia coli activate the hut and nif operons of Klebsiella pneumoniae. Gene 19(1):109–116

Yang ZY, Ledbetter R, Shaw S, Pence N, Tokmina-Lukaszewska M, Eilers B, Guo Q, Pokhrel N, Cash VL, Dean DR, Antony E, Bothner B, Peters JW, Seefeldt LC (2016) Evidence that the pi release event is the rate-limiting step in the nitrogenase catalytic cycle. Biochemistry 55(26): 3625–3635. https://doi.org/10.1021/acs.biochem.6b00421

Zhang H, Prithiviraj B, Souleimanov A, D'Aoust F, Charles TC, Driscoll BT, Smith DL (2002) The effect of temperature and genistein concentration on lipo-chitooligosaccharide (LCO) production by wild-type and mutant strains of *Bradyrhizobium japonicum*. Soil Biol Biochem 34(8): 1175–1180

Chapter 3
Interactions of Rhizobia with Nonleguminous Plants: A Molecular Ecology Perspective for Enhanced Plant Growth

Sourav Debnath, Nandita Das, Dinesh Kumar Maheshwari, and Piyush Pandey

Abstract Rhizobia are known for its symbiotic association with the leguminous plants, which have role in biological nitrogen fixation in root nodules. However, its association with nonlegumes has received relatively lesser attention. With the progress in technology and research strategies, the molecular ecological perspective of rhizobial interaction with nonlegumes has recently gained much progress. Rhizobia are now known to form symbiosis with nonlegumes without forming true nodules, and yet promote the growth of nonlegumes through direct and indirect mechanisms. Plant growth-promoting traits such as production of phytohormones, siderophore, ACC deaminase activity, phosphate solubilization, and improving the nutrient uptake by modulating the root structure are the PGPR mechanisms described for rhizobia. Recently, rhizobia have also been reported to modulate the rhizospheric bacterial community structure that helps plants to adapt to a new or hostile environment. The rhizobia can also mediate biocontrol through antibiosis, parasitism, or competition which inhibits plant pathogens, induces systemic resistance in the host plant, and also releases exopolysaccharides for improving root adhering soil in the plants. The research on cell-to-cell communication for this unique synergistic interaction with nonlegumes, such as rice and wheat plants, has revealed interesting facts, which may be used for better plant growth. Therefore, the application of rhizobia as PGPR and further use as a biofertilizer, stress regulators, and biocontrol agents for nonleguminous plants need more intervention from the perspective of its interaction with nonlegumes, which has been addressed in this article. Also, the importance of rhizobia with the perspective of molecular ecology,

S. Debnath · N. Das · P. Pandey (✉)
Department of Microbiology, Assam University, Silchar, India

D. K. Maheshwari
Department of Botany and Microbiology, Gurukula Kangri (Deemed to be University), Haridwar, Uttarakhand, India

© The Author(s), under exclusive license to Springer Nature Singapore Pte Ltd. 2022

D. K. Maheshwari et al. (eds.), *Nitrogen Fixing Bacteria: Sustainable Growth of Non-legumes*, Microorganisms for Sustainability 36,
https://doi.org/10.1007/978-981-19-4906-7_3

genomics attributes of rhizobia colonizing nonlegumes, and possible rhizobial engineering have been included.

Keywords Rhizobia · PGPR · Nonlegumes · OMICS · Nitrogenase · Nitrogen fixation

3.1 Introduction

The demand for food has been increasing at an exaggerating rate worldwide. For such a demanding process, the farmers apply chemical fertilizers, insecticides, herbicides, etc. more than their recommended level for enhancing the production. These applied chemicals, in turn, affect soil health and increase a load of contaminants into the environment, Consequently, affecting the health of humans and other organisms. Therefore, a sustainable approach must be adopted to ensure effective management of all the resources in an agriculture system that reduces the impact of the chemicals while maintaining the fertility of the soil. Presently, the trend in the agricultural sector is to explore the alternatives for the harmful chemicals and focus on organic and inorganic fertilizers (Haggag and Wafaa 2002), which is a daunting task (Ray et al. 2000; Bera et al. 2006). Plant growth-promoting rhizobacteria (PGPR) are a group of beneficial microbes which are involved in symbiotic and nonsymbiotic beneficial traits to improve the growth and yield of legumes as well as nonlegumes (Antoun et al. 1998; García-Fraile et al. 2012; Ahmad et al. 2013; Khaitov et al. 2016; Ziaf et al. 2016). Thus, the use of microbes as biofertilizers for sustainable agriculture is hereby utmost necessary considering their beneficial traits and mode of action (Nosheen et al. 2021).

Rhizobia are soil bacteria belonging to family *Rhizobiaceae* which are gram-negative, chemo-organotroph, or chemolithotroph in nature (Werner 1992), and are capable of fixing atmospheric nitrogen popularly known as biological nitrogen fixation (BNF) (Franche et al. 2009). Some of the well-known genera of rhizobia are *Rhizobium, Sinorhizobium, Mesorhizobium, Bradyrhizobium, Agrobacterium, Azorhizobium, Allorhizobium,* etc. (Rao et al. 2018) which possess host-specific ability to establish symbiosis with leguminous plants (Mehboob et al. 2012). However, rhizobia also possess the ability to associate with nonlegumes without forming true nodules which are nonspecific (Reyes and Schmidt 1979). This leads to speculations and further work on the working mechanism of the well-established fact that increases the yield upon their inoculation.

Rhizobia are known to promote the growth of many plants including various crops and grasses (Machado et al. 2016; Borges et al. 2019). Yet various factors govern the successful nature of the inoculants. Rhizobia meditates the growth of nonlegume plants through its direct and indirect mechanisms or a combination of both. These include PGP traits such as IAA production, siderophore activity, and ACC deaminase activity to name a few including biocontrolling property as well as by influencing other beneficial microbes in the vicinity for better growth of the plant (Shakhawat Hossain and Mårtensson 2008).

On the other hand, certainly incompatible rhizobia might have a deleterious effect on certain crops (Perrine et al. 2001). Therefore, it is important to determine the specificity of a particular strain and understand the underlying interaction before selecting it as a PGPR.

3.2 Rhizobia and Nonlegume Interaction

Rhizobia are known for their ability to form root nodules in the leguminous plants, by which they fix atmospheric nitrogen and provide nourishment to the plants (Schloter et al. 1997), there had been early reports for their interaction with the nonlegumes (Reyes and Schmidt 1979; Chabot et al. 1996). The rhizobia possess the ability to survive as well as to colonize the roots of the nonlegume plants (Antoun and Prevost 2000; Bhattacharjee et al. 2008). In fact, bacterial associations with plants are of two types, i.e., close and loose. This may be endophytic, phyllosphere, or rhizospheric (Weyens et al. 2009). This colonizing ability of the bacteria brings about stimulating or inhibiting effects (Höflich et al. 1994; Antoun et al. 1998). The rhizobia enter the nonlegume through cracks present in the root epidermis and colonize the cortex within the xylem (Sabry et al. 1997) and between the root intercellular spaces (Reddy et al. 1997). The roots of a particular plant and rhizobia interact with each other, while this interaction results in enhancement of the growth and yield of the plant (Lemanceau 1992; Yanni et al. 1997). Therefore, those specific and nonspecific interactions make rhizobia a potential endophyte or rhizobacteria for the nonlegumes (Sessitsch et al. 2002). There are various studies which suggests rhizobia as endophytes in nonleguminous plants, e.g., *Rhizobium laguerreae* in spinach (Jiménez-Gómez et al. 2018), *Rhizobium phaseoli*, *Sinorhizobium americanum,* and *Azospirillum brasilense* in maize (Gómez-Godínez et al. 2019), *Rhizobium* species in cotton plant (Qureshi et al. 2019), and *Rhizobium alamii* in *Brassica napus* (Tulumello et al. 2021).

Rhizobial endosymbiosis with other nonlegumes such as *Parasponia* has also been reported (Sytsma et al. 2002). Different *Rhizobium* species are associated with the nodulation process of *Parasponia* (Trinick and Galbraith 1980; Trinick and Hadobas 1989) with diverse genes for Nod factor biosynthesis (Op den Camp et al. 2012). The structure of such nodules is like lateral roots, and is formed following the typical flavonoid-dependent mechanism (Chapman and Muday 2021). It was reported that Nod factors lysin-motif (LysM) domain proteins are important for the symbiosis of nodulation and mycorrhization in *P. andersonii* (Op den Camp et al. 2011).

Rhizobia can flourish in both legumes as well as nonlegumes (Pena-Cabriales and Alexander 1983). There are reports of the appearance of nodule-like structures in nonlegumes (Ridge et al. 1992; Trinick and Hadobas 1995; Naidu et al. 2004). Rhizobial colonization in rice and wheat seedling has been reported by Shimshick and Hebert (1979), while the effectiveness of rhizobial competence was determined by Wiehe and Höflich (1995) in maize. Many such reports of rhizobial colonization in nonlegumes were reported by Wiehe et al. (1994), Schloter et al. (1997), Reddy et al. (1997), and Sabry et al. (1997). Along with endophytic colonization, the ascending migration toward stem, leaves, and leaf sheath has been reported by Chi et al. (2005). The survival and multiplication of rhizobia in the rhizospheric region of wheat, corn, rape, etc. (Wiehe and Höflich 1995), and lettuce (Pena and Reyes 2007) are well studied. Moreover, the presence of rhizobia has been reported from the epidermis of sorghum and millet plants, after inoculation (Matiru et al. 2005). Perrine-Walker et al. (2007) detected the presence of rhizobia and their ability to colonize rice plants.

Rhizobia are also known to secrete different kinds of metabolites which ensure the development of nonleguminous plants. Such compounds provide stabilizations and protection to the plant. These compounds include cytokinins (Noel et al. 1996), abscisic acid (Minamisawa et al. 1996), indole-3-acetic acid (Pandey and Maheshwari 2007; Venieraki et al. 2011), gibberellic acid (Humphry et al. 2007), ethylene (Boiero et al. 2007), ACC-deaminase (Glick et al. 1994), antibiotics (Bhattacharya et al. 2013), etc. These metabolites are produced through the interaction of rhizobia and the nonlegume which results in better tolerance of stress, growth, and yield (Mehboob et al. 2012). In contrast, sometimes overproduction of certain metabolites may also harm the plant. Production of bacteriocin was reported from *Sinorhizobium meliloti* which inhibits the growth of rice (Perrine-Walker et al. 2009). Similarly, a high concentration of auxin and nitrate by rhizobia was reported to inhibit nonleguminous plants (Perrine-Walker et al. 2007). The PGP of endophytic *Bradyrhizobium* sp. strain SUTN9-2 isolated from rice plants was examined. The expression of genes involved in IAA (*nit*) and ACC deaminase (*acdS*) synthesis was contradictory with the results of quantitative analysis of IAA and ACC deaminase. This inconsistency suggested that IAA and ACC deaminase generated by SUTN9-2 have no direct effect on rice development, but those other components arising from IAA and ACC deaminase activities may have their role. Furthermore, SUTN9-2 enhanced the expression of genes involved in nitrogen-fixing (*nifH* and *nifV*) in rice tissues (Greetatorn et al. 2019). Hara et al. (2019) discovered that the functional N_2-fixing *Bradyrhizobia* (TM122 and TM124) found in sorghum roots were phylogenetically related to photosynthetic *B. oligotrophicum* S58T and non-nodulating *Bradyrhizobium* sp. S23321. In terms of the G+C content of the *nifDK* genes, *nifV*, and possibly *nif* gene regulation, the *nif* genes of "Free-living diazotrophs" TM122, TM124, S58T, and S23321 differ significantly from those on the symbiosis islands of nodule-forming *Bradyrhizobium* sp.

The successful nature of the rhizobial and nonleguminous plant association depends on many factors. Along with the bacterial strain, the type of plant, culture condition, microflora, quality of soil, and various biotic-abiotic factors contribute to the success of the inoculum (Lynch 1990a, b; O'sullivan and O'Gara 1992; Antoun et al. 1998; Biswas et al. 2000; Hilali et al. 2001; Dobbelaere et al. 2003; Depret et al. 2004; Mehboob et al. 2008; Hussain et al. 2009). Depending upon these factors, rhizobia have been divided into three groups depending upon their growth-promotional, inhibitory ability, and nonassociating nature (Prayitno et al. 1999; Perrine et al. 2001, 2005). The development of competent rhizobial strains by the plant, soil, and environment is key (Mehboob et al. 2012). On the basis of these reports, it may be concluded that just like the rhizobial-legume interaction, rhizobial and nonlegume interaction is also much important for green and sustainable agriculture.

3.2.1 Molecular Interaction of Rhizobia in Nonlegumes

The molecular aspect of rhizobial inoculation has been extensively explored in *Parasponia andersonii* and rice plants. The recruitment of LysM-Type Mycorrhizal Receptor, which is responsible for the symbiotic association with *Rhizobium*, is the fundamental mechanism of Parasponia-Rhizobia interaction (Op den Camp et al. 2011). A class of LysM-type receptors namely *MtNFP/LjNFR5* is reported from *Parasponia* and the functional analysis of this gene revealed a dual symbiotic function in *P. andersonii* (Streng et al. 2011). Comparative transcriptomics of *P. andersonii* revealed 290 symbiotic genes which are similar to a legume *Medicago truncatula* that is responsible for its nodule-enhanced expression profile. Some important genes are Nodule Inception (Nin) And *Rhizobium*-Directed Polar Growth (RPG), known for their importance for nitrogen-fixing root nodules. These set of genes along with a putative ortholog of the NFP/NFR5-type LysM receptor for *Rhizobium* LCO Signaling molecules namely NFP2 in *Parasponia* are critical in forming the nodules which separate it from other plants of its category (van Velzen et al. 2018; Dupin et al. 2020).

In rice plants, however, rhizobial invasion occurs mostly through pores in the epidermis and fissures formed during the development of lateral roots (Reddy et al. 1997). This infection process is nod-gene independent, nonspecific, and does not include infection thread development. Naringenin, a flavonoid, has been shown to enhance this form of rhizobial colonization in rice plants (Webster et al. 1997). Perrine et al. (2001) reported the involvement of specific plasmids carried by rhizobial strains affecting the growth and development of rice seedlings. Piromyou et al. (2015) investigated the effect of *Bradyrhizobium* inoculation in rice seedlings and reported strong expression of *peces*, *rhcJ*, *virD4*, exopolysaccharide production (*fliP*), and glutathione-*S*-transferase (*gst* genes). Wu et al. (2018) reported the growth-promotional and signaling potential of *Sinorhizobium meliloti* in rice seedlings, which resulted in increased gene expression, which is responsible for accelerated cell division and cell expansion. Transcriptomic analysis revealed that differentially expressed genes (DEG) are involved in upregulation of phytohormone production, photosynthetic efficiency, glucose metabolism, cell division, and cell-wall expansion. Moreover, the inoculation of *Bradyrhizobium* sp. in rice plants revealed colonization, enlargement of bacterial cells, increased DNA content, and nitrogen fixation. Some factors in rice extract induced the expression of cell cycle and nitrogen fixation genes. The transcriptomic analysis revealed encoding a class of oxidoreductases that act with oxygen atoms and may play a role in maintaining an appropriate level of oxygen for nitrogenase activity, followed by GroESL chaperonins, which are required for nitrogenase functioning. The expression of the antimicrobial peptide transporter (*sapDF*) was also increased, leading to cell differentiation (Greetatorn et al. 2020).

3.3 Methods to Detect N_2 Fixation by Rhizobia in Nonlegumes

There are methods by which we can identify the activity of nitrogen fixers in nonlegumes. One indirect method is to detect the *nifH* DNA in the tissues having DNA of endophytes, which indicates the occupancy of N_2-fixating bacteria. The expression of *nifH* genes stipulates the probability of active N_2 fixation by diazotrophs. It is done with the help of Rt-PCR where soft stem tissues of plants like sugarcane are being used to detect any signs of *nifH* expression (Thaweenut et al. 2011). RNA is isolated and reverse transcripted into cDNA in this method (Thaweenut et al. 2011). Using the product of RT-PCR as a template, the fragments of *nifH* are amplified through nested PCR with Taq DNA polymerase. The efficiency of the *nifH* PCR primer has been re-examined in different laboratories (Gaby et al. 2018) and a new modified annealing temperature was set at 58 °C to determine the largest diversity of *nifH* templates.

The second way is to detect the diazotrophic rhizobia by metaproteomics. For this, the first step is to obtain the bacterial cell-enriched fraction. The bacterial cells are extracted from the root tissues of rice plants through different centrifugation steps followed by a density gradient centrifugation followed by proteins extraction. A metaproteomic analysis based on metagenome analysis on the roots of rice plant was used to determine the peptide abundances of the proteins involved in methane oxidation (particulate/soluble methane monooxygenase (pMMO/sMMO), methanol dehydrogenase (MxaFI), formaldehyde dehydrogenase (FAD), formate dehydrogenase (FDH)) and N_2 fixation (NifH, NifD, NifK, VnfD). This was followed up by Nanoliquid chromatography (LC)–electrospray ionization–tandem mass spectrometry (MS/MS) analyzed using an LTQ ion-trap MS coupled with a multidimensional high-performance LC Paradigm MS2 chromatograph and a nanospray electrospray ionization device. The tryptic peptide spectra were recorded in an m/z range of 450–180. The MS/MS data were explored against the rice root microbiome database that was constructed using metagenome data targeting the same rice root samples (Bao et al. 2014).

3.4 Genomic Attributes of Rhizobia Colonizing Nonlegumes

Genomics is the study of genes and genomes that focuses on the structure, function, evolution, mapping, epigenomic, mutagenomic, and aspects of genome editing (Muthamilarasan et al. 2019). Genomics plays an important role in elucidating genetic variation, which may enhance the performance or the efficiency of the strains resulting in improved crop production. The rhizobial genomes that are studied, largely belong to α and β class of Proteobacteria. The average and median genome sizes of rhizobia were reported to be 3.65 Mb and 3.46 Mb, respectively (Dicenzo et al. 2016) which are nearly two-three times larger than other bacterial groups. The rhizobial genomes reflect their ability to adapt in complex conditions, where limited and diverse types of nutrients are available to the rhizobia (Dicenzo et al. 2016). Mostly, the genomes are multipartite, which are split into two or more large self-replicating fragments (replicons). The replicons vary from100 to >2000 kb in size (Geddes et al. 2020). Though the majority of the research works have been associated with the rhizobia of legume crops, there are some genomic data available for the rhizobia in the nonleguminous group which enable us to understand the role of molecular machinery other than nodule formation.

de Souza et al. (2015) reported the genome of *Rhizobium* sp. UR51a isolated from roots of rice plants which is associated with plant growth-promoting traits such as siderophore, IAA production along with biological nitrogen fixation. The genome analyses revealed the genes for siderophore aerobactin uptake (*fhuABCD*), genes for biosynthesis of auxin, genes for antioxidant enzymes, antibiotic, and toxic compounds resistance genes. Flores-Félix et al. (2021) isolated *Rhizobium laguerreae* PEPV16 strain from root nodules of *Phaseolus vulgaris* and performed genomic analysis. The beneficial traits identified through the analysis have led its application to other vegetables such as carrot and lettuce, subsequently enhancing their growth. The analysis revealed the genomes possess genes related to *N*-acyl-homoserine lactone (AHL) and biosynthesis of cellulose, genes for quorum sensing, and formation of biofilm. Moreover, the genes related to PGP traits such as phosphate solubilization, indole acetic acid production, siderophore biosynthesis, and nitrogen fixation were also reported from the genome. The content of genes related to amino acids and other associated genes were also present. For the production of cellulose, the presence of *bcsA* and *bcsB* genes were reported. Also, a third gene (*celC*) encoding an endonuclease enzyme, CelC2 has been reported to be associated with the biosynthesis of cellulose, and the formation of biofilm. A gene encoding an *N*-acyl-L-homoserine lactone (AHL) synthase has been reported to be associated with quorum sensing. For the colonization which mediates the formation of biofilm and attachment to plant surface, many associated genes for motility, chemotaxis, and biosynthesis of EPS have been reported. Moreover, genes that benefit PGP such as phosphate solubilization-related genes that carry out the phosphate solubilization from organic compounds. A siderophore-producing gene that encodes acetyltransferase that is similar to the *vbsA* gene responsible for the biosynthesis of vicibactin, a siderophore produced in other rhizobial groups is also reported from the genome.

3.5 Mechanisms of Growth Promotion of Nonlegumes by Rhizobia

Hiltner (1904) termed the soil around the roots as the rhizosphere, where the microbial population is very high (Bodelier et al. 1997). This region is rich in compounds such as amino acids, sugars, vitamins, organic acids, auxins, flavonoids, etc. which are released by the plants. The microbes get attracted by these compounds which are also known as root exudates utilized to the microbial population for their multiplication (Lynch and Whipps 1990; Dakora and Phillips 2002; Somers et al. 2004; Dardanelli et al. 2008, 2010; Raaijmakers et al. 2009). This interaction between plants' roots and bacteria leads all the exchanges between them and governs beneficial, deleterious, and neutral processes. In other words, those compounds act as chemo-attractants and help the microbial population to communicate with the plants, resulting in successful interaction (Bolton et al. 1986; Dardanelli et al. 2008, 2010). As a result, the competent bacteria which multiply and colonize the rhizosphere are known as rhizobacteria (Antoun and Kloepper 2001). These rhizobacteria often possess beneficial traits which enhance the growth of plants, also known as plant growth-promoting rhizobacteria (PGPR) (Kloepper 1978). These groups of bacteria possess different modes of action; some provide direct nourishment by synthesizing beneficial compounds or through indirect mechanisms helping plants to withstand deleterious effects or pathogen crisis (Glick et al. 1995). Rhizobia are also considered as PGPRs (Chandra et al. 2007), which associate themselves with leguminous as well as nonleguminous plants (Höflich et al. 1994; Noel et al. 1996; Yanni et al. 1997; Antoun et al. 1998; Rodríguez and Fraga 1999; Sessitsch et al. 2002). Some of the well-known rhizobial PGPRs belong to genera *Allorhizobium, Azorhizobium, Bradyrhizobium, Mesorhizobium, Rhizobium,* and *Sinorhizobium* (Mehboob et al. 2012). These rhizobia benefit the plants in many ways (Fig. 3.1), some of which are mentioned below.

3.5.1 *Direct Mechanisms*

The direct mechanism of PGPR shown by various bacterial genera includes phytohormone production, mineral solubilization, nitrogen fixation, siderophore, and HCN production. These mechanisms highly influence the plant growth and result in better crop yield.

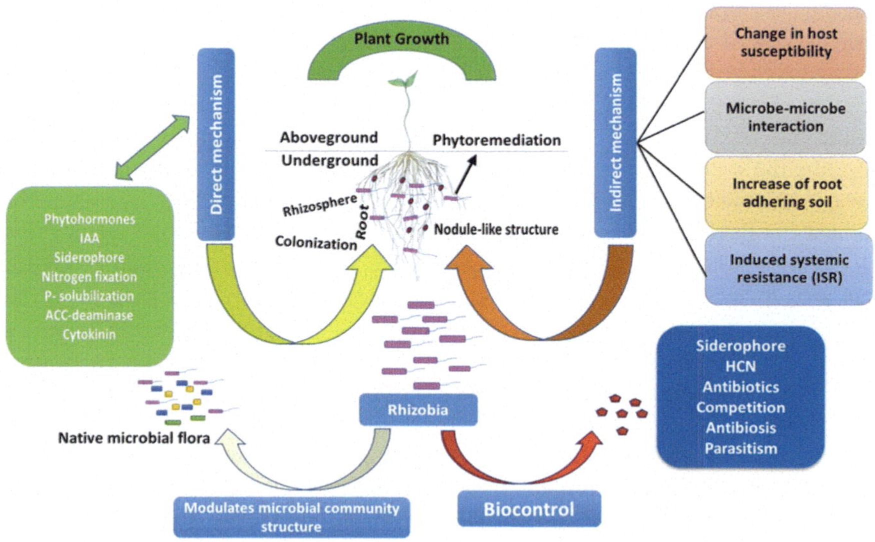

Fig. 3.1 Various mechanisms of rhizobia by which they benefit a nonlegume plant

3.5.1.1 Production of Important Compounds

Rhizobia produces lower molecular weight plant hormones (phytohormones) which are known to regulate important physiological and developmental processes during the growth of the plant (Chiwocha et al. 2003). These compounds affect the process of flowering, aging, root and stem development, fruit coloration, formation and shredding of leaves, and many other processes. Some of the important phytohormones are auxins, cytokinins, gibberellins, abscisic acid, indole-3-acetic acid (IAA), and ethylene (Zahir and Arshad 2004; Khalid et al. 2006). The production of these important compounds is an important characteristic of rhizobia (Phillips and Torrey 1970; Hirsch et al. 1997; Law and Strijdom 1988; Atzorn et al. 1988; Minamisawa et al. 1996), and also benefits the nonleguminous category (Biswas et al. 2000; Yanni et al. 2001; Hafeez et al. 2004; Matiru and Dakora 2005a; Mishra et al. 2006; Chandra et al. 2007; Humphry et al. 2007; Pena and Reyes 2007).

The Nod factors produced by rhizobia which are essential in forming nodules in leguminous plants (Buhian and Bensmihen 2018), also play an important role in nonleguminous crops. These Nod factors help in rapid and transient alkalinization of cells of tobacco (Baier et al. 1999), tomato (Staehelin et al. 1994), and restore division of cell and embryonic development in carrot (De Jong et al. 1993), increasing root mass and length (Smith et al. 2002), enhance photosynthate production and yield of grain when sprayed over the surface of leaves (Smith et al. 2001, 2002). It has also been reported to restore cell division and embryogenesis in the plants when auxins and cytokinins are absent (Dyachok et al. 2000). Moreover, in maize and cotton, Nod factors induce the germination of seeds and pitches for early seedling development, at low temperatures. Nod factors also promote colonization of legumes as well as nonlegumes by AM fungi (Xie et al. 1995).

Besides rhizobia produce some signaling compounds such as lumichrome which stimulates growth of plants (Yang et al. 2002; Beveridge et al. 2003; Dakora 2003; Matiru and Dakora 2005b). This compound is also known to help host plants in surviving the water stress by decreasing the leaf stomatal conductance and reduction of water loss via transpiration through the leaves (Phillips et al. 1999). Rhizobia also produce riboflavin which possesses a significant role in plant-microbe interactions (McCormick 1989). It can be further converted to lumichrome, which promotes plant growth.

3.5.1.2 Production of Enzymes

Ethylene is a hormone that promotes the ripening of fruit, breaks the dormancy of seed, and promotes the formation of root hairs (Dolan 2001). However, its overproduction inhibits the growth of the plant (Li et al. 2018). *Rhizobium* sp. produces 1-aminocyclopropane-1-carboxylate (ACC) deaminase, which is known to reduce the ethylene levels in plants, by hydrolyzing ACC (the precursor of ethylene) (Walsh et al. 1981; Yang and Hoffman 1984) into ammonia and α-ketobutyrate, then absorbing them as a source of nitrogen and carbon (Honma and Shimomura 1978; Klee et al. 1991). Rhizobia with ACC deaminase activity possess longer roots (Glick et al. 1999) and are known to resist the ethylene stress imposed of heavy metals (Burd et al. 2000), attack of pathogens (Wang et al. 2000), drought stress (Arshad et al. 2008; Zahir et al. 2008), salinity (Mayak et al. 2004; Nadeem et al. 2007; Zahir et al. 2009), and water stress (Grichko and Glick 2001). Thus, impart indirect benefit to the plants.

3.5.1.3 Production of Siderophore

Siderophores are chelating compounds that are produced by bacteria and supply iron to the plants which is necessary for the synthesis of chlorophyll and also present as co-factors (Rout and Sahoo 2005). It solubilizes ferric iron from the soil and transports it readily into the cells (Neilands 1993). Siderophores contribute the majority of the available iron supply to the plants from the rhizospheric soil (Masalha et al. 2000). Different strains of rhizobia are known to possess siderophore activity in nonlegumes. *Rhizobium meliloti* (Schwyn and Neilands 1987; Arora et al. 2001), *S. meliloti*, *R. leguminosarum* bv. *viciae*, *R. leguminosarum* bv. *trifolii*, *R. leguminosarum* bv. *phaseoli*, *R. tropici* (Chabot et al. 1993; Carson et al. 2000), *Rhizobium* sp. (Deryło et al. 1994; Antoun et al. 1998), and *Bradyrhizobium* (Plessner et al. 1993; Jadhav et al. 1994; Dudeja et al. 1997; Antoun et al. 1998) to name a few which produce siderophore for the acquisition of Fe^{3+} chelation in the iron-deficient environment (Guerinot 1991; Carson et al. 1992; Reigh and O'Connell 1993; Guerinot 1994; Arora et al. 2001).

3.5.1.4 Solubilization and Uptake of Nutrients

Phosphorus is an important nutrient for plants which is available in soil in two forms, organic and inorganic. Organic phosphates are phosphomonoesters, phosphodiesters (phospholipids and nucleic acids), and phosphotriesters (Rodriguez and Fraga 1999). Inorganic forms are apatite, hydroxapatite, and oxyapatite (Rodriguez and Fraga 1999; Fernández et al. 2007) which are insoluble. Its deficiency can lead to limited plant growth and low yield (Fernández et al. 2007). Phosphorous remains unavailable for plants due to their immovable nature and depends on oil type as well as pH. Some rhizobia possess the phosphate solubilization ability (both organic and inorganic) which in turn supplies phosphate to the plant (Abd-Alla 1994; Antoun et al. 1998; Dazzo et al. 2000; Alikhani et al. 2007; Afzal and Bano 2008). It was reported that *R. meliloti* possesses phosphate solubilization activity in nonlegumes to enhance their growth (Egamberdiyeva et al. 2004).

Similarly, supply of other important nutrients such as N, P, K, Ca, Mg, Zn, Na, Mo, and Fe by *Rhizobium*, *R. leguminosarum* bv. *trifolii*, *Bradyrhizobium* (Khokhar and Qureshi 1998; Biswas et al. 2000; Yanni et al. 2001), K^+ and Ca^+ in cotton by *R. leguminosarum* bv. *Trifolii* (Hafeez et al. 2004), and N, K, Na, Zn, Fe, and Cu in wheat by *Rhizobium* (Amara and Dahdoh 1995) are some important examples of nutrient supply by rhizobia in nonlegumes.

3.5.1.5 Amelioration of Different Plant-Stress Conditions

Rhizobial inoculation to nonleguminous plants has yielded promising results in stress amelioration (Silva et al. 2020) as rhizobia help in combating different types of biotic and abiotic stresses. Rhizobial inoculation has resulted in countering water stress in the host plant as reported by several workers (Figueiredo et al. 1999; Alami et al. 2000; Tulumello et al. 2021). Rhizobial inoculation alters the stomatal conductance and transpiration (Matiru and Dakora 2005b), improving photosynthetic capacity (Chi et al. 2005), and also known to alter the morphology of roots which helps in absorbing the nutrients from the soil and also resists drought conditions. Pesticides affects the growth of the plant by disturbing the normal root functioning altering root architecture, sites of rhizobial infection, ammonia transformation, and exchange of compounds between plants and microbes, and also by affecting the microbial population and diversity (al-ani et al. 2019). Kanade et al. (2010) reported the use of rhizobia from the fenugreek plant for the degrading of malathion. Though in other reports, the field results were not found to be very satisfactory and require more research (Gopalakrishnan et al. 2015).

3.5.2 Indirect Mechanism

This involved the functional role of rhizobacteria in inhibiting the phytopathogens causing disease in plants.

3.5.2.1 Biocontrol

Biocontrol is the phenomenon by which microbes play an important role to eliminate or reducing the effect of pathogens by secreting various kinds of compounds such as antibiotics, HCN, cell-wall lytic enzymes such as chitinase and glucanase (Chakraborty and Purkayastha 1984; Deshwal et al. 2003; Chandra et al. 2007). Rhizobia possess antagonistic activity against pathogens and also change the level of host susceptibility against a particular pathogen. Different mechanisms are being exhibited by the rhizobia such as competition, antibiosis, or parasitism to eliminate the pathogen. The competition of nutrients between the bacteria and pathogen may also result in the elimination of the pathogen. *Rhizobium* spp. suppress the disease-causing pathogen by the production of lytic enzymes, antibiotics, and ISR (Volpiano et al. 2019). Siderophore activity plays an important role in starving the pathogen from acquiring iron (Carrillo and Vazquez 1992; Arora et al. 2001). Arora et al. (2001) reported the action of siderophore-producing rhizobia against *Macrophomina phaseolina*, a disease-causal fungus in more than 500 angiosperm plants. In antibiosis, the rhizobia produce compounds called antibiotics which act as an eliminator to the pathogen. *R. leguminosarum* bv. *trifolii* produces trifolitoxin (Schwinghamer and Belkengren 1968; Breil et al. 1993) which is potent enough against many plant and animal pathogens (Triplett et al. 1994). Parasitism includes the elimination of the pathogen with the help of enzymes. For instance, chitinase and glucanase break the cell wall of pathogenic fungi. *R. leguminosarum*, *S. meliloti,* and *B. japonicum* are known to be used against genera *Macrophmina*, *Rhizoctonia,* and *Fusarium* (Ehteshamul-Haque and Ghaffar 1993; Özkoç and Deliveli 2001). *S. meliloti* and *R. trifolii* are reported to inhibit *F. oxysporum*, and rot/knot disease of the root of sunflower and tomato plants (Antoun et al. 1978; Siddiqui et al. 2000; Shaukat and Siddiqui 2003), *R. leguminosarum* bv. *viciae* is known to control *Pythium* that causes damping-off of sugar beet (Bardin et al. 2004), *M. loti* inhibits the growth of *Sclerotinia sclerotiorum* (Chandra et al. 2007), *B. japonicum* controls root rot of mustard and sunflower and may decrease the sporulation of *Phytophthora megasperma*, *Pythium ultimum*, *Fusaruim oxysporum,* and *Ascochyta imperfecta* (Tu 1978, 1979; Ehteshamul-Haque and Ghaffar 1992, 1993; Siddiqui et al. 2000). Long back, *R. meliloti* was reported to control root-knot phytoparasitic nematode in okra (Parveen and Ghaffar 1991; Parveen et al. 1993; Ehteshamul-Haque et al. 1996).

3.5.2.2 Change in Host Susceptibility

The microbes often induce resistance in plants (Van Loon 2007), and the process by which resistance is incurred in the plants is known as induced systemic resistance (ISR). Rhizobia can limit the effect of the pathogen through the induction of plant defense mechanisms (Abdel-Aziz et al. 1996). ISR system is adopted by rhizobia for controlling many fungal pathogens of nonlegumes such as sunflower, okra, and soybean (Ehteshamul-Haque and Ghaffar 1993; Nautiyal 1997). Rhizobia have been reported to produce several biostimulatory agents (Yanni et al. 2001; Peng et al. 2002; Mishra et al. 2006; Singh et al. 2006), eliciting ISR in the plants. *Rhizobium etli* was reported to induce ISR in the roots of potato through a special transduction pathway that protects against *Globodera pallida* (Reitz et al. 2000). *R. leguminosarum* bv. *phaseoli* and *R. leguminosarum* bv. *trifolii* inoculations induce increased synthesis of phenolic compounds in rice plants which mediates ISR and provides bioprotection to the plants against pathogens (Mishra et al. 2006). *Mesorhizobium* sp. Showed increased growth and defense against *Sclerotium rolfsii* infection (Singh et al. 2014).

3.5.2.3 Microbe-Microbe Interaction

The qualities of rhizobia as PGPR can further be enhanced with the addition of one or more bacterial cultures, thus a consortium with other PGPR can prove much beneficial. It was reported that using multiple cultures of PGPR promote the yield of nonlegumes like sorghum (Alagawadi and Gaur 1988), rice barley (Belimov et al. 1995; Höflich et al. 1994), rice (Yanni et al. 1997), maize (Chabot et al. 1993), and wheat (Galal 2003). Nitrogen-fixing bacteria like rhizobia along with other PGPRs are highly beneficial to the crop (Şahin et al. 2004). Sheikh et al. (2006) studied the beneficial traits of using *R. meliloti* and *B. thuringiensis* in okra plants which resulted in better plant growth and performance against fungal pathogens. Han and Lee (2005) reported better growth of lettuce while using co-inoculation of *Serratia* sp. And *Rhizobium* together. Moreover, in degrading soil environments, use of AM fungi, rhizobia, and other PGP strains have been very successful in uplifting the quality of soil (Requena et al. 1997). Also, inoculation of rhizobia can modulate the rhizospheric microbial community, thus improving the soil health and thus growth of the plant (Xu et al. 2020).

3.5.2.4 Increase of Root Adhering Soil

Root adhering soil (RAS) is very important to plants as this region provides water and other nutrients. Two types of such soil exist namely loosely adhering and closely adhering. The soil around the root is much important to the plant as it supports the plant (Dobbelaere et al. 2003). This is the region where the microbial activity is much higher, results in an exchange of several beneficial compounds. Rhizobia-producing exopolysaccharides (EPS) are of great importance which increase soil aggregation (Martens and Frankenberger 1993), and also trap moisture, and other essential nutrients (Alami et al. 2000). Thus, EPS improves RAS and contributes to soil aggregation (Kaci et al. 2005).

3.6 Nitrogen Fixation in Nonlegumes

BNF in the nonleguminous plants by symbiotic rhizobia has been relatively less studied. Fixation of nitrogen by different rhizobia which form exogenous or endogenous symbiosis in nonleguminous plants has been reported by some of the scientists. Werner (1992) reported *Rhizobium* genus to form nodule-like structures in *Parasponia* and similarly fix N_2 as in leguminous plants. *Rhizobium parasponium* and *Bradyrhizobium* were reported to form nodules in oilseed plants (Cocking et al. 1992). Structures like nodules, galls, or root outgrowths have been observed in many nonleguminous plants such as rice, oilseed, *Arabidopsis thaliana* (Al-Mallah et al. 1989, 1990; Bender et al. 1990; Rolfe and Bender 1990; Jing et al. 1990, 1992; Li et al. 1991; Ridge et al. 1992; Spencer et al. 1994; De Bruijn et al. 1995; Trinick and Hadobas 1995). Velázquez et al. (2005) reported the presence of both symbiosis and pathogenicity-related genes *Rhizobium rhizogenesi,* which help to form nodule-like structures in different plants. *Rhizobium* inoculation enables nitrogen fixation in wheat was reported by Chen et al. (1991), Yu and Kennedy (1995), and Cocking et al. (1995). *Azorhizobium caulinodans* was reported to increase dry weight and nitrogen content resulting from nitrogenase activity when inoculated in wheat, further validating BNF in nonlegumes (Sabry et al. 1997). Nitrogenase activity was observed after inoculation of *A. caulinodans* in rice plants (Naidu et al. 2004). It was suggested that the endophytic nature of particular rhizobia should be active for effective nitrogen fixation with nonlegumes. Diverse genera like *Azoarcus* sp., *Burkholderia* sp., *Gluconacetobacter diazotrophicus,* and *Herbaspirillum* sp. Were reported to have the nitrogen-fixing ability as endophytes (Vessey 2003). Verma et al. (2004) reported higher nitrogen fixation in rice plants inoculated with *Ochrobactrum* sp. Moreover, various attempts have been made by using the latest techniques to incorporate the BNF by rhizobia in nonleguminous plants through genetic engineering but with limited success (Saikia and Jain 2007).

3.7 Application of Rhizobia with Nonlegumes

Rhizobia as a PGPR have multiple practical applications associated with it. Rhizobia are mostly known for its biofertilizer property, biocontrol ability, phytoremediation, and stress regulating properties (Kumari et al. 2019). Biofertilizer increases the growth of the plant through multiple mechanisms such as nitrogen fixation, releasing compound which helps in the growth of the plant, by increasing the availability of nutrients (Cocking 2003). The biofertilizer supplies or mobilizes the important compounds with minimal resources. These properties were reported from rhizobia while using it as a biofertilizer (Bardin et al. 2004; Chi et al. 2005). These biofertilizers are cost-effective and environment-friendly alternative to chemical fertilizers. Rhizobia are used as commercial biofertilizers in various nonlegumes for enhancing their growth and yield (Perrine et al. 2001; Hussain et al. 2009). Such rhizobial biofertilizer strains have been known to compete with the pathogen (Arora et al. 2001), secrete metabolites such as antibiotics (Deshwal et al. 2003), produce enzymes for cell wall lysis (Özkoç and Deliveli 2001), siderophore activity (Deshwal et al. 2003), HCN production (Chandra et al. 2007), and also reported inducing ISR (Singh et al. 2006). Many PGPR strains including rhizobia are reported for their biocontrol ability (Reitz et al. 2000; Bardin et al. 2004; Chandra et al. 2007). *B. japonicum, R. meliloti,* and *R. leguminosarumi* are used against *M. phaseolina, R. solani, Fusarium solani,* and *F solani* (Ehteshamul-Haque and Ghaffar 1993); *M. loti* against white rot disease of *Brassica campestris* (Chandra et al. 2007); *R. leguminosarum* bv. *Phaseoli* and *R. leguminosarum* bv. *Trifolii* against *R. solani* in rice plants (Mishra et al. 2006).

PGPR are also known for its usefulness in phytoremediation (Khan et al. 2009; Glick 2010). Apart from plant-microbe interactions, phytoremediation largely depends on several abiotic and biotic factors such as soil physicochemical properties, nutrient availability, water content, type, and concentration of contaminants (Thijs et al. 2017). Efficient phytoremediation depends on the growth and survival of both plant and active rhizospheric microbiome in polluted soil. Heavy contamination restricts the microbial population due to its toxic nature (Cook and Hesterberg 2013). It becomes more potent when used in conjunction with a plant, increasing the availability and mobility of pollutants, and also acidifies the targeted contaminants, along with phosphate solubilization and release of chelating agents in addition to enhancing plant growth (Abou-Shanab et al. 2003; Höflich et al. 1994; Noel et al. 1996; Yanni et al. 1997; Dazzo et al. 2000; Arora et al. 2001; Özkoç and Deliveli 2001; Dakora 2003; Matiru et al. 2005; Van loon 2007). Fagorzi et al. (2018) emphasized the advantage of using rhizobia in phytoremediation techniques of heavy metals.

Drought stress is one of the most important limiting factors for plant growth which can ultimately affect agricultural crop yields (García et al. 2017; Khan et al. 2018). Drought tolerance can be regulated by the production of ethylene, ACC deaminase, IAA, cytokinin, EPS, and antioxidant production (Joshi et al. 2019). Due to high salt concentration, soil become dry and thus plants are unable to uptake the water and also a high level of salt toxicity for plant cell (Kumar et al. 2019). Salt-resistant rhizobial strains can survive under osmotic stress (Irshad et al. 2021). Recent research on PGPR suggested that some of the strains can produce heat-/cold-resistant proteins which can enhance the thermal tolerance in plants (Ali et al. 2009). Alexandre and Oliveira (2013) discussed the physiology of rhizobia under thermal stress. There are several reports on ACC-deaminase producing root-nodulating rhizobia such as *Rhizobium leguminosarum* and *Mesorhizobium loti*

(Belimov et al. 2001, 2005; Ma et al. 2003; Sullivan et al. 2002) helping the plant to cop stress. These beneficial rhizobia are being used in different nonlegume crops as mentioned in Table 3.1.

Currently, various rhizobial biofertilizers are commercially available in the Indian as well as global market. The formulation of biofertilizers can be solid carrier-based (organic and inorganic), liquid-based (with or without additives), synthetic polymer-based, and metabolite-based formulations. The solid carrier materials are coal, coconut shell, wheat straw, cellulose, charcoal, etc. Using solid carrier-based formulation provides easy storage, application, and handling of the biofertilizers. Whereas liquid-based formulations are more useful for the legume plants during their sowing in large fields (Arora et al. 2017). A brief summary on crops like rice, wheat, and maize is further discussed.

*3.7.1 Rice (*Oryza sativa*)*

Rhizobia are known to improve the growth and yield of rice plants. There are several reports of rhizobial inoculation enhancing the growth of rice plants (Peng et al. 2002; Yanni et al. 2001; Chaintreuil et al. 2000; Matiru and Dakora 2004; Singh et al. 2005; Bhattacharjee et al. 2008; Senthilkumar et al. 2008). Naidu et al. (2004) reported the increased growth and yield of rice after rhizobial inoculation. Colonization of rice was checked by Chi et al. (2005), who reported increased root and shoot biomass followed by a rate of photosynthesis, stomatal conductance, transpiration rate, efficiency in water utilization, and increased area of flag leaves when inoculated with rhizobia. Singh et al. (2005) reported increased biomass and grain yield of rice due to the application of three rhizobial strains. These rhizobial strains are potent enough to colonize the rice plants and exhibit different PGP characteristics (Yanni et al. 1997). Biswas et al. (2000) studied rhizobial isolates from different legumes and their application in rice plants, resulting increased grain (8–22%), the yield of straw (4–19%), nutrients N, P, K (10–28%), and Fe uptake (15–64%). Rhizobial strains significantly contributed to the increased vigor of rice seedlings, growth physiology, and modulate root morphology (Mehboob et al. 2012).

*3.7.2 Wheat (*Triticum aestivum*)*

Rhizobia colonize endophytically in wheat and result in various growth and yield promotion (Sabry et al. 1997; Biederbeck et al. 2000). Webster et al. (1997) reported *A. caulinodans* inoculation elicits lateral roots in the wheat plants. *R. leguminosarum* bv. *Trifolii* is reported to increase shoot length in the wheat (Höflich 2000). Anyia et al. (2004) observed inoculation of *A. caulinodans* enhances increased grain yield and total biomass by 34% and 49%, respectively and also larger leaf surface area. Amara and Dahdoh (1995) discussed *Rhizobium* inoculation resulted in a high yield

Table 3.1 Some of the rhizobia and their mode of action in nonleguminous crops

Host plant	Rhizobia	Mechanism	References
Rice	*Bradyrhizobium* sp.	Plant growth	Chaintreuil et al. (2000)
	R. leguminosarum	IAA production	Biswas et al. (2000), Dazzo et al. (2000)
	Rhizobium leguminosarum	Auxin and nitrate production, and root colonization	Perrine et al. (2001)
	Bradyrhizobium sp.	Plant growth promotion	Peng et al. (2002)
	Rhizobium sp.	Indole-3-acetic acid, gibberellin production, and root colonization	Chi et al. (2005)
	Rhizobium leguminosarum	Biocontrol/phenolics production	Mishra et al. (2006)
	Rhizobium sp.	N_2-fixation and root colonization	Singh et al. (2006)
	Rhizobium leguminosarum bv. *trifolii*	N_2-fixation	Perrine-Walker et al. (2007)
	Rhizobium phaseoli, Mesorhizobium cicer	Plant growth promotion	Hussain et al. (2009)
	Bradyrhizobium	Plant growth promotion	Mia and Shamsuddin (2009)
	Rhizobium sp.	Rhizosphere, root colonization, and N_2-fixation	Vargas et al. (2009)
	Sinorhizobium meliloti	Nutrient uptake and indole-3-acetic acid production	Chi et al. (2010)
	Azorhizobium caulinodans	Indole-3- acetic acid, cytokinins production, and nitrogenase activity	Senthilkumar et al. (2009)
Wheat	*Rhizobium leguminosarum*	Plant growth promotion	Hilali et al. (2001)
	Rhizobium sp.	EPS production	Kaci et al. (2005)
	Rhizobium leguminosarum	Phosphate solubilization	Afzal and Bano (2008)
	Rhizobium leguminosarum	Production of indole-3- acetic acid and nutrient solubilization	Etesami et al. (2009)
Maize	*Bradyrhizobium japonicum*	Plant growth promotion	Prévost et al. (2000)
	Rhizobium	Low nutrient solubilization	El-Tarabily et al. (2006)
	Mesorhizobium ciceri, Rhizobium leguminosarum, Rhizobium phaseoli	Plant growth promotion	Mehboob et al. (2008)

(continued)

Table 3.1 (continued)

Host plant	Rhizobia	Mechanism	References
	Bradyrhizobium	Plant growth promotion	Roesch et al. (2008)
Barley	*Mesorhizobium mediterraneum*	Phosphate solubilization	Peix et al. (2001)
	Rhizobium radiobacter	Indole-3-acetic acid and gibberellic acid production	Humphry et al. (2007)
	Bradyrhizobium japonicum	Lipo-chitooligosaccharides and gibberellin production	Miransari and Smith (2009)
Brassica campestris/ napus	*Rhizobium leguminosarum*	Indole-3- acetic acid and cytokinin production	Noel et al. (1996)
	Rhizobium alamii	Plant growth	Tulumello et al. (2021)
Sunflower	*Rhizobium sp.*	EPS production	Alami et al. (2000)
	Mesorhizobium loti	Biocontrol, production of hydrocyanic acid, indole-3-acetic acid, and phosphate solubilization	Chandra et al. (2007)
Sorghum	*Bradyrhizobium japonicum, Sinorhizobium meliloti*	Indole-3- acetic production and nutrient solubilization	Matiru et al. (2005)
Cotton	*Rhizobium leguminosarum*	Indole-3-acetic acid production	Hafeez et al. (2004)
	Rhizobium sp.	Plant growth and yield	Qureshi et al. 2019
Raddish	*Rhizobium, Bradyrhizobium*	Plant growth	Antoun et al. (1998)
Canola	*R. leguminosarum*	Plant growth	Noel et al. (1996)
Potato	*Rhizobium etli*	Biocontrol	Reitz et al. (2000)
Tomato	*Bradyrhizobium japonicum*	Plant growth	Carletti et al. (1994)
Lettuce	*Rhizobium sp.*	Phosphate solubilization, siderophores and auxins production	Chabot et al. (1993)
	Rhizobium sp.	Indole-3-acetic acid production and P-solubilization	Pena and Reyes (2007)
Switchgrass	*Bradyrhizobium* spp. *Rhizobium helanshanense*	Plant growth promotion	Bahulikar et al. (2014)
Sugarcane	*Rhizobium daejonense Sinorhizobium fredii*	Plant growth promotion	Thaweenut et al. (2011)
Sweet potato	*Bradyrhizobium*	Plant growth promotion	Reiter et al. (2003)
	Sinorhizobium meliloti Bradyrhizobium japonicum Rhizobium leguminosarum	Plant growth promotion	Terakado-Tonooka et al. (2008)

of grains as compared to control. Kaci et al. (2005) studied the inoculation of *Rhizobium* in wheat, increased shoot dry mass (85%), root dry mass (56%), root adhering soil (RAS) dry mass (dm) per root dm (RAS/RT) up to 137%, and aggregate water stability in RAS with its EPS-producing property. Similarly, Afzal and Bano (2008) reported rhizobia along with other PGPR considerably enhance the grain yield of wheat.

*3.7.3 Maize (*Zea mays L.*)*

Rhizobia are also reported to increase the yield of maize. Though they do not contribute to the nitrogen-fixing element (Höflich et al. 1994), inoculation of *R. etli* has resulted in increased dry matter (Martínez-Romero et al. 2000). Chabot et al. (1998) reported rhizobial inoculation under P-deficient and P-rich soils has resulted in better growth of maize. Höflich (2000) reported *R. leguminosarum* bv. *Trifolii* strain promotes the growth of maize in both greenhouse and field trials. Shakhawat Hossain and Mårtensson (2008) reported rhizobial inoculation enhanced shoot and root dry weight of maize plants. Mehboob et al. (2008) reported inoculation of *Rhizobium phaseoli* has resulted in increased root length, shoot length, and seedling biomass as compared to uninoculated control. Rhizobia with multiple PGP traits have to increase the dry matter of shoots after inoculation (Chabot et al. 1993).

3.7.4 Other Crops

Other than above crops, the application of rhizobia as PGPR has also been tested in cotton plants with *R. meliloti,* which resulted in increased yield (Egamberdiyeva et al. 2004). Hafeez et al. (2004) reported increased seedling emergence, shoot dry weight, biomass, and nitrogen uptake after inoculation with various rhizobia strains. *B. japonicum, A. caulinodan, Rhizobium, Rhizobium, S. meliloti, R. leguminosarum* bv. *Viceae,* and *R. leguminosarum* bv. Viceae have been reported to promote the growth and yield of sorghum, millet, and sudangrass (Matiru et al. 2005). Chabot et al. (1993) examined increased growth of lettuce after application of rhizobial strains. Noel et al. (1996) observed inoculation of *R. leguminosarum* resulted in increased growth of lettuce. Along with growth promotion, biocontrol activity of rhizobia has also been reported from *B. japonicum* and *R. leguminosarumi* against *M. phaseolina, R. solani,* and *Fusarium* spp. Causing disease in sunflower and okra plants (Ehteshamul-Haque and Ghaffar 1993). Sheikh et al. (2006) used *R. meliloti* and *B. thuringiensis* against *M. phaseolina, R. solani,* and *Fusarium* spp. In okra plants. Moreover, EPS-producing *Rhizobium* strain plays a role in PGP, mediates water stress, and also supplies water in sunflower plants (Alami et al. 2000). Peix et al. (2001) reported *Mesorhizobium mediterraneum* enhances the growth of barley, while Humphry et al. (2007) observed the effect of *R. radiobacter* strain in barley plants. Application of *B. japonicum* in radish induces plant dry matter (Antoun et al. 1998). Chandra et al. (2007) reported enhanced seed germination, early vegetative growth, and yield of Indian mustard (*Brassica campestris*) by *M. loti*. It was also reported that the use of multiple strains of PGPR is more beneficial than using single culture of rhizobia for growth promotion (Akintokun and Taiwo 2016).

3.8 Rhizobial Bioengineering

The competitiveness of rhizobia in various types of soil can be by increasing their multiplication in the specific environment or through modifying the signal mechanism of the competitive microbes which in turn disrupts the normal functioning of the introduced microbes (Savka et al. 2002). As we know for a successful interaction, the soil of a particular environment, associated microbes, and the plant are interlinked. Altering, one of them can be beneficial for the colonization of the target rhizobia. The genetic aspect is always important which governs the competitive nature of the target bacteria. Several studies have underlined the causative genes, their deficit leads to limited or less competitiveness. However, the study of genes that might increase the competitive nature is yet to be determined (Geetha and Joshi 2013). Some successful techniques for manipulating the genes are to construct chimeric *Nif HDK* operon under NifHc promoter and expression in PHB negative mutants of *R. etli* (Peralta et al. 2004), to develop an acid-tolerant *R. leguminosarum* bv. *Trifolii* strain (Chen et al. 1991), to express the ACC deaminase gene in *S. meliloti* (Ma et al. 2004), overexpression of *putA* gene (Van Dillewijn et al. 2001), overexpression of trehalose 6-phosphate synthase gene (Suárez et al. 2008), overexpression of *rosR* and *pssR* genes (Janczarek et al. 2009), heterologous expression of ferrichrome siderophore receptor *fegA* and *fhuA* genes (Joshi et al. 2008; Geetha et al. 2009; Joshi et al. 2009), and overproduction of the adhesion rap1 (Mongiardini et al. 2009). Also introducing the property to utilize diverge nature of siderophore into the bacterial inoculants further enhances the root colonization ability and biofilm formation. Though the *nif*H genes are critical for competitiveness, the genes of iron up-taking are equally important. Through genome analysis, it was established that TonB-dependent siderophore receptors are important in iron uptake and are not adequately present naturally in the rhizobia (Joshi et al. 2009). Among rhizobia, *Bradyrhizobium* possesses the most TonB receptor and hence their accumulation and competitive nature are higher than other rhizobia groups (Hume and Shelp 1990). Also, some FhuA homologs are present in the inner membrane, possess similar functioning to FhuE (rhodotorulic acid and coprogen receptor) and IutA (aerobactin receptor) (Streeter 1994). The receptors work in combination with FhuBCD (ferrichrome system), suggest the transport of ferric siderophores through the inner membrane is more specific than the outer membrane, resulting in a lesser number of periplasmic and cytoplasmic membrane proteins present in the inner membrane (Stevens et al. 1999). Thus, the increase of repertoire of outer membrane siderophore receptors could enable rhizobial isolates to enhance iron uptake and colonization in different environments (Geetha and Joshi 2013). The BNF can be made more efficient by accelerating the delivery of electrons required for catalyzing the biochemical reaction performed by nitrogenase enzyme. This is by overexpressing the set of *nif* and *fix* groups of genes (Goyal et al. 2021). Moreover, the structurally similar genes such as *Nod* and *Myc* factors are responsible for activating the signaling pathway during mycorrhizal symbiosis in various crops (Maillet et al. 2011). The modulation of nod factors for activating the mycorrhizal symbiosis signaling pathway which activates the modified nodulation-related genes has been reviewed in nonlegumes (Rogers and Oldroyd 2014). As such, a transgenic rice plant exhibiting root deformation similar to initial nodule formation in legumes through expressing legume-specific nodulation (*Nod*) factor receptor protein genes suitably responded to the rhizobial Nod factors (Altúzar-Molina et al. 2020) but more alteration is to be paid in carrying out the similar work on the crops.

3.9 Challenges and Limitations

Though in many instances, rhizobia act as a potential PGPR and enhance the quality of applied crops, sometimes it also turns harmful to the plant. Though such phenomenon may be caused due to noncompatibility of the plant with an interacting microbe or the applied inoculant may lead to overproduction of certain harmful compounds. This phenomenon leads to deleterious effects on the plant (Alström 1991). Some PGP traits such as IAA, HCN, etc. are proved better for the plants when released in low concentration, but are harmful to the plant at supra-optimal concentration (Antoun et al. 1998; Alström and Burns 1989; O'Sullivan and O'Gara 1992). Perrine et al. (2001) reported the harmful nature of auxin and nitrate when available in high concentrations. Further, the growth inhibitors produced by the rhizobial strain proved harmful to the plant (El-Tarabily et al. 2006). Other factors, such as the plant-microbe or microbe-microbe interaction, where the inoculated PGPR may not be competent enough to bend with the native microbial flora led to undesired results (Antoun et al. 1998). It is also stated that the soil, pH, and environmental factors also play multifarious role in the plant-microbe interaction (Lynch 1990a, b; O'Sullivan and O'Gara 1992; Hilali et al. 2001).

To evaluate rhizobia as PGPR, and to develop it on a mass scale, requires a considerable amount of time and require various steps. To develop an effective biofertilizer, we must aim to evaluate the developmental processes, the policymakers, associated industries, research, and tie-ups with educational institutions. All should work collaboratively and must be implemented as per guidelines. The field-oriented research carried out must be readily made available to the public domain. The commercialization of the outcome of the conducted work should be more encouraged and technology be transferred to the industries. There are some limitations and associated disadvantages which are suggested below.

3.9.1 Limitation in Field Application

Rhizobial application as PGPR in greenhouse or laboratory trials showed optimistic outcomes. But the growth conditions in greenhouses can be controlled and adjustable to the favorable growing requirement of the crop throughout the season (Paulitz and Bélanger 2001). Thus, achieving such controlled field trials is not possible as several biotic and abiotic factors influence crop developments. Also, the abundance of indigenous microorganisms is more pronounced in field soil which can alter or affect the proliferation of applied PGPR strains. Knowledge and research are required for the successful application of rhizobia in the field. The proper timing of inoculation, types of crops, mutual interactions between host plant and microbes, bioformulation of rhizobial strains, the concentration of inoculum applied, and management of crops can ensure the growth support, augmentation, and bioactivity of PGPR in field practices (Bowen and Rovira 1999; Gardener and Fravel 2002; Mansouri et al. 2002). However, recent approaches such as rhizosphere engineering and improved carrier techniques can overcome the limitations of rhizobial field applications (Date 2001; Yardin et al. 2000).

3.9.2 Selection and Characterization

Major challenges in rhizobial product application are the screening of potential microbial strains and its bioformulation process (Kumari et al. 2019). For the selection and screening of the most promising strains, plant adaptions to particular soil types, root exudates, and surrounding ecological environmental status play a vital role (Bowen and Rovira 1999). Various approaches include the use of enrichment medium for the selection of need-based indigenous N-fixing bacteria from the rhizosphere. Another application of the spermosphere model is where plant root exudates use as a sole nutrient source for the proliferation of rhizosphere bacteria (Joshi et al. 2019). The selection of microbial populations based on their phosphate solubilizing, siderophore, and antibiotic production abilities (Weller et al. 2002; Silva et al. 2003) with other beneficial traits are most desirable.

3.9.3 Limitations in Commercialization

Slow growth in commercialization is due to a lack of knowledge among farmers. The field trainers and farmers must be educated about the beneficial role of rhizobial inoculants, its bioformulation, and its economical acceptability to the diverse genera (Kumari et al. 2019). Several factors are to be considered before the commercialization of the PGPR. These include large-scale production of strains, shelf-life compatibility, temperature tolerance, eco-friendly economic which does not impart toxicity or pathogenicity to human and animal should be measured before marketing (Joshi et al. 2019).

3.10 Rhizobia and Omics Technologies

Didier Raoult and Jean-Christophe Lagier coined the word culturomics to describe an approach for bringing more bacterial isolates from environmental microbiomes into laboratory culturing (Lagier et al. 2018). PCR amplification of the ubiquitous 16S ribosomal RNA (rRNA) has been used to identify bacterial isolates in conjunction with these culture techniques (Turner et al. 2013). Despite its significance, "culturomics" has many limitations, the most notable of which is the still limited ability for cultivating some bacterial taxa. Now a days, the culturome (strains that can be cultured in the laboratory) does not represent the entire microbiome (Martiny 2019; Steen et al. 2019). The genus *Rhizobium* is found in the core microbiome of many plants (Oberholster et al. 2018; Pérez-Jaramillo et al. 2019). Besides next-generation sequencing (NGS), the classification, platforms like Illumina and PacBio are essential for analyzing the genomes of *Rhizobium* species (Ormeno-Orrillo et al. 2015; González et al. 2019). Some studies have already used PacBio to generate genomes of novel species, such as *Rhizobium jaguaris* CCGE525T isolated from *Calliandra grandiflora* nodules (Servín-Garcidueñas et al. 2019), or to complete genome sequences, such as *Rhizobium* sp. strain 11515TR from tomato rhizosphere (Montecillo et al. 2018). Irar et al. (2014), on the other hand, described a proteomic approach to the nodule response to drought in *Pisum sativum*. Plants were inoculated with *R. leguminosarum* strains and cultivated in "normal well-irrigated" conditions and the other was impacted by a drought. The results showed a total of 18 proteins expressed during a period of drought: *Rhizobium leguminosarum* encodes 11 genes, and *Pisum sativum* encodes seven nodule proteins. These proteins have such a relation to RNA-binding proteins, flavonoid metabolism, and sulfur metabolism. All of the data gave a new goal for improving legume drought tolerance. Despite the relevance of these techniques, the scientists used model organisms such as *Sinorhizobium* or *Bradyrhizobium* species for their research. By using nuclear magnetic resonance, researchers were able to detect the exo-metabolomes generated by *Rhizobium etli* CFN42T, *Rhizobium leucaenae* CFN299T, *Rhizobium tropici* CIAT899T, and *Rhizobium phaseoli* Ch24-10 from free-living culture (Montes-Grajales et al. 2019), except the culture supernatant of *R. tropici* CIAT 899T none of them contained ornithine. This chemical has been linked to symbiotic efficiency as well as resilience to stress conditions like acidity (Rojas-Jiménez et al. 2005; Vences-Guzmán et al. 2011).

3.11 *Rhizobium* in Microbiome of Nonlegumes

The omics-based research revealed that the order
is a keystone taxon in a variety of settings, including forests, agricultural land, Arctic and Antarctic ecosystems, polluted soils, and plant-associated microbiota (Banerjee et al. 2018; LeBlanc and Crouch 2019). These habitats identify *Rhizobium* as a keystone taxon in the core microbiomes of several plant crops rhizospheres, including tropical crops, e.g., sunflower and sorghum (Bulgarelli et al. 2015; Yeoh et al. 2017; Oberholster et al. 2018), as well as their well-known presence and functions in the legume nodule microbiome (Velázquez et al. 2019; Zheng et al. 2020). In long-term experiments, several genera from the order Rhizobiales that are closely related to *Rhizobium*, such as *Agrobacterium*, *Bradyrhizobium*, and *Devosia* have been identified to be part of the maize rhizospheric core microbiome (Walters et al. 2018). Members of the Rhizobiaceae family and certain other Rhizobiales members appeared to be part of the heritable component of the maize rhizosphere microbiome. Several reports have been published in recent years about the occurrence of *Rhizobium* and related taxa in the rhizosphere, endosphere, and phyllosphere of nonleguminous crops. This is due to the interest in the investigation of agricultural microbiomes with the goal of discovering native rhizobial and nonrhizobial bacteria that may be endophytes to create benefits in nonlegume crops, being friendly with the indigenous microbiomes (Menéndez and Paço 2020).

Further, nonleguminous crops inhabit *Rhizobium*, also fix nitrogen within legume nodules, and other endophytic diazotrophs (Yoneyama et al. 2017, 2019). Using *nifH* gene amplification and cloning from various sources, some studies reported the presence of *Rhizobium* sp. in the roots and stems of maize plants grown in fields (Roesch et al. 2008), *R. etli* in the roots of one cultivar of sorghum grown with low and high nitrogen fertilizer doses (Rodrigues Coelho et al. 2008), while, *R. leguminosarum* applied in sweet potato tubers (Terakado-Tonooka et al. 2008), *R. helanshanense* in switchgrass roots and shoots (Bahulikar et al. 2014), and *R. daejeonense* in sugarcane stems and roots in Japan and Brazil (Thaweenut et al. 2011). Lay et al. (2018) used NGS approaches to compare the rhizosphere and endosphere of canola, pea, and wheat grown on the Canadian prairies. On the other hand, *R. leguminosarum* was detected in varying degrees of abundance in the endospheres and rhizospheres of the three crops; however, similar members of the Rhizobiaceae family, such as *Agrobacterium* sp., were associated with the endospheres of canola and wheat, but not in case of pea (Lay et al. 2018). Essel et al. (2019) investigated the selection of appropriate agronomic procedures for isolation of rhizobia from rhizospheric soils of rotationally farmed wheat and pea. This indicates that *Rhizobium* is more prevalent in soils that are closely linked to the roots, revealing the specialized functioning of genus *Rhizobium* with crops. *Rhizobium* was identified as a prominent OTU among other diazotrophs in rice fields (Jha et al. 2020). Other related OTUs, such as unclassified *Rhizobiales* and unclassified Rhizobiaceae, as well as other rhizobia OTUs, were also detected with a lower prevalence. The inclusion of a *R. leguminosarum* strain as an inoculant with or without a low dosage of urea fertilizers lowered the OTU richness; *Rhizobium* remained a relevant OTU, but other α-Proteobacteria OTUs were less prevalent. Nonetheless, the beneficial effects of inoculation and inoculation + low dose of N showed enhanced rice growth and yield, implying that the communities are not negatively affected by selective dosage of chemical fertilizers and adaptive fertilizer adaptive nature of rhizobia explored.

The majority of the nonlegume researched are cereals, although, work also conducted on the microbiomes of vegetable plants, trees, and shrubs. Member of genus *Rhizobium* and related genera were reported from those microbiomes which indicates their relevance in plant growth promotion and biocontrol measures. *Rhizobium* spp. were found in bulk and rhizospheric soils of cucumber plants (Jia et al. 2019). Marasco et al. (2013) identified several *Rhizobium* species in grapevine roots, both in the rhizosphere and in the interior tissues, using DGGE rather than amplicon sequencing or metagenomics. Members of the *Allorhizobium–Rhizobium/ ParaRhizobium–Rhizobium* complex were only discovered in *Xylella*-infected and *Xylella*-uninfected olive trees of the variety "Leccino" (tolerant to *Xylella* infection). This was relevant after using NGS in the phyllosphere and endosphere of leaves and branches (Vergine et al. 2019). *Rhizobium* was detected in the resistant cultivar but not in the susceptible cultivar, implying that this taxon may have a role in this cultivar's resistance to infections. Recently, Wang et al. (2020) identified *Rhizobium* as a key bacterial genus in the microbiome of rice root and shoot.

3.12 Conclusion and Future Aspect

The rhizobia can benefit the nonlegumes as well as the legume plants. The compounds released or secreted by rhizobia are beneficial to both the category of plants alter their environment with the help of these compounds. With the advent of new technology, the plant-microbe interaction is better understood and more research allows us to predict the exact requirement of both the plant and microbe. With the positive interaction, the microbe may fix atmospheric N_2, release phytohormones, increasing the immunity of the plant against different stress. It also allows the plant to blend in a new environment, altering rhizospheric microflora. The goal is to achieve and identify beneficial communities which not only save time but are also cost-effective. Therefore, with the new technologies, more research has to be done emphasizing the genetic aspect, molecular biology, and ecology of the rhizobia and better understanding of nonleguminous plants for improving the productivity, to attain useful rhizobia for sustainable agriculture. The futuristic focus should be to understand the signaling mechanisms between rhizobia and nonlegume plants and the process of colonization, to exhibit synergistic effect between host plant and rhizobia, to genetically modify the partners for better co-operation, the use of crop-specific promoters per the environment or soil type, selecting mutant types with better growth traits. Also focus should be there to use of multiple beneficial nitrogen-fixing strains benefited to diverse germ plasm of nonlegume crops so as to achieve sustainable goal in agroecological practices.

Acknowledgment SD, ND and PP are grateful to Department of Biotechnology, Ministry of Science and Technology, Govt of India, for financial support.

Conflict of interest Author(s) declares no conflict of interest.

References

Abd-Alla MH (1994) Use of organic phosphorus by *Rhizobium leguminosarum* biovarviceae phosphatases. Biol Fertil Soil 18(3):216–218

Abdel-Aziz RA, Radwan SMA, Abdel-Kader MM, Barakat MIA (1996) Biocontrol of faba bean root-rot using VA mycorrhizae and its effect on biological nitrogen fixation. Egypt J Microbiol (Egypt)

Abou-Shanab RA, Angle JS, Delorme TA, Chaney RL, Van Berkum P, Moawad H et al (2003) Rhizobacterial effects on nickel extraction from soil and uptake by *Alyssum murale*. New Phytol 158(1):219–224

Afzal A, Bano A (2008) *Rhizobium* and phosphate-solubilizing bacteria improve the yield and phosphorus uptake in wheat (*Triticum aestivum*). Int J Agric Biol 10(1):85–88

Ahmad E, Khan M, Zaidi A (2013) ACC deaminase producing *Pseudomonas putida* strain PSE3 and *Rhizobium leguminosarum* strain RP2 in synergism improves growth, nodulation and yield of pea grown in alluvial soils. Symbiosis 61(2):93–104

Akintokun AK, Taiwo MO (2016) Comparison of single culture and the consortium of growth-promoting rhizobacteria from three tomato (Lycopersicon esculentum Mill) varieties. Adv Plants Agric Res 5(1):00167

Alagawadi AR, Gaur AC (1988) Associative effect of *Rhizobium* and phosphate-solubilizing bacteria on the yield and nutrient uptake of chickpea. Plant Soil 105(2):241–246

Alami Y, Achouak W, Marol C, Heulin T (2000) Rhizosphere soil aggregation and plant growth promotion of sunflowers by an exopolysaccharide-producing *Rhizobium* sp. strain isolated from sunflower roots. Appl Environ Microbiol 66(8):3393–3398

Alexandre A, Oliveira S (2013) Response to temperature stress in rhizobia. Crit Rev Microbiol 39(3):219–228

Ali SZ, Sandhya V, Grover M, Kishore N, Rao LV, Venkateswarlu B (2009) *Pseudomonas* sp. strain AKM-P6 enhances tolerance of sorghum seedlings to elevated temperatures. Biol Fertil Soil 46(1):45–55

Alikhani HA, Saleh-Rastin N, Antoun H (2007) Phosphate solubilization activity of rhizobia native to Iranian soils. In: Velázquez E, Rodríguez-Barrueco C (eds) First international meeting on microbial phosphate solubilization. Developments in plant and soil sciences, vol 102. Springer, Dordrecht, pp 35–41

Al-Mallah MK, Davey MR, Cocking EC (1989) Formation of nodular structures on rice seedlings by rhizobia. J Exp Bot 40(4):473–478

Al-Mallah MK, Davey MR, Cocking EC (1990) Enzyme treatment, PEG, biotin and mannitol, stimulate nodulation of white clover by *Rhizobium trifolii*. J Plant Physiol 137(1):15–19

Alström S (1991) Induction of disease resistance in common bean susceptible to halo blight bacterial pathogen after seed bacterization with rhizosphere pseudomonads. J Gen Appl Microbiol 37(6):495–501

Alström S, Burns RG (1989) Cyanide production by rhizobacteria as a possible mechanism of plant growth inhibition. Biol Fertil Soil 7(3):232–238

Altúzar-Molina A, Lozano L, Ortíz-Berrocal M, Ramírez M, Martínez L, de Lourdes Velázquez-Hernández M et al (2020) Expression of the legume-specific nod factor receptor proteins alters developmental and immune responses in rice. Plant Mol Biol Reporter 38(2):262–281

Amara MA, Dahdoh MSA (1995) Effect of inoculation with plant-growth promoting rhizobacteria, PGPR on yield and uptake of nutrients by wheat grown on sandy soil. In: 5th. National Congress on Bio-Agriculture in Relation to Environment, Cairo (Egypt) 20–21 Nov 1995

Antoun H, Kloepper JW (2001) Plant growth-promoting rhizobacteria (PGPR). In: Brenner S, Miller JH (eds) Encyclopedia of genetics. Academic Press, New York, pp 1477–1480

Antoun H, Prevost D (2000) PGPR activity of *Rhizobium* with non-leguminous plants. In: Proceedings of the 5th international PGPR workshop. Villa Carlos Paz, Córdoba, Argentina, p 62

Antoun H, Bordeleau LM, Gagnon C (1978) Antagonisme entre *Rhizobium meliloti* et *Fusarium oxysporum* en relation avec l'efficacité symbiotique. Can J Plant Sci 58(1):75–78

Antoun H, Beauchamp CJ, Goussard N, Chabot R, Lalande R (1998) Potential of *Rhizobium* and *Bradyrhizobium* species as plant growth-promoting rhizobacteria on non-legumes: effect on radishes (*Raphanus sativus* L.). In: Hardarson G, Broughton WJ (eds) Molecular microbial ecology of the soil. Developments in plant and soil sciences, vol 83. Springer, Dordrecht, pp 57–67

Anyia AO, Archambault DJ, Slaski JJ (2004) Growth promoting effects of the diazotroph *Azorhizobium caulinodans* on Canadian wheat cultivars. In: Proceedings of the 4th International Crop Science Congress, Brisbane, Australia, pp 201–202

Arora NK, Kang SC, Maheshwari DK (2001) Isolation of siderophore-producing strains of *Rhizobium meliloti* and their biocontrol potential against *Macrophomina phaseolina* that causes charcoal rot of groundnut. Curr Sci 81:673–677

Arora NK, Verma M, Mishra J (2017) Rhizobial bioformulations: past, present and future. In: Mehnaz S (ed) Rhizotrophs: plant growth promotion to bioremediation. Microorganisms for sustainability, vol 2. Springer, Singapore, pp 69–99

Arshad M, Shaharoona B, Mahmood T (2008) Inoculation with *Pseudomonas* spp. containing ACC-deaminase partially eliminates the effects of drought stress on growth, yield, and ripening of pea (*Pisum sativum* L.). Pedosphere 18(5):611–620

Atzorn R, Crozier A, Wheeler CT, Sandberg G (1988) Production of gibberellins and indole-3-acetic acid by *Rhizobium phaseoli* in relation to nodulation of *Phaseolus vulgaris* roots. Planta 175(4):532–538

Bahulikar RA, Torres-Jerez I, Worley E, Craven K, Udvardi MK (2014) Diversity of nitrogen-fixing bacteria associated with switchgrass in the native tallgrass prairie of Northern Oklahoma. Appl Environ Microbiol 80(18):5636–5643

Baier R, Schiene K, Kohring B, Flaschel E, Niehaus K (1999) Alfalfa and tobacco cells react differently to chitin oligosaccharides and *Sinorhizobium meliloti* nodulation factors. Planta 210(1):157–164

Banerjee S, Schlaeppi K, van der Heijden MG (2018) Keystone taxa as drivers of microbiome structure and functioning. Nat Rev Microbiol 16(9):567–576

Bao Z, Okubo T, Kubota K, Kasahara Y, Tsurumaru H, Anda M et al (2014) Metaproteomic identification of diazotrophic methanotrophs and their localization in root tissues of field-grown rice plants. Appl Environ Microbiol 80(16):5043–5052

Bardin SD, Huang HC, Pinto J, Amundsen EJ, Erickson RS (2004) Biological control of *Pythium* damping-off of pea and sugar beet by *Rhizobium leguminosarum* bv. viceae. Can J Bot 82(3): 291–296

Belimov AA, Kojemiakov AP, Chuvarliyeva CN (1995) Interaction between barley and mixed cultures of nitrogen fixing and phosphate-solubilizing bacteria. Plant Soil 173(1):29–37

Belimov AA, Safronova VI, Sergeyeva TA, Egorova TN, Matveyeva VA, Tsyganov VE et al (2001) Characterization of plant growth-promoting rhizobacteria isolated from polluted soils and containing 1-aminocyclopropane-1-carboxylate deaminase. Can J Microbiol 47(7):642–652

Belimov AA, Hontzeas N, Safronova VI, Demchinskaya SV, Piluzza G, Bullitta S, Glick BR (2005) Cadmium-tolerant plant growth-promoting bacteria associated with the roots of Indian mustard (*Brassica juncea* L. Czern.). Soil Biol Biochem 37(2):241–250

Bender GL, Preston L, Barnard D, Rolfe BG (1990) Formation of nodule-like structures on the roots of the non-legumes rice and wheat. In: Nitrogen fixation: Achievements and objectives. Chapman and Hall, New York, p 825

Bera R, Seal A, Bhattacharyya P, Das TH, Sarkar D, Kangjoo K (2006) Targeted yield concept and a framework of fertilizer recommendation in irrigated rice domains of subtropical India. J Zhejiang Univ Sci B 7(12):963–968

Beveridge CA, Gresshoff PM, Rameau C, Turnbull CG (2003) Additional signalling compounds are required to orchestrate plant development. J Plant Growth Regul 22(1):15–24

Bhattacharjee RB, Singh A, Mukhopadhyay SN (2008) Use of nitrogen-fixing bacteria as biofertiliser for non-legumes: prospects and challenges. Appl Microbiol Biotechnol 80(2): 199–209

Bhattacharya C, Deshpande B, Pandey B (2013) Isolation and characterization of *Rhizobium* sp. form root of legume plant (*Pisum sativum*) and its antibacterial activity against different bacterial strains. Int J Agric Food Sci 3(4):138–141

Biederbeck VO, Lupwayi NZ, Hanson KG, Rice WA, Zentner RP (2000) Effect of long-term rotation with lentils on rhizosphere ecology and on endophytic rhizobia in wheat. In: Book of abstracts, 17th North American Conference on Symbiotic Nitrogen Fixation 23:28–29

Biswas JC, Ladha JK, Dazzo FB (2000) Rhizobia inoculation improves nutrient uptake and growth of lowland rice. Soil Sci Soc Am J 64(5):1644–1650

Bodelier PL, Wijlhuizen AG, Blom CW, Laanbroek HJ (1997) Effects of photoperiod on growth of and denitrification by *Pseudomonas chlororaphis* in the root zone of *Glyceria maxima*, studied in a gnotobiotic microcosm. Plant Soil 190(1):91–103

Boiero L, Perrig D, Masciarelli O, Penna C, Cassán F, Luna V (2007) Phytohormone production by three strains of *Bradyrhizobium japonicum* and possible physiological and technological implications. Appl Microbiol Biotechnol 74(4):874–880

Bolton GW, Nester EW, Gordon MP (1986) Plant phenolic compounds induce expression of the *Agrobacterium tumefaciens* loci needed for virulence. Science 232(4753):983–985

Borges CS, de Sá ES, Muniz AW, Osorio Filho BD (2019) Potential use of *Rhizobium* for vegetable crops growth promotion. Afr J Agric Res 14(8):477–483

Bowen GD, Rovira AD (1999) The rhizosphere and its management to improve plant growth. Adv Agron 66:1–102

Breil BT, Ludden PW, Triplett EW (1993) DNA sequence and mutational analysis of genes involved in the production and resistance of the antibiotic peptide trifolitoxin. J Bacteriol 175(12):3693–3702

Buhian WP, Bensmihen S (2018) Mini-review: nod factor regulation of phytohormone signaling and homeostasis during rhizobia-legume symbiosis. Front Plant Sci 9:1247

Bulgarelli D, Garrido-Oter R, Münch PC, Weiman A, Dröge J, Pan Y et al (2015) Structure and function of the bacterial root microbiota in wild and domesticated barley. Cell Host Microbe 17(3):392–403

Burd GI, Dixon DG, Glick BR (2000) Plant growth-promoting bacteria that decrease heavy metal toxicity in plants. Can J Microbiol 46(3):237–245

Carletti S, Caceres ER, Llorent B (1994) Growth promotion by PGPR on different plant species growing in hydroponic conditions. In: Improving plant productivity with rhizosphere bacteria. Proc. 3rd international workshop on plant growth-promoting rhizobacteria, Adelaide, Australia

Carrillo GC, Vazquez MDRG (1992) Comparative study of siderophore-like activity of *Rhizobium phaseoli* and *Pseudomonas fluorescens*. J Plant Nutr 15(5):579–590

Carson KC, Holliday S, Glenn AR, Dilworth MJ (1992) Siderophore and organic acid production in root nodule bacteria. Arch Microbiol 157(3):264–271

Carson KC, Meyer JM, Dilworth MJ (2000) Hydroxamate siderophores of root nodule bacteria. Soil Biol Biochem 32(1):11–21

Chabot R, Antoun H, Cescas MP (1993) Stimulation de la croissance du maïs et de la laitue romaine par des microorganismes dissolvant le phosphore inorganique. Can J Microbiol 39(10):941–947
Chabot R, Antoun H, Cescas MP (1996) Growth promotion of maize and lettuce by phosphate-solubilizing *Rhizobium leguminosarum* biovar. phaseoli. Plant Soil 184(2):311–321
Chabot R, Beauchamp CJ, Kloepper JW, Antoun H (1998) Effect of phosphorus on root colonization and growth promotion of maize by bioluminescent mutants of phosphate-solubilizing *Rhizobium leguminosarum* biovar phaseoli. Soil Biol Biochem 30(12):1615–1618
Chaintreuil C, Giraud E, Prin Y, Lorquin J, Bâ A, Gillis M et al (2000) Photosynthetic bradyrhizobia are natural endophytes of the African wild rice *Oryza breviligulata*. Appl Environ Microbiol 66(12):5437–5447
Chakraborty U, Purkayastha RP (1984) Role of rhizobitoxine in protecting soybean roots from Macrophomina phaseolina infection. Can J Microbiol 30(3):285–289
Chandra S, Choure K, Dubey RC, Maheshwari DK (2007) Rhizosphere competent *Mesorhizobium loti* MP6 induces root hair curling, inhibits *Sclerotinia sclerotiorum* and enhances growth of Indian mustard (*Brassica campestris*). Braz J Microbiol 38(1):124–130
Chapman JM, Muday GK (2021) Flavonols modulate lateral root emergence by scavenging reactive oxygen species in *Arabidopsis thaliana*. J Biol Chem 296
Chen H, Richardson AE, Gartner E, Djordjevic MA, Roughley RJ, Rolfe BG (1991) Construction of an acid-tolerant *Rhizobium leguminosarum* biovar trifolii strain with enhanced capacity for nitrogen fixation. Appl Environ Microbiol 57(7):2005–2011
Chi F, Shen SH, Cheng HP, Jing YX, Yanni YG, Dazzo FB (2005) Ascending migration of endophytic rhizobia, from roots to leaves, inside rice plants and assessment of benefits to rice growth physiology. Appl Environ Microbiol 71(11):7271–7278
Chi F, Yang P, Han F, Jing Y, Shen S (2010) Proteomic analysis of rice seedlings infected by *Sinorhizobium meliloti* 1021. Proteomics 10(9):1861–1874
Chiwocha SD, Abrams SR, Ambrose SJ, Cutler AJ, Loewen M, Ross AR, Kermode AR (2003) A method for profiling classes of plant hormones and their metabolites using liquid chromatography-electrospray ionization tandem mass spectrometry: an analysis of hormone regulation of thermodormancy of lettuce (*Lactuca sativa* L.) seeds. Plant J 35(3):405–417
Cocking EC (2003) Endophytic colonization of plant roots by nitrogen-fixing bacteria. Plant Soil 252(1):169–175
Cocking EC, Srivastava JS, Kothari SL, Davey M (1992) Invasion of nonlegume plants by diazotrophic bacteria. In: Khush G, Bennett J (eds) Nodulation and nitrogen fixation in rice: potentials and prospects, pp 119–121
Cocking EC, Kothari SL, Batchelor CA, Jain S, Webster G, Jones J et al (1995) Interaction of rhizobia with non-legume crops for symbiotic nitrogen fixation nodulation. In: Fendrik I, del Gallo M, Vanderleyden J, de Zamaroczy M (eds) *Azospirillum* VI and related microorganisms. NATO ASI series, vol 37. Springer, Berlin, pp 197–205
Cook RL, Hesterberg D (2013) Comparison of trees and grasses for rhizoremediation of petroleum hydrocarbons. Int J Phytoremediation 15(9):844–860
Dakora FD (2003) Defining new roles for plant and rhizobial molecules in sole and mixed plant cultures involving symbiotic legumes. New Phytol 158(1):39–49
Dakora FD, Phillips DA (2002) Root exudates as mediators of mineral acquisition in low-nutrient environments. In: Food security in nutrient-stressed environments: exploiting plants' genetic capabilities, pp 201–213
Dardanelli MS, de Cordoba FJF, Espuny MR, Carvajal MAR, Díaz MES, Serrano AMG et al (2008) Effect of *Azospirillum brasilense* coinoculated with *Rhizobium* on *Phaseolus vulgaris* flavonoids and Nod factor production under salt stress. Soil Biol Biochem 40(11):2713–2721
Dardanelli MS, Manyani H, González-Barroso S, Rodríguez-Carvajal MA, Gil-Serrano AM, Espuny MR et al (2010) Effect of the presence of the plant growth-promoting rhizobacterium (PGPR) *Chryseobacterium balustinum* Aur9 and salt stress in the pattern of flavonoids exuded by soybean roots. Plant Soil 328(1):483–493
Date RA (2001) Advances in inoculant technology: a brief review. Aust J Exp Agric 41(3):321–325

Dazzo FB, Yanni YG, Rizk R, de Bruijn FJ, Rademaker J, Squartini A, Corich V, Mateos P, Martínez-Molina E, Velázquez E, Biswas JC, Hernandez RJ, Ladha JK, Hill J, Weinman J, Rolfe BG, Vega-Hernández M, Bradford JJ, Hollingsworth RI, Ostrom P, Marshall E, Jain T, Orgambide G, Philip-Hollingsworth S, Triplett E, Malik KA, Maya-Flores J, Hartmann A, Umali-Garcia M, Izaguirre-Mayoral ML (2000) Progress in multinational collaborative studies on the beneficial association between Rhizobium leguminosarum bv. trifolii and rice. In: Ladha JK, Reddy PM (eds) The quest for nitrogen fixation in rice. Los Baños, The Philippines, IRRI Press, pp 167–189

De Bruijn FJ, Jing Y, Dazzo FB (1995) Potential and pitfalls of trying to extend symbiotic interactions of nitrogen-fixing organisms to presently non-nodulated plants, such as rice. In: Management of biological nitrogen fixation for the development of more productive and sustainable agricultural systems. Springer, Dordrecht, pp 225–240

De Jong AJ, Heidstra R, Spaink HP, Hartog MV, Meijer EA, Hendriks T et al (1993) *Rhizobium* lipooligosaccharides rescue a carrot somatic embryo mutant. Plant Cell 5(6):615–620

Depret G, Houot S, Allard MR, Breuil MC, Nouaïm R, Laguerre G (2004) Long-term effects of crop management on *Rhizobium leguminosarum* biovar viciae populations. FEMS Microbiol Ecol 51(1):87–97

Deryło M, Choma A, Puchalski B, Suchanek W (1994) Siderophore activity in *Rhizobium* species isolated from different legumes. Acta Biochim Pol 41(1):7–11

Deshwal VK, Dubey RC, Maheshwari DK (2003) Isolation of plant growth-promoting strains of *Bradyrhizobium* (*Arachis*) sp. with biocontrol potential against *Macrophomina phaseolina* causing charcoal rot of peanut. Curr Sci:443–448

de Souza R, Sant'Anna FH, Ambrosini A, Tadra-Sfeir M, Faoro H, Pedrosa FO, Souza EM, Passaglia LM (2015) Genome of Rhizobium sp. UR51a, isolated from rice cropped in Southern Brazilian fields. Genome Announc 3(2):e00249-15

DiCenzo GC, Zamani M, Milunovic B, Finan TM (2016) Genomic resources for identification of the minimal N2-fixing symbiotic genome. Environ Microbiol 18(8):2534–2547

Dobbelaere S, Vanderleyden J, Okon Y (2003) Plant growth-promoting effects of diazotrophs in the rhizosphere. Crit Rev Plant Sci 22(2):107–149

Dolan L (2001) The role of ethylene in root hair growth in *Arabidopsis*. J Plant Nutr Soil Sci 164(2): 141–145

Dudeja SS, Suneja S, Khurana AL (1997) Iron acquisition system and its role in legume-*Rhizobium* symbiosis. Indian J Microbiol 37:1–12

Dupin SE, Geurts R, Kiers ET (2020) The non-legume *Parasponia andersonii* mediates the fitness of nitrogen-fixing rhizobial symbionts under high nitrogen conditions. Front Plant Sci 10:1779

Dyachok JV, Tobin AE, Price NPJ, Von Arnold S (2000) Rhizobial Nod factors stimulate somatic embryo development in *Picea abies*. Plant Cell Rep 19(3):290–297

Ehteshamul-Haque S, Ghaffar A (1992) Use of *Bradyrhizobium japonicaum* and fungicides in the control of root rot disease of sun flower. In: Proceedings of Status of Plant Pathology in Pakistan, Department of Botany, University of Karachi, Karachi, pp 261–266

Ehteshamul-Haque S, Ghaffar A (1993) Use of rhizobia in the control of root rot diseases of sunflower, okra, soybean and mungbean. J Phytopathol 138(2):157–163

Ehteshamul-Haque S, Abid M, Sultana V, Ara J, Ghaffar A (1996) Use of organic amendments on the efficacy of biocontrol agents in the control of root rot and root knot disease complex of okra. Nematol Mediterr 24(1):13–16

El-Tarabily KA, Soaud AA, Saleh ME, Matsumoto S (2006) Isolation and characterisation of sulfur-oxidising bacteria, including strains of *Rhizobium*, from calcareous sandy soils and their effects on nutrient uptake and growth of maize (*Zea mays* L.). Aust J Agric Res 57(1):101–111

Essel E, Xie J, Deng C, Peng Z, Wang J, Shen J et al (2019) Bacterial and fungal diversity in rhizosphere and bulk soil under different long-term tillage and cereal/legume rotation. Soil Till Res 194:104302

Etesami H, Alikhani HA, Jadidi M, Aliakbari A (2009) Effect of superior IAA producing rhizobia on N, P, K uptake by wheat grown under greenhouse condition. World Appl Sci J 6:1629–1633

Fagorzi C, Checcucci A, DiCenzo GC, Debiec-Andrzejewska K, Dziewit L, Pini F et al (2018) Harnessing rhizobia to improve heavy-metal phytoremediation by legumes. Genes 9(11):542

Fernández LA, Zalba P, Gómez MA, Sagardoy MA (2007) Phosphate-solubilization activity of bacterial strains in soil and their effect on soybean growth under greenhouse conditions. Biol Fertil Soil 43(6):805–809

Figueiredo MVB, Vilar JJ, Burity HA (1999) Alleviation of water stress effects in cowpea by *Bradyrhizobium* spp. inoculation. Plant Soil 207(1):67–75

Flores-Félix JD, Velázquez E, Martínez-Molina E, González-Andrés F, Squartini A, Rivas R (2021) Connecting the lab and the field: Genome analysis of phyllobacterium and *Rhizobium* strains and field performance on two vegetable crops. Agronomy 11(6):1124

Franche C, Lindström K, Elmerich C (2009) Nitrogen-fixing bacteria associated with leguminous and non-leguminous plants. Plant Soil 321(1):35–59

Gaby JC, Rishishwar L, Valderrama-Aguirre LC, Green SJ, Valderrama-Aguirre A, Jordan IK, Kostka JE (2018) Diazotroph community characterization via a high-throughput nifH amplicon sequencing and analysis pipeline. Appl Environ Microbiol 84(4):e01512–e01517

Galal YGM (2003) Assessment of nitrogen availability to wheat (*Triticum aestivum* L.) from inorganic and organic N sources as affected by *Azospirillum brasilense* and *Rhizobium* leguminosarum inoculation. Egypt J Microbiol 38:57–73

García JE, Maroniche G, Creus C, Suárez-Rodríguez R, Ramirez-Trujillo JA, Groppa MD (2017) In vitro PGPR properties and osmotic tolerance of different *Azospirillum* native strains and their effects on growth of maize under drought stress. Microbiol Res 202:21–29

García-Fraile P, Carro L, Robledo M, Ramírez-Bahena MH, Flores-Félix JD, Fernández MT, Mateos PF, Rivas R, Igual JM, Martínez-Molina E, Peix A, Velázquez E (2012) *Rhizobium* promotes non-legumes growth and quality in. PLoS One 7:38122

Gardener BBM, Fravel DR (2002) Biological control of plant pathogens: research, commercialization, and application in the USA. Plant Health Prog 3(1):17

Geddes BA, Kearsley J, Morton R, Finan TM (2020) The genomes of rhizobia. Adv Bot Res 94: 2013–2249

Geetha SJ, Joshi SJ (2013) Engineering rhizobial bioinoculants: a strategy to improve iron nutrition. Sci World J

Geetha R, Desai AJ, Archana G (2009) Effect of the expression of *Escherichia coli* fhuA gene in *Rhizobium* sp. IC3123 and ST1 in planta: its role in increased nodule occupancy and function in pigeon pea. Appl Soil Ecol 43(2–3):185–190

Glick BR (2010) Using soil bacteria to facilitate phytoremediation. Biotechnol Adv 28(3):367–374

Glick BR, Jacobson CB, Schwarze MM, Pasternak JJ (1994) 1-Aminocyclopropane-1-carboxylic acid deaminase mutants of the plant growth-promoting rhizobacterium *Pseudomonas putida* GR12-2 do not stimulate canola root elongation. Can J Microbiol 40(11):911–915

Glick BR, Karaturovíc DM, Newell PC (1995) A novel procedure for rapid isolation of plant growth-promoting pseudomonads. Can J Microbiol 41(6):533–536

Glick BR, Holguin G, Patten CL, Penrose DM (1999) Biochemical and genetic mechanisms used by plant growth-promoting bacteria. World Scientific

Gómez-Godínez LJ, Fernandez-Valverde SL, Romero JCM, Martínez-Romero E (2019) Metatranscriptomics and nitrogen fixation from the rhizoplane of maize plantlets inoculated with a group of PGPRs. Syst Appl Microbiol 42(4):517–525

González V, Santamaría RI, Bustos P, Pérez-Carrascal OM, Vinuesa P, Juárez S et al (2019) Phylogenomic *Rhizobium* species are structured by a continuum of diversity and genomic clusters. Front Microbiol 10:910

Gopalakrishnan S, Sathya A, Vijayabharathi R, Varshney RK, Gowda CL, Krishnamurthy L (2015) Plant growth-promoting rhizobia: challenges and opportunities. 3 Biotech 5(4):355–377

Goyal RK, Schmidt MA, Hynes MF (2021) Molecular biology in the improvement of biological nitrogen fixation by rhizobia and extending the scope to cereals. Microorganisms 9(1):125

Greetatorn T, Hashimoto S, Sarapat S, Tittabutr P, Boonkerd N, Uchiumi T, Teaumroong N (2019) Empowering rice seedling growth by endophytic Bradyrhizobium sp. SUTN 9-2. Lett Appl Microbiol 68(3):258–266
Greetatorn T, Hashimoto S, Maeda T, Fukudome M, Piromyou P, Teamtisong K et al (2020) Mechanisms of rice endophytic Bradyrhizobial cell differentiation and its role in nitrogen fixation. Microb Environ 35(3):ME20049
Grichko VP, Glick BR (2001) Amelioration of flooding stress by ACC deaminase-containing plant growth-promoting bacteria. Plant Physiol Biochem 39(1):11–17
Guerinot ML (1991) Iron uptake and metabolism in the rhizobia/legume symbioses. In: Iron nutrition and interactions in plants. Springer, Dordrecht, pp 239–249
Guerinot ML (1994) Microbial iron transport. Annu Rev Microbiol 48(1):743–772
Hafeez FY, Safdar ME, Chaudhry AU, Malik KA (2004) Rhizobial inoculation improves seedling emergence, nutrient uptake and growth of cotton. Aust J Exp Agric 44(6):617–622
Haggag WM, Wafaa MH (2002) Sustainable agriculture management of plant diseases. J Biol Sci 2(4):280–284
Han HS, Lee KD (2005) Plant growth-promoting rhizobacteria effect on antioxidant status, photosynthesis, mineral uptake and growth of lettuce under soil salinity. Res J Agric Biol Sci 1(3):210–215
Hara S, Morikawa T, Wasai S, Kasahara Y, Koshiba T, Yamazaki K, Fujiwara T, Tokunaga T, Minamisawa K (2019) Identification of nitrogen-fixing Bradyrhizobium associated with roots of field-grown sorghum by metagenome and proteome analyses. Front Microbiol 10:407
Hilali A, Prévost D, Broughton WJ, Antoun H (2001) Effets de l'inoculation avec des souches de *Rhizobium leguminosarum* biovar trifolii sur la croissance du blé dans deux sols du Maroc. Can J Microbiol 47(6):590–593
Hiltner L (1904) Uber nevere erfahrungen und probleme auf dem gebiet der boden bakteriologie und unter besonderer beurchsichtigung der grundungung und broche. Arbeit Deut Landw Ges Berlin 98:59–78
Hirsch AM, Fang Y, Asad S, Kapulnik Y (1997) The role of phytohormones in plant-microbe symbioses. Plant Soil 194(1):171–184
Höflich G (2000) Colonization and growth promotion of non-legumes by *Rhizobium* bacteria. In: Bell CR, Brylinsky M, Johnson-Green P (eds) Microbial biosystems: new frontiers. Proceedings of the 8th international symposium on microbial ecology. Atlantic Canada Society for Microbial Ecology, Halifax, pp 827–830
Höflich G, Wiehe W, Kühn G (1994) Plant growth stimulation by inoculation with symbiotic and associative rhizosphere microorganisms. Experientia 50(10):897–905
Honma M, Shimomura T (1978) Metabolism of 1-aminocyclopropane-1-carboxylic acid. Agric Biol Chem 42(10):1825–1831
Hume DJ, Shelp BJ (1990) Superior performance of the hup–*Bradyrhizobium japonicum* strain 532C in Ontario soybean field trials. Can J Plant Sci 70(3):661–666
Humphry DR, Andrews M, Santos SR, James EK, Vinogradova LV, Perin L et al (2007) Phylogenetic assignment and mechanism of action of a crop growth-promoting *Rhizobium* radiobacter strain used as a biofertiliser on graminaceous crops in Russia. Antonie van Leeuwenhoek 91(2): 105–113
Hussain MB, Mehboob I, Zahir ZA, Naveed M, Asghar HN (2009) Potential of *Rhizobium* spp. for improving growth and yield of rice (*Oryza sativa* L.). Soil Environ 28(1):49–55
Irar S, González EM, Arrese-Igor C, Marino D (2014) A proteomic approach reveals new actors of nodule response to drought in split-root grown pea plants. Physiol Plant 152(4):634–645
Irshad A, Rehman RNU, Abrar MM, Saeed Q, Sharif R, Hu T (2021) Contribution of *Rhizobium*–legume symbiosis in salt stress tolerance in *Medicago truncatula* evaluated through photosynthesis, antioxidant enzymes, and compatible solutes accumulation. Sustainability 13(6):3369
Jadhav RS, Thaker NV, Desai A (1994) Involvement of the siderophore of cowpea *Rhizobium* in the iron nutrition of the peanut. World J Microbiol Biotechnol 10(3):360–361

Janczarek M, Jaroszuk-Ściseł J, Skorupska A (2009) Multiple copies of rosR and pssA genes enhance exopolysaccharide production, symbiotic competitiveness and clover nodulation in *Rhizobium leguminosarum* bv. trifolii. Antonie Van Leeuwenhoek 96(4):471–486

Jha PN, Gomaa AB, Yanni YG, El-Saadany AEY, Stedtfeld TM, Stedtfeld RD et al (2020) Alterations in the endophyte-enriched root-associated microbiome of rice receiving growth-promoting treatments of urea fertilizer and *Rhizobium* biofertilizer. Microb Ecol 79(2):367–382

Jia HT, Liu JY, Shi YJ, Li DL, Wu FZ, Zhou XG (2019) Characterization of cucumber rhizosphere bacterial community with high-throughput amplicon sequencing. Allelopathy J 47(1):103–112

Jiménez-Gómez A, Flores-Félix JD, García-Fraile P, Mateos PF, Menéndez E, Velázquez E, Rivas R (2018) Probiotic activities of *Rhizobium laguerreae* on growth and quality of spinach. Sci Rep 8(1):1–10

Jing Y, Li G, Jin G, Shan X, Zhang B, Guan C, Li J (1990) Rice root nodules with acetylene reduction activity. In: Gresshoff PM, Roth LE, Stacey G, Newton WE (eds) Nitrogen fixation achievements and objectives, p 829

Jing Y, Li G, Shan X (1992) Development of nodule-like structure on rice roots. In: Khush GS, Bennett J (eds) Nodulation and nitrogen fixation in rice, pp 123–126

Joshi F, Chaudhari A, Joglekar P, Archana G, Desai A (2008) Effect of expression of *Bradyrhizobium japonicum* 61A152 fegA gene in *Mesorhizobium* sp., on its competitive survival and nodule occupancy on *Arachis hypogea*. Appl Soil Ecol 40(2):338–347

Joshi FR, Desai DK, Archana G, Desai AJ (2009) Enhanced survival and nodule occupancy of pigeon pea nodulating *Rhizobium* sp. ST1 expressing *fegA* gene of *Bradyrhizobium japonicum* 61A152. J Biol Sci 9:40–51

Joshi AU, Andharia KN, Patel PA, Kotadiya RJ, Kothari RK (2019) Plant growth-promoting rhizobacteria: mechanism, application, advantages and disadvantages. In: Green biotechnology. Day Publishing House: Division of Astral International Pvt. Ltd., New Delhi, pp 13–40

Kaci Y, Heyraud A, Barakat M, Heulin T (2005) Isolation and identification of an EPS-producing *Rhizobium* strain from arid soil (Algeria): characterization of its EPS and the effect of inoculation on wheat rhizosphere soil structure. Res Microbiol 156(4):522–531

Kanade SN, Shaikh SM, Ade AB, Khilare VC (2010) Degradation of Malathion by *Rhizobium* isolated from fenugreek (*Trigonella foenumgraecum*). J Biotechnol Bioinform 1:240–242

Khaitov B, Kurbonov A, Abdiev A, Adilov M (2016) Effect of chickpea in association with *Rhizobium* to crop productivity and soil fertility. Eurasian J Soil Sci 5(2):105–112

Khalid A, Arshad M, Zahir ZA (2006) Phytohormones: microbial production and applications. In: Biological approaches to sustainable soil system, pp 207–220

Khan MS, Zaidi A, Wani PA, Oves M (2009) Role of plant growth-promoting rhizobacteria in the remediation of metal contaminated soils. Environ Chem Lett 7(1):1–19

Khan N, Zandi P, Ali S, Mehmood A, Adnan Shahid M, Yang J (2018) Impact of salicylic acid and PGPR on the drought tolerance and phytoremediation potential of *Helianthus annus*. Front Microbiol 2507

Khokhar SN, Qureshi A (1998) Interaction of *Azorhizobium caulinodans* with different rice cultivars for increased N_2-fixation. In: Nitrogen fixation with non-legumes. Springer, Dordrecht, pp 91–93

Klee HJ, Hayford MB, Kretzmer KA, Barry GF, Kishore GM (1991) Control of ethylene synthesis by expression of a bacterial enzyme in transgenic tomato plants. Plant Cell 3(11):1187–1193

Kloepper JW (1978) Plant growth-promoting rhizobacteria on radishes. In: Proceedings of the 4th international conference on plant pathogenic bacteria, Station de Pathologie Vegetale et Phytobacteriologie, INRA, Angers, France, 2, pp 879–882

Kumar A, Patel JS, Meena VS, Srivastava R (2019) Recent advances of PGPR-based approaches for stress tolerance in plants for sustainable agriculture. Biocatal Agric Biotechnol 20:101271

Kumari B, Mallick MA, Solanki MK, Solanki AC, Hora A, Guo W (2019) Plant growth-promoting rhizobacteria (PGPR): modern prospects for sustainable agriculture. In: Plant health under biotic stress. Springer, Singapore, pp 109–127

Lagier JC, Dubourg G, Million M, Cadoret F, Bilen M, Fenollar F et al (2018) Culturing the human microbiota and culturomics. Nat Rev Microbiol 16(9):540–550
Law IJ, Strijdom BW (1988) Inoculation of cowpea and wheat with strains of *Bradyrhizobium* sp. that differ in their production of indole acetic acid. S Afr J Plant Soil 6(3):161–166
Lay CY, Bell TH, Hamel C, Harker KN, Mohr R, Greer CW et al (2018) Canola root–associated microbiomes in the Canadian Prairies. Front Microbiol 9:1188
LeBlanc N, Crouch JA (2019) Prokaryotic taxa play keystone roles in the soil microbiome associated with woody perennial plants in the genus Buxus. Ecol Evol 9:11102–11111
Lemanceau P (1992) Effets bénéfiques de rhizobactéries sur les plantes: exemple des *Pseudomonas* spp fluorescents. Agronomie 12(6):413–437
Li WX, Kodama O, Akatsuka T (1991) Role of oxygenated fatty acids in rice phytoalexin production. Agric Biol Chem 55(4):1041–1047
Li W, Nishiyama R, Watanabe Y, Van Ha C, Kojima M, An P, Tian L, Tian C, Sakakibara H, Tran LS (2018) Effects of overproduced ethylene on the contents of other phytohormones and expression of their key biosynthetic genes. Plant Physiol Biochem 128:170–177
Lynch J (1990a) Substrate flow in the rhizosphere. Plant Soil 129:1–10
Lynch JM (1990b) The rhizosphere. Wiley Interscience, Chichester
Lynch JM, Whipps JM (1990) Substrate flows in the rhizosphere. Plant Soil 129:1–10
Ma W, Sebestianova SB, Sebestian J, Burd GI, Guinel FC, Glick BR (2003) Prevalence of 1-aminocyclopropane-1-carboxylate deaminase in *Rhizobium* spp. Antonie Van Leeuwenhoek 83(3):285–291
Ma W, Charles TC, Glick BR (2004) Expression of an exogenous 1-aminocyclopropane-1-carboxylate deaminase gene in *Sinorhizobium meliloti* increases its ability to nodulate alfalfa. Appl Environ Microbiol 70(10):5891–5897
Machado RG, de Sá ELS, Hahn L, Oldra S, Mangrich dos Passos JF, Osorio Filho BD et al (2016) Rhizobia symbionts of legume forages native to south brazil as promoters of cultivated grass growing. Int J Agric Biol 18(5)
Maillet F, Poinsot V, André O, Puech-Pagès V, Haouy A, Gueunier M et al (2011) Fungal lipochitooligosaccharide symbiotic signals in arbuscular mycorrhiza. Nature 469(7328):58–63
Mansouri H, Petit A, Oger P, Dessaux Y (2002) Engineered rhizosphere: the trophic bias generated by opine-producing plants is independent of the opine type, the soil origin, and the plant species. Appl Environ Microbiol 68(5):2562–2566
Marasco R, Rolli E, Vigani G, Borin S, Sorlini C, Ouzari H et al (2013) Are drought-resistance promoting bacteria cross-compatible with different plant models? Plant Signal Behav 8(10): e26741
Martens DA, Frankenberger WT (1993) Soil saccharide extraction and detection. Plant Soil 149(1): 145–147
Martínez-Romero E, Wang ET, López-Merino A, Caballero-Mellado J, Rogel MA, Gándara B et al (2000) Ribosomal gene-based phylogenies on trial: the case of *Rhizobium* and related genera. Biol Plant Microb Interact 2:59–64
Martiny AC (2019) High proportions of bacteria are culturable across major biomes. ISME J 13(8): 2125–2128
Masalha J, Kosegarten H, Elmaci Ö, Mengel K (2000) The central role of microbial activity for iron acquisition in maize and sunflower. Biol Fertil Soil 30(5):433–439
Matiru VN, Dakora FD (2004) Potential use of rhizobial bacteria as promoters of plant growth for increased yield in landraces of African cereal crops. Afr J Biotechnol 3(1):1–7
Matiru VN, Dakora FD (2005a) Xylem transport and shoot accumulation of lumichrome, a newly recognized rhizobial signal, alters root respiration, stomatal conductance, leaf transpiration and photosynthetic rates in legumes and cereals. New Phytol 165(3):847–855
Matiru VN, Dakora FD (2005b) The rhizosphere signal molecule lumichrome alters seedling development in both legumes and cereals. New Phytol 166(2):439–444

Matiru VN, Jaffer MA, Dakora FD (2005) Rhizobial infection of African landraces of sorghum (*Sorghum bicolor* L.) and finger millet (*Eleucine coracana* L.) promotes plant growth and alters tissue nutrient concentration under axenic conditions. Symbiosis

Mayak S, Tirosh T, Glick BR (2004) Plant growth-promoting bacteria confer resistance in tomato plants to salt stress. Plant physiol Biochem 42(6):565–572

McCormick DB (1989) Two interconnected B vitamins: riboflavin and pyridoxine. Physiol Rev 69(4):1170–1198

Mehboob I, Zahir ZA, Mahboob A, Shahzad SM, Jawad A, Arshad M (2008) Preliminary screening of *Rhizobium* isolates for improving growth of maize seedlings under axenic conditions. Soil Environ 27:64–71

Mehboob I, Naveed M, Zahir ZA, Ashraf M (2012) Potential of rhizobia for sustainable production of non-legumes. In: Crop production for agricultural improvement. Springer, Dordrecht, pp 659–704

Menéndez E, Paço A (2020) Is the application of plant probiotic bacterial consortia always beneficial for plants? Exploring synergies between rhizobial and non-rhizobial bacteria and their effects on agro-economically valuable crops. Life 10(3):24

Mia MAB, Shamsuddin ZH (2009) Enhanced emergence and vigor seedling production of rice through growth-promoting bacterial inoculation. Res J Seed Sci 2(4):96–104

Minamisawa K, Ogawa KI, Fukuhara H, Koga J (1996) Indolepyruvate pathway for indole-3-acetic acid biosynthesis in *Bradyrhizobium elkanii*. Plant Cell Physiol 37(4):449–453

Miransari M, Smith D (2009) Rhizobial lipo-chitooligosaccharides and gibberellins enhance barley (*Hordeum vulgare* L.) seed germination. Biotechnology 8(2):270–275

Mishra RP, Singh RK, Jaiswal HK, Kumar V, Maurya S (2006) *Rhizobium*-mediated induction of phenolics and plant growth promotion in rice (*Oryza sativa* L.). Curr Microbiol 52(5):383–389

Mongiardini EJ, Pérez-Giménez J, Althabegoiti MJ, Covelli J, Quelas JI, López-García SL, Lodeiro AR (2009) Overproduction of the rhizobial adhesin RapA1 increases competitiveness for nodulation. Soil Biol Biochem 41(9):2017–2020

Montecillo AD, Raymundo AK, Papa IA, Aquino GMB, Rosana ARR (2018) Complete genome sequence of *Rhizobium* sp. strain 11515TR, isolated from tomato rhizosphere in the Philippines. Microbiol Resour Announ 7(7):e00903–e00918

Montes-Grajales D, Esturau-Escofet N, Esquivel B, Martinez-Romero E (2019) Exo-metabolites of *Phaseolus vulgaris*-nodulating rhizobial strains. Metabolites 9(6):105

Muthamilarasan M, Singh NK, Prasad M (2019) Multi-omics approaches for strategic improvement of stress tolerance in underutilized crop species: a climate change perspective. Adv Genet 103: 1–38

Nadeem SM, Zahir ZA, Naveed M, Arshad M (2007) Preliminary investigations on inducing salt tolerance in maize through inoculation with rhizobacteria containing ACC deaminase activity. Can J Microbiol 53(10):1141–1149

Naidu VSGR, Panwar JDS, Annapurna K (2004) Effect of synthetic auxins and *Azorhizobium caulinodans* on growth and yield of rice. Indian J Microbiol 44:211–213

Nautiyal CS (1997) A method for selection and characterization of rhizosphere-competent bacteria of chickpea. Curr Microbiol 34(1):12–17

Neilands JB (1993) Siderophores. Arch Biochem Biophys 302(1):1–3

Noel TC, Sheng C, Yost CK, Pharis RP, Hynes MF (1996) *Rhizobium leguminosarum* as a plant growth-promoting rhizobacterium: direct growth promotion of canola and lettuce. Can J Microbiol 42(3):279–283

Nosheen S, Ajmal I, Song Y (2021) Microbes as biofertilizers, a potential approach for sustainable crop production. Sustainability 13(4):1868

O'Sullivan DJ, O'Gara F (1992) Traits of fluorescent Pseudomonas spp. involved in suppression of plant root pathogens. Microbiol Rev 56(4):662–676

Oberholster T, Vikram S, Cowan D, Valverde A (2018) Key microbial taxa in the rhizosphere of sorghum and sunflower grown in crop rotation. Sci Total Environ 624:530–539

Op den Camp R, Streng A, De Mita S, Cao Q, Polone E, Liu W, Ammiraju JS, Kudrna D, Wing R, Untergasser A, Bisseling T (2011) LysM-type mycorrhizal receptor recruited for rhizobium symbiosis in nonlegume *Parasponia*. Science 331(6019):909–912

Op den Camp RH, Polone E, Fedorova E, Roelofsen W, Squartini A, Op den Camp HJ et al (2012) Nonlegume *Parasponia andersonii* deploys a broad *Rhizobium* host range strategy resulting in largely variable symbiotic effectiveness. Mol Plant Microbe Interact 25(7):954–963

Ormeno-Orrillo E, Servín-Garcidueñas LE, Rogel MA, González V, Peralta H, Mora J et al (2015) Taxonomy of rhizobia and agrobacteria from the Rhizobiaceae family in light of genomics. Syst Appl Microbiol 38(4):287–291

Özkoç İ, Deliveli MH (2001) In vitro inhibition of the mycelial growth of some root rot fungi by *Rhizobium leguminosarum* biovar phaseoli isolates. Turkish J Biol 25(4):435–445

Pandey P, Maheshwari DK (2007) Two-species microbial consortium for growth promotion of Cajanus cajan. Curr Sci 25:1137–1142

Parveen S, Ghaffar A (1991) Effect of microbial antagonists in the control of root-rot of tomato. Pak J Bot 23(2):179–182

Parveen S, Ehteshamul-Haque S, Ghaffar A (1993) Biological control of Meloidogyne javanica on tomato and okra in soil infested with *Fusarium oxysporum*. Pak J Nematol 11(2):151–156

Paulitz TC, Bélanger RR (2001) Biological control in greenhouse systems. Annu Rev Phytopathol 39(1):103–133

Peix A, Rivas-Boyero AA, Mateos PF, Rodriguez-Barrueco C, Martınez-Molina E, Velazquez E (2001) Growth promotion of chickpea and barley by a phosphate solubilizing strain of *Mesorhizobium mediterraneum* under growth chamber conditions. Soil Biol Biochem 33(1): 103–110

Pena-Cabriales JJ, Alexander M (1983) Growth of *Rhizobium* in unamended soil. Soil Sci Soc Am J 47(1):81–84

Pena HB, Reyes I (2007) Nitrogen fixing bacteria and phosphate solubilizers isolated in lettuce (Lactuca sativa L.) and evaluated as plant growth promoters. Interciencia 32(8):560–565

Peng S, Biswas JC, Ladha JK, Gyaneshwar P, Chen Y (2002) Influence of rhizobial inoculation on photosynthesis and grain yield of rice. Agron J 94(4):925–929

Peralta H, Mora Y, Salazar E, Encarnación S, Palacios R, Mora J (2004) Engineering the *nifH* promoter region and abolishing poly-β-hydroxybutyrate accumulation in *Rhizobium etli* enhance nitrogen fixation in symbiosis with *Phaseolus vulgaris*. Appl Environ Microbiol 70(6):3272–3281

Pérez-Jaramillo JE, de Hollander M, Ramírez CA, Mendes R, Raaijmakers JM, Carrión VJ (2019) Deciphering rhizosphere microbiome assembly of wild and modern common bean (*Phaseolus vulgaris*) in native and agricultural soils from Colombia. Microbiome 7(1):1–16

Perrine FM, Prayitno J, Weinman JJ, Dazzo FB, Rolfe BG (2001) *Rhizobium* plasmids are involved in the inhibition or stimulation of rice growth and development. Funct Plant Biol 28(9):923–937

Perrine FM, Hocart CH, Hynes MF, Rolfe BG (2005) Plasmid-associated genes in the model microsymbiont *Sinorhizobium meliloti* 1021 affect the growth and development of young rice seedlings. Environ Microbiol 7(11):1826–1838

Perrine-Walker FM, Prayitno J, Rolfe BG, Weinman JJ, Hocart CH (2007) Infection process and the interaction of rice roots with rhizobia. J Exp Bot 58(12):3343–3350

Perrine-Walker FM, Hynes MF, Rolfe BG, Hocart CH (2009) Strain competition and agar affect the interaction of rhizobia with rice. Can J Microbiol 55(10):1217–1223

Phillips DA, Torrey JG (1970) Cytokinin production by *Rhizobium* japonicum. Physiol Plant 23(6): 1057–1063

Phillips DA, Joseph CM, Yang GP, Martínez-Romero E, Sanborn JR, Volpin H (1999) Identification of lumichrome as a *Sinorhizobium* enhancer of alfalfa root respiration and shoot growth. Proc Natl Acad Sci U S A 96(22):12275–12280

Piromyou P, Songwattana P, Greetatorn T, Okubo T, Kakizaki KC, Prakamhang J et al (2015) The type III secretion system (T3SS) is a determinant for rice-endophyte colonization by non-photosynthetic *Bradyrhizobium*. Microb Environ 30(4):291–300

Plessner O, Klapatch T, Guerinot ML (1993) Siderophore utilization by Bradyrhizobium japonicum. Appl Environ Microbiol 59(5):1688–1690

Prayitno J, Stefaniak J, McIver J, Weinman JJ, Dazzo FB, Ladha JK et al (1999) Interactions of rice seedlings with bacteria isolated from rice roots. Funct Plant Biol 26(6):521–535

Prévost D, Saddiki S, Antoun H (2000) Growth and mineral nutrition of corn inoculated with effective strains of *Bradyrhizobium japonicum*. In: Proceedings of the 5th international PGPR workshop. Villa Carlos Paz, Córdoba, Argentina

Qureshi MA, Shahzad H, Saeed MS, Ullah S, Ali MA, Mujeeb F, Anjum MA (2019) Relative potential of *Rhizobium* species to enhance the growth and yield attributes of cotton (*Gossypium hirsutum* L.). Eurasian J Soil Sci 8(2):159–166

Raaijmakers JM, Paulitz TC, Steinberg C, Alabouvette C, Moënne-Loccoz Y (2009) The rhizosphere: a playground and battlefield for soilborne pathogens and beneficial microorganisms. Plant Soil 321(1):341–361

Rao D, Mohanty S, Acharya C, Atoliya N (2018) Rhizobial taxonomy-current status. IUNFC Newslett 3:1–3

Ray PK, Jana AK, Maitra DN, Saha MN, Chaudhury J, Saha S, Saha AR (2000) Fertilizer prescriptions on soil test basis for jute, rice and wheat in a Typic Ustochrept. J Indian Soc Soil Sci 48(1):79–84

Reddy PM, Ladha JK, So RB, Hernandez RJ, Ramos MC, Angeles OR et al (1997) Rhizobial communication with rice roots: induction of phenotypic changes, mode of invasion and extent of colonization. Plant Soil 194(1):81–98

Reigh G, O'Connell M (1993) Siderophore-mediated iron transport correlates with the presence of specific iron-regulated proteins in the outer membrane of *Rhizobium meliloti*. J Bacteriol 175(1): 94–102

Reiter B, Bürgmann H, Burg K, Sessitsch A (2003) Endophytic *nifH* gene diversity in African sweet potato. Can J Microbiol 49(9):549–555

Reitz M, Rudolph K, Schroder I, Hoffmann-Hergarten S, Hallmann J, Sikora R (2000) Lipopolysaccharides of *Rhizobium etli* strain G12 act in potato roots as an inducing agent of systemic resistance to infection by the cyst nematode *Globodera pallida*. Appl Environ Microbiol 66(8): 3515–3518

Requena N, Jimenez I, Toro M, Barea JM (1997) Interactions between plant-growth-promoting rhizobacteria (PGPR), arbuscular mycorrhizal fungi and *Rhizobium* spp. in the rhizosphere of *Anthyllis cytisoides*, a model legume for revegetation in mediterranean semi-arid ecosystems. New Phytol 136(4):667–677

Reyes VG, Schmidt EL (1979) Population densities of *Rhizobium japonicum* strain 123 estimated directly in soil and rhizospheres. Appl Environ Microbiol 37(5):854–858

Ridge RW, Bender GL, Rolfe BG (1992) Nodule-like structures induced on the roots of wheat seedlings by the addition of the synthetic auxin 2, 4-dichlorophenoxyacetic acid and the effects of microorganisms. Funct Plant Biol 19(5):481–492

Rodrigues Coelho MR, De Vos M, Carneiro NP, Marriel IE, Paiva E, Seldin L (2008) Diversity of *nifH* gene pools in the rhizosphere of two cultivars of sorghum (*Sorghum bicolor*) treated with contrasting levels of nitrogen fertilizer. FEMS Microbiol Lett 279(1):15–22

Rodríguez H, Fraga R (1999) Phosphate solubilizing bacteria and their role in plant growth promotion. Biotechnol Adv 17(4-5):319–339

Roesch LFW, Camargo FA, Bento FM, Triplett EW (2008) Biodiversity of diazotrophic bacteria within the soil, root and stem of field-grown maize. Plant Soil 302(1):91–104

Rogers C, Oldroyd GE (2014) Synthetic biology approaches to engineering the nitrogen symbiosis in cereals. J Exp Bot 65(8):1939–1946

Rojas-Jiménez K, Sohlenkamp C, Geiger O, Martínez-Romero E, Werner D, Vinuesa P (2005) A ClC chloride channel homolog and ornithine-containing membrane lipids of *Rhizobium* tropici CIAT899 are involved in symbiotic efficiency and acid tolerance. Mol Plant Microbe Interact 18(11):1175–1185

Rolfe BG, Bender GL (1990) Evolving a *Rhizobium* for non-legume nodulation. In: Nitrogen fixation. Springer, Boston, MA, pp 779–780
Rout GR, Sahoo S (2005) Role of iron in plant growth and metabolism. Rev Agric Sci 3:1–24
Sabry SR, Saleh SA, Batchelor CA, Jones J, Jotham J, Webster G et al (1997) Endophytic establishment of *Azorhizobium caulinodans* in wheat. Proc R Soc Lond B: Biol Sci 264(1380):341–346
Şahin F, Çakmakçi R, Kantar F (2004) Sugar beet and barley yields in relation to inoculation with N_2-fixing and phosphate solubilizing bacteria. Plant Soil 265(1):123–129
Saikia SP, Jain V (2007) Biological nitrogen fixation with non-legumes: an achievable target or a dogma? Curr Sci:317–322
Savka MA, Dessaux Y, Oger P, Rossbach S (2002) Engineering bacterial competitiveness and persistence in the phytosphere. Mol Plant Microbe Interact 15(9):866–874
Schloter M, Wiehe W, Assmus B, Steindl H, Becke H, Höflich G, Hartmann A (1997) Root colonization of different plants by plant-growth-promoting *Rhizobium leguminosarum* bv. *trifolii* R39 studied with monospecific polyclonal antisera. Appl Environ Microbiol 63(5): 2038–2046
Schwinghamer EA, Belkengren RP (1968) Inhibition of rhizobia by a strain of *Rhizobium trifolii*: Some properties of the antibiotic and of the strain. Arch Mikrobiol 64(2):130–145
Schwyn B, Neilands JB (1987) Universal chemical assay for the detection and determination of siderophores. Anal Biochem 160(1):47–56
Senthilkumar M, Madhaiyan M, Sundaram SP, Sangeetha H, Kannaiyan S (2008) Induction of endophytic colonization in rice (*Oryza sativa* L.) tissue culture plants by *Azorhizobium caulinodans*. Biotechnol Lett 30(8):1477–1487
Senthilkumar M, Madhaiyan M, Sundaram SP, Kannaiyan S (2009) Intercellular colonization and growth-promoting effects of *Methylobacterium* sp. with plant-growth regulators on rice (*Oryza sativa* L. Cv CO-43). Microbiol Res 164(1):92–104
Servín-Garcidueñas LE, Guerrero G, Rogel-Hernández MA, Martínez-Romero E (2019) Genome sequence of *Rhizobium jaguaris* CCGE525T, a strain isolated from *Calliandra grandiflora* nodules from a rain forest in Mexico. Microbiol Resour Announ 8(9):e01584–e01518
Sessitsch A, Reiter B, Pfeifer U, Wilhelm E (2002) Cultivation-independent population analysis of bacterial endophytes in three potato varieties based on eubacterial and actinomycetes-specific PCR of 16S rRNA genes. FEMS Microbiol Ecol 39(1):23–32
Shakhawat Hossain M, Mårtensson A (2008) Potential use of *Rhizobium spp*. to improve fitness of non-nitrogen-fixing plants. Acta Agric Scand Sect B Soil Plant Sci 58(4):352–358
Shaukat SS, Siddiqui IA (2003) The influence of mineral and carbon sources on biological control of charcoal rot fungus, *Macrophomina phaseolina* by fluorescent pseudomonads in tomato. Lett Appl Microbiol 36(6):392–398
Sheikh LI, Dawar S, Zaki MJ, Ghaffar A (2006) Efficacy of *Bacillus thuringiensis* and *Rhizobium meliloti* with nursery fertilizers in the control of root infecting fungi on mung bean and okra plants. Pak J Bot 38(2):465
Shimshick EJ, Hebert RR (1979) Binding characteristics of N_2-fixing bacteria to cereal roots. Appl Environ Microbiol 38(3):447–453
Siddiqui IA, Ehteshamul-Haq S, Zaki MJ, Ghaffar A (2000) Effect of urea on the efficacy of *Bradyrhizobium* sp. and *Trichoderma harzianum* in the control of root infecting fungi in mungbean and sunflower. Sarhad J Agric (Pak)
Silva HSA, Romeiro RDS, Mounteer A (2003) Development of a root colonization bioassay for rapid screening of rhizobacteria for potential biocontrol agents. J Phytopathol 151(1):42–46
Silva FB, Winck B, Borges CS, Santos FL, Bataiolli RD, Backes T et al (2020) Native rhizobia from southern Brazilian grassland promote the growth of grasses. Rhizosphere 16:100240
Singh R, Kumar V, Sharma S, Behl RK, Singh BP, Narula N (2005) Performance and persistence of green fluorescent protein (gfp) marked *Azotobacter chroococcum* in sterilized and unsterilized wheat rhizospheric soil. J Appl Environ Biol 11:751–755

Singh RK, Mishra RP, Jaiswal HK, Kumar V, Pandey SP, Rao SB, Annapurna K (2006) Isolation and identification of natural endophytic rhizobia from rice (*Oryza sativa* L.) through rDNA PCR-RFLP and sequence analysis. Curr Microbiol 52(5):345–349

Singh A, Jain A, Sarma BK, Upadhyay RS, Singh HB (2014) Rhizosphere competent microbial consortium mediates rapid changes in phenolic profiles in chickpea during *Sclerotium rolfsii* infection. Microbiol Res 169(5-6):353–360

Smith SE, Dickson S, Smith FA (2001) Nutrient transfer in arbuscular mycorrhizas: how are fungal and plant processes integrated? Funct Plant Biol 28(7):685–696

Smith DL, Prithiviraj B, Zhang F (2002) Rhizobial signals and control of plant growth. In: Nitrogen fixation: global perspectives. CABI Publishing, Wallingford, pp 327–330

Somers E, Vanderleyden J, Srinivasan M (2004) Rhizosphere bacterial signalling: a love parade beneath our feet. Crit Rev Microbiol 30(4):205–240

Spencer D, James EK, Ellis GJ, Shaw JE, Sprent JI (1994) Interaction between rhizobia and potato tissues. J Exp Bot 45(10):1475–1482

Staehelin C, Granado J, Müller J, Wiemken A, Mellor RB, Felix G, Boller T (1994) Perception of *Rhizobium* nodulation factors by tomato cells and inactivation by root chitinases. Proc Natl Acad Sci U S A 91(6):2196–2200

Steen AD, Crits-Christoph A, Carini P, DeAngelis KM, Fierer N, Lloyd KG, Cameron Thrash J (2019) High proportions of bacteria and archaea across most biomes remain uncultured. ISME J 12:3126–3130

Stevens JB, Carter RA, Hussain H, Carson KC, Dilworth MJ, Johnston AW (1999) The fhu genes of *Rhizobium leguminosarum*, specifying siderophore uptake proteins: *fhuDCB* are adjacent to a pseudogene version of *fhuA*. Microbiology 145(3):593–601

Streeter JG (1994) Failure of inoculant rhizobia to overcome the dominance of indigenous strains for nodule formation. Can J Microbiol 40(7):513–522

Streng A, op den Camp R, Bisseling T, Geurts R (2011) Evolutionary origin of *Rhizobium* Nod factor signaling. Plant Signal Behav 6(10):1510–1514

Suárez R, Wong A, Ramírez M, Barraza A, Orozco MDC, Cevallos MA et al (2008) Improvement of drought tolerance and grain yield in common bean by overexpressing trehalose-6-phosphate synthase in rhizobia. Mol Plant Microbe Interact 21(7):958–966

Sullivan JT, Trzebiatowski JR, Cruickshank RW, Gouzy J, Brown SD, Elliot RM et al (2002) Comparative sequence analysis of the symbiosis island of *Mesorhizobium loti* strain R7A. J Bacteriol 184(11):3086–3095

Sytsma KJ, Morawetz J, Pires JC, Nepokroeff M, Conti E, Zjhra M et al (2002) Urticalean rosids: circumscription, rosid ancestry, and phylogenetics based on rbcL, trnL-F, and ndhF sequences. Am J Bot 89(9):1531–1546

Terakado-Tonooka J, Ohwaki Y, Yamakawa H, Tanaka F, Yoneyama T, Fujihara S (2008) Expressed *nifH* genes of endophytic bacteria detected in field-grown sweet potatoes (*Ipomoea batatas* L.). Microb Environ 23(1):89–93

Thaweenut N, Hachisuka Y, Ando S, Yanagisawa S, Yoneyama T (2011) Two seasons' study on *nifH* gene expression and nitrogen fixation by diazotrophic endophytes in sugarcane (*Saccharum* spp. hybrids): expression of *nifH* genes similar to those of rhizobia. Plant Soil 338(1):435–449

Thijs S, Sillen W, Weyens N, Vangronsveld J (2017) Phytoremediation: state-of-the-art and a key role for the plant microbiome in future trends and research prospects. Int J Phytoremed 19(1): 23–38

Trinick MJ, Galbraith J (1980) The *Rhizobium* requirements of the non-legume *Parasponia* in relationship to the cross-inoculation group concept of legumes. New Phytol 86(1):17–26

Trinick MJ, Hadobas PA (1989) Biology of the *Pavasponia-Bradyrhizobium* symbiosis. In: Nitrogen fixation with non-legumes. Springer, Dordrecht, pp 25–33

Trinick MJ, Hadobas PA (1995) Formation of nodular structures on the non-legumes *Brassica napus*, *B. campestris*, *B. juncea* and *Arabidopsis thaliana* with *Bradyrhizobium* and *Rhizobium* isolated from *Parasponia* spp. or legumes grown in tropical soils. Plant Soil 172(2):207–219

Triplett EW, Breil BT, Splitter GA (1994) Expression of tfx and sensitivity to the rhizobial peptide antibiotic trifolitoxin in a taxonomically distinct group of alpha-proteobacteria including the animal pathogen *Brucella abortus*. Appl Environ Microbiol 60(11):4163–4166
Tu JC (1978) Protection of soybean from severe Phytophthora root rot by *Rhizobium*. Physiol Plant Pathol 12(2):233–240
Tu JC (1979) Evidence of differential tolerance among some root rot fungi to rhizobial parasitism *in vitro*. Physiol Plant Pathol 14:171–177
Tulumello J, Chabert N, Rodriguez J, Long J, Nalin R, Achouak W, Heulin T (2021) *Rhizobium* alamii improves water stress tolerance in a non-legume. Sci Total Environ 797:148895
Turner TR, James EK, Poole PS (2013) The plant microbiome. Genome Biol 14:209
Van Loon LC (2007) Plant responses to plant growth-promoting rhizobacteria. In: New perspectives and approaches in plant growth-promoting. Rhizobacteria research, pp 243–254
Van Dillewijn P, Soto MJ, Villadas PJ, Toro N (2001) Construction and environmental release of a *Sinorhizobium meliloti* strain genetically modified to be more competitive for alfalfa nodulation. Appl Environ Microbiol 67(9):3860–3865
van Velzen R, Holmer R, Bu F, Rutten L, van Zeijl A, Liu W et al (2018) Comparative genomics of the nonlegume *Parasponia* reveals insights into evolution of nitrogen-fixing *Rhizobium* symbioses. Proc Natl Acad Sci U S A 115(20):E4700–E4709
Vargas LK, Lisboa BB, Schlindwein G, Granada CE, Giongo A, Beneduzi A, Passaglia LMP (2009) Occurrence of plant growth-promoting traits in clover-nodulating rhizobia strains isolated from different soils in Rio Grande do Sul state. Rev Brasil Ciên Solo 33(5):1227–1235
Velázquez E, Peix A, Zurdo-Piñiro JL, Palomo JL, Mateos PF, Rivas R et al (2005) The coexistence of symbiosis and pathogenicity-determining genes in *Rhizobium* rhizogenes strains enables them to induce nodules and tumors or hairy roots in plants. Mol Plant Microbe Interact 18(12):1325–1332
Velázquez E, Carro L, Flores-Félix JD, Menéndez E, Ramírez-Bahena MH, Peix A (2019) Bacteria-inducing legume nodules involved in the improvement of plant growth, health and nutrition. In: Microbiome in plant health and disease. Springer, Singapore, pp 79–104
Vences-Guzmán MÁ, Guan Z, Ormeño-Orrillo E, González-Silva N, López-Lara IM, Martínez-Romero E et al (2011) Hydroxylated ornithine lipids increase stress tolerance in *Rhizobium* tropici CIAT899. Mol Microbiol 79(6):1496–1514
Venieraki A, Dimou M, Vezyri E, Kefalogianni I, Argyris N, Liara G, Pergalis P, Chatzipavlidis I, Katinakis P (2011) Characterization of nitrogen-fixing bacteria isolated from field-grown barley, oat, and wheat. J Microbiol 49(4):525–534
Vergine M, Meyer JB, Cardinale M, Sabella E, Hartmann M, Cherubini P et al (2019) The Xylella fastidiosa-resistant olive cultivar "Leccino" has stable endophytic microbiota during the olive quick decline syndrome (OQDS). Pathogens 9(1):35
Verma SC, Singh A, Chowdhury SP, Tripathi AK (2004) Endophytic colonization ability of two deep-water rice endophytes, *Pantoea* sp. and *Ochrobactrum* sp. using green fluorescent protein reporter. Biotechnol Lett 26(5):425–429
Vessey JK (2003) Plant growth-promoting rhizobacteria as biofertilizers. Plant Soil 255(2): 571–586
Volpiano CG, Lisboa BB, Granada CE, São José JFB, de Oliveira AMR, Beneduzi A (2019) Microbiome in plant health and disease
Walsh C, Pascal RA Jr, Johnston M, Raines R, Dikshit D, Krantz A, Honma M (1981) Mechanistic studies on the pyridoxal phosphate enzyme 1-aminocyclopropane-1-carboxylate deaminase from *Pseudomonas* sp. Biochemistry 20(26):7509–7519
Walters WA, Jin Z, Youngblut N, Wallace JG, Sutter J, Zhang W et al (2018) Large-scale replicated field study of maize rhizosphere identifies heritable microbes. Proc Natl Acad Sci U S A 115(28):7368–7373
Wang C, Knill E, Glick BR, Défago G (2000) Effect of transferring 1-aminocyclopropane-1-carboxylic acid (ACC) deaminase genes into *Pseudomonas fluorescens* strain CHA0 and its

gacA derivative CHA96 on their growth-promoting and disease-suppressive capacities. Can J Microbiol 46(10):898–907

Wang M, Eyre AW, Thon MR, Oh Y, Dean RA (2020) Dynamic changes in the microbiome of rice during shoot and root growth derived from seeds. Front Microbiol 2183

Webster G, Gough C, Vasse J, Batchelor CA, O'callaghan KJ, Kothari SL et al (1997) Interactions of rhizobia with rice and wheat. In: Opportunities for biological nitrogen fixation in rice and other non-legumes. Springer, Dordrecht, pp 115–122

Weller DM, Raaijmakers JM, Gardener BBM, Thomashow LS (2002) Microbial populations responsible for specific soil suppressiveness to plant pathogens. Annu Rev Phytopathol 40(1): 309–348

Werner D (1992) Symbiosis of plants and microbes (No. SB731 W49). Chapman & Hall, London

Weyens N, van der Lelie D, Taghavi S, Vangronsveld J (2009) Phytoremediation: plant–endophyte partnerships take the challenge. Curr Opin Biotechnol 20(2):248–254

Wiehe W, Höflich G (1995) Survival of plant growth-promoting rhizosphere bacteria in the rhizosphere of different crops and migration to non-inoculated plants under field conditions in north-east Germany. Microbiol Res 150(2):201–206

Wiehe W, Hecht-Buchholz CH, Hoflich G (1994) Electron microscopic investigations on root colonization of *Lupinus albus* and *Pisum sativum* with two associative plant growth-promoting rhizobacteria, *Pseudomonas fluorescens* and *Rhizobium leguminosarum* bv. trifolii. Symbiosis

Wu Q, Peng X, Yang M, Zhang W, Dazzo FB, Uphoff N et al (2018) Rhizobia promote the growth of rice shoots by targeting cell signaling, division and expansion. Plant Mol Biol 97(6):507–523

Xie ZP, Staehelin C, Vierheilig H, Wiemken A, Jabbouri S, Broughton WJ et al (1995) Rhizobial nodulation factors stimulate mycorrhizal colonization of nodulating and nonnodulating soybeans. Plant Physiol 108(4):1519–1525

Xu H, Yang Y, Tian Y, Xu R, Zhong Y, Liao H (2020) *Rhizobium* inoculation drives the shifting of rhizosphere fungal community in a host genotype-dependent manner. Front Microbiol 3135

Yang SF, Hoffman NE (1984) Ethylene biosynthesis and its regulation in higher plants. Annu rev Plant Physiol 35(1):155–189

Yang G, Bhuvaneswari TV, Joseph CM, King MD, Phillips DA (2002) Roles for riboflavin in the *Sinorhizobium*-alfalfa association. Mol Plant Microbe Interact 15(5):456–462

Yanni YG, Rizk RY, Corich V, Squartini A, Ninke K, Philip-Hollingsworth S et al (1997) Natural endophytic association between *Rhizobium leguminosarum* bv. trifolii and rice roots and assessment of its potential to promote rice growth. In: Opportunities for biological nitrogen fixation in rice and other non-legumes. Springer, Dordrecht, pp 99–114

Yanni YG, Rizk RY, Abd El-Fattah FK, Squartini A, Corich V, Giacomini A et al (2001) The beneficial plant growth-promoting association of *Rhizobium leguminosarum* bv. trifolii with rice roots. Funct Plant Biol 28(9):845–870

Yardin MR, Kennedy IR, Thies JE (2000) Development of high-quality carrier materials for field delivery of key microorganisms used as bio-fertilisers and bio-pesticides. Radiat Phys Chem 57(3–6):565–568

Yeoh YK, Dennis PG, Paungfoo-Lonhienne C, Weber L, Brackin R, Ragan MA et al (2017) Evolutionary conservation of a core root microbiome across plant phyla along a tropical soil chronosequence. Nat Commun 8(1):1–9

Yoneyama T, Terakado-Tonooka J, Minamisawa K (2017) Exploration of bacterial N_2-fixation systems in association with soil-grown sugarcane, sweet potato, and paddy rice: a review and synthesis. Soil Sci Plant Nutr 63(6):578–590

Yoneyama T, Terakado-Tonooka J, Bao Z, Minamisawa K (2019) Molecular analyses of the distribution and function of diazotrophic rhizobia and methanotrophs in the tissues and rhizosphere of non-leguminous plants. Plants 8(10):408

Yu D, Kennedy IR (1995) Nitrogenase activity (C2H2 reduction) of *Azorhizobium* in 2, 4-D-induced root structures of wheat. Soil Biol Biochem 27(4-5):459–462

Zahir ZA, Arshad M (2004) Perspectives in agriculture. Adv Agron 81:97–98

Zahir ZA, Munir A, Asghar HN, Shaharoona B, Arshad M (2008) Effectiveness of rhizobacteria containing ACC deaminase for growth promotion of peas (*Pisum sativum*) under drought conditions. J Microbiol Biotechnol 18(5):958–963

Zahir ZA, Ghani U, Naveed M, Nadeem SM, Asghar HN (2009) Comparative effectiveness of *Pseudomonas* and *Serratia* sp. containing ACC-deaminase for improving growth and yield of wheat (*Triticum aestivum* L.) under salt-stressed conditions. Arch Microbiol 191(5):415–424

Zheng Y, Liang J, Zhao DL, Meng C, Xu ZC, Xie ZH, Zhang CS (2020) The root nodule microbiome of cultivated and wild halophytic legumes showed similar diversity but distinct community structure in Yellow River Delta saline soils. Microorganisms 8(2):207

Ziaf K, Latif U, Amjad M, Shabir MZ, Asghar W, Ahmed S et al (2016) Combined use of microbial and synthetic amendments can improve radish (*Raphanus sativus*) yield. J Environ Agric Sci 6: 10–15

Chapter 4
Biotechnological Solutions to Improve Nitrogenous Nutrition in Nonlegume Crops

Hassan Etesami and Byoung Ryong Jeong

Abstract In many developing countries around the world, nitrogen availability greatly limits crop production. On the other hand, nitrogen (N) fertilization in industrialized countries has become unsustainable and lead to environmental consequences. It has therefore become necessary to find alternatives to chemical nitrogen fertilizers to ensure a secure, sustainable food production. Plants are unable to directly utilize the freely available N_2 in the atmosphere, necessitating the chemical application of N fertilization. However, certain archaea and bacteria possess the ability to convert atmospheric N_2 to ammonia, which can directly be utilized by plants for various biological processes. This opens engineering possibilities to improve the N nutrition in nonlegume plants, such as (1) applying nitrogenase to plant cells; (2) introducing legume symbiosis for nonlegumes; and (3) imparting the ability to associate with N_2-fixing bacteria and/or other plant growth-promoting bacteria (PGPB) in nonlegumes. These are challenging biotechnological approaches, but the groundwork upon which these solutions may be implemented have been laid out by recent advances in the field. This chapter attempts to review and collect important up-to-date information on biotechnological solutions to improve N nutrition in nonlegume crops.

Keywords Biotechnology · Biological nitrogen fixation · Cereal crops · *nif* gene · Nitrogenase

H. Etesami (✉)
Soil Science Department, College of Agriculture and Natural Resources, University of Tehran, Tehran, Iran
e-mail: hassanetesami@ut.ac.ir

B. R. Jeong
Division of Applied Life Science (BK21 Four), Department of Horticulture, Graduate School, Gyeongsang National University, Jinju, Republic of Korea

Institute of Agriculture and Life Science, Gyeongsang National University, Jinju, Republic of Korea

© The Author(s), under exclusive license to Springer Nature Singapore Pte Ltd. 2022

D. K. Maheshwari et al. (eds.), *Nitrogen Fixing Bacteria: Sustainable Growth of Non-legumes*, Microorganisms for Sustainability 36,
https://doi.org/10.1007/978-981-19-4906-7_4

4.1 Introduction

The global population is experiencing an exponential growth and society's need for food, including protein, is increasing. Therefore, providing solutions based on sustainable development and paying attention to maintaining environmental health are essential to increase agricultural production. Nitrogen (N) is an essential element for plants, influencing their growth and forming an integral part of protein structure, constituting about 2% of the dry weight of a plant (Santi et al. 2013). Although molecular nitrogen (N_2) forms more than 78% of the Earth's atmosphere, the stability of the triple bond between the two nitrogen atoms makes it unavailable for plants to absorb in this form. Instead, soil N is absorbed by plants as ammonium and nitrates through the roots.

Adding chemical N fertilizers to agricultural soils is a common method to provide sufficient N for agricultural products and increase their production (Rosenblueth et al. 2018; Westhoff 2009) and on the other hand has a high potential for pollution (e.g., eutrophication of aquatic systems and atmospheric pollution and deterioration of the quality of the soil and water) (Rockstrom et al. 2009). In addition, N fertilizer manufacture is very energy-intensive, with as much as six times the input than that needed to for either P or K fertilizer production (Da Silva et al. 1978). Fossil fuels are used in the energy-intensive processes during the production of chemical N fertilizers, the use of which is estimated to be approximately 1–2% of the total global energy supply and accounts for an equivalent proportion of greenhouse gas production (Van Deynze et al. 2018). Furthermore, plants have been reported to effectively utilize less than 30% of the fertilizers provided, and the remainder ends up in soils and bodies of water (Priyadarshini et al. 2021). Studies with maize, rice, and wheat indicate that typically, plants are only able to utilize less than 50% of the N from fertilization (Anas et al. 2020; Ladha et al. 2016). Improper, unprincipled chemical N fertilization harms the environmental and human health, examples of which include degeneration of the ozone layer and production of greenhouse gases (Erisman et al. 2015; Glendining et al. 2009; Ladha and Reddy 2003; Stokstad 2016). Annually, approximately 100 million tons of N is introduced into freshwater, marine, and soil environments (Rockström et al. 2009; Galloway et al. 2008).

Finding alternative N sources for agricultural uses has been an ongoing research topic, as the current use of chemical N sources poses diverse environmental problems and threats to human health, resources to produce N fertilizers become scarce as petroleum reserves used in the Haber-Bosch process decline, and plants are unable to efficiently utilize N from chemical fertilizers. An alternative N source for agriculture that may potentially be attractive is biological N_2 fixation (BNF) (Ladha and Reddy 1995; Beatty and Good 2011; Rogers and Oldroyd 2014). BNF is responsible for 30–50% of all N in crop fields (Martinez-Romero 2006; Rosenblueth et al. 2018) and shows promise of replacing traditional chemical N fertilizers (Olivares et al. 2013; Dent and Cocking 2017; Good 2018). Furthermore, in mixed intercropping systems like the wheat-soybean system and seasonal crop rotation, the fixed nitrogen can be transferred to nonlegumes (Fustec et al. 2010). Biological N_2 fixation is exclusive to certain types of prokaryotic organisms (archaea and bacteria) that are able to produce the enzyme nitrogenase (Franche et al. 2009). Nitrogenase acts as a catalyst to reduce N_2 to NH_3 (or ammonium), to fix N_2 at normal temperature and pressure, and depends on high ATP and reductant levels (Seefeldt et al. 2009). The resulting ammonium is then converted to nitrogenous compounds (amino acids, etc.) required by the cell or, in the case of symbiotic diazotrophs, by the host plant. In this way, with the help of diazotrophs, this life-giving element is continuously injected into the soil system, making it possible for other organisms to continue living.

Legumes are usually the only participant in the agriculturally important N_2-fixing symbioses (Werner et al. 2014). For hundreds of years, legumes have been used to introduce N into agricultural systems without chemical fertilization. Therefore, for a majority of agriculturally important nonlegumes like maize, rice, and wheat, NFB is not directly available. Nitrogen availability often limits cereal crop production. It is predicted that the projected food necessity in 2050 will not be met by cereal crop production, based on the genetics and trends in the cereal crop (nonlegumes) management (Ray et al. 2013). Cereal crop production historically has involved using large amounts of chemical N fertilizers, and as such, application of NFB for cereal crops is certainly desirable (Rosenblueth et al. 2018). The potential of reducing the need for chemical N fertilizers has made application of nitrogen-fixing bacteria to nonlegumes, especially cereals, a topic of great research interest for over a century (Bennett et al. 2020; Roesch et al. 2008; Triplett 1996; Beatty and Good 2011; Mus et al. 2016). A 50-year global investigation of maize, rice, and wheat concluded that up to 24% of the total N in these crops originate from symbiotic nitrogen fixation of nonlegumes, and suggest that a substantial proportion of the total N intake of cereal crops is via associative nitrogen fixation (Ladha et al. 2016). Furthermore, in certain environments, sugarcane at least partly depends on diazotrophic endophytes for its N nutrition (Urquiaga et al. 2012; Luo et al. 2016; Sevilla et al. 2001). The 15 nitrogen dilution experiments in a study involving Miscanthus × giganteus have demonstrated that the bioenergy feedstock acquires about 16% of the total N from the atmosphere (Keymer and Kent 2014). Van Deynze et al. (2018) have observed in a study that in maize grown in nitrogen-scarce environments acquire 29–82% of the nitrogen from the atmosphere. These examples demonstrate that while the total proportion may be small, some monocots are able to associate with diazotrophs to obtain utilizable amounts of fixed nitrogen from the atmosphere (Bennett et al. 2020). The model C4 grass *Setaria viridis* inoculated with diazotrophs has been observed to be able to acquire most of its fixed N_2 via associative nitrogen fixation (Pankievicz et al. 2015).

Efforts have been made with a focus on overexpressing the genes involved in the transport of ammonium and nitrates in the roots of maize, rice, and wheat over the past decades to improve the nitrogen use and assimilation efficiencies, although to differing degrees of success (Li et al. 2020). Another historical research focus has been on the ammonium tolerance in the plants (Song et al. 2021, 2022). Since nitrogenase was transferred from *Klebsiella pneumoniae* to *Escherichia coli* in the 1970s, engineering cereal crops to be self-sustainable on nitrogen via N_2 fixation has been a major research objective (Geddes et al. 2015). With a shared goal of transferring fixed nitrogen to cereal crops, the aforementioned objective of N_2-fixation-based self-sustainability of cereal crops has shed light on several new approaches. A very significant contribution to agriculture could be made by biotechnology by adapting symbiotic biological nitrogen fixation to nonlegumes, but the task has been recognized as a major challenge for research for many years (Conway 2019). Scientists have continually researched strategies to make nonlegumes self-sustainable in terms of N_2 consumption. Modifying the nitrogenase expression in plant organelles (transferring nitrogenase into crops), applying endophytic diazotrophs to nonlegumes to help fix N_2, and engineering nonlegumes' perception of rhizobia and the subsequent nodule formation (developing the root nodular symbiosis in nonlegumes) are some of these strategies (Ladha and Reddy 1995, 2003; Beatty and Good 2011; Santi et al. 2013; Curatti and Rubio 2014; Geddes et al. 2015; Oldroyd and Dixon 2014; Oldroyd et al. 2011). This book chapter aims to investigate and report on the progress made to fix nitrogen in nonlegumes, in hopes to help shed light on how to achieve one of the most important research goals in agricultural sciences.

4.2 Types of Biological Nitrogen Fixers

Biological nitrogen fixation (BNF) is the process that converts N_2 into ammonia, and diazotrophs are the bacteria that fix the nitrogen. The term diazotroph originates from diazo and troph which denotes "two nitrogens" (or dinitrogen) and "pertaining to food", respectively. Diazotrophs are categorized into three groups on the basis of their dependence on plants to provide carbon and energy for N_2 fixation: Free-living, Associative, and Symbiotic, which are discussed below.

4.2.1 Free-Living Diazotrophs

Free-living diazotrophs are N_2-fixing bacteria that can independently fix N_2 without the cooperation of a host plant. Autotrophic and heterotrophic bacteria execute free-living nitrogen fixation, respectively obtaining their energy from photosynthesis and organic matter decomposition. Free-living diazotrophs are relatively rare in the rhizosphere, although in certain cases they may account for the majority of the rhizospheric nitrogen content. Free-living diazotrophs are categorized into physiologically, phylogenetically diverse groups, which include Alpha-proteobacteria (*Bradyrhizobium*, *Rhodospirillum*, *Beijerinckia*, *Rhodobacteria*, *Rhodopseudomonas*, and *Rhizobium*), Beta-proteobacteria (*Nitrosospira* and *Burkholderia*), cyanobacteria (*Anabaena* and *Nostoc*), and Gamma-proteobacteria (*Azotobacter*, *Klebsiella*, *Pseudomonas*, and *Xanthomonas*), firmicutes (*Clostridium* and *Paenibacillus*) (Mahmud et al. 2020; Priyadarshini et al. 2021).

4.2.2 Associative Diazotrophs

Associative diazotrophs are N_2-fixing bacteria that make simple physical contact with the plant but do not form a common, visible biological organ in the plant. Such diazotrophs loosely associate with the root surfaces of plants and likely transfer the fixed nitrogen through death and mineralization. Endophytic diazotrophs, on the other hand, can colonize the insides of the host plant in the parenchyma, dead cells, and intercellular spaces, for instance. Such endophytic diazotrophs do not cause any apparent damage but invade the plant tissues and result in defense responses of the host plant. Associative and/or endophytic diazotrophs constitute diverse microbial genera (e.g., *Azospirillum*, *Gluconacetobacter*, *Herbaspirillum*, *Moraxella*, *Brevibacillus*, *Burkholderia*, *Klebsiella*, *Pseudomonas*, *Enterobacter*, *Nostoc*, *Anabaena*, *Azoarcus*, and *Bradyrhizobium*) and have been observed to be associated with various economically important plants, such as rice, maize, wheat, sugarcane, kallar grasssugar beet, coffee, potato, sorghum, tomato, and oilseed rape (Priyadarshini et al. 2021; Engelhard et al. 2000; Vaishampayan et al. 2001; Eskin et al. 2014; James et al. 2002; de Almeida et al. 2009; Compant et al. 2008; Reyna-Flores et al. 2018; Etesami 2019). Except for a few strains that are considered facultative endophytes, bacteria from the genus *Azospirillum* are primarily associative diazotrophs (Döbereiner et al. 1995). *Azoarcus* spp., *Herbaspirillum seropedicae*, and *Gluconacetobacter diazotrophicus* are the most researched obligate endophytic diazotrophs, and seem to limit their habit to the interior of plant tissues. Associative and endophytic colonization are both developed by heterocystous nitrogen-fixing cyanobacteria. The cyanobacteria *Anabaena azollae* develops a symbiotic relationship with the aquatic fern *Azolla,* whose exclusive nitrogen source in lowland paddy is the cyanobacteria (Rai et al. 2019). *Nostoc* is another cyanobacteria that dwell in microaerobic environments of host plants such as liverworts, hornworts, and cycads (Rai et al. 2000). Free-living diazotrophs are rare in the rhizosphere. Knowledge on diverse diazotrophs have been enhanced with the recent advances in gene-sequencing technology. Using nitrogenase as a marker, some studies have surveyed DNAs to identify the microbiota that fix nitrogen from the metagenome (Gaby et al. 2018). However, the mere presence of certain genes does not guarantee that the microorganism can fix nitrogen. Metagenomics therefore should be accompanied by additional omics approaches to be able to identify functional diazotrophs with a certainty (Pankievicz et al. 2019).

4.2.3 Symbiotic Diazotrophs

Symbiotic diazotrophs are N_2-fixing bacteria that are closely related to the plant and are able to fix N_2 by forming a specialized organ (nodules) common to a plant. Such endosymbiotic diazotrophs include rhizobia that belong to the phylum proteobacteria alpha subgroup that associates with the nonlegume *Parasponia* species (family Cannabaceae), and legumes (family Fabaceae) (Desbrosses and Stougaard 2011), and actinomycete family's *Frankia* sp. members of that associate with a various plants of eight actinorhizal plant families. Furthermore, it has been discovered that nitrogen-fixing cyanobacteria (mainly *Nostoc* sp.) colonize different plant organs, either extracellularly in *Azolla*, Cycadaceae, hornworts, and liverworts, or intracellularly in Gunneraceae. Host plants and diazotrophs form a system where the plant receives the benefit of the fixed nitrogen provided by the diazotroph, and the diazotroph receives carbon and other nutrients from the plant partner. Additionally, the endophytic or symbiotic plant structures colonized by diazotrophs may help protect nitrogenase from oxygen exposure. Multiple evidences exist to corroborate successful symbiotic relationships between different bacteria and nonlegumes. A well-known example is the symbiosis between actinorhizal plants and *Frankia* for N_2 fixation (Santi et al. 2013). Nodule formation is known to occur in the symbiosis between Parasponia and Rhizobia, where Parasponia is the only nonlegume where rhizobia effectively drive N_2 fixation by colonizing the insides of nodules (Akkermans et al. 1978).

4.3 Nitrogenase Enzyme Complex

Nitrogenase catalyzes the reduction of dinitrogen (N_2) to ammonia (NH_3), which depends on ATP. Nitrogen fixation is a dynamic process with high energy demands (Rosenblueth et al. 2018). The biological reduction of N_2 (inert) into the NH_3 (reactive) occurs as follows, under microaerobic conditions:

$$N_2 + 8\,H^+ + 8\,e^- + 16\,Mg - ATP \rightarrow 2\,NH_3 + H_2 + 16\,P + 16\,Mg - ADP$$

The overall reaction enthalpy of forming ammonia with molecular nitrogen and hydrogen is negative ($\Delta H° = -45.2$ KJ mol^{-1} NH_3); without catalysis, it is challenging to overcome the energy requirement for activation (EA°= 420 KJ mol^{-1}). Hence, nitrogenase requires a great amount of chemical energy release from ATP hydrolysis (1 mol of reduced N_2 with 16 mol of ATP) and reducing agents, like ferredoxin in vivo or dithionite in vitro. Nitrogenase is present in specialized bacteria and archaea called diazotrophs but is not found in eukaryotes. The free-living, associative, and symbiotic diazotrophs use nitrogenase to reduce N_2 to NH_3. Among different prokaryotic species (cyanobacteria, green sulfur bacteria, methanobacteria, and proteobacteria), genes have been observed to be horizontally transferred (Ivleva et al. 2016). Nitrogenase is a complex enzyme and consists of two enzymes (Fig. 4.1): dinitrogenase reductase, which contains iron (an Fe protein), and dinitrogenase, which contains molybdenum and iron (a MoFe protein) (Seefeldt et al. 2009; Addo and Dos Santos 2020). Nitrogenase is distinguished by the fact that ATP-dependent MoFeP-FeP interactions drive conformational changes that synchronize the proton and electron transfer processes associated with substrate

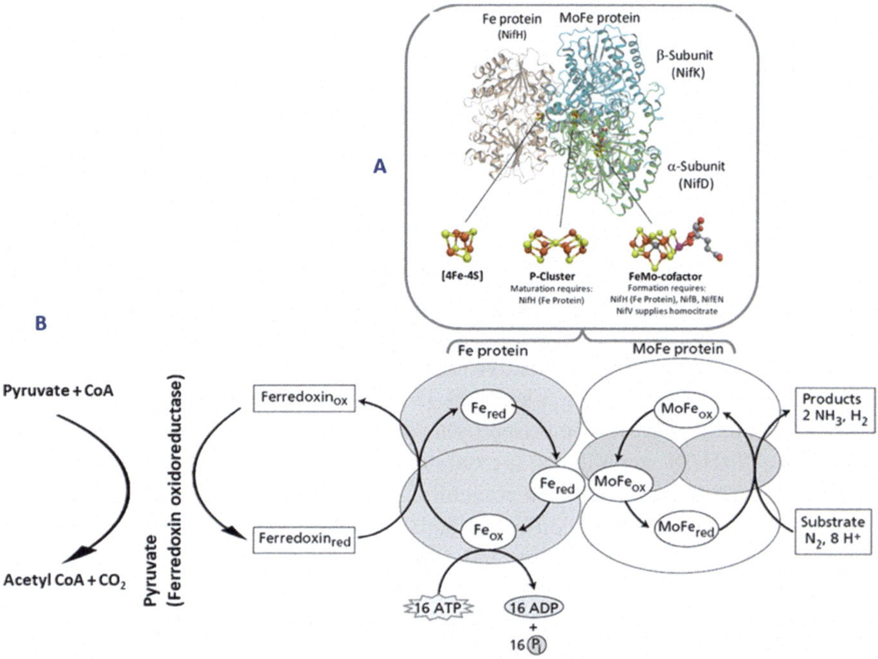

Fig. 4.1 Schematic representation of Mo-dependent nitrogenase and its associated metal-containing cofactors. Fe protein subunits (encoded by *nifH*) are shown in light brown, the MoFe protein a subunit (encoded by *nifD*) is shown in green, and the MoFe protein b subunit (encoded by *nifK*) is shown in blue. Atoms in metal-containing cofactors are indicated as: Fe (rust), S (yellow), Mo (magenta), C (gray), and O (red) (**a**). The schematic diagram of the nitrogenase complex, showing the flow of reducing power and substrates in enzymatic nitrogen fixation. The process is catalyzed by a cytoplasmic nitrogenase complex consisting of two enzymes: one enzyme is dinitrogenase, which contains molybdenum and iron (a MoFe protein); the other enzyme is dinitrogenase reductase, an iron-containing enzyme (an Fe protein). The source of the electrons for N_2 reduction is usually the reduced form of the Fe-S protein ferredoxin (Fd_{red}) which has a very negative E_O' value. Anaerobic or microaerophilic bacteria can provide Fd_{red} from oxidation of pyruvic acid by pyruvate: ferredoxin oxidoreductase. Aerobic bacteria reduce NAD^+ to NADH during pyruvic acid oxidation and thus must use the proton motive force to power reversed electron transport allowing NADH to reduce Fd (**b**)

reduction. The N_2 binding and reduction take place on the molybdenum-iron protein (MoFeP), which is a heterotetramer ($\alpha_2\beta_2$) encoded by *nifD* and *nifK*. The dinitrogenase reductase (FeP) provides electrons to the MoFeP, and is a homodimer (α_2) encoded by *nifH,* where each subunit contains one Mg·ATP-binding site. Reduced flavodoxin II provides electrons to FeP, and the FeP provides obligate electrons to MoFeP, and in the FeMo-co inside each MoFe subunit, the substrate reduction occurs. About 16 mol of ATP are required for 1 mol of N_2 in the overall catalysis, which produces 2 mol of ammonia and reduction equivalents provided by the reduced ferredoxin (Seefeldt et al. 2012; Curatti and Rubio 2014). The genes *nifD* , *nifK*, and *nifH* comprise the same operon and are frequently found in the form of *nifHDK* (Dixon and Kahn 2004). Additionally, several other genes are involved in the operon, such as *nifF* and *nifJ* that encode the electron transport proteins, and *nifBEN* that participates in the *nifA* biosynthesis, which is the iron-molybdenum cofactor of dinitrogenase. Moreover, analyzing the biochemistry and genetics has demonstrated that many additional *nif* genes (*nifA*, *nifB*, *nifE*, *nifQ*, *nifN*, *nifS nifV*, *nifW*, *nifX*, and *nifZ*) help regulate the nif genes, electron transport maturation processes, as well as the assembly and biosynthesis of the FeMo-cofactor biosynthesis (Masepohl et al. 2004; Lee et al. 2000). The nif regulon is composed of various operons that include nitrogenase-encoding genes, proteins associated with electron transfer, and regulator genes, which therefore regulates the nitrogen fixation (Shin et al. 2016). In *Azotobacter vinelandii* and *Rhodobacter capsulatus*, which carry iron and vanadium at their active sites, twos vanadium and iron-only nitrogenases have additionally been identified (Mus et al. 2018). The AnfHDK and VnfHDK subunits that are comprised by these enzymes are homologous to the Mo nitrogenase NifHDK subunits but are only expressed in low Molybdenum environments. Iron-iron or vanadium-iron cofactors, as well as additional components with unknown functions, like the AnfG and VnfG subunits, are contained at the active sites of these nitrogenases (Dixon and Kahn 2004). Reductants, like ferredoxin, flavodoxin, or sodium dithionite, are needed in BNF to deliver electrons to reduce N_2. In principle, N_2 reduction to NH_3 requires six electrons, but H_2 is also generated in the coupled process (Newton 2007). Therefore, most diazotrophs receive eight electrons from reduced ferredoxin. The reductase-nitrogenase electron transfer is also coupled with ATP hydrolysis by reductase.

4.4 Biotechnological Approaches to Develop Nitrogen-Fixing Nonlegumes

Historically, nonlegumes such as cereals have long interacted with associative N_2-fixing microbes. Such examples can be found in rice (Roger and Ladha 1992; Etesami and Alikhani 2016; Etesami et al. 2014; Chi et al. 2005), several sugarcane varieties (Boddey and Dobereiner 1995), and some maize cultivars (Garcia de Salomone and Döbereiner 1996). On the other hand, even with artificial inoculation with diazotrophs like *Azospirillum*, some cereals like certain wheat or maize cultivars have proven to be a tough target host for N_2-fixing microbes (Garcia de Salamone et al. 1996). In nonlegumes, the fixed nitrogen provided by N_2-fixing bacteria is insufficient and not as good as levels provided by chemical fertilizers or found in legumes in symbiosis with rhizobium; therefore, researchers have studied for decades to identify mechanisms that can effectively introduce biologically fixed atmospheric nitrogen for use in nonlegumes (Rosenblueth et al. 2018). Numerous projects are funded to genetically modify nonlegumes such that they can form nodules with nitrogen-fixing rhizobia, fix nitrogen themselves, or enhance colonization by diazotrophs (Rosenblueth et al. 2018), although many projects have not yet demonstrated success. A commercially viable symbiosis between cereals and rhizobia root nodules has yet to be established. A lack of dedicated chemicals and machinery necessary for legume-rhizobium association are attributable for the inability of nonlegumes to establish endosymbiotic relationship with rhizobia. Another critical setback for nonlegumes in establishing root nodule symbiosis is the absence of the transcription activator of nodule formation genes (ARN and NIN) (Priyadarshini et al. 2021). Major breakthroughs have been made in diazotroph genomics, nitrogen fixation genetics, and legume-rhizobia symbiotic processes have been made lately, which have enabled new, practical approaches aimed to establish a systematic symbiosis between legumes and rhizobia (Beatty and Good 2011). As observed in *Parasponia,* the only nonlegume capable of biologically fixing N_2 through root nodule symbiosis (Akkermans et al. 1978), it has been discovered that rhizobia can also enter plants through natural cracks, unlike in legumes where rhizobia infect through the infection threads in the root hairs (Behm et al. 2014). This presents the potential for all plants to intercellularly host rhizobia. Also, plant growth hormones are well known to facilitate nodule formation. Researchers have been trying to exploit and merge existing technology and knowledge to enable root nodule symbioses in cereals. 2,4-D is synthetic auxin that is a well-known herbicide, and has been observed to promote nodulation in soybean, which otherwise does not form nodules (Akao et al. 1991). 2,4-D inflicts are believed to injure the root, which enables rhizobia to penetrate and establish itself, resulting in the implementation of the symbiosis (Azam 2002). Cellulase and pectolyase, examples of cell-wall degrading enzymes, may be able to similarly help nonlegumes establish symbiosis with rhizobia (Cocking et al. 1990). The first barrier for cereals to notch up N_2 fixation may potentially be overcome by such chemicals and enzymes. However, 2,4-D-induced nodules histologically from those are stimulated by rhizobia (Francisco and Akao 1994). In contrast to legume nodules that have peripheral vasculature, nonlegume nodules induced by 2,4-D have central vasculature (Francisco and Akao 1994). The N_2-fixing nonlegume *Parasponia* has a central vascular structure, which suggests that the vasculature type does not impair the N_2-fixing ability in cereals. Because cereals possess significant homologies with *Parasponia*, it is plausible for cereals to be able to establish nodules to fix N_2 that are like those in *Parasponia*, even if such nodules bear little resemblance to those found in legumes. Paranodules, the nodule-like structures induced by 2, 4-D, are anatomically and structurally unfit for N_2 fixation, although such paranodules were able to successfully fix N_2 in *Azospirillum* and *Nostoc* (Francisco and Akao 1994). Because cereals are unable to naturally form nodules that can host rhizobia, transforming chloroplasts could be an alternative to successfully express *nif* and other associated genes to help cereals shelter rhizobia (Priyadarshini et al. 2021). The following sections discuss strategies explored by scientists to make cereals self-reliant on fixed nitrogen (Fig. 4.2).

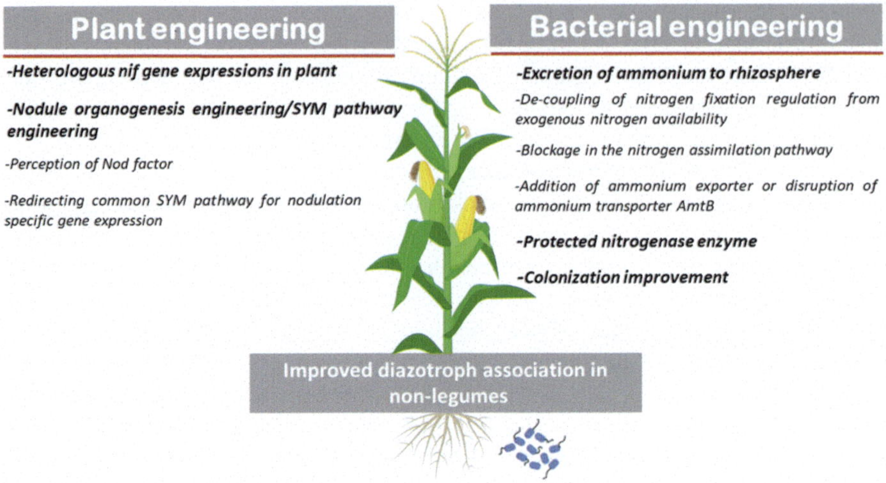

Fig. 4.2 Biotechnological interventions to establish nonlegume-rhizobium interaction. (For more details see the text and Priyadarshini et al. (2021))

4.4.1 Engineering Nitrogenase to Function in Nonlegume Cells

It is suggested by research of evolutionary genomics that to transfer nitrogen-fixing capabilities from legumes to nonlegumes, relatively few genetic elements are involved (Bailey-Serres et al. 2019). New synthetic engineering tools have recently applied to nitrogenase biosynthesis. Early successes in these approaches to deliver nitrogen to cereals, transferring nitrogenase and other such traits to microorganisms in close association with cereals is a logical strategy (Geddes et al. 2015). Researchers are considering incorporating and expressing the genetic machinery of bacteria to encode and support a function nitrogen system, to assemble an active nitrogenase system in plants. A minimum set of three genes must be created by consumerization of bacterial genetic units to transfer nitrogenase to plants (Deng et al. 2019). The extreme sensitivity of nitrogenase to oxygen, and the complexity of the nitrogenase biosynthesis (e.g., a coordinated expression of at least 16 N_2-fixing genes in plant cells) make it challenging to introduce nitrogenase-encoding bacterial *nif* genes into nonlegumes (Li and Chen 2020; Temme et al. 2012). Nitrogen fixation also has great energy demands (Curatti and Rubio 2014; Seefeldt et al. 2012). The common core set of genes and gene products necessary for functionally biosynthesizing nitrogenase has been identified by extensive studies in biochemistry and genetics (Rubio and Ludden 2008). Active nitrogenase expression in plants in potential subcellular low-oxygen environments (micro air pockets) offered by mitochondria and plastids makes transferring nitrogenase to plants feasible (Curatti and Rubio 2014). Since chloroplasts and mitochondria can meet the energy requirements for nitrogenase in plant cells, they are viewed as suitable sites for nitrogen fixation.

However, because nitrogenase is extremely sensitive to and is inactivated by oxygen, the oxygen produced during photosynthesis by chloroplasts may prove lethal to maintaining the integrity of the nitrogenase enzyme complex. Therefore, photosynthesis and nitrogen fixation must be separated if expressing functional nitrogenase in chloroplasts, either temporally by confining *nif* expression only to when photosynthesis is not happening, or spatially by limiting *nif* expression to tissues not involved in photosynthesis, such as the roots (Rosenblueth et al. 2018). López-Torrejón et al. (2016) used yeast, an organism that does not photosynthesize, as a proof of concept to engineer *nif*H, *nif*M, *nif*S, and *nif*U into the eukaryotic cell from *Azotobacter vinelandii* and demonstrated that if NifH polypeptide and *nif*M maturase are targeted to the mitochondrial milieu, active nitrogenase Fe protein can be produced. They further demonstrated that because *nif*H can acquire and incorporate endogenously generated mitochondrial Fe-S clusters, *nif*S and *nif*U, the *nif*H-specific Fe-S cofactor synthesizing protein components, need not be transferred into the mitochondria for the generation of an active Fe protein. Burén et al. (2017) targeted a minimum set of nine *nif* genes in *A. vinelandii* (*nif*B, *nif*D, *nif*E, *nif*H, *nif*K, *nif*M, *nif*N, *nif*S, and *nif*U) into the mitochondria and demonstrated that the *nif*DK tetramer was successfully formed, which is an essential first step to bring together a functional nitrogenase in a eukaryotic cell. Attempts have also been made to transfer the *nif* gene in plants. *nif*H and *nif*M were expressed in the chloroplasts of tobacco and generated functional *nif*H albeit with low activity (Ivleva et al. 2016). Allen et al. (2017) demonstrated recently that the complete range of biosynthetic and catalytic nitrogenase (nif) proteins can be feasibly expressed in tobacco leaves as transit peptide-*nif* fusions targeting mitochondria. However, studies of tobacco and yeast demonstrated that the *nif*D polypeptide is susceptible to degradation in eukaryotic cells (Burén et al. 2017; Allen et al. 2017), and therefore, the amino acid sequence needs to be optimized for stability such that the catalytic activity remains uncompromised. The readers are referred to the excellent array of recently published review articles for a comprehensive account of how to transfer *nif* genes to eukaryotes (Curatti and Rubio 2014; Burén and Rubio 2018).

4.4.2 Engineering the Legume Symbiosis into Nonlegumes

In this approach, the development of root-nodule symbioses (RNS) in nonlegumes, like that found in legumes, is considered (Rogers and Oldroyd 2014). Exudates excreted by legumes contain specific plant chemicals that trigger only the compatible nearby rhizobia. The arbuscular mycorrhizal symbiosis is also affected by some components of the legume symbiotic signaling (SYM). It has also been observed that flavonoids and strigolactones released by cereals induce specific signals in arbuscular mycorrhizal fungi (AMF), which help initiate the symbiosis between AMF and cereals (Steinkellner et al. 2007). Since cereals already contain the SYM pathway for arbuscular mycorrhizal associations, the pathway can be activated by engineering the association to perceive the rhizobial signaling molecules, and engineering the activation outputs to the nodule-like root organs with limited oxygen to fix nitrogen (Mus et al. 2016). Legumes and rice, and possibly other cereals have similar genetic constituents that are import in initializing the AM symbiosis (AMS) development (Gutjahr et al. 2008). The same genetic components are crucial in legumes in aiding the initial stages of RNS development. The common symbiosis pathway (CSP) is constituted of these genetic elements that promote both AMS and RNS development (Markmann and Parniske 2009). The conserved genetic constituents of the CSP are being used in current research with cereals to form a basis on which genetic networks can be extended to build a complete signaling pathway to support RNS in cereals like that found in legumes (Rogers and Oldroyd 2014; Davidson et al. 2015; Mus et al. 2016). Latest research in phylogenomics also suggest that a species that associates with arbuscular mycorrhizal fungi can be converted into a nitrogen-fixing symbiont with a small set of genes (Griesmann et al. 2018; van Velzen et al. 2018). Generally, symbiotic N_2 fixation is very complex, which requires execution and regulation of multiple events in the host plant as well as the rhizobia. Therefore, adapting the existing developmental and signaling mechanisms to establish a suitable environment for nitrogenase activity in the new cereal nodules is necessary to successfully engineer a N_2-fixing symbiosis in cereals (Mus et al. 2016; Oldroyd and Dixon 2014; Goyal et al. 2021).

4.4.3 Engineering Nonlegumes to Associate with N_2-fixing Bacteria and/or Other PGPB

As mentioned above, diazotrophs are found among Alpha-proteobacteria, Beta-proteobacteria, Cyanobacteria, Firmicutes, and Gamma-proteobacteria. However, diazotrophs are observed to not be the dominant bacteria in rhizospheres. Therefore, it may be possible to increase nitrogen fixation by increasing the diazotroph populations. Two nonmutually exclusive, but distinct approaches are suggested to enhance the existing interactions between bacteria and nonlegumes: increasing the colonization of plants by highly efficient N_2-fixing microbes (e.g., developing cereals that promote diazotroph growth), and engineering into bacteria that already closely associate with cereals to the transfer of efficient nitrogen fixation (Geddes et al. 2015). Endophytic bacterial population exists in too low density in nonlegume tissues to fix sufficient nitrogen, and therefore systems should be established to increase the colonization of diazotrophic endophytes for improved nitrogen fixation in nonlegumes. Therefore, it is important to find and engineer bacteria that specifically associate with cereals to enrich the microbiomes of cereals with associative and endophytic N_2 fixers and develop competent plant varieties. To this end, improving the chances of the inoculating diazotroph to selectively colonize the crop plant is important. This is especially crucial as newly introduced bacterial strains to the rhizosphere are generally dominated by the native microbes. This problem may be solved by engineering plants to produce specific metabolites that lead to a "biased rhizosphere" that favors the growth and development of the new bacterial strain to be introduced into (Rossbach et al. 1994). Historically, researchers have investigated how the bacterial population selection in the rhizosphere is affected by new nutritional resources (Oger et al. 1997; Savka and Farrand 1997). Pea root mucilage, for example, is the sole carbon source for some *Burkholderia* sp., *Pseudomonas* sp., and *Rhizobium* sp. (Knee et al. 2001). This "biased rhizosphere" approach necessarily will involve identifying appropriate target genes, signals, and receptors, to effectively construct a rhizosphere to encourage the growth of the newly introduced diazotroph (Rossbach et al. 1994). The traits involved for plant colonization are poorly understood and hundreds of genes are likely involved. Engineering into bacteria, the ability to colonize and associate with plants is daunting, and can be avoided if associative bacteria or pre-existing endophytes are employed as the basis for enhancing or transferring the N_2-fixing ability via synthetic biology. Direct engineering of plants has been favored over engineering bacteria that are already native in an ecological niche to help cereals fix nitrogen (Geddes et al. 2015). Molecular approaches independent of the culture have recently shown that bacterial nitrogenase genes are expressed in plants, and some rhizobia are found in cereals (Rosenblueth et al. 2018). In the end, multiple approaches should be integrated to transfer nitrogen fixation to cereals. One such strategy would involve developing in nonlegumes the ability to perceive rhizobial signaling molecules, formation of nodule-like root organ in an oxygen-limited setting, and infection of the newly formed root nodules with N_2-fixing bacteria. In this new niche, the ideal symbionts would be associative or endophytic organisms already engineered for efficient nitrogen fixation and transfer to plants (Geddes et al. 2015).

To enhance the competitiveness of diazotrophs, nonlegume plants may be selected or modified in such a way that the growth of certain diazotrophs is favored. For example, a specialized carbon source may be utilized to strengthen the competition for carbon for a given population of nitrogen-fixing microbes, in order to establish signals between those microbes and cereals for effective colonization (Mus et al. 2016). Transgenic plants produce opine molecules that are known to boost the rhizosphere with bacteria that catabolize opine molecules; however, this is accompanied with the risk of filling the rhizosphere with chemical compounds produced by pathogenic organisms (Mondy et al. 2014; Oger et al. 1997; Savka and Farrand 1997). A rare group of chemicals produced by rhizobia inside legume nodules that are exuded into the rhizosphere are referred to as rhizopines. Of these, 3-*O*-methyl-scyllo-inosamine 2 (3-*O*-MSI) and scyllo-inosamine 1 (SIA) are thought to be suitable for ideal chemical signaling between rhizospheric bacteria and plants. However, little success has been made in engineering plants to produce rhizopines (Murphy et al. 1995; Savka et al. 2013; Wexler et al. 1995; Gordon et al. 1996). Rhizobia receive their energy from rhizopines (carbon and nitrogen). M*os*ABC is responsible for rhizopine synthesis, and *moc*CABRDEF is responsible for rhizopine catabolsim, which have been found in the rhizobium *Sinorhizobium meliloti* L5-30 (Murphy et al. 1987). Continued pursuits in the rhizospheric engineering of cereal crops have led to recent success in transferring rhizospine biosynthesis into barley (Hordeum vulgare) (Geddes et al. 2019). It is also known that even if the soil life of diazotrophs is short, regular diazotroph inoculation, as is common for legumes, can provide enough bacterial cells for plants (Rosenblueth et al. 2018).

Free-living, nitrogen-fixing bacteria commonly assimilate to and be used by bacteria for their own growth, instead of excreting nitrogen compounds to the host plant with ammonium, as occurs in nodules. Through nitrogen fixation, genetically modified bacteria were observed to improve plant growth. Mutants of some diazotrophs that excrete ammonium were effective at supplying N to their host plants (Setten et al. 2013; Rosenblueth et al. 2018). *Azospirillum* that excreted nitrogen, for example, was shown to improve the nitrogen supply to wheat plants (Van Dommelen et al. 2009). Similarly, *Azotobacter*, *Azospirillum*, *Kosakonia*, and *Pseudomonas* mutants (Setten et al. 2013; Zhang et al. 2012; Geddes et al. 2015; Ambrosio et al. 2017; Bageshwar et al. 2017) were observed to promote plant growth. Ammonium-excreting mutants of *Azoarcus*, *Herbaspirillum*, or *Paraburkholderia* should also be tested to check whether they enhance plant growth through nitrogen fixation. Setten et al. (2013) have engineered *Pseudomonas protegens* Pf-5, a root-colonizing nondiazotrophic endophyte, by transferring a DNA stretch from *P. stutzeri* with 52 genes including the *nif* gene cluster (Vermeiren et al. 1999). The modified *P. protegens* strain constitutively fixed nitrogen and released an abundance of ammonium to its surroundings, even in the presence of combined nitrogen. Fox et al. (2016) demonstrated in greenhouse tests, increased maize and wheat yields when inoculated with *Pseudomonas protegens* Pf-5, and ^{15}N isotope dilution analysis confirmed that nitrogen fixation in the roots was clearly responsible for this positive effect.

All plant roots secrete exudates (Badri and Vivanco 2009), which comprise a broad spectrum of high-molecular-weight (polysaccharides and proteins) and low-molecular-weight (amino acids, organic acids, phenolic compounds, and sugars) compounds (Huang et al. 2014). A wide range of functions are performed by these exudates, which include improving the nutrient uptake, increasing abiotic stress tolerances, structuring the soil-plant microbiome, and suppressing diseases (Badri and Vivanco 2009; Huang et al. 2014; Hirsch et al. 2003; Dennis et al. 2010). Mucilage derived from the root cap is usually a viscoelastic root exudate released to the rhizosphere from the root cap cells. Mucilage is composed primarily by amino acids, alcohols, fatty acids, polysaccharides, and organic acids, and is categorized chemically as a high-molecular weight (HMW) carbohydrate (Naveed et al. 2017; van Veelen et al. 2018). About 5–21% of a plant's photosynthates are known to be released by the roots as amino acids, soluble sugars, or secondary metabolites, which in turn recruit and support rhizospheric microbial communities (Huang et al. 2014; Badri and Vivanco 2009; Badri et al. 2013; Chaparro et al. 2013). A sufficient supply of organic acids, sugars, as well as a low-nitrogen, low-oxygen environment are generally considered essential to support microbial communities that fix nitrogen (Bennett et al. 2020). For plants, mucilage is known to be an essential determinant of the nitrogen fixation and researchers have suggested that for nonlegumes, mucilage secretion may play a more general role in sheltering the microbial diazotroph communities (Van Deynze et al. 2018; Bennett et al. 2020; Johansson and Bergman 1992; Forni and Caiola 1993). Based on their findings with the aerial root mucilage of maize landrace native to the Sierra Mixe indigenous maize landrace (Bennett et al. 2020), they also proposed a diazotrophic microbial community model that is supported by mucilage. In their suggested model, Bennett et al. (2020) explained how mucilage can support the general requirements of a community of nitrogen-fixing microbes. Mucilage refers to the wealth of sugars that can potentially act as an energy source for diazotrophs (Bennett et al. 2020; Van Deynze et al. 2018; Osborn et al. 1999; Chaboud 1983). Root mucilage contains fucose, glucose, galactose, xylose, and arabinose. According to Van Deynze et al. (2018), the high levels of nitrogen fixation (29–82% N) in indigenous landraces of maize grown in nitrogen-depleted fields were attributed to the abundant production of sugars (41% fucose, 36% galactose, 14% arabinose, 3% xylose, and 3% mannose): the aerial-plant-root-associated rich mucilage supported a complex nitrogen-fixing microbiome. They concluded that the nitrogen fixation and subsequent delivery to maize plants may be at least partially attributed to mucilage (Van Deynze et al. 2018). The monosaccharide composition of mucilage may help signal the associative diazotrophic bacteria that are able to disintegrate the polysaccharides of the mucilage complex, and provide the released monosaccharides that support nitrogen fixation and growth promotion (Bennett et al. 2020). It is likely that the polysaccharides of mucilage provide the optimal environment-specific microbiota, by serving as an energy source, and helping bacteria establish themselves in the mucilage-root environment by supporting their metabolism and essential mechanisms of colonization (Bennett et al. 2020). A majority of the root exudate diversity is provided by the low-molecular-weight compounds, while a larger mass proportion is taken up by the high-molecular-weight compounds (Bennett et al. 2020). It is likely that mucilage will be a utilizable feature of cereal crops (e.g., wheat, barley, maize, and sorghum) (Bennett et al. 2020; Sinha Roy et al. 2002; Carter et al. 2019; Werker and Kislev 1978; Li et al. 2014). The polysaccharide structure of mucilage details the terminal positions and proportions of arabinose, fucose, and xylose residues; this suggests that enzymatic release of the mucilage could feed microbiota, like diazotrophs, that reside in the mucilage (Bennett et al. 2020). The microbial genes that encode the minimum gene set for molybdenum nitrogenase to fix nitrogen (*nif* genes) are known to be contained in the mucilage (Van Deynze et al. 2018; Dos Santos et al. 2012). Molybdenum nitrogenase is likely an essential member for nitrogen fixation in mucilage of nonlegumes; two additional nitrogenases, iron-only nitrogenase and vanadium nitrogenase, may also contribute to the mucilage nitrogen fixation (Bennett et al. 2020). The wealth of sugars in the mucilage polysaccharides can be used by diazotrophs to fuel nitrogen fixation, which are arabinose, fucose, and/or xylose (Bennett et al. 2020). Diazotrophs in the mucilage possess and are able to express genes that are necessary for importing and metabolizing components of the mucilage polysaccharides, in order to produce ATP which serves as an energy source for the energy-intensive nitrogenase activities (Bennett et al. 2020).

As previously mentioned, to support diazotrophs, the microbial community's ability to reduce the oxygen levels is critical. A delicate balance in the oxygen level management is necessary for nitrogen fixation by aerobic bacteria, because high ATP levels as a fuel are usually produced through aerobic respiration, but high oxygen levels inhibit nitrogenase activities (Hunt and Layzell 1993; Marchal and Vanderleyden 2000). At a depth of 8 mm, it was observed that root mucilage maintained oxygen levels below 5%, which suggests that a microaerobic environment that supports nitrogenase activities could be sustained with mucilage (Van Deynze et al. 2018). A similar oxygen depletion level was observed in a 0.2% agar medium; the reduction of atmospheric oxygen diffusion into the mucilage matrix, mediated by the mucilage, may result in these low oxygen levels. Oxygen levels were observed to be similarly depleted by embedded diazotrophs in an exopolysaccharide pellicle, which enable aerobic nitrogen fixation, and in bacterial biofilm aggregates (Wessel et al. 2014; Wang et al. 2017). Nitrogen fixation is a highly regulated molecular mechanism because it requires great energy levels. Transcriptional and/or post-transcriptional mechanisms are used by diazotrophs to deactivate nitrogenase activities when nitrogen is present in the environment (Halbleib and Ludden 2000). Very low nitrate and ammonium levels were observed in Sierra Mixe maize mucilage by Bennett et al. (2020), when the mucilage nitrogenase levels were very high and the nitrogenase activity was uninhibited. These low nitrogen levels likely help enrich diazotrophs in microbial community of the mucilage, as nitrogen-fixing bacteria are exposed to a significant advantage in such a high-carbon, low-oxygen, and low-nitrogen environment. However, it has yet to be clearly determined how nitrogen is transferred to the host plant from bacteria. The model suggested by Bennett et al. (2020) provides a general framework on which diazotrophic activities can be evaluated for cereal crops and optimize the functionalities of the mucilage by specifically structuring the microbial communities associated with the mucilage and/or through genetic selection to potentially increase nitrogen fixation in cereals. Further research is necessary to determine whether genetic engineering or breeding can transfer the aerial root mucilage trait can be transferred to conventional corn varieties and other cereal crops also possess the same trait.

One of the promising sustainable ways for reducing application of chemical N fertilizer and increasing N use efficiency (NUE) is by providing plant growth-promoting bacteria (PGPB) to nonlegumes (Roy et al. 2015; Adesemoye and Kloepper 2009; Etesami 2019; Etesami and Maheshwari 2018). In addition, the combined use of N_2-fixing bacteria and PGPB can improve the root growth and help cereals better utilize nutrients and environmental resources. To this end, systems that can be effectively colonized by PGPB should be designed such that the diazotrophs selectively colonize the targeted plants, since the diazotroph and plant variety all substantially influence the N fixation (Boujenna and del Moral 2021).

The other much short-term and simpler approach to improve N nutrition in nonlegumes is to use naturally occurring nonrhizobial, N_2-fixing endophytic bacteria (commonly known as associative N_2 fixation) that can colonize the root systems of nonlegumes intracellularly and fix N_2 without any need for nodulation (Dent and Cocking 2017; Cocking et al. 2006; Bargaz et al. 2018; Etesami 2019). Bacterial endophytes are a plant-associated bacteria that can nonpathogenically colonize various plant tissues internally (with no visible disease symptoms in plants) (Kaga et al. 2009; Hallmann et al. 1997; Gaiero et al. 2013; Compant et al. 2010; Sessitsch et al. 2012). Endophytic bacteria are considered to be a rhizobacterial subpopulation (Compant et al. 2010). Contrasted to other bacteria (i.e., rhizosphere, rhizoplane, and phyllosphere bacteria), the bacterial endophytes might establish more intimate relationships with the host plant and provide fixed N without any loss to the plant. In addition, because there is no competition between bacterial endophytes and non-endophytic microorganisms (rhizosphere, rhizoplane, and phyllosphere bacteria) in the endorhizosphere, and because carbon sources are provided with low-pressure oxygen oscillations (James et al. 2002; Etesami 2019), endorhizospheric bacteria contribute much more extensively to N fixation, than rhizospheric bacteria (Etesami 2019). Different nonlegume plant tissues are well-known to host different nitrogen-fixing bacterial genera, such as *Enterobacter, Microbacterium, Klebsiella, Beijerinckia, Citrobacter, Herbaspirillum, Bacillus, Alcaligenes, Rhizobium, Azospirillum, Penibacillus, Azotobacter, Agrobacterium, Sphingomonas, Methylosinus* sp., *Corynebacterium, Azoarcus, Gluconacetobacter, Clostridium, Methanosarcina, Burkholderia,* and *Paenibacillus* (Reinhold-Hurek et al. 2007; Hongrittipun et al. 2014; Yanni and Dazzo 2010; Gupta et al. 2012; Ji et al. 2014; Prayitno and Rolfe 2010; Etesami 2019; Yoneyama et al. 2017; Bargaz et al. 2018). Utilizing such nonsymbiotic N_2-fixing bacteria may help reduce N fertilization for agriculture by producing 22–50 kg of N per hectare, and contribute to reducing damage to the environment (Elbeltagy et al. 2001; Etesami 2019; Ladha et al. 2016). In earlier reports, an increase of 21.2–23% in yield and K, N, and P levels in rice plants inoculated with N_2-fixing bacteria (i.e., *Anabaena oscillarioides* CR3, *Brevundimonas diminuta* PR7, *Ochrobactrum anthropi* PR10, and *Pseudomonas* sp.) was observed, when compared to the use of recommended rate of NPK fertilizers was observed (Rana et al. 2015; Mäder et al. 2011).

The results of some reports indicate that nitrogen fixation is not the only mechanism by N_2-fixing bacteria to stimulate growth and yield of rice plants (Saharan and Nehra 2011; Bhattacharjee et al. 2008). These bacteria can also improve growth, yield, and nutrient absorption (N, P, K, Zn, etc.) in rice plants through other mechanisms such as production of auxin (IAA, etc.), ACC (1–aminocyclopropane–1–carboxylate) deaminase, siderophores, and exopolysaccharides (extracellular polymeric substance), as well as insoluble inorganic solubilization (Etesami and Alikhani 2016; Etesami 2019; Etesami and Maheshwari 2018; Estrada et al. 2013; Ji et al. 2014; de Souza et al. 2013). In other words, the N in nonlegumes like as rice can be either due to nitrogen fixation or due to an increase in the uptake of soil nitrogen by the plant (Yanni et al. 1997; Prayitno et al. 1999; Elbeltagy et al. 2001; Oliveira et al. 2002).

In addition to N availability, moisture availability (soil water) is well known to be a key factor determining crop yield and NUE (Bänziger et al. 1999; Huang et al. 2018). To paraphrase, plants cannot extract nutrients (especially, nutrients that are absorbed by the mass flow process like nitrate, sulfate, calcium, and magnesium) from the soil without sufficient water (Huang et al. 2018). In saturated or partially saturated environments, it was observed that exopolysaccharide-producing bacteria associated with plants increased the soil water retention, diminished soil water evaporation, and decreased the hydraulic conductivity, which helps make more water available for a longer time for plants (Zheng et al. 2018) and consequently increases the water use efficiency of plants (Roberson and Firestone 1992). The water holding capacity of bacterial exopolysaccharides is great (a polymer matrix is 97% water); bacterial exopolysaccharides can change the pore space connectivity and the soil matrix structure, as well as modify the viscosity and surface tension of water (Zheng et al. 2018; Roberson and Firestone 1992; Bozorg et al. 2015; Volk et al. 2016; Kroener et al. 2018). Bacterial exopolysaccharides can also rectify rhizospheric soil structural properties (i.e., increase the rhizospheric soil aggregation, decrease the percentage of soil aggregates, increase the ratio of root-adhering soil to the root tissue, form a hydrophilic biofilm around roots, etc.) (Etesami and Maheshwari 2018; Kaushal and Wani 2016).

Root architecture is also known as an important trait that affects N uptake efficiency and plays a pivotal role in extracting available N from the soil. The capacity of the root to absorb nutrient (i.e., N) depends on the extent of root expansion and its absorption surface area. Some bacteria can increase plant root system development and thereby increase the absorption of nutrients (improved N use efficiency) by producing IAA and ACC deaminase (stress ethylene level-reducing enzyme) (Etesami and Maheshwari 2018). When soil nutrients (i.e., P and Fe) are scarcely available, limits are usually placed on biological nitrogen fixation (BNF). Therefore, sufficient Fe and P levels should be maintained to increase the nitrogen fix efficiency during BNF (Etesami and Beattie 2017; Schulze and Drevon 2005; Alkama et al. 2012). Phosphate-solubilizing bacteria (PSB) and siderophore-producing bacteria (SPB) are able to respectively increase the P and Fe availability to plants grown with low P and Fe levels (Etesami 2019; Etesami and Maheshwari 2018).

Based on the above statements, if the purpose of inoculating nonlegume plants with nitrogen-fixing bacteria (NFB) is to supply the plant nitrogen, it is recommended that other bacteria with the ability to produce ACC deaminase, IAA, exopolysaccharides, and siderophores, as well as the ability to solubilize phosphates (or fix nitrogen and promote various plant growth traits) are also inoculated into the plants (multistrain inoculation). This bacterial consortium can improve nitrogen uptake by rice plant via different mechanisms (Fig. 4.3). Combined use of such bacteria can be effective at achieving maximum N uptake, N-fertilizer savings, and nonlegume growth.

There are reports showing that simultaneous inoculation (multistrain inoculation) of nonlegumes like rice with several superior bacteria has a significantly greater effect on the nutrient uptake (decreased dependency on exogenous nitrogen supply) and consequently plant growth, compared to inoculation with a single superior bacterium strain (Etesami and Alikhani 2016; Roy et al. 2015; Nguyen et al. 2003; Cong et al. 2011; Uthiraselvam et al. 2012; Hasan et al. 2014; Roy and Srivastava 2010; Williams and Kennedy 2002; Malik et al. 2002; Hegazi et al. 1998). For example, Etesami and Alikhani (2016) observed in a study that supplementing 75% of the recommended N-fertilizization rate with endophytic (*Pseudomonas fluorescens* REN1) and rhizospheric (*Pseudomonas putida* REN5) bacteria as a multistrain inoculation with several plant growth-promoting traits resulted in

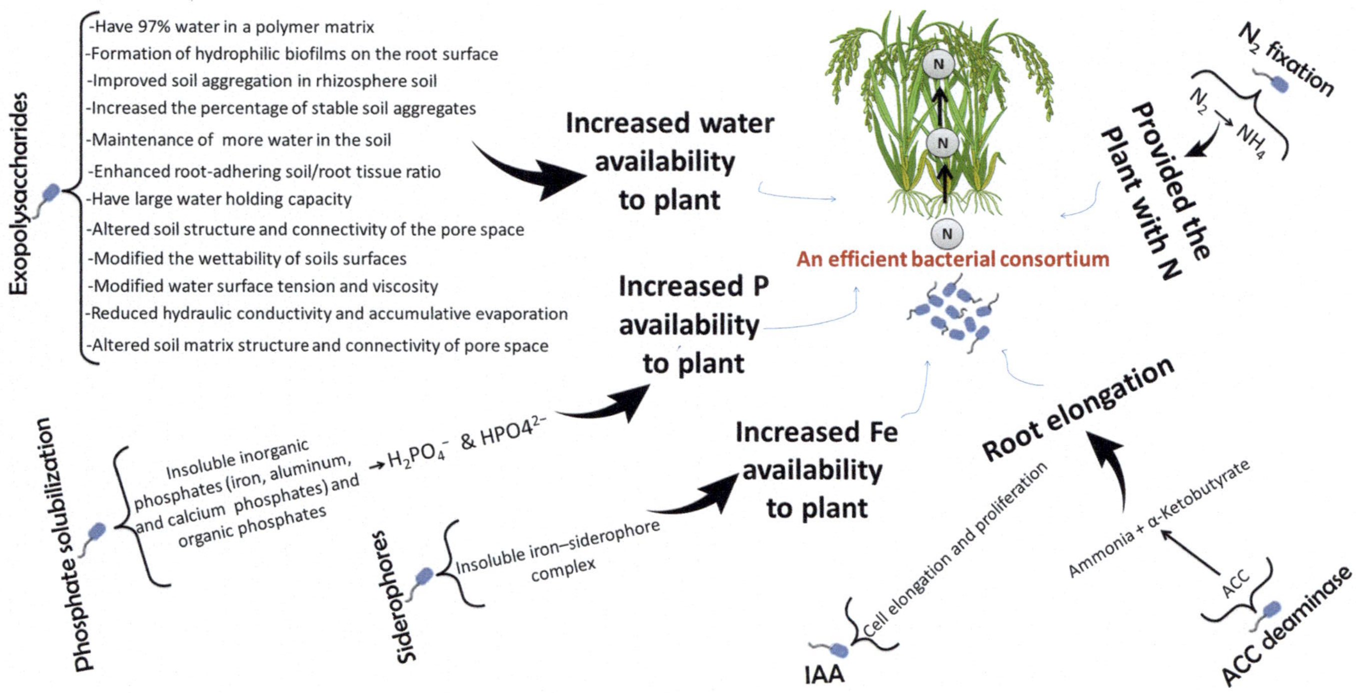

Fig. 4.3 An efficient bacterial consortium for improving water use efficiency and nitrogen (N) use efficiency in nonlegume plant

enhanced N content and growth indices in rice compared to single-strain inoculation, which statistically were similar to when the full fertilizer rate (313 kg urea ha^{-1}) was employed without these isolates. In other words, in this study multistrain inoculation with these bacterial strains diminished N-fertilizer application by up to 25%. It is well known that bacteria, when used in combination, may produce synergy or a particular bacterium may serve the "helper" role to boost the other bacterium's performance. It is found that these bacteria contribute to effects that can be complemented by diverse functions (i.e., nutrient (N, P, etc.) provision, inhibitory product removal, and biochemically/physically induced mutual stimulation) (Etesami and Maheshwari 2018; Roy et al. 2015; Bashan and Holguin 1997). Since most soils under cultivation of cereal crops, including rice, are deficient in N, and rice production is highly dependent on N application, and nitrogen-fixing bacteria cannot replace nitrogen fertilizer in rice plant (they can be used as a supplement to chemical N fertilizer) (Etesami and Alikhani 2016), a combined use of sufficient N nutrition and nonsymbiotic nitrogen-fixing bacteria would help improve rice yields (Bargaz et al. 2018; de Souza et al. 2016; Biswas et al. 2000; Yanni and Dazzo 2010; Duarah et al. 2011; Khorshidi et al. 2011; Khan et al. 2017; Etesami 2019).

4.5 Conclusions and Future Prospects

The N availability is a major bottleneck for crop growth. For a reliable quality and yields, sufficient N fertilization is necessary for plants. However, excessive chemical N fertilization resulted in not only a worldwide severe N pollution, but also shrunk plants' N use efficiency. Alternatives to chemical N fertilization need to be found, to securely and sustainably produce food. One such sustainable N nutrition may be achieved through biological nitrogen fixation in plants and may shift the dependence of N provision from industrial settings. Symbiotic nitrogen fixation is currently limited to legumes to a significant degree, multiple nonlegume rhizospheric diazotrophs exist that have been observed to enhance N nutrition in nonlegumes. Interventions made biotechnologically have also importantly helped provide nitrogen-fixing capacity to nonlegumes. Constructing N_2-fixing nonlegumes including cereals, which normally use large amounts of chemical fertilizers, is an enormous challenge biotechnologically, which if successful, would revolutionize agricultural systems worldwide. Great efforts have been made to establish nitrogen-fixing ability in nonlegumes, especially cereals, as they comprise a major proportion of the global food supply. The mechanisms with which symbiotic systems are formed between microbes and plants will help researchers successfully transfer the nitrogen-fixing ability to nonlegumes in the future. Introducing nitrogenase for nonlegumes, and transferring the legume symbiosis to nonlegumes are both complex technological challenges, but if/when successful, possess the potential to revolutionize crop production. While there will undoubtedly be obstacles to launch the two approaches, researchers must not limit efforts in fear of the complexity of these engineering problems to enable nonlegumes to fix nitrogen (Oldroyd and Dixon 2014). In these potential nitrogen-fixing cereals, even minute increases in the available N levels would lead to a substantial increase in the yields of low-input agricultural systems in developing countries. Manipulating host plants and soil diazotrophs may also be able to solve the overuse problem of synthetic N fertilizers in the short term. Genetically modifying cereal crops to be able to fix nitrogen is a complex challenge, but the current approaches being attempted are presenting exciting possibilities for successful implementation of such crops in the foreseeable future. This will enable the world to greatly benefit from less extensive chemical fertilization, while nitrogen-fixing nonlegumes are unlikely to pose any substantial harm to the environment. In addition to nitrogen, agricultural inputs like water and phosphorus may limit agricultural productivity. Plant cultivars and mycorrhiza with high-use efficiencies of phosphate use need to also be considered when researching to develop nitrogen-fixing cereals. To achieve this, the authors reckon that effective programs to control human population growth, and a more efficient crop management are required in addition to genetically modifying plants and utilizing microbes (Rosenblueth et al. 2018). These approaches integrated together then could research to realize the dream of self-supporting, nitrogen-fixing cereals. Associative nitrogen fixation in nonlegumes, especially cereals like maize, rice, and wheat, biased/ targeted rhizosphere, artificial symbioses, and endophytic nitrogen fixation in nonlegume plants should be considered further and researched into to help realize the goal of self-supporting nitrogen-fixing nonlegumes (Mahmud et al. 2020).

Acknowledgments We wish to thank University of Tehran for providing the necessary facilities for this study.

Conflict of Interest Author(s) declares no conflict of interest.

References

Addo MA, Dos Santos PC (2020) Distribution of nitrogen-fixation genes in prokaryotes containing alternative nitrogenases. ChemBiochem 21(12):1749–1759

Adesemoye AO, Kloepper JW (2009) Plant–microbes interactions in enhanced fertilizer-use efficiency. Appl Microbiol Biotechnol 85(1):1–12

Akao S, Nakata S, Yoneyama T (1991) Formation of nodules on non-nodulating soybean T201 with treatment of 2,4-D. Plant Soil, 138, 207–212

Akkermans ADL, Abdulkadir S, Trinick MJ (1978) N2-fixing root nodules in Ulmaceae: *Parasponia* or (and) *Trema* spp.? Plant Soil 49(3):711–715

Alkama N, Ounane G, Drevon JJ (2012) Is genotypic variation of H^+ efflux under P deficiency linked with nodulated-root respiration of N2-Fixing common-bean (*Phaseolus vulgaris* L.)? J Plant Physiol 169(11):1084–1089

Allen RS, Tilbrook K, Warden AC, Campbell PC, Rolland V, Singh SP, Wood CC (2017) Expression of 16 nitrogenase proteins within the plant mitochondrial matrix. Front Plant Sci 8:287

Ambrosio R, Ortiz-Marquez JCF, Curatti L (2017) Metabolic engineering of a diazotrophic bacterium improves ammonium release and biofertilization of plants and microalgae. Metabol Eng 40:59–68

Anas M, Liao F, Verma KK, Sarwar MA, Mahmood A, Chen Z-L, Li Q, Zeng X-P, Liu Y, Li Y-R (2020) Fate of nitrogen in agriculture and environment: agronomic, eco-physiological and molecular approaches to improve nitrogen use efficiency. Biol Res 53(1):1–20

Azam F (2002) Nodulation in cereals as a means to decreasing their dependence on nitrogenous fertilizers—an achievable target or a dogma. Pak J Biol Sci 5(1):122–127

Badri DV, Vivanco JM (2009) Regulation and function of root exudates. Plant Cell Environ 32(6): 666–681

Badri DV, Chaparro JM, Zhang R, Shen Q, Vivanco JM (2013) Application of natural blends of phytochemicals derived from the root exudates of *Arabidopsis* to the soil reveal that phenolic-related compounds predominantly modulate the soil microbiome. J Biolgic Chem 288(7): 4502–4512

Bageshwar UK, Srivastava M, Pardha-Saradhi P, Paul S, Gothandapani S, Jaat RS, Shankar P, Yadav R, Biswas DR, Kumar PA (2017) An environmentally friendly engineered Azotobacter strain that replaces a substantial amount of urea fertilizer while sustaining the same wheat yield. Appl Environ Microbiol 83(15):e00590-00517

Bailey-Serres J, Parker JE, Ainsworth EA, Oldroyd GED, Schroeder JI (2019) Genetic strategies for improving crop yields. Nature 575(7781):109–118

Bänziger M, Edmeades GO, Lafitte HR (1999) Selection for drought tolerance increases maize yields across a range of nitrogen levels. Crop Sci 39(4):1035–1040

Bargaz A, Lyamlouli K, Chtouki M, Zeroual Y, Dhiba D (2018) Soil microbial resources for improving fertilizers efficiency in an integrated plant nutrient management system. Front Microbiol 9:1606

Bashan Y, Holguin G (1997) *Azospirillum*–plant relationships: environmental and physiological advances (1990–1996). Can J Microbiol 43(2):103–121

Beatty PH, Good AG (2011) Future prospects for cereals that fix nitrogen. Science 333(6041): 416–417

Behm JE, Geurts R, Kiers ET (2014) Parasponia: a novel system for studying mutualism stability. Trend Plant Sci 19(12):757–763

Bennett AB, Pankievicz VCS, Ané J-M (2020) A model for nitrogen fixation in cereal crops. Trend Plant Sci 25(3):226–235

Bhattacharjee RB, Singh A, Mukhopadhyay SN (2008) Use of nitrogen-fixing bacteria as biofertiliser for non-legumes: prospects and challenges. Appl Microbiol Biotechnol 80(2): 199–209

Biswas J, Ladha J, Dazzo F (2000) Rhizobia inoculation improves nutrient uptake and growth of lowland rice. Soil Sci Soc Am J 64(5):1644–1650

Boddey RM, Dobereiner J (1995) Nitrogen fixation associated with grasses and cereals: recent progress and perspectives for the future. In: Ahmad N (ed) Nitrogen economy in tropical soils: proceedings of the international symposium on nitrogen economy in tropical soils, held in Trinidad, W.I., January 9–14, 1994, pp 241–250

Boujenna A, del Moral LFG (2021) Biotechnological approaches to develop nitrogen-fixing cereals: a review. Span J Agric Res 19(4):e08R01

Bozorg A, Gates ID, Sen A (2015) Using bacterial bioluminescence to evaluate the impact of biofilm on porous media hydraulic properties. J Microbiol Method 109:84–92

Burén S, Rubio LM (2018) State of the art in eukaryotic nitrogenase engineering. FEMS Microbiol Lett 365(2):fnx274

Burén S, Young EM, Sweeny EA, Lopez-Torrejón G, Veldhuizen M, Voigt CA, Rubio LM (2017) Formation of nitrogenase NifDK tetramers in the mitochondria of *Saccharomyces cerevisiae*. ACS Syn Biol 6(6):1043–1055

Carter AY, Ottman MJ, Curlango-Rivera G, Huskey DA, D'Agostini BA, Hawes MC (2019) Drought-tolerant barley: II. Root tip characteristics in emerging roots. Agronomy 9(5):220

Chaboud A (1983) Isolation, purification and chemical composition of maize root cap slime. Plant Soil 73(3):395–402

Chaparro JM, Badri DV, Bakker MG, Sugiyama A, Manter DK, Vivanco JM (2013) Root exudation of phytochemicals in *Arabidopsis* follows specific patterns that are developmentally programmed and correlate with soil microbial functions. PLoS One 8(2):e55731

Chi F, Shen S-H, Cheng H-P, Jing Y-X, Yanni YG, Dazzo FB (2005) Ascending migration of endophytic rhizobia, from roots to leaves, inside rice plants and assessment of benefits to rice growth physiology. Appl Environ Microbiol 71(11):7271–7278. https://doi.org/10.1128/AEM.71.11.7271-7278.2005

Cocking EC, Al-Mallah MK, Benson E, Davey MR (1990) Nodulation of non-legumes by rhizobia. In: Gresshoff PM, Roth LE, Stacey G, Newton WE (eds) Nitrogen fixation. Springer, Boston, MA, pp 813–823

Cocking EC, Stone PJ, Davey MR (2006) Intracellular colonization of roots of Arabidopsis and crop plants by Gluconacetobacter diazotrophicus. In Vitro Cell Dev Biol-Plant 42:74–82

Compant S, Nowak J, Coenye T, Clément C, Ait Barka E (2008) Diversity and occurrence of Burkholderia spp. in the natural environment. FEMS Microbiol Rev 32(4):607–626

Compant S, Clément C, Sessitsch A (2010) Plant growth-promoting bacteria in the rhizo-and endosphere of plants: their role, colonization, mechanisms involved and prospects for utilization. Soil Biol Biochem 42(5):669–678

Cong PT, Dung TD, Hien NT, Choudhury ATMA, Rose MT, Kecskes ML, Deaker R, Kennedy IR (2011) Effects of a multistrain biofertilizer and phosphorus rates on nutrition and grain yield of paddy rice on a sandy soil in Southern Vietnam. J Plant Nutr 34(7):1058–1069

Conway G (2019) The doubly green revolution: food for all in the twenty-first century. Cornell University Press, Ithaca, NY

Curatti L, Rubio LM (2014) Challenges to develop nitrogen-fixing cereals by direct nif-gene transfer. Plant Sci 225:130–137

Da Silva JG, Serra GE, Moreira JR, Conçalves JC, Goldemberg J (1978) Energy balance for ethyl alcohol production from crops. Science 210:903–906

Davidson FW, Whitney HG, Tahlan K (2015) Genome Sequences of *Klebsiella variicola* isolates from dairy animals with bovine mastitis from Newfoundland, Canada. Genom Announce 3(5): e00938-00915. https://doi.org/10.1128/genomeA.00938-15

de Almeida CV, Andreote FD, Yara R, Tanaka FAO, Azevedo JL, de Almeida M (2009) Bacteriosomes in axenic plants: endophytes as stable endosymbionts. World J Microbiol Biotechnol 25(10):1757–1764

de Souza R, Beneduzi A, Ambrosini A, Da Costa PB, Meyer J, Vargas LK, Schoenfeld R, Passaglia LM (2013) The effect of plant growth-promoting rhizobacteria on the growth of rice (*Oryza sativa* L.) cropped in southern Brazilian fields. Plant Soil 366(1-2):585–603

de Souza R, Schoenfeld R, Passaglia LM (2016) Bacterial inoculants for rice: effects on nutrient uptake and growth promotion. Arch Agron Soil Sci 62(4):561–569

Deng Y, Wu T, Wang M, Shi S, Yuan G, Li X, Chong H, Wu B, Zheng P (2019) Enzymatic biosynthesis and immobilization of polyprotein verified at the single-molecule level. Nat Commun 10(1):1–11

Dennis PG, Miller AJ, Hirsch PR (2010) Are root exudates more important than other sources of rhizodeposits in structuring rhizosphere bacterial communities? FEMS Microbiol Ecol 72(3): 313–327

Dent D, Cocking E (2017) Establishing symbiotic nitrogen fixation in cereals and other non-legume crops: the Greener Nitrogen Revolution. Agric Food Sec 6(1):1–9

Desbrosses GJ, Stougaard J (2011) Root nodulation: a paradigm for how plant-microbe symbiosis influences host developmental pathways. Cell Host Microbe 10(4):348–358

Dixon R, Kahn D (2004) Genetic regulation of biological nitrogen fixation. Nat Rev Microbiol 2(8): 621–631

Döbereiner J, Baldani VLD, Reis VM (1995) Endophytic occurrence of diazotrophic bacteria in non-leguminous crops. In: Fendrik I, del Gallo M, Vanderleyden J, de Zamaroczy M (eds)

Azospirillum VI and related microorganisms. NATO ASI series, vol 37. Springer, Berlin. https://doi.org/10.1007/978-3-642-79906-8_1

Dos Santos PC, Fang Z, Mason SW, Setubal JC, Dixon R (2012) Distribution of nitrogen fixation and nitrogenase-like sequences amongst microbial genomes. BMC Genomics 13(1):1–12

Duarah I, Deka M, Saikia N, Boruah HPD (2011) Phosphate solubilizers enhance NPK fertilizer use efficiency in rice and legume cultivation. 3 Biotech 1(4):227–238

Elbeltagy A, Nishioka K, Sato T, Suzuki H, Ye B, Hamada T, Isawa T, Mitsui H, Minamisawa K (2001) Endophytic colonization and in planta nitrogen fixation by a *Herbaspirillum* sp. isolated from wild rice species. Appl Environ Microbiol 67(11):5285–5293

Engelhard M, Hurek T, Reinhold-Hurek B (2000) Preferential occurrence of diazotrophic endophytes, *Azoarcus* spp., in wild rice species and land races of *Oryza sativa* in comparison with modern races. Environ Microbiol 2(2):131–141

Erisman JW, Galloway JN, Dise NB, Sutton MA, Bleeker A, Grizzetti B, Leach AM, De Vries W (2015) Nitrogen: too much of a vital resource. Sci Brief WWF, Netherlands

Eskin N, Vessey K, Tian L (2014) Research progress and perspectives of nitrogen-fixing bacterium, *Gluconacetobacter diazotrophicus*, in monocot plants. Int J Agron 2014:208383

Estrada GA, Baldani VLD, de Oliveira DM, Urquiaga S, Baldani JI (2013) Selection of phosphate-solubilizing diazotrophic *Herbaspirillum* and *Burkholderia* strains and their effect on rice crop yield and nutrient uptake. Plant Soil 369(1-2):115–129

Etesami H (2019) Plant growth promotion and suppression of fungal pathogens in rice (*Oryza sativa* L.) by plant growth-promoting bacteria. In: Maheshwari DK, Dheeman S (eds) Field crops: sustainable management by PGPR. Springer, Cham, pp 351–383

Etesami H, Alikhani HA (2016) Co-inoculation with endophytic and rhizosphere bacteria allows reduced application rates of N-fertilizer for rice plant. Rhizosphere 2:5–12

Etesami H, Beattie GA (2017) Plant-microbe interactions in adaptation of agricultural crops to abiotic stress conditions. In: Kumar V, Kumar M, Sharma S, Prasad R (eds) Probiotics and plant health. Springer, Singapore, pp 163–200

Etesami H, Maheshwari DK (2018) Use of plant growth promoting rhizobacteria (PGPRs) with multiple plant growth-promoting traits in stress agriculture: action mechanisms and future prospects. Ecotoxicol Environ Saf 156:225–246

Etesami H, Hosseini HM, Alikhani HA, Mohammadi L (2014) Bacterial biosynthesis of 1-aminocyclopropane-1-carboxylate (ACC) deaminase and indole-3-acetic acid (IAA) as endophytic preferential selection traits by rice plant seedlings. J Plant Growth Regul 33(3):654–670

Forni C, Caiola MG (1993) *Azolla*: an efficient N_2-fixing association with three components. Plant Biosys 127(3):422–427

Fox AR, Soto G, Valverde C, Russo D, Lagares A Jr, Zorreguieta Á, Alleva K, Pascuan C, Frare R, Mercado-Blanco J (2016) Major cereal crops benefit from biological nitrogen fixation when inoculated with the nitrogen-fixing bacterium *Pseudomonas protegens* Pf-5 X940. Environ Microbiol 18(10):3522–3534

Franche C, Lindström K, Elmerich C (2009) Nitrogen-fixing bacteria associated with leguminous and non-leguminous plants. Plant Soil 321(1):35–59

Francisco PB, Akao S (1994) The 2, 4-D-induced wheat para-nodules are modified lateral roots with structure enhanced by rhizobial inoculation. Plant Soil 159(2):121–129

Fustec J, Lesuffleur F, Mahieu S, Cliquet J-B (2010) Nitrogen rhizodeposition of legumes. A review. Agron Sustain Dev 30(1):57–66

Gaby JC, Rishishwar L, Valderrama-Aguirre LC, Green SJ, Valderrama-Aguirre A, Jordan IK, Kostka JE (2018) Diazotroph community characterization via a high-throughput *nif*H amplicon sequencing and analysis pipeline. Appl Environ Microbiol 84(4):e01512-01517

Gaiero JR, McCall CA, Thompson KA, Day NJ, Best AS, Dunfield KE (2013) Inside the root microbiome: bacterial root endophytes and plant growth promotion. Am J Bot 100(9): 1738–1750

Galloway JN, Townsend AR, Erisman JW, Bekunda M, Cai Z, Freney JR, Martinelli LA, Seitzinger SP, Sutton MA (2008) Transformation of the nitrogen cycle: recent trends, questions, and potential solutions. Science 320(5878):889–892

Garcia de Salamone IE, Döbereiner J, Urquiaga S, Boddey RM (1996) Biological nitrogen fixation in Azospirillum strain-maize genotype associations as evaluated by the ^{15}N isotope dilution technique. Biol Fertil Soil 23(3):249–256

Garcia de Salomone I, Döbereiner J (1996) Maize genotype effects on the response to Azospirillum inoculation. Biol Fertil Soil 21(3):193–196

Geddes BA, Ryu M-H, Mus F, Garcia Costas A, Peters JW, Voigt CA, Poole P (2015) Use of plant-colonizing bacteria as chassis for transfer of N_2-fixation to cereals. Curr Opin Biotechnol 32: 216–222. https://doi.org/10.1016/j.copbio.2015.01.004

Geddes BA, Paramasivan P, Joffrin A, Thompson AL, Christensen K, Jorrin B, Brett P, Conway SJ, Oldroyd GED, Poole PS (2019) Engineering transkingdom signalling in plants to control gene expression in rhizosphere bacteria. Nat Commun 10(1):1–11

Glendining MJ, Dailey AG, Williams AG, Van Evert FK, Goulding KWT, Whitmore AP (2009) Is it possible to increase the sustainability of arable and ruminant agriculture by reducing inputs? Agric Sys 99(2–3):117–125

Good A (2018) Toward nitrogen-fixing plants. Science 359(6378):869–870

Gordon DM, Ryder MH, Heinrich K, Murphy PJ (1996) An experimental test of the rhizopine concept in Rhizobium meliloti. Appl Environ Microbiol 62(11):3991–3996

Goyal RK, Schmidt MA, Hynes MF (2021) Molecular biology in the improvement of biological nitrogen fixation by rhizobia and extending the scope to cereals. Microorganisms 9(1):125

Griesmann M, Chang Y, Liu X, Song Y, Haberer G, Crook MB, Billault-Penneteau B, Lauressergues D, Keller J, Imanishi L (2018) Phylogenomics reveals multiple losses of nitrogen-fixing root nodule symbiosis. Science 361(6398):eaat1743

Gupta G, Panwar J, Akhtar MS, Jha PN (2012) Endophytic nitrogen-fixing bacteria as biofertilizer. In: Lichtfouse E (ed) Sustainable agriculture reviews, vol 11. Springer, Dordrecht, pp 183–221

Gutjahr C, Banba M, Croset V, An K, Miyao A, An G, Hirochika H, Imaizumi-Anraku H, Paszkowski U (2008) Arbuscular mycorrhiza–specific signaling in rice transcends the common symbiosis signaling pathway. Plant Cell 20(11):2989–3005. https://doi.org/10.1105/tpc.108.062414

Halbleib CM, Ludden PW (2000) Regulation of biological nitrogen fixation. J Nutr 130(5): 1081–1084

Hallmann J, Quadt-Hallmann A, Mahaffee W, Kloepper J (1997) Bacterial endophytes in agricultural crops. Can J Microbiol 43(10):895–914

Hasan M, Bano A, Hassan SG, Iqbal J, Awan U, Rong-ji D, Khan KA (2014) Enhancement of rice growth and production of growth-promoting phytohormones by inoculation with Rhizobium and other rhizobacteria. World Appl Sci J 31(10):1734–1743

Hegazi N, Fayez M, Amin G, Hamza M, Abbas M, Youssef H, Monib M (1998) Diazotrophs associated with non-legumes grown in sandy soils. In: Malik KA, Mirza MS, Ladha JK (eds) Nitrogen fixation with non-legumes. Developments in plant and soil sciences, vol 79. Springer, Dordrecht, pp 209–222. https://doi.org/10.1007/978-94-011-5232-7_24

Hirsch AM, Bauer WD, Bird DM, Cullimore J, Tyler B, Yoder JI (2003) Molecular signals and receptors: controlling rhizosphere interactions between plants and other organisms. Ecology 84(4):858–868

Hongrittipun P, Youpensuk S, Rerkasem B (2014) Screening of nitrogen-fixing endophytic bacteria in *Oryza sativa* L. J Agric Sci 6(6):66

Huang X-F, Chaparro JM, Reardon KF, Zhang R, Shen Q, Vivanco JM (2014) Rhizosphere interactions: root exudates, microbes, and microbial communities. Botany 92(4):267–275

Huang S, Zhao C, Zhang Y, Wang C (2018) Nitrogen use efficiency in rice. In: Amanullah, Fahad S (eds) . Nitrogen in agriculture-updates, InTech, Croatia, pp 189–208

Hunt S, Layzell DB (1993) Gas exchange of legume nodules and the regulation of nitrogenase activity. Annl Rev Plant Biol 44(1):483–511

Ivleva NB, Groat J, Staub JM, Stephens M (2016) Expression of active subunit of nitrogenase via integration into plant organelle genome. PLoS One 11(8):e0160951

James EK, Gyaneshwar P, Mathan N, Barraquio WL, Reddy PM, Iannetta PP, Olivares FL, Ladha JK (2002) Infection and colonization of rice seedlings by the plant growth-promoting bacterium *Herbaspirillum seropedicae* Z67. Mol Plant-Microbe Interact 15(9):894–906. https://doi.org/10.1094/mpmi.2002.15.9.894

Ji SH, Gururani MA, Chun S-C (2014) Isolation and characterization of plant growth-promoting endophytic diazotrophic bacteria from Korean rice cultivars. Microbiol Res 169(1):83–98

Johansson C, Bergman B (1992) Early events during the establishment of the Gunnera/Nostoc symbiosis. Planta 188(3):403–413

Kaga H, Mano H, Tanaka F, Watanabe A, Kaneko S, Morisaki H (2009) Rice seeds as sources of endophytic bacteria. Microbe Environ 2009:E09113. https://doi.org/10.1264/jsme2.ME09113

Kaushal M, Wani SP (2016) Rhizobacterial-plant interactions: strategies ensuring plant growth promotion under drought and salinity stress. Agric Ecosys Environ 231:68–78

Keymer DP, Kent AD (2014) Contribution of nitrogen fixation to first year Miscanthus × giganteus. GCB Bioenergy 6(5):577–586

Khan MMA, Haque E, Paul NC, Khaleque MA, Al-Garni SMS, Rahman M, Islam MT (2017) Enhancement of growth and grain yield of rice in nutrient deficient soils by rice probiotic bacteria. Rice Sci 24(5):264–273

Khorshidi YR, Ardakani MR, Ramezanpour MR, Khavazi K, Zargari K (2011) Response of yield and yield components of rice (*Oryza sativa* L.) to *Pseudomonas flouresence* and *Azospirillum lipoferum* under different nitrogen levels. Am Euras J Agric Environ Sci 10:387–395

Knee EM, Gong F-C, Gao M, Teplitski M, Jones AR, Foxworthy A, Mort AJ, Bauer WD (2001) Root mucilage from pea and its utilization by rhizosphere bacteria as a sole carbon source. Mol Plant-Microbe Interact 14(6):775–784

Kroener E, Holz M, Zarebanadkouki M, Ahmed M, Carminati A (2018) Effects of mucilage on rhizosphere hydraulic functions depend on soil particle size. Vadose Zone J 17(1):1–11. https://doi.org/10.2136/vzj2017.03.0056

Ladha JK, Reddy PM (1995) Extension of nitrogen fixation to rice—necessity and possibilities. GeoJournal 35(3):363–372

Ladha JK, Reddy PM (2003) Nitrogen fixation in rice systems: state of knowledge and future prospects. Plant Soil 252(1):151–167

Ladha JK, Tirol-Padre A, Reddy CK, Cassman KG, Verma S, Powlson DS, Van Kessel C, Richter DB, Chakraborty D, Pathak H (2016) Global nitrogen budgets in cereals: a 50-year assessment for maize, rice and wheat production systems. Sci Rep 6(1):1–9

Lee S, Reth A, Meletzus D, Sevilla M, Kennedy C (2000) Characterization of a major cluster of nif, fix, and associated genes in a sugarcane endophyte, *Acetobacter diazotrophicus*. J Bacteriol 182(24):7088–7091

Li Q, Chen S (2020) Transfer of nitrogen fixation (*nif*) genes to non-diazotrophic hosts. ChemBioChem 21(12):1717–1722

Li R, Han Y, Lv P, Du R, Liu G (2014) Molecular mapping of the brace root traits in sorghum (*Sorghum bicolor* L. Moench). Breed Sci 64(2):193–198

Li M, Xu J, Gao Z, Tian H, Gao Y, Kariman K (2020) Genetically modified crops are superior in their nitrogen use efficiency—a meta-analysis of three major cereals. Sci Rep 10(1):1–9

López-Torrejón G, Jiménez-Vicente E, Buesa JM, Hernandez JA, Verma HK, Rubio LM (2016) Expression of a functional oxygen-labile nitrogenase component in the mitochondrial matrix of aerobically grown yeast. Nat Commun 7(1):1–6

Luo T, Ou-Yang XQ, Yang LT, Li YR, Song XP, Zhang GM, Gao YJ, Duan WX, An Q (2016) *Raoultella* sp. strain L03 fixes N_2 in association with micropropagated sugarcane plants. J Basic Microbiol 56(8):934–940

Mäder P, Kaiser F, Adholeya A, Singh R, Uppal HS, Sharma AK, Srivastava R, Sahai V, Aragno M, Wiemken A (2011) Inoculation of root microorganisms for sustainable wheat–rice and wheat–black gram rotations in India. Soil Biol Biochem 43(3):609–619

Mahmud K, Makaju S, Ibrahim R, Missaoui A (2020) Current progress in nitrogen-fixing plants and microbiome research. Plants 9(1):97

Malik KA, Mirza MS, Hassan U, Mehnaz S, Rasul G, Haurat J, Bally R, Normand P (2002) The role of plant-associated beneficial bacteria in rice-wheat cropping system. In: Kennedy IR, Choudhury ATMA (eds) Biofertilisers in action: a report for the rural industries research and development corporation. RIRDC Publication, Barton, pp 73–83. https://www.agrifutures.com.au/wp-content/uploads/publications/02-086.pdf#page=82. Accessed 11 Apr 2022

Marchal K, Vanderleyden J (2000) The "oxygen paradox" of dinitrogen-fixing bacteria. Biol Fertil Soil 30(5):363–373

Markmann K, Parniske M (2009) Evolution of root endosymbiosis with bacteria: how novel are nodules? Trend Plant Sci 14(2):77–86. https://doi.org/10.1016/j.tplants.2008.11.009

Martinez-Romero E (2006) Dinitrogen-fixing prokaryotes. In: The prokaryotes, vol 2. Springer, New York, p 793

Masepohl B, Drepper T, Klipp W (2004) Nitrogen fixation in the photosynthetic purple bacterium Rhodobacter capsulatus. In: Klipp W, Masepohl B, Gallon JR, Newton WE (eds) Genetics and regulation of nitrogen fixation in free-living bacteria. Nitrogen fixation: origins, applications, and research progress, vol 2. Springer, Dordrecht, pp 141–173. https://doi.org/10.1007/1-4020-2179-8_7

Mondy S, Lenglet A, Beury-Cirou A, Libanga C, Ratet P, Faure D, Dessaux Y (2014) An increasing opine carbon bias in artificial exudation systems and genetically modified plant rhizospheres leads to an increasing reshaping of bacterial populations. Mol Ecol 23(19):4846–4861

Murphy PJ, Heycke N, Banfalvi Z, Tate ME, De Bruijn F, Kondorosi A, Tempé J, Schell J (1987) Genes for the catabolism and synthesis of an opine-like compound in *Rhizobium meliloti* are closely linked and on the Sym plasmid. Proc Nat Acad Sci U S A 84(2):493–497

Murphy PJ, Wexler W, Grzemski W, Rao JP, Gordon D (1995) Rhizopines—their role in symbiosis and competition. Soil Biol Biochem 27(4-5):525–529

Mus F, Crook MB, Garcia K, Garcia Costas A, Geddes BA, Kouri ED, Paramasivan P, Ryu M-H, Oldroyd GED, Poole PS (2016) Symbiotic nitrogen fixation and the challenges to its extension to nonlegumes. Appl Environ Microbiol 82(13):3698–3710

Mus F, Alleman AB, Pence N, Seefeldt LC, Peters JW (2018) Exploring the alternatives of biological nitrogen fixation. Metallomics 10(4):523–538

Naveed M, Brown LK, Raffan AC, George TS, Bengough AG, Roose T, Sinclair I, Koebernick N, Cooper L, Hackett CA (2017) Plant exudates may stabilize or weaken soil depending on species, origin and time. Euro J Soil Sci 68(6):806–816

Newton WE (2007) Physiology, biochemistry, and molecular biology of nitrogen fixation. In: Bothe H, Ferguson SJ, Newton WE (eds) Biology of the nitrogen cycle. Elsevier, pp 109–129

Nguyen HT, Deaker R, Kennedy IR, Roughley RJ (2003) The positive yield response of field-grown rice to inoculation with a multi-strain biofertiliser in the Hanoi area, Vietnam. Symbiosis 35(1):231–245

Oger P, Petit A, Dessaux Y (1997) Genetically engineered plants-producing opines alter their biological environment. Nat Biotechnol 15(4):369–372

Oldroyd GED, Dixon R (2014) Biotechnological solutions to the nitrogen problem. Curr Opin Biotechnol 26:19–24

Oldroyd GED, Murray JD, Poole PS, Downie JA (2011) The rules of engagement in the legume-rhizobial symbiosis. Annl Rev Genet 45:119–144

Olivares J, Bedmar EJ, Sanjuán J (2013) Biological nitrogen fixation in the context of global change. Mol Plant-Microbe Interact 26(5):486–494

Oliveira ALM, Urquiaga S, Döbereiner J, Baldani JI (2002) The effect of inoculating endophytic N_2-fixing bacteria on micropropagated sugarcane plants. Plant Soil 242(2):205–215

Osborn HMI, Lochey F, Mosley L, Read D (1999) Analysis of polysaccharides and monosaccharides in the root mucilage of maize (Zea mays L.) by gas chromatography. J Chromatogr A 831(2):267–276

Pankievicz VCS, do Amaral FP, Santos KFDN, Agtuca B, Xu Y, Schueller MJ, Arisi ACM, Steffens MBR, de Souza EM, Pedrosa FO (2015) Robust biological nitrogen fixation in a model grass–bacterial association. Plant J 81(6):907–919

Pankievicz V, Irving TB, Maia LGS, Ané J-M (2019) Are we there yet? The long walk towards the development of efficient symbiotic associations between nitrogen-fixing bacteria and non-leguminous crops. BMC Biol 17(1):1–17

Prayitno J, Rolfe B (2010) Characterization of endophytic diazotroph bacteria isolated from rice. HAYATI J Biosci 17(2):73–78

Prayitno J, Stefaniak J, McIver J, Weinman JJ, Dazzo FB, Ladha JK, Barraquio W, Yanni YG, Rolfe BG (1999) Interactions of rice seedlings with bacteria isolated from rice roots. Funct Plant Biol 26(6):521–535

Priyadarshini P, Choudhury S, Tilgam J, Bharati A, Sreeshma N (2021) Nitrogen fixing cereal: a rising hero towards meeting food security. Plant Physiol Biochem 167:912–920

Rai AN, Söderbäck E, Bergman B (2000) Tansley review no. 116 cyanobacterium–plant symbioses. New Phytol 147(3):449–481

Rai AN, Singh AK, Syiem MB (2019) Plant growth-promoting abilities in cyanobacteria. In: Mishra AK, Tiwari DN, Rai AN (eds) Cyanobacteria. Elsevier, pp 459–476

Rana A, Kabi SR, Verma S, Adak A, Pal M, Shivay YS, Prasanna R, Nain L (2015) Prospecting plant growth-promoting bacteria and cyanobacteria as options for enrichment of macro-and micronutrients in grains in rice–wheat cropping sequence. Cogent Food Agric 1(1):1037379

Ray DK, Mueller ND, West PC, Foley JA (2013) Yield trends are insufficient to double global crop production by 2050. PLoS One 8(6):e66428

Reinhold-Hurek B, Krause A, Leyser B, Miché L, Hurek T (2007) the rice apoplast as a habitat for endophytic N_2-fixing bacteria. In: Sattelmacher B, Horst WJ (eds) The apoplast of higher plants: compartment of storage, transport and reactions. Springer, Dordrecht, pp 427–443

Reyna-Flores F, Barrios-Camacho H, Dantán-González E, Ramírez-Trujillo JA, Lozano Aguirre Beltrán LF, Rodríguez-Medina N, Garza-Ramos U, Suárez-Rodríguez R (2018) Draft genome sequences of endophytic isolates of *Klebsiella variicola* and *Klebsiella pneumoniae* obtained from the same sugarcane plant. Genom Announc 6(12):e00147-00118

Roberson EB, Firestone MK (1992) Relationship between desiccation and exopolysaccharide production in a soil *Pseudomonas* sp. Appl Environ Microbiol 58(4):1284–1291

Rockström J, Steffen W, Noone K, Persson Å, Chapin FS, Lambin EF, Lenton TM, Scheffer M, Folke C, Schellnhuber HJ (2009) A safe operating space for humanity. Nature 461(7263): 472–475

Rockstrom J, Steffen W, Noone K, Persson A, Chapin Iii FS, Lambin EF, Lenton TM, Scheffer M, Folke C, Schellnhuber HJ (2009) A safe operating space for humanity: identifying and quantifying planetary boundaries that must not be transgressed could help prevent human activities from causing unacceptable environmental change, argue Johan Rockstrom and colleagues. Nature 461(7263):472–476

Roesch LFW, Camargo FAO, Bento FM, Triplett EW (2008) Biodiversity of diazotrophic bacteria within the soil, root and stem of field-grown maize. Plant Soil 302(1):91–104. https://doi.org/10.1007/s11104-007-9458-3

Roger P-A, Ladha JK (1992) Biological N_2 fixation in wetland rice fields: estimation and contribution to nitrogen balance. In: Ladha JK, George T, Bohlool BB (eds) Biological nitrogen fixation for sustainable agriculture. Developments in plant and soil sciences, vol 49. Springer, Dordrecht, pp 41–55

Rogers C, Oldroyd GED (2014) Synthetic biology approaches to engineering the nitrogen symbiosis in cereals. J Exp Biol 65(8):1939–1946

Rosenblueth M, Ormeño-Orrillo E, López-López A, Rogel MA, Reyes-Hernández BJ, Martínez-Romero JC, Reddy PM, Martínez-Romero E (2018) Nitrogen fixation in cereals. Front Microbiol 9:1794. https://doi.org/10.3389/fmicb.2018.01794

Rossbach S, McSpadden B, Kulpa D, Rasul G, Ganoof M, De Bruijn FJ (1994) Use of rhizopine synthesis and catabolism genes to monitor soil bacteria and to create biased rhizospheres. Mol Ecol 3:610–611

Roy ML, Srivastava RC (2010) Single and co-inoculation effects of different biofertilizers on growth, in vivo nitrate reductase activity and soluble protein in *Oryza sativa* L. J Appl Biosci 36(1):101–104

Roy M, Saha S, Das J, Srivastava RC (2015) Technologies of microbial inoculation in rice—a review. Agric Rev 36(2):125–132

Rubio LM, Ludden PW (2008) Biosynthesis of the iron-molybdenum cofactor of nitrogenase. Annl Rev Microbiol 62:93–111. https://doi.org/10.1146/annurev.micro.62.081307.162737

Saharan BS, Nehra V (2011) Plant growth-promoting rhizobacteria: a critical review. Life Sci Med Res 2011:LSMR-21. https://agris.fao.org/agris-search/search.do?recordID=DJ2012070623. Accessed 11 Apr 2022

Santi C, Bogusz D, Franche C (2013) Biological nitrogen fixation in non-legume plants. Ann Bot 111(5):743–767

Savka MA, Farrand SK (1997) Modification of rhizobacterial populations by engineering bacterium utilization of a novel plant-produced resource. Nat Biotechnol 15(4):363–368

Savka MA, Dessaux Y, McSpadden Gardener BB, Mondy S, Kohler PRA, de Bruijn FJ, Rossbach S (2013) The "biased rhizosphere" concept and advances in the omics era to study bacterial competitiveness and persistence in the phytosphere. Mol Microbial Ecol Rhizosph 1:1145–1161

Schulze J, Drevon J-J (2005) P-deficiency increases the O_2 uptake per N_2 reduced in alfalfa. J Exp Bot 56(417):1779–1784

Seefeldt LC, Hoffman BM, Dean DR (2009) Mechanism of Mo-dependent nitrogenase. Ann Rev Biochem 78:701–722

Seefeldt LC, Hoffman BM, Dean DR (2012) Electron transfer in nitrogenase catalysis. Curr Opin Chem Biol 16(1-2):19–25

Sessitsch A, Hardoim P, Döring J, Weilharter A, Krause A, Woyke T, Mitter B, Hauberg-Lotte L, Friedrich F, Rahalkar M (2012) Functional characteristics of an endophyte community colonizing rice roots as revealed by metagenomic analysis. Mol Plant-Microbe Interact 25(1):28–36

Setten L, Soto G, Mozzicafreddo M, Fox AR, Lisi C, Cuccioloni M, Angeletti M, Pagano E, Díaz-Paleo A, Ayub ND (2013) Engineering Pseudomonas protegens Pf-5 for nitrogen fixation and its application to improve plant growth under nitrogen-deficient conditions. PLoS One 8(5): e63666

Sevilla M, Burris RH, Gunapala N, Kennedy C (2001) Comparison of benefit to sugarcane plant growth and 15N2 incorporation following inoculation of sterile plants with Acetobacter diazotrophicus wild-type and nif mutant strains. Mol Plant-Microbe Interact 14(3):358–366

Shin W, Islam R, Benson A, Joe MM, Kim K, Gopal S, Samaddar S, Banerjee S, Sa T (2016) Role of diazotrophic bacteria in biological nitrogen fixation and plant growth improvement. Korean J Soil Sci Fertil 49(1):17–29

Sinha Roy S, Mittra B, Sharma S, Das TK, Babu CR (2002) Detection of root mucilage using an anti-fucose antibody. Ann Bot 89(3):293–299

Song J, Yang J, Jeong BR (2021) Growth, quality, and nitrogen assimilation in response to high ammonium or nitrate supply in cabbage (*Brassica campestris* L.) and lettuce (*Lactuca sativa* L.). Agronomy 11(12):2556

Song J, Yang J, Jeong BR (2022) Root GS and NADH-GDH play important roles in enhancing the ammonium tolerance in three bedding plants. Int J Mol Sci 23(3):1061

Steinkellner S, Lendzemo V, Langer I, Schweiger P, Khaosaad T, Toussaint JP, Vierheilig H (2007) Flavonoids and strigolactones in root exudates as signals in symbiotic and pathogenic plant-fungus interactions. Molecules 12(7):1290–1306. https://doi.org/10.3390/12071290

Stokstad E (2016) The nitrogen fix. Science 353(6305):1225–1227

Temme K, Zhao D, Voigt CA (2012) Refactoring the nitrogen fixation gene cluster from Klebsiella oxytoca. Proc Natl Acad Sci U S A 109(18):7085–7090

Triplett EW (1996) Diazotrophic endophytes: progress and prospects for nitrogen fixation in monocots. Plant Soil 186(1):29–38

Urquiaga S, Xavier RP, de Morais RF, Batista RB, Schultz N, Leite JM, Maia e Sá J, Barbosa KP, de Resende AS, BJR A (2012) Evidence from field nitrogen balance and 15N natural abundance data for the contribution of biological N_2 fixation to Brazilian sugarcane varieties. Plant Soil 356(1):5–21

Uthiraselvam M, Ravikumar S, Abideen S, Selvam MB, Fathima SA (2012) Effect of multiple inoculation of magnetotactic bacteria on along with nitrogen fixers and phosphate solubilisers on the growth and yield of agriculture crop *Oryza Sativa*. J Microbiol Biotechnol Res 2(5): 758–765

Vaishampayan A, Sinha RP, Hader DP, Dey T, Gupta AK, Bhan U, Rao AL (2001) Cyanobacterial biofertilizers in rice agriculture. Bot Rev 67(4):453–516

Van Deynze A, Zamora P, Delaux P-M, Heitmann C, Jayaraman D, Rajasekar S, Graham D, Maeda J, Gibson D, Schwartz KD (2018) Nitrogen fixation in a landrace of maize is supported by a mucilage-associated diazotrophic microbiota. PLoS Biol 16(8):e2006352

Van Dommelen A, Croonenborghs A, Spaepen S, Vanderleyden J (2009) Wheat growth promotion through inoculation with an ammonium-excreting mutant of Azospirillum brasilense. Biol Fertil Soil 45(5):549–553

van Veelen A, Tourell MC, Koebernick N, Pileio G, Roose T (2018) Correlative visualization of root mucilage degradation using X-ray CT and MRI. Front Environ Sci 6. https://doi.org/10.3389/fenvs.2018.00032

van Velzen R, Holmer R, Bu F, Rutten L, van Zeijl A, Liu W, Santuari L, Cao Q, Sharma T, Shen D (2018) Comparative genomics of the nonlegume *Parasponia* reveals insights into evolution of nitrogen-fixing rhizobium symbioses. Proc Natl Acad Sci U S A 115(20):E4700–E4709

Vermeiren H, Willems A, Schoofs G, De Mot R, Keijers V, Hai W, Vanderleyden J (1999) The rice inoculant strain *Alcaligenes faecalis* A15 is a nitrogen-fixing *Pseudomonas stutzeri*. Syst Appl Microbiol 22(2):215–224

Volk E, Iden SC, Furman A, Durner W, Rosenzweig R (2016) Biofilm effect on soil hydraulic properties: experimental investigation using soil-grown real biofilm. Water Resour Res 52(8): 5813–5828

Wang D, Xu A, Elmerich C, Ma LZ (2017) Biofilm formation enables free-living nitrogen-fixing rhizobacteria to fix nitrogen under aerobic conditions. ISME J 11(7):1602–1613

Werker E, Kislev M (1978) Mucilage on the root surface and root hairs of Sorghum: heterogeneity in structure, manner of production and site of accumulation. Ann Bot 42(4):809–816

Werner GDA, Cornwell WK, Sprent JI, Kattge J, Kiers ET (2014) A single evolutionary innovation drives the deep evolution of symbiotic N_2-fixation in angiosperms. Nat Commun 5(1):1–9

Wessel AK, Arshad TA, Fitzpatrick M, Connell JL, Bonnecaze RT, Shear JB, Whiteley M (2014) Oxygen limitation within a bacterial aggregate. MBio 5(2):e00992–e00914

Westhoff P (2009) The economics of biological nitrogen fixation in the global economy. In: Emerich DW, Krishnan HB (eds) Nitrogen fixation in crop production. Agronomy Monograph 52. Madison, WI, pp 309–328

Wexler M, Gordon D, Murphy PJ (1995) The distribution of inositol rhizopine genes in Rhizobium populations. Soil Biol Biochem 27(4-5):531–537

Williams RL, Kennedy IR (2002) A model for testing the effectiveness of biofertiliser for Australian rice production. In: Kennedy IR, Choudhury ATMA (eds) Biofertilizers in action: a report for the rural industries research and devlopment coporation. RIRDC Publications, pp 112–114

Yanni YG, Dazzo FB (2010) Enhancement of rice production using endophytic strains of *Rhizobium leguminosarum* bv. trifolii in extensive field inoculation trials within the Egypt Nile delta. Plant Soil 336(1–2):129–142

Yanni YG, Rizk RY, Corich V, Squartini A, Ninke K, Philip-Hollingsworth S, Orgambide G, De Bruijn F, Stoltzfus J, Buckley D (1997) Natural endophytic association between *Rhizobium leguminosarum* bv. trifolii and rice roots and assessment of its potential to promote rice growth. In: Ladha JK, de Bruijn FJ, Malik KA (eds) Opportunities for biological nitrogen fixation in rice and other non-legumes. Developments in plant and soil sciences, vol 75. Springer, Dordrecht, pp 99–114

Yoneyama T, Terakado-Tonooka J, Minamisawa K (2017) Exploration of bacterial N_2-fixation systems in association with soil-grown sugarcane, sweet potato, and paddy rice: a review and synthesis. Soil Sci Plant Nutr 63(6):578–590

Zhang T, Yan Y, He S, Ping S, Alam KM, Han Y, Liu X, Lu W, Zhang W, Chen M (2012) Involvement of the ammonium transporter AmtB in nitrogenase regulation and ammonium excretion in *Pseudomonas stutzeri* A1501. Res Microbiol 163(5):332–339

Zheng W, Zeng S, Bais H, LaManna JM, Hussey DS, Jacobson DL, Jin Y (2018) Plant growth-promoting rhizobacteria (PGPR) reduce evaporation and increase soil water retention. Water Res Res 54(5):3673–3687

Chapter 5
Contribution of Nitrogen-Fixing Bacteria in Rice Cultivation: Past, Present, and Future

Thilini A. Perera and Shamala Tirimanne

Abstract Potential of nitrogen-fixing bacteria had been known by the farmers as early as the 1900s. From early days, legumes were incorporated into the rice soils to be mineralized and absorbed into rice plants. A major re-focus toward the usage of nitrogen-fixing microbes came into play after the realization of the negative consequences that occurred due to excess inorganic nitrogen fertilizer usage during green revolution. Initial research included the identification of rhizosphere nitrogen fixers that can colonize non-legume rice plants, observing their behavior, and measuring the amounts of nitrogen fixed. Then, it was targeted for induction of root nodules in rice roots, engineering rice plants with nitrogen-fixing genes, and activation of the nitrogenase enzyme inside the rice plant. With time, it was realized that nitrogen fixation is an extremely complex process to be induced in a non-leguminous plant as it is a high energy-demanding process occurring in low-oxygen conditions. A decade ago, the scientists were under the impression that genetic engineering of a rice plant to fix nitrogen is not an achievable target. But at present, with the improved understanding of the molecular biology of nitrogen fixation and the use of molecular tools and technology, the genes and the QTL regions involved in nitrogen fixation that are crucial for BNF are identified and producing a biologically nitrogen-fixing rice plant is not as impossible as it seems to be a decade ago. Nitrogen fixer application as bio-fertilizers has also yielded positive results, but this process is not as rapid as legume–Rhizobium symbiosis. Therefore, there will be research continuing around the world until a completely nitrogen-fixing rice plant is created. More research studies related to the genetic regulation, the factors involved in bacterial colonization of the rice plants, and molecular levels studies needs to be taken place.

Keywords Rice production · Nitrogen fixation · Rhizosphere · Biofertilizers · Organic agriculture

T. A. Perera · S. Tirimanne (✉)
Department of Plant Sciences, Faculty of Science, University of Colombo, Colombo, Sri Lanka
e-mail: thilini@pts.cmb.ac.lk; shamala@pts.cmb.ac.lk

© The Author(s), under exclusive license to Springer Nature Singapore Pte Ltd. 2022
D. K. Maheshwari et al. (eds.), *Nitrogen Fixing Bacteria: Sustainable Growth of Non-legumes*, Microorganisms for Sustainability 36,
https://doi.org/10.1007/978-981-19-4906-7_5

5.1 Introduction

Rice is one of the worlds' most consumed cereal crops. It is a staple for more than half of the world's population providing 80% of their food requirement. According to literature, rice (*Oryza sativa*) had been domesticated from the wild grass *Oryza rufipogon* 10,000–14,000 years ago. Reports from China have shown evidence of rice cultivation even before 8000 years. Rice is grown mostly by the poor people in the world, especially in the Asian regions. According to statistics, in the years 2020/2021 global rice production had been 496 million metric tons utilizing about 165 million hectares of land. As stated by IRRI, globally 21% of per capita energy and 15% per capita protein are provided by rice (IRRI, Knowledge Bank). It is also estimated that by the year 2050, the demand for rice will be 584 million tons (Samal et al. 2022).

5.2 Nitrogen as a Nutrient for Rice

To feed the rapidly increasing world population, with decreasing arable lands, increase in productivity of crops per unit land area has become a fundamental need. Several initiatives have been taken to face these challenges, especially during the era of green revolution. Main measures include the use of genetically improved rice varieties and the use of synthetic fertilizers, insecticides, and pesticides for rice cultivation. According to Evenson and Gollin (2003), high-yielding rice varieties were a major accomplishment of the era of green revolution. Doubling of the population leading to the upsurge of food demand was taken care of by these modern varieties produced. These improved varieties were highly responsive to the synthetic fertilizers used.

Among the nutrients that are required for rice plant growth and development, nitrogen is termed as the "*sine quo none*" or the absolute necessity in high-yielding agriculture (Ladha et al. 2016) as nitrogen has a direct, positive correlation with the growth and development of rice. Moreover, nitrogen is identified as the most frequent yield-limiting nutrient (Chauhan et al. 2017). There is no phase of growth of the life cycle of rice that is not affected by nitrogen. During the vegetative growth phase, nitrogen is highly involved in the increase in plant height, tiller number, leaf size (Dobermann and Fairhurst 2000), number of stalk (total tillers/hill) (Chaturvedi 2005; Ghanbari-malidarreh 2011), dry matter content of a stem per hill (Chaturvedi 2005; Youseftabar et al. 2012), root volume (Anil et al. 2014), and many other characteristics. During vegetative growth, sink organs assimilate nitrogen. The assimilated nitrogen is used for amino acid, protein, and enzyme synthesis, building up the plant architecture and components required for photosynthetic machinery (Hirel et al. 2007). During reproductive and ripening phases of a rice plant, accumulated nitrogen is remobilized to the seeds. The roots and the shoots now behave as nitrogen source tissues, and the proteins stored in source tissues are hydrolyzed releasing amino acids that are transported to the seeds.

5.3 Detrimental Consequences Associated with Nitrogen Fertilizers

Urea is the most commonly used nitrogen fertilizer and is an expensive commodity. It is manufactured by the "Haber Bosch" process, and it consumes a large amount of petroleum energy. Therefore, the urea prices in the global market always fluctuate with the fluctuations of the global petroleum market. The issues associated with the COVID-19 pandemic have led to an unexpected urea price hike during the last few months of the year 2021. The price of a metric ton of urea that had been around USD 452 in July 2021 has increased to USD 890 in November 2021[1] making it a major issue to the developing countries.

Although expensive, since it has become an essential commodity for rice cultivation, many Asian countries have introduced urea fertilizer subsidy schemes. In Sri Lanka, in the year 2018, a 50 kg bag of urea with the price of about Rs. 3500 in the global market was provided to the farmers for a subsidized cost of Rs. 500.00 while the government bore the cost of Rs. 3000.00,[2] resulting in major economic issues that apply especially to other developing countries as well.

When nitrogen fertilizers are added to the soil, only 30% is absorbed by the plant and the rest is lost from the system through leaching, volatilization, and denitrification. Environmental and health issues of the nitrogen fertilizer usage are explained briefly in Fig. 5.1 (Perera and Tirimanne 2021). Eutrophication of water bodies, groundwater contamination, greenhouse gas emissions, global warming, and acid rains are some of the hazardous environmental consequences. Methemoglobinemia in infants, skin cancers, and gastric cancers are some of the detrimental health issues associated with the urea fertilizer usage.

Due to the detrimental issues stated above, the scientific community had been experimenting on alternatives that can replace nitrogen fertilizer usage in rice cultivation. There, the usage of nitrogen-fixing bacteria has been identified as a promising method that can contribute immensely to the reduction in nitrogen fertilizer in rice cultivation,[3] given that the challenges associated with the non-legume nitrogen fixation can be overcome.

---

[1] https://www.indexmundi.com/commodities/?commodity=urea.

[2] Agriculture Ministry to provide fertilizer at old prices. https://news.lk/economy/item/22652-agriculture-ministry-to-provide-fertilizer-at-old-prices. Accessed May 20, 2021.

[3] Essential nutrients in rice production. http://www.knowledgebank.irri.org/ericeproduction/IV.1_Essential_nutrients.htm. Accessed May 8, 2020.

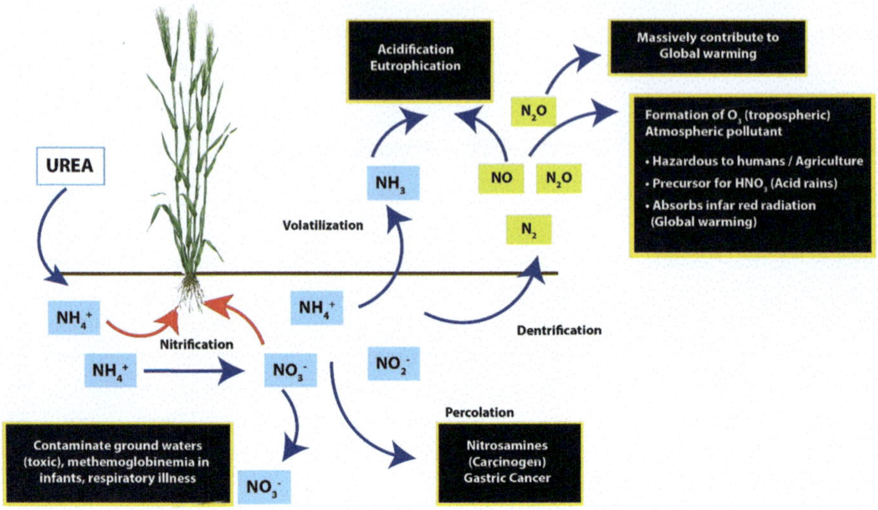

Fig. 5.1 Environmental and health issues of the nitrogen fertilizer usage (Perera and Tirimanne 2021)

5.4 Nitrogen Fixation and Nitrogenase

Seventy-nine percent of the atmosphere around us consists of dinitrogen (N_2). Even though this is available for free, rice plants cannot utilize this gaseous nitrogen directly for their growth and development. But the plants of the family Leguminosae have the ability to take in (or fix) this freely available atmospheric nitrogen through a process called biological nitrogen fixation (BNF), in association with nitrogen-fixing bacteria. Through nitrogen fixation, the atmospheric di-nitrogen can be converted (reduced) into ammonia, a form that can be utilized by living organisms (plants and animals) for bio-organic matter synthesis (Cheng 2008). It has been estimated that annually 200 million tons of nitrogen are fixed in the world (Rascio et al. 2008).

Overall, the nitrogen fixation reaction is given in the following equation:

$$N_2 + 8\,e^- + 8\,H + +16\,MgATP \rightarrow 2\,NH_3 + H_2 + 16\,MgADP + 16\,Pi$$

This reaction is catalyzed by a complex metalloenzyme, nitrogenase. This enzyme has a conserved structure and mechanistic features (Rees and Howard 2000; Lawson and Smith 2002). This is considered as a high energy-consuming process as each electron transfer of every catalytic cycle of nitrogenase enzyme, and two MgATPs are hydrolyzed (Dixon and Kahn 2004).

This phenomenal conversion, which greatly benefits agriculture, is carried out by certain groups of symbiotic, associative, and free-living bacteria, which belong to kingdoms of Bacteria and Archaea but not Eukaryotes. In Bacteria, BNF is found in subdivisions of Proteobacteria, green sulfur bacteria, Firmibacteria, Actinobacteria, and Cyanobacteria, but is restricted to methanogens in Archaea (Dixon and Kahn 2004).

Biological nitrogen-fixing organisms could be either aerobic (e.g., *Azotobacter*), anaerobic (e.g., *Clostridium*), facultative anaerobic (e.g., *Klebsiella*), anoxygenic heterotrophs (e.g., *Rhodobacter*), or oxygenic heterotroph (e.g., *Anabaena*). Phototrophs and chemolithotrophs (e.g., *Leptospirillum ferrooxidans*) have also been found to fix N_2 (Dixon and Kahn 2004).

5.5 Major Groups of Nitrogen Fixers

Two major groups of nitrogen fixers are of importance and are termed symbiotic and free-living nitrogen fixers.

5.5.1 Symbiotic Nitrogen Fixers

Symbiotic nitrogen fixers develop a symbiotic relationship with the eukaryotic host, which can transfer the fixed ammonia to the eukaryotic host. There the eukaryotic host provides the biological niche and support for the bacteria through better feeding (Lee et al. 2008). The symbiotic relationship between the plants of family Leguminosae and the bacteria rhizobia (Lee et al. 2008) is of significant importance to agriculture.

In the leguminous plants, root nodule formation involves a legume host, factors released by the legume, a micro-symbiont, and the factors released by the micro-symbiont. Nodule formation is a highly synchronized, highly regulated process where molecular signals travel both ways, from the plant to the rhizobium and vice versa (Kulasooriya 2008).

5.5.1.1 Regulation of Root Nodule Formation

When compatible bacteria encounter a suitable leguminous plant, a highly specific molecular dialog between the plant and the microbe leads to a nodule formation (Kulasooriya 2008; Schultze and Kondorosi 1998). As the first step of the nodulation process, the plant host secretes flavonoids, chalcones, and conjugated isoflavonoids. On the reception of these signals, the specific bacteria receive the signal and activate their NodD regulatory proteins and otherwise silent *nod* genes (Franche et al. 2009; Kulasooriya 2008). This is a very specific process, and every bacterial strain has a highly specific host range (Young and Johnston 1989).

5.5.1.2 Nitrogenase Enzyme and Associated Genes

Generally, a bacterium comprises a single circular chromosome and one or more extra chromosomal DNA or plasmids. A thorough study on the genetics behind nitrogen fixation had been done on the facultative anaerobe, *Klebsiella oxytoca* strain M5a1 (initially *Klebsiella pneumoniae*) (Franche et al. 2009). Rhizobial *nif* genes are recorded to show structural similarity to *Klebsiella nif* genes (Arnold et al. 1988).

A specific type of gene called the bacterial *nod* genes is involved in regulating the root nodule formation in the legume. The *nif* genes of the bacteria are responsible for regulating nitrogen fixation. Several other types of genes from the bacteria (e.g., *exo, lps, ndv*) are involved in regulating and helping root nodule formation (Kulasooriya 2008).

According to the review by MacLean et al. (2007), nitrogen fixation and nodule formation are considered as plasmid-borne functions; hence, the genes and the protein they code for as well as the integrated properties could be easily lost or gained (Franche et al. 2009).

The genes responsible for nodule induction (*nod* genes) are carried in the large sym-plasmid (pSym) of rhizobia and in the chromosome in *Azorhizobium caulinodans* and *Bradyrhizobium japonicum* as they do not contain plasmids (Fischer 1994).

5.5.1.3 Prevention of Oxygen Damage of Nitrogenase

The nitrogenase enzyme that catalyzes molecular dinitrogen to ammonia is highly conserved in free-living and symbiotic diazotrophs and is highly sensitive to oxygen. Nitrogenase enzyme is composed of two metalloproteins, metalloprotein 1, or the Mo-Fe protein, and metalloprotein 2, the Fe protein. While Mo-Fe protein is composed of two non-identical subunits, the component 2 or the Fe protein is one with two identical subunits (Franche et al. 2009). The common form of nitrogenase is the Mo-nitrogenase, and it contains a Fe-Mo-co, a prosthetic group with molybdenum. Both Fe protein and the Mo-Fe proteins in the nitrogenase enzyme are highly oxygen-sensitive (Einsle et al. 2002; Seefeldt et al. 2004).

This high oxygen sensitivity of nitrogenase affects the nitrogen-fixing ability of the bacteria, and it is compelled to protect the enzyme from irreversible oxygen damage. Several physiological adaptations can be found in diazotrophs in regard to nitrogenase protection. These include the avoidance of oxygen (by anaerobic growth), excess oxygen consumption through respiration, evolving oxygen diffusion barriers, and enzyme compartmentalization (Dixon and Kahn 2004). In the legume–root symbiosis, one of the strategies is the nodule cortex acting as an oxygen diffusion barrier. The second strategy is the reversible binding of the nodulin leghaemoglobin to oxygen, leading to lower frequency diffusion of oxygen. Even the action of *nif* genes is controlled by the presence of oxygen. A diazotroph needs to respond to the concentration of fixed nitrogen and the external oxygen and also needs to provide sufficient energy for nitrogen fixation. All the diazotrophs have a common regulatory principle but vary somewhat according to the host physiology and the type of the microbe (Dixon and Kahn 2004).

5.5.2 *Non-symbiotic Nitrogen Fixers*

From the time the scientists understood the importance of inducing BNF in cereals, there had been many attempts to achieve it from as early as 1917, but as explained above, it was realized that it is extremely difficult to induce nitrogen fixation in non-legume plants.

After understanding the relationship between the legume plants and rhizobia for nitrogen fixation and the discovery of genes encoding for nitrogen-fixating enzyme component/s, it was realized that making a nitrogen-fixing crop plant, rice, wheat, or maize might 1 day become a reality (Pankievicz et al. 2019).

Any new attempt on non-legume nitrogen fixation faces the following challenges. Firstly, there should be sufficient energy available as reduction of di-nitrogen to ammonia is a high energy-demanding process requiring at least 16 ATP per dinitrogen fixed. Secondly, the process should happen in an environment protected from oxygen. Finally, there should be efficient nutrient exchange between the plant and the microbe (Pankievicz et al. 2019).

It was understood that a successful non-symbiotic nitrogen fixation can occur, if there are sufficient numbers of diazotrophic bacteria in the rhizosphere or in the plant and the fixed nitrogen is transferred to the plant. There is evidence of this nitrogen transfer happening in many systems and evidence to show that the bacteria are living in and around the root system. Whether bacteria directly provide fixed nitrogen to the host plant was a question. James et al. (2000) suggest that this nitrogen is transferred upon the death and mineralization of the symbiotic diazotroph but not as directly as in legume–rhizobium symbiosis.

In non-legumes, BNF happens as associative, endosymbiotic, and endophytic nitrogen fixation. The nitrogen fixers involved contribute in reducing nitrogen fertilizer use in agriculture, increased plant nutrient content, and soil health reclamation (Mahmud et al. 2020).

5.6 History and Early Evidences of Possible Nitrogen Fixation in Non-legumes

Discovery of the possibilities of rhizobium forming nitrogen-fixing nodules (Cocking et al. 1992) in the non-legume plant *Parasponia andersonii* (Trinick 1979) showed the possibility of induction of nitrogen fixation in non-legumes naturally.

Al-mallah et al. (1989) have shown that, after the degradation of the cell walls by a mixture of enzymes (cellulase and pectolyase that are known for complete removal of rice root cell walls) and inoculation of the roots with Rhizobia or Bradyrhizobia in the presence of polyethylene glycol (PEG), induces nodular structures in the rice plant roots which are similar to *Parasponia* nodules. Other than the monocots rice and wheat, the treatment with enzymes had been able to induce more prominent nodules in the dicotyledon, oil seed rape (*Brassica napus*), which is similar to *Parasponia*. It was also suggested that, given the proper conditions, non-legumes nodulate with rhizobia (Sprent 2001; Santi et al. 2013). The findings above strengthen this statement. Cocking et al. (1990) found that, the dicot oil seed rape can be nodulated even without the enzymatic treatment. This led to the focus of research on the possibility of nodulating the cereals, rice, wheat, and maize by *Parasponia* nodulating rhizobia (Cocking et al. 1992). When sterilized indica rice seeds were inoculated with *Parasponia* nodulating Bradyrhizobia and grown in a medium without nitrogen, short, thick lateral roots are formed on the rice seedlings. The membrane-bound bacteria containing vesicles were found in the cytoplasm of the cortical cells and in the microfibrils between the cells (Cocking et al. 1990, 1992).

Rhizosphere-associated nitrogen fixation can happen in three ways (Giraud et al. 2007). In the absence of *nod* A, B, and C genes, first the rhizobia enter the plant through crack entry invasion (Bonaldi et al. 2010). Therefore, this is a nod factor-independent process. Xylem parenchyma is invaded via the cortical cells. Experiments conducted by Perera et al. (2017) have shown evidence for these findings recently. They have shown that the rice root cortex, lateral root cracks, and the xylem region were colonized by clusters of the bacterium *Azorhizobium caulinodans* labeled with green fluorescent protein (*gfp*) (Fig. 5.2).

In low-oxygen or micro-aerobic conditions, rhizobia can show free-living nitrogen fixation, e.g., *Bradyrhizobium* spp., *Azorhizobium* spp., and *Burkholderia* spp. All three species can form symbiotic associations with leguminous plants and fix nitrogen symbiotically but also can fix nitrogen in free-living conditions as well (Alazard et al. 1988; Dreyfus et al. 1983).

In a non-symbiotic system, e.g., rhizosphere-associated nitrogen fixation, carbon and energy for the root growth are taken by the outside environment. The nitrogen fixed by the bacterium is released after the lysis of the cells (James 2000; White et al. 2012). The bacterium is essentially free-living (Barber and Lynch 1977; Delwiche and Wijler 1956) dependent on the carbohydrate released by the plant root and the

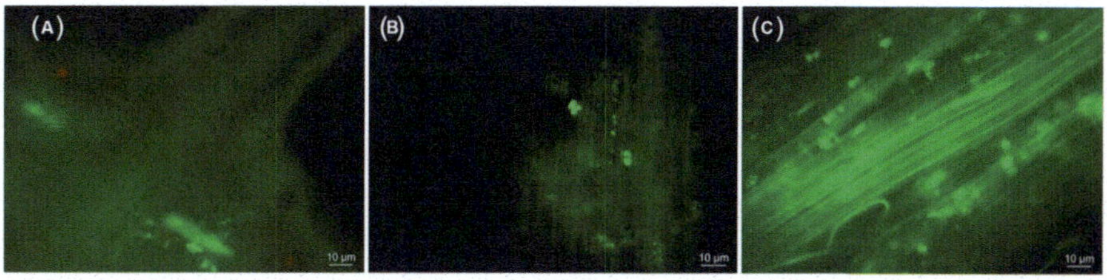

Fig. 5.2 Epifluorescent micrographs of rice roots showing colonization of *Azorhizobium caulinodans*: 15 days after application of plants treated with naringenin 1 × 10^{-4} M and *A. caulinodans* (×40). Colonization of the *Azorhizobium caulinodans* ORS 571 (**a**) in lateral root cracks, (**b**) in the xylem region, and (**c**) in the cortical region. (Extracted from Perera et al. (2017))

digestion of soil organic matter and rice rhizosphere secretions. These non-legume systems are generated in energy-sufficient but nitrogen-deficient conditions.

Associated nitrogen fixation by rice is of utmost significance (Mahmud et al. 2020). In any non-symbiotic nitrogen providing system (for example, rhizosphere associative nitrogen fixation), the following criteria need to be met in order to provide nitrogen to the non-symbiotic system.

Basically, in these systems, carbon and energy for the bacterial growth are gained from the environment (Barber and Lynch 1977). Root excretes carbohydrates and the bacteria are attracted toward them, which stimulate bacterial growth. Nitrogen-fixing bacteria fix nitrogen, which is released after the lysis of the bacterial cells, and this is the most common method as predicted by James 2000. Physiologically, oxygen should be deficient but sufficient energy should be present. Free-living nitrogen-fixing bacteria living in the rhizosphere represent a small fraction of the rhizosphere ecosystem. These bacteria basically belong to the groups alphaproteobacteria (*Rhizobia* spp., *Bradyrhizobia* spp., *Rhodobacter* spp.), betaproteobacteria (*Burkholderia* spp., *Nitrosospira* spp.), gammaproteobacteria (*Pseudomonas* spp., *Xanthomonas* spp.), firmicutes, and cyanobacteria (Mahmud et al. 2020).

5.7 Use of Nitrogen-Fixing Bacteria for Rice Cultivation

5.7.1 Use of Nitrogen-Fixing Bacteria for Rice Cultivation in the Past (1970–2000)

In the past, to make use of nitrogen-fixing bacteria in the rice systems, methods such as crop rotation and co-culturing with legumes had been practiced. These methods are still being used by the farmers around the world. During the 1990s, the commendable efforts of International Rice Research Institute (IRRI), through their dream project of BNF in rice, several discoveries were made that can be practiced directly and provide the basis and evidence for many other researchers to carry out their research related to BNF.

In Egyptian rice fields, rice rotation with berseem clover (*Trifolium alexandrinum*) has greatly benefitted rice cultivation by being able to replace a portion of nitrogen fertilizer requirement (Yanni et al. 1997). Examination of the possibility of colonization of rice plants by the clover plant endosymbiont, *Rhizobium leguminosarum* bv. *trifolii*, has yielded positive results. The bacterium has been able to migrate into the rice root interior leading to significant increase in the shoot and root growth, increment in the grain yield, and increment in nitrogen use efficiency (Yanni et al. 1997).

Experiments done with the intention of looking at the possibilities of rice–*Rhizobium* symbiosis have revealed that the primary mode of infection of rice plants is through epidermal cracks and fissures raised by lateral root emergence. This infection was *nod* gene-independent, non-specific, and no infection threads had been formed during the infection. Rhizobia majorly colonize intercellular air spaces and lysed host cells (Reddy et al. 1997).

Yoshida and Ancajas (1973) has reported that flooded rice soils fix a higher amount of nitrogen compared to upland soils. In the flooded paddy soils, the root–soil interface was identified as the site of nitrogen fixation (Kimura et al. 1979; Yoshida and Ancajas 1973) with highest nitrogenase activity (Yoshida and Ancajas 1973).

In an experiment conducted at IRRI, it was revealed that photosynthetic cyanobacteria and heterotrophic diazotrophs can fix nitrogen in the rice systems. Root secretions of cyanobacterial origin in the rhizosphere are utilized (Yoneyama et al. 2017). A positive nitrogen balance was observed, indicating a significant nitrogen fixation in the rice fields (Ladha et al. 2016).

App et al. (1980) have reported a positive nitrogen balance in flooded rice fields. Nitrogen balance is the difference between nitrogen inputs and nitrogen output in an ecosystem. The reason for the positive nitrogen balance is predicted as a result of photo-autotrophic and nitrogen-fixing agents present in the soil. This can be taken as an example of natural nitrogen fertility of the flooded rice fields. App et al. (1980) have also reported that, together with P and Fe, even though blue-green algae (cyanobacteria) do not increase the nitrogen balance, inoculating *Azolla* spp. has improved the nitrogen balance in the rice fields.

5.7.1.1 Dream Project of BNF in Rice

A serious consideration of biological nitrogen fixation is rice was put forward in early 1990, as one of the major projects in the decade by the International Rice Research Institute (IRRI). The dream project of biological nitrogen fixation in rice started in 1992 (Sofi and Wani 2007). Based on expert recommendations, IRRI stated four major approaches to achieve the objective, and first was to improve endophytic associations between rice and nitrogen-fixing bacteria. Number of research teams from around the world started working independently and in collaboration to study the probable endophytic diazotrophic relationships with rice with the potential of developing endosymbiotic relationships with rice (Sofi and Wani 2007).

Among the endophytic diazotrophs studied, *Alcaligenes* sp. (Bennet and Ladha 1992), Serratia marcescens (James et al. 2000), and *Azorhizobium caulinodans* were identified as aggressive colonizers of rice. A study on interaction between various rhizobial strains and rice has found out that root exudates from rice do not often get involved in root deformations, *nod* factor inductions, formation of thick lateral roots, or attachments to the plant. Rhizobia primarily enter the rice plant by cracks of epidermal cells or fissures caused by developing lateral roots. Endophytic colonization is restricted only to intercellular spaces (Al-mallah et al. 1989; Reddy et al. 1997).

Second approach was the engineering of a rice plant capable of forming legume-like symbiosis and nodules with rhizobia. Identification of diazotrophic (*nod, nif* and *fix* genes) genes and plant-specific (noduline) genes related to the process and identification of rice-specific promoters have shown the possibility of realizing the second approach. Engineering rice plant was approached by the transformation of rice leaf and the transformation of rice root (Parakaran 1997). It was realized that engineering the rice plant to fix nitrogen will be extremely difficult as it will require a large number of genes to work together in a foreign environment. Engineering of rice plant capable of N_2 fixation requires coordinated and regulated expression of almost 16 *nif* genes, 8 housekeeping *nif* genes assembled in an appropriate cellular location, additional genes to keep nitrogenase in an active form, promoters to activate the 16 *nif* genes, and appropriate mechanisms to protect the sensitive nitrogenase from oxygen (Sofi and Wani 2007). Achieving the nitrogen-fixing transgenic system completely failed due to the complexity of the process. "Genetic engineering through biotechnology has little or no success in achieving the induction of symbiosis between cereals and diazotrophs" (Saikia and Jain 2007). Although this was thought initially, recent evidence suggests that this is something not impossible (Refer section below).

The third major approach was transforming rice to ensure expression of nitrogenase and protect the nitrogenase system from oxygen damage. Since nitrogenase is extremely oxygen-sensitive, it is extremely important to develop mechanisms that can protect nitrogenase inside the eukaryotic cell. The reasons were, in the chloroplast, the genes that are expressed in prokaryotic fashion and polycistronic mRNAs are translated. There it was proposed that the process of photosynthesis and nitrogen fixation can coexist, if the oxygen damage for nitrogenase can be stopped. The fourth major approach was enhancing nitrogen use efficiency in rice.

5.7.2 Current Developments (2000–2021)

At present, everyone is well aware of the negative consequences of the use of nitrogen fertilizers. Focus toward probable methods of induction of nitrogen fixation in non-legumes such as rice is therefore intensified to great heights. A large number of studies concerning the molecular biology of nitrogen fixation, bio-fertilizer preparation, metagenomic studies, improvement of molecular diagnostics, combining nitrogen-fixing bacterial inoculum with compost or other organic methods, endophytic nitrogen fixation, complete genome sequencing of bacterial genomes, use of Internet of things (IoT), and other smart tools in agriculture and nitrogen management are carried out. The results of these research studies are available and can be used to improve the plant–diazotroph interactions.

5.7.2.1 Nitrogen-Fixing Biofertilizers with Improved Efficiency

Since the complexity of the nitrogen fixation process was well studied and understood, several options related to making use of nitrogen fixers in rice cultivation were taken into consideration. During this time, a major focus was given toward the production of nitrogen-fixing biofertilizers to be added to the paddy fields.

Biofertilizers are defined as natural fertilizers that contain active or latent soil microbial strains for enhancing plant nutrient uptake and soil productivity via their metabolic processes such as N-fixing and phosphate solubilization (Banayo et al. 2012; Simarmata et al. 2016). According to Banayo et al. (2012), the concept of bio-fertilizers has developed after the discovery of beneficial effects on plants from the microbes dwelling on the plant body.

For proper functioning of a biofertilizer, at the time of biofertilizer preparation, the micro-organisms should be selected wisely. The microorganisms chosen should facilitate the requirement. For example, if N and P need to be provided to the plant, the biofertilizer can be incorporated with N-fixing and P-solubilizing bacteria. Care should be taken when microbes are selected for the process, and they should be able to coexist and also should be able to compete with the rhizobia already residing in the rhizosphere and survive themselves.

It was understood that, rather than adding micro-organisms as mono-cultures, multi-bacteria complexes are more efficient in surviving in the soils and carrying out nitrogen fixation. Other than fixing nitrogen and providing it to the plants, microbial consortia can improve the surrounding soil of the plant resulting in healthier and productive soil. Another group of important microorganisms that can contribute positively to non-legume cultivation are the plant growth-promoting rhizobacteria (PGPRs). Enrichment of the soil microbial reserve will improve the nutrient absorption efficiency.

Combined application of bio-fertilizers with efficient nitrogen fertilizer application methods would improve the overall productivity of rice fields. Biofilms are another important form of microbial communities that can contribute massively to the improvement of soil. Perera et al. (2015) and Perera and Tirimanne (2021) report that through the incorporation of a biofilm comprised of *Azorhizobium caulinodans* and the fungi *Aspergillus* spp., into the rice soils with only 50% of the nitrogen recommendation for rice, yield similar to application of 100% nitrogen recommendation can be obtained.

5.7.2.2 Focus of Bill and Melinda Gates Foundation on Nitrogen-Fixing Non-legumes

Bill and Melinda Gates Foundation's focus on nitrogen fixation in non-legumes (Beatty and Good 2011) is one of the important recent advances in this regard. Three approaches that have been discussed in the Gates foundation meeting are of importance.

First focus was on developing root nodule symbiosis (*similar to the second approach by IRRI*) in cereals. Several recent findings contributed positively to the first approach. Discovery of Myc factors that are structurally similar to Nod factors involved in legume rhizobium symbiosis is of major importance. It has been discovered that Myc factors are involved in cross-talk between 70% and 90% of terrestrial plants and arbuscular mycorrhizal endosymbiotic fungi (Maillet et al. 2011). Moreover, the discoveries of crack entry invasion of diazotrophs to non-legumes (Sect. 5.7.1) and finding that the hormones that are involved in nodule formation are common to all the plants were also positive indicators of possibility of root nodule symbiosis in cereals (Madsen et al. 2010).

The second approach discussed at the Gates Foundation meeting (2011) was about the possible preparation of biofertilizers from the nodule-independent associations that are being discovered (Beatty and Good 2011). Although there are several formations of biofertilizers present, their mode of action and efficiency needs further research and clarification (Beatty and Good 2011).

The third approach of the Gates foundation meeting deals with the introduction of nitrogenase into plant organelles through the introduction of nitrogenase enzyme-encoding bacterial genes into the plants and letting plants fix their own nitrogen. Although earlier it was thought impossible due to the higher number of genes and factors involved in the formation of nitrogenase and the other metal cofactor (Fe-Mo-co), the recent discoveries show that it requires only three proteins (Lill and Mühlenhoff 2008; Rubio and Ludden 2008). Moreover, to analyze nitrogenase maturation, novel in vitro assay tools are now available. These are positive indicators of the possibilities of making an engineering rice plant a feasible one (Beatty and Good 2011). Mitochondria and chloroplast are suggested as the places that are suitable for nitrogen fixation as both-high energy and low-oxygen environments can be provided.

Rhizosphere nitrogen fixers who are free-living need to be competent in striving successfully and tolerating the different soil types, plant cultivars, and agricultural practices. To be beneficial to the plant, it is important to research on the available PGPRs and their competency with the available soil type and the environmental condition. For this, a kind of matching is vital. A well-fitting microbe will be able to contribute to the reduction in urea fertilizer usage (Igiehon and Babalola 2018). Hence, it is extremely important to study the microorganisms associated with different types of crop plants. Understanding the microbiome requires studying and identifying them correctly. Latest methodologies such as DNA analysis, metagenomics profiling of the entire microbiome, and next-generation sequencing

(NGS) are possible approaches. NGS techniques have assisted largely in determining the accurate microbial population among the plants. NGS has assisted the identification of these microbes more than any other method (Igiehon and Babalola 2018).

5.8 Future Directions

Large amount of research related to symbiotic nitrogen fixation has revealed major requirements of efficient symbiotic nitrogen fixation systems. These findings are of extreme importance when it comes to improving the probable contribution that nitrogen-fixing microbes can do for paddy cultivations. It has been clearly understood by now that symbiotic nitrogen fixation is an extremely efficient active method when compared to the contribution from free-living passive nitrogen fixation.

In the future, much research will be targeted on engineering non-legumes such as rice, to nodulate and fix nitrogen symbiotically. Although this has been a target in the past, it was decided unachievable. A great deal of information is available now, and therefore, these data will assist positively in achieving this crucial task of producing a nitrogen-fixing rice plant. Huisman and Geurts (2020) categorize the available information in four themes.

First, the availability of variety of blueprints is related to non-legume nitrogen fixation. Ten lineages in the related taxonomic orders Fabales, Fagales, Cucurbitales, and Rosales perform nitrogen fixation (Soltis et al. 1995; Doyle 2011; Li et al. 2015). Evolutionary studies describe this as the nitrogen-fixing clade even though there are many lineages of non-nodulating species. The nodulating and non-nodulating plants have a shared evolutionary origin (Soltis et al. 1995; van Velzen et al. 2019).

Second is the identified core symbiosis genes. With the help of evolutionary genomic studies, it has been discovered that only few genetic elements are required to induce non-legume nitrogen fixation, unlike it was thought in the past. According to reports by Yang et al. (2018) and Bailey-Serres et al. (2019), nitrogen fixation in non-legumes such as rice can be achieved through genetic engineering of a few genetic components and also by exploring bacterial genetic units can be used to create a minimal set of three genes, necessary for the transfer of nitrogen fixation. As explained earlier, nitrogenase enzyme synthesis and nitrogen fixation are extremely complicated processes demanding lower amounts of oxygen and higher energy. But through extensive research, a common core set of genes/gene products required for functional nitrogenase biosynthesis has been identified (Rubio and Ludden 2008). To provide the low oxygen environment that is required for nitrogenase biosynthesis, plastids and mitochondria are being identified as probable locations. Both could provide the ATP and electrons required for nitrogenase to function, but they differ in their internal O_2 levels and their ability to incorporate ammonium into amino acids.

Third theme is the information and understanding of how the nodulation of the ancient arbuscular mycorrhizal (AM) signaling pathway is used. Fourth is the identified communication between generic plant development programs and nodulation.

5.9 Conclusions

Application of biological nitrogen fixation particularly using bio-fertilizers, in the form of symbiotic or free-living bacteria, has gained momentum and yielded positive results, but certain limitation under this process are also advocated besides the success of legume–rhizobium symbiosis. Therefore, it is high time to resolve limitation of nitrogen fixation in other crops like rice, which can feed the world. Continuing researches across the world is solicited until we can find an impressive and acceptable solution to this problem. Expansion of research in the scope of genetic regulation and controlling limiting factors for successful colonization in rice's rhizosphere are important approach that needs to be studied.

Acknowledgement Not applicable.

Conflicts of Interest The author(s) declares(declare) no conflicts of interest.

Disclaimer(s) The author(s) is(are) aware of any ethical concern for using and reproducing Figs. 5.1 and 5.2 in this publication from their previous publications.

References

Alazard D, Ndoye I, Dreyfus B (1988) *Sesbania rostrata* and other stem-nodulated legumes. In: Bothe H et al (eds) Nitrogen fixation: hundred years. Fischer, New York, pp 765–769

Al-Mallah MK, Favey MR, Cocking EC (1989) Formation of nodular structures in rice seedlings by *Rhizobia*. J Exp Bot 40:473–478

Anil K, Yakadri M, Jayasree G (2014) Influence of nitrogen levels and times of application on growth parameters of aerobic rice. Int J Plant Animal Env Sci 4(3):231–234

App AA, Watanabe I, Alexander M, Ventura W, Daez C, Santiago T, De Datta SK (1980) Nonsymbiotic nitrogen fixation associated with the rice plant in flooded soils. Soil Sci 130(5): 283–289

Arnold W, Rump A, Klipp W, Priefer UB, Pühler A (1988) Nucleotide sequence of a 24 206-base-pair DNA fragment carrying the entire nitrogen fixation gene cluster of *Klebsiella pneumoniae*. J Mol Biol 203(3):715–738

Bailey-Serres J, Parker JE, Ainsworth EA, Oldroyd GE, Schroeder JI (2019) Genetic strategies for improving crop yields. Nature 575(7781):109–118

Banayo NP, Cruz PC, Aguilar EA, Badayos RB, Haefele SM (2012) Evaluation of bio-fertilizers in irrigated rice: effects on grain yield at different fertilizer rates. Agriculture 2(1):73–86

Barber D, Lynch JJ (1977) Microbial growth in the rhizosphere. Soil Biol Biochem 9:305–308

Beatty PH, Good AG (2011) Future prospects for cereals that fix nitrogen. Science 333(6041): 416–417

Bennet J, Ladha JK (1992) Introduction: feasibility of nodulation and nitrogen fixation in rice—potential and prospects. In: Khush GS, Bennett J (eds) Nodulation and nitrogen fixation in rice: potential and Prospects. International Rice Research Institute, Los Banos, pp 1–17

Bonaldi K, Gourion B, Fardoux J, Hannibal L, Cartieaux F, Boursot M, Vallenet D, Chaintreuil C, Prin Y, Nouwen N, Giraud E (2010) Large-scale transposon mutagenesis of photosynthetic *Bradyrhizobium* sp. strain ORS278 reveals new genetic loci putatively important for nod-independent symbiosis with *Aeschynomene indica*. Mol Plant Microbe Interact 23(6): 760–770

Chaturvedi I (2005) Effect of nitrogen fertilizers on growth, yield and quality of hybrid rice (*Oryza sativa*). J Cent Eur Agric 6(4):611–618

Chauhan BS, Jabran K, Mahajan G (eds) (2017) Rice production worldwide, vol 247. Springer, Cham

Cheng Q (2008) Perspectives in biological nitrogen fixation research. J Integr Plant Biol 50(7): 786–798

Cocking EC, Al-Mallah MK, Benson E, Davey MR (1990) Nodulation of non-legumes by rhizobia. In: Gresshoff PM et al (eds) Nitrogen fixation. Springer, New York, pp 813–823

Cocking EC, Srivastava JS, Kothari SL, Davey MR (1992) Invasion of non-legume plants by diazotrophic bacteria. In: Khush G, Bennett J (eds) Nodulation and nitrogen fixation in rice: potentials and prospects. International Rice Research Institute, Los Banos, pp 119–121

Delwiche C, Wijler JJ (1956) Non-symbiotic nitrogen fixation in soil. Plant Soil 7:113–129

Dixon R, Kahn D (2004) Genetic regulation of biological nitrogen fixation. Nat Rev Microbiol 2(8): 621–631

Dobermann A, Fairhurst TH (2000) Nutrient disorders and nutrient management. Potash & Phosphate Institute (PPI), Potash & Phosphate Institute of Canada (PPIC) and International Rice Research Institute (IRRI)

Doyle JJ (2011) Phylogenetic perspectives on the origins of nodulation. Mol Plant Microbe Interact 24(11):1289–1295

Dreyfus BL, Elmerich C, Dommergues YR (1983) Free-living *Rhizobium* strain able to grow on N_2 as the sole nitrogen source. Appl Environ Microbiol 45(2):711–713

Einsle O, Tezcan FA, Andrade SL, Schmid B, Yoshida M, Howard JB, Rees DC (2002) Nitrogenase MoFe-protein at 1.16 Å resolution: a central ligand in the FeMo-cofactor. Science 297(5587):1696–1700

Evenson RE, Gollin D (2003) Crop variety improvement and its effect on productivity: the impact of international agricultural research. Cabi Publication, Cambridge, p 548

Fischer HM (1994) Genetic regulation of nitrogen fixation in rhizobia. Microbiol Rev 58(3): 352–386

Franche C, Lindström K, Elmerich C (2009) Nitrogen-fixing bacteria associated with leguminous and non-leguminous plants. Plant Soil 321(1–2):35–59

Ghanbari-Malidarreh A (2011) Silicon application and nitrogen on yield and yield components in rice (*Oryza sativa* L.), in two irrigation systems. World Acad Sci Eng Technol 5(2):40–47

Giraud E, Moulin L, Vallenet D, Barbe V, Cytryn E, Avarre J-C, Jaubert M et al (2007) Legumes symbioses: absence of *Nod* genes in photosynthetic bradyrhizobia. Science 316:1307–1312

Hirel B, Le Gouis J, Ney B, Gallais A (2007) The challenge of improving nitrogen use efficiency in crop plants: towards a more central role for genetic variability and quantitative genetics within integrated approaches. J Exp Bot 58(9):2369–2387

Huisman R, Geurts R (2020) A roadmap toward engineered nitrogen-fixing nodule symbiosis. Plant Commun 1(1):100019

Igiehon NO, Babalola OO (2018) Rhizosphere microbiome modulators: contributions of nitrogen fixing bacteria towards sustainable agriculture. Int J Environ Res Public Health 15(4):574

James EK (2000) Nitrogen fixation in endophytic and associative symbiosis. Field Crop Res 65(2–3):197–209

James EK, Gyaneshwar P, Barraquio WL, Mathan N, Ladha JK (2000) Endophytic bacteria associated with rice. In: Ladha JK, Reddy PM (eds) The quest for nitrogen fixation in rice. IRRI, Los Banos, pp 119–140

Kimura M, Panichsakpatana S, Wada H, Takai Y (1979) Influences of organic debris and rice root on the nitrogen fixation in the submerged soil. Soil Sci Plant Nutr 25:637–640

Kulasooriya SA (2008) Biological nitrogen fixation, fundamentals and utilization. Peradeniya Science Publication, Peradeniya

Ladha JK, Tirol-Padre A, Reddy CK, Cassman KG, Verma S, Powlson DS et al (2016) Global nitrogen budgets in cereals: a 50-year assessment for maize, rice and wheat production systems. Sci Rep 6(1):1–9

Lawson DM, Smith BE (2002) Molybdenum nitrogenase: a crystallographic and mechanistic view. In: Sigel A, Sigel H (eds) Ions in biological systems. Marcel Dekker, New York, pp 75–119

Lee KB, De Backer P, Aono T, Liu CT, Suzuki S, Suzuki T et al (2008) The genome of the versatile nitrogen fixer *Azorhizobium caulinodans* ORS571. BMC Genomics 9(1):1–14

Li HL, Wang W, Mortimer PE, Li RQ, Li DZ, Hyde KD, Chen ZD (2015) Large-scale phylogenetic analyses reveal multiple gains of actinorhizal nitrogen-fixing symbioses in angiosperms associated with climate change. Sci Rep 5(1):1–8

Lill R, Mühlenhoff U (2008) Maturation of iron-sulfur proteins in eukaryotes: mechanisms, connected processes, and diseases. Annu Rev Biochem 77:669–700

MacLean AM, Finan TM, Sadowsky MJ (2007) Genomes of the symbiotic nitrogen-fixing bacteria of legumes. Plant Physiol 144(2):615–622

Madsen LH, Tirichine L, Jurkiewicz A, Sullivan JT, Heckmann AB, Bek AS et al (2010) The molecular network governing nodule organogenesis and infection in the model legume *Lotus japonicus*. Nat Comm 1(1):1–12

Mahmud K, Makaju S, Ibrahim R, Missaoui A (2020) Current progress in nitrogen fixing plants and microbiome research. Plants 9(1):97

Maillet F, Poinsot V, André O, Puech-Pagès V, Haouy A, Gueunier M et al (2011) Fungal lipochitooligosaccharide symbiotic signals in arbuscular mycorrhiza. Nature 469(7328):58–63

Pankievicz V, Irving TB, Maia LG, Ané JM (2019) Are we there yet? The long walk towards the development of efficient symbiotic associations between nitrogen-fixing bacteria and non-leguminous crops. BMC Biol 17(1):1–17

Parakaran J (1997) BNF rice: the choice of new generations. Paper presented in plant genetic seminar on nitrogen fixation, University of Wisconsin Madison, Wisconsin, USA

Perera TA, Tirimanne S (2021) Role of microbial communities in sustainable rice cultivation. In: Role of microbial communities for sustainability. Springer, Singapore, pp 75–119

Perera E, Gnaneswaran N, Staines C, Win AK, Sinclair R (2015) Incidence and prevalence of non-melanoma skin cancer in Australia: a systematic review. Aus J Dermatol 56(4):258–267

Perera TA, Tirimanne TL, Seneviratne G, Kulasooriya SA (2017) *Azorhizobium caulinodans* ORS 571–*Aspergillus* spp. biofilm in the presence of flavonoid naringenin: an extremely effective association for rice root colonization with a definite future as a nitrogen bio-fertilizer

Rascio N, La Rocca N, Sven Erik J, Brian F (2008) Biological nitrogen fixation. In: Encyclopedia of ecology

Reddy PM, Ladha JK, So RB, Hernandez RJ, Ramos MC, Angeles OR et al (1997) Rhizobial communication with rice roots: induction of phenotypic changes, mode of invasion and extent of colonization. Plant Soil 194(1–2):81–98

Rees DC, Howard JB (2000) Nitrogenase: standing at the crossroads. Curr Opin Chem Biol 4(5): 559–566

Rubio LM, Ludden PW (2008) Biosynthesis of the iron-molybdenum cofactor of nitrogenase. Ann Rev Microbiol 62

Saikia SP, Jain V (2007) Biological nitrogen fixation with non-legumes: an achievable target or a dogma? Curr Sci 92:317–322

Samal P, Babu SC, Mondal B, Mishra SN (2022) The global rice agriculture towards 2050: an inter-continental perspective. Outlook Agric 00307270221088338

Santi C, Bogusz D, Franche C (2013) Biological nitrogen fixation in non-legume plants. Ann Bot 111(5):743–767

Schultze M, Kondorosi A (1998) Regulation of symbiotic root nodule development. Annu Rev Genet 32(1):33–57

Seefeldt LC, Dance IG, Dean DR (2004) Substrate interactions with nitrogenase: Fe versus Mo. Biochemistry 43(6):1401–1409

Simarmata T, Turmuktini T, Fitriatin BN, Setiawati MR (2016) Application of bioameliorant and biofertilizers to increase the soil health and rice productivity. HAYATI J Biosci 23(4):181–184

Sofi P, Wani S (2007) Prospects of nitrogen fixation in rice. Asian J Plant Sci 6(1):203–213

Soltis DE, Soltis PS, Morgan DR, Swensen SM, Mullin BC, Dowd JM, Martin PG (1995) Chloroplast gene sequence data suggest a single origin of the predisposition for symbiotic nitrogen fixation in angiosperms. Proc Natl Acad Sci 92(7):2647–2651

Sprent JI (2001) Nodulation in legumes. Royal Botanic Gardens, Oxford University Press, Kew

Trinick MJ (1979) Structure of nitrogen-fixing nodules formed by *Rhizobium* on roots of *Parasponia andersonii* Planch. Can J Microbiol 25(5):565–578

van Velzen R, Doyle JJ, Geurts R (2019) A resurrected scenario: single gain and massive loss of nitrogen-fixing nodulation. Trends Plant Sci 24(1):49–57

White JF, Crawford H, Torres MS, Mattera R, Irizarry I, Bergen MA (2012) Proposed mechanism for nitrogen acquisition by grass seedlings through oxidation of symbiotic bacteria. Symbiosis 57:161–171

Yang Z, Han Y, Ma Y, Chen Q, Zhan Y, Lu W, Lin M (2018) Global investigation of an engineered nitrogen-fixing Escherichia coli strain reveals regulatory coupling between host and heterologous nitrogen-fixation genes. Sci Rep 8(1):1–13

Yanni YG, Rizk RY, Corich V, Squartini A, Ninke K, Philip-Hollingsworth S (1997) Natural endophytic association between *Rhizobium leguminosarum* bv. trifolii and rice roots and assessment of its potential to promote rice growth. Plant Soil 194(1–2):99–114

Yoneyama T, Terakado-Tonooka J, Minamisawa K (2017) Exploration of bacterial N_2-fixation systems in association with soil-grown sugarcane, sweet potato, and paddy rice: a review and synthesis. Soil Sci Plant Nutr 63(6):578–590

Yoshida T, Ancajas RR (1973) Nitrogen-fixing activity in upland and flooded rice fields. Soil Sci Soc Am J 37:42–46

Young JPW, Johnston AWB (1989) The evolution of specificity in the legume-*Rhizobium* symbiosis. Trends Ecol Evol 4(11):341–349

Youseftabar S, Fallah A, Daneshiyan J (2012) Effect of split application of nitrogen fertilizer on growth and yield of hybrid rice (GRH1). Aus J Basic Appl Sci 6(6):1–5

Chapter 6
Nitrogen-Fixing Archaea and Sustainable Agriculture

Amr M. Mowafy, Engy Atef Abou El-ftouh, Mohammed Y. Sdiek, Shreef Abdelfattah Abdelshafi, Asmaa A. Sallam, Mona S. Agha, and Walaa R. Abou Zeid

Abstract Whether free-living or symbiotic, diazotrophic microorganisms are the main and only source of nitrogen fixation and utilization in different biotic systems. The fixed nitrogen is an important component in various anabolic processes of many macromolecules, which are important to the cell in terms of structure and function. The most famous and specialized in nitrogen fixation are the root nodule bacteria, but other types of free-living bacteria and Archaea can fix nitrogen and enrich their environment with this important element. Archaea are a significant division of life forms, abundant in both severe and normal habitats; little attention has been paid to them as an integrated component of various metabolic processes of the plant microbiome. The employment of these microorganisms in agriculture replacing or at least decreasing the input of chemical fertilizers is one of the most important proposals for sustainable agriculture particularly for nonleguminous plants. This chapter is focusing on Archaea as plant growth promoters with special emphasis on their role in nitrogen fixation.

A. M. Mowafy (✉)
Botany Department, Faculty of Science, Mansoura University, Mansoura, Egypt

Department of Biological Sciences, Faculty of Science, New Mansoura University, New Mansoura city, Egypt
e-mail: ammr79@mans.edu.eg

E. A. A. El-ftouh · M. Y. Sdiek · S. A. Abdelshafi
Biotechnology and Its Application Program, Faculty of Science, Mansoura University, Mansoura, Egypt

A. A. Sallam · M. S. Agha
Botany Department, Faculty of Science, Mansoura University, Mansoura, Egypt
e-mail: asmaaalisallam@mans.edu.eg; mona.agha@mans.edu.eg

W. R. Abou Zeid
Cotton and Crops Acarology Department, Plant Protection Research Institute, Agriculture Research Center, Dokki, Giza, Egypt
e-mail: dr-walaaroushdy@arc.sci.eg

© The Author(s), under exclusive license to Springer Nature Singapore Pte Ltd. 2022
D. K. Maheshwari et al. (eds.), *Nitrogen Fixing Bacteria: Sustainable Growth of Non-legumes*, Microorganisms for Sustainability 36,
https://doi.org/10.1007/978-981-19-4906-7_6

Keywords Archaea · Methanogenic euryarchaeota · Rhizosphere · Nitrogen fixation

6.1 Introduction

Nitrogen (N) is a considerable essential macronutrient affecting growth, metabolism, and yield of plants. It represents 2% of the plant's dry weight (Miller and Cramer 2005). It is a major component of amino acids which are the building blocks of both structural and functional proteins. It is also critical for nitrogen bases, DNA, RNA, alkaloids, vitamins, amides, coenzymes, hormones, and many other vital components that constitute cell life and activity. The plant's different physiological processes are significantly affected by nitrogen supply. The biosynthesis of chlorophyll and stem as well as root growth is all affected by the available nitrogen dose. Additionally, its sufficient supply improves fruit quality and the protein content of fodder plants. Synergistically, it encourages the uptake and utilization of other nutrients including potassium and phosphorous (Bloom 2015; Hemerly 2016) as shown in Fig. 6.1.

For higher plants, the introduced organic and inorganic forms of nitrogen in soil could be utilized after uptake by root in the form of nitrate (NO_3^-) or ammonium (NH_4^+). Most of the ammonium is incorporated into organic compounds in the root, whereas nitrate is mostly translocated through the xylem to the shoot system or stored in the vacuoles of both root and stem cells, and it might be incorporated into storage organs. In order to be assimilated into organic polymers, nitrate must be reduced into ammonia, a step that is as vital as CO_2 assimilation via photosynthesis (Hecht and Mohr 1990). Plants cannot utilize N or the 98% of planetary N that is immobilized in the geosphere (Rosswall 1983). To meet the increasing demands of the growing population requirements of food, the synthetic N produced by Haber–Bosch process, which is considered as the most important world invention of the

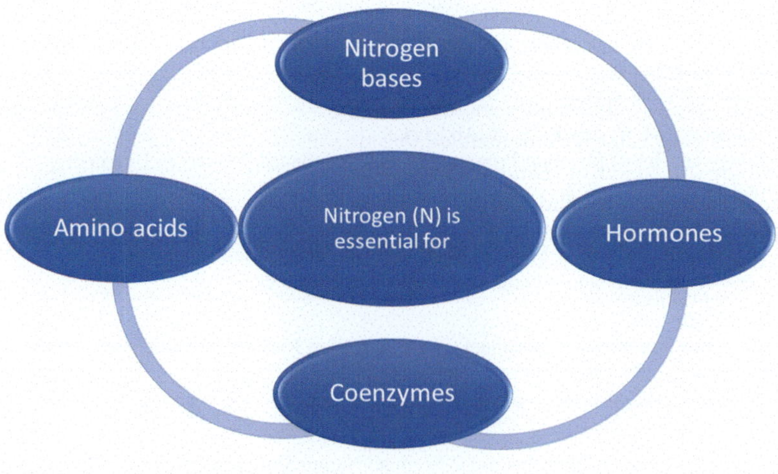

Fig. 6.1 Importance of nitrogen element

20th century (Dent and Cocking 2017). This has been intensively added to crop leading to the increase in the yield (Galloway et al. 2008; Kaur et al. 2017).

The intensive use of synthetic fertilizers led to a contamination that might reach to a toxic level not only for soil but also for the groundwater resulting in eutrophication in the environment besides the consumption of the limited energy resources during manufacturing (Miller and Cramer 2005; Kaur et al. 2017). In North America, the intensive use of N fertilizers is responsible for 75% of nitrous oxide emission in addition to the contamination of 1.5 million drinking well water with nitrate (Dent and Cocking 2017) indicating that nitrogen deposition around the world due to N fertilizers is not usually used efficiently by crops (Peoples et al. 1995a, b).

These disadvantages clear up the requirement of crops with high yields and lower consumption of nitrogen fertilizers in the call of a "second green revolution" by Norman E. Borlaug (Dent and Cocking 2017). Symbiotic nitrogen fixation, which is positioned as a major part of biological nitrogen fixation, is an important alternative source of nitrogen supply for sustainable agriculture. It has been estimated that 200–300 kg nitrogen could be fixed per hectare when legume crop and pasture species are associated (Peoples et al. 1995a, b). As an alternative to chemicals, the use of beneficial microbes is growing as biostimulant, bioprotectant, and a sustainable source for different plant nutrients such as nitrogen, phosphorus, potassium, zinc, and iron. The use and application of plant growth-promoting bacteria (PGPB), as well as fungi, are well-formulated, and more attention thus is given to Archaea as plant growth promotors.

6.2 General Information About Archaea

Archaea are one of the three main life domains beside bacteria and eukarya which are assigned according to the rRNA and protein (Dave et al. 2006). It is derived from Greek "Archaios" that means primitive or ancient (Wu et al. 2022). It has been discovered about 30 years ago as a domain of extreme environments being able to coexist in different environments with bacteria & eukarya, as well it contains the only organisms being able to run methanogenesis and so-called methanogens. It tends to be more similar to eukarya than bacteria based on their mechanisms, but according to other aspects such as the chromosomes organization and the size, it is more similar to bacteria. But with that whole similarity they have a significant characteristic being absent in all other organisms as till now all discovered archaeal membrane phospholipids are isoprenoid ethers linked to glycerol 1 phosphate (G1P) in contrast to bacterial and eukaryal fatty acid ester-containing membranes linked to glycerol 3 phosphate (G3P). There are also some characteristics that distinguish archaea from other domains like presence of special flagellins and absence of murein of the cell wall (Borrel et al. 2020), having many RNA polymerases with more than eight polypeptides. Among archaeal members, there are *Pyrolobus fumarii*, which can live up to 113 °C equivalent to 235 °F and was found in hydrothermal vents. And *Picrophilus*, isolated from soils of high acidity, which is the most known organism could tolerate high acidity as they can grow at around pH 0. The methanogens can produce methane as a byproduct and are anaerobic, so they can be found in the guts of animals including humans, the hot springs, and the marshes (Wu et al. 2022).

6.3 Archaea as Plant Growth Promotors

Our understanding of the Archaea's diversity and metabolic capacity in a variety of contexts has been greatly transformed. Archaea represents a large, wealthy, and unknown entity to some extent, wrongly famous for existence only in harsh environmental niches despite its spread and recent isolation from different environments. Woese and coworkers were the first to be credited with distinguishing between bacteria and Archaea through the comparative study of 16S rRNA sequences analysis leading to split the "prokaryotic domain" eubacteria and archaebacteria (Woese and Fox 1977). Archaebacteria or as it is commonly known as Archaea are a significant component of Earth's planets and may have a role in the C-cycle and N-cycle. Archaea differs from bacteria in that their membranes contain isoprene lipids attached to glycerol-1-phosphate through ether bonds, they lack peptidoglycan in their cell walls, and their proteins resemble those of eukaryotes (Wenli et al. 2021). Until recently, Archaea was believed to be restricted to extreme environmental niches, but recent metagenomic studies have proven its spread in many mesophilic environments as shown in Fig. 6.2, especially the rhizosphere and endosphere regions of different plants (Alori et al. 2020; Naitam and Kaushik 2021).

Nitrogen fixation along with others that directly affect plant growth including the ability to solubilize phosphate and secrete phytohormones like IAA and chelate iron due to siderophores productions makes them potential biotechnological candidates

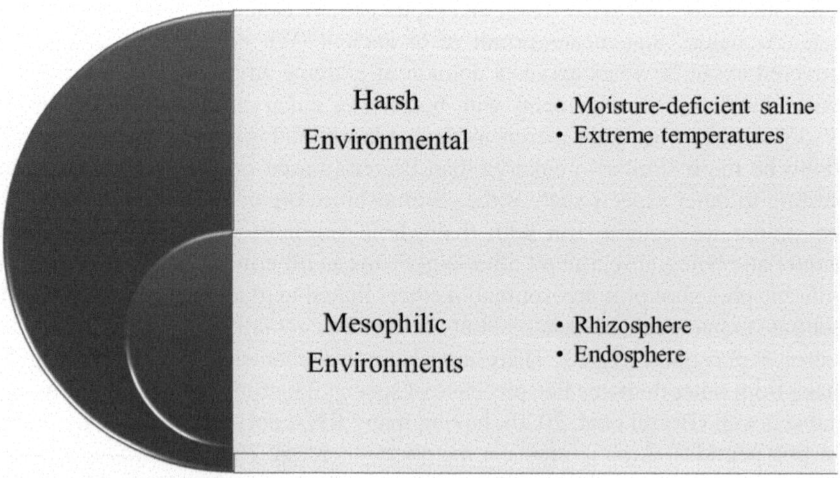

Fig. 6.2 Archaea occurrence in different environmental conditions

Table 6.1 Some archaea and their potential roles in plant promotion

Archaea	Role	References
Candidatus, Nitrosocosmicusfranklandus C13	Ammonium oxidation	Prudence et al. (2019)
Asgard archaea	Nutrient (nitrogen and sulfur) cycling, heavy metals (arsenic and copper) extraction	MacLeod et al. (2019)
Nitrosocosmicusoleophilus MY3	Nutrient supply (ammonia oxidation), biocontrol against pathogenic organisms	Song et al. (2019)
Thaumarchaeota, Crenarchaeota, Euryarchaeota	Siderophore production	Dave et al. (2006)
Natrialba, Natrinema, Halolamina, Halosarcina	Phosphorus solubilization, nitrogen fixation, siderophore production and indole acetic acid production	Yadav et al. (2017)
Crenarchaeota, Euarchaea	N. transformation (nitrification)	Dubey et al. (2016)
Thaumarchaeota, Euryarchaeota	CO_2 fixation and glycogen degradation	Taffner et al. (2019)

for plant growth promotion particularly for their adaptability for various ecosystems (Yadav et al. 2017) as shown in Table 6.1.

Several phosphate-solubilizing halophilic Archaea have been isolated from the rhizosphere of wild grasses growing in halo saline Indian desert using a set of organic acids to lower pH (Yadav et al. 2015). Besides, some of them showed the ability to solubilize potassium and zinc. The ability of Archaea to produce IAA was first discovered in the thermophilic *Sulfolobus acidocaldarius* and then followed by other halophiles (White 1987; Yadav et al. 2019). Siderophores, the iron chelators, are produced by several types of Archaea as such the haloalkaliphilic Archaea (Dave et al. 2006; Patil et al. 2016; Alori et al. 2020). The ability of Archaea to thrive under extreme conditions such as moisture-deficient saline regimes make them the best candidates to assist plants to overcome abiotic stresses (Taffner et al. 2019) in addition to their ability to interplay with fungi at the rhizosphere region (Taffner et al. 2018). Plant defense response mechanisms against biotic stress caused by phytopathogens are activated by Archaea according to some reports (Song et al. 2019; Wang et al. 2019).

Recently, it has been found that *Arabidopsis thaliana*-induced systemic resistance (ISR) against *Pectobacterium carotovorum* and *Pseudomonas syringae* is triggered by the ammonia-oxidizing *Nitrosocosmicusoleophilus* (Song et al. 2019). Hg-methylating Archaea in the paddy field of rice are suggested to play an important role in alleviating Hg stress on the plant (Ma et al. 2019). Indirectly, Archaea superoxide dismutase when expressed in *Arabidopsis* made the plant more resistant to heat and light (Im et al. 2009).

6.3.1 Nitrogen Fixation by Archaea

The process of biological nitrogen fixation (BNF) is the reduction of atmospheric nitrogen to ammonia (De Bruijn 2015), a process that could be done by some bacterial genera and Archaea those might be free-living diazotrophs (*Rhodobacter* and *Azotobacter*), as well as those, are living in close (*Azospirillum*) or in deep association with host plants (*Rhizobium*) or endophytic in nature (*Bacillus*) (Leigh 2000; Kaschuk and Hungria 2017). Among nitrogen-fixing diazotrophs, cyanobacteria are a distinct group with a potential ability to commit oxygen-labile nitrogen fixation and oxygen-evolving photosynthesis within the same cell (Mitsui et al. 1986), a description that has been turned over after discovering the ability of some rhizobial species to perform photosynthesis via a specialized photosynthetic system (Fleischman and Kramer 1998).

The process of BNF is of considerable interest in agriculture for its inputs of available nitrogen forms for plants unable to make symbiotic relationship with nitrogen fixers. Thus BNF allow the nitrogen cycle to maintain the life of non-nitrogen fixing organisms. It has been reported that the most important agent in nitrogen fixation is that obtained through the symbiotic relationship (80 % of the total nitrogen biologically fixed) between legumes and *Rhizobium* (Herridge et al. 2008). Symbiotic nitrogen fixation has been estimated to provide at least 70 million metric tons of nitrogen per year as stated by Brockwell and Bottomley (1995). Therefore, the association between legumes and their specific *Rhizobia* offers cheap nitrogen control than synthetic N fertilizers because the nitrogen pool becomes slowly available to nonlegume species when crop rotation is well designed and controlled (Pandey et al. 2000).

In Archaea, nitrogen fixation was initially discovered in *Methanosarcina barkeri* and *Methanococcus thermolithotrophicus* and was restricted to methanogenic euryarchaeota. Bacterial nitrogen fixation has been a prominent study area in recent years due to the importance of the N-cycle in agriculture. On the other hand, Archaea may undertake a variety of reductive N-cycle reactions, including assimilatory activities like nitrate assimilation and N_2 fixation, as well as dissimilatory reactions like nitrate respiration and denitrification. However, Archaea's nitrogen metabolism is far less well understood in comparison to that of bacteria (Cabello et al. 2004).

Ammonia-oxidizing Archaea (AOA) and bacteria (AOB) are the two major drivers of the biological nitrogen cycle. Metabolic pathways are often shared by Archaea and bacteria, and nearly all genes involved in this process are present in these domains (Sun et al. 2021). The capacity to fix nitrogen is located in a wide range of phylogenetically different bacteria, but seems to be limited to *Methanococcus thermolithotrophicus, Methanobacterium bryantii, Methanosarcina barkeri, and Methanospirillum hungatei;* those are mostly halophilic and thermophilic in habitat although they could be mesophilic and psychrophilic (Leigh 2000; Mehta et al. 2003; Boyd et al. 2011; Dhakephalkar et al. 2019).

M. thermolithotrophicus is regarded as the only known organism with the ability to fix nitrogen at more than 60 °C suggesting the presence of different fixation mechanisms. However, biochemical and genetic studies demonstrated that Archaea nitrogen fixation is evolutionarily relevant to bacteria nitrogen fixation and operated by almost the similar mechanism. At least six *nif* genes (*nif*-H, *nif*-D, *nif*-K, *nif*-E, *nif*-N, and *nif*-X) discovered in bacteria are also found in diazotrophic methanogens. The majority of nitrogenases found in methanogens are most likely molybdenum-type enzymes with few exceptions of vanadium and iron nitrogenases. Gene organization and regulation, on the other hand, differ from that in bacteria. *Methanococcus maripaludis* has a single operon that contains all six known methanogen *nif* genes, as well as two homologues of the bacterial nitrogen sensor-regulator *gln*B (Leigh 2000). The iron protein, or dinitrogenase reductase, and the molybdenum iron protein, or dinitrogenase, together make up nitrogenase, the enzyme complex that catalyzes nitrogen fixation (Turk et al. 2011). The iron protein genes (*nif*H, *vnf*H, and *anf*H) are highly conserved among diverse prokaryotes, and the *nif*H phylogenetic tree closely mimics that of 16S rRNA, and this conserved nature makes it an ideal molecular tool to discriminate between different diazotrophs. Generally, the *nif*H gene is grouped into four clusters in which cluster II is for *Methanogens* and bacterial *anf*H (Mehta et al. 2003). Nitrogenase reductase from *M. barkeri* was found to be homotetramer rather than a homodimer, and the activity was lower compared to that of bacteria (Leigh 2000). Nitrogen-fixing methanogens were found to dominate subtropical wetlands and the stagnant water as such of rice fields making these organisms very suitable candidates forest land reclamation and rice agriculture (Bae et al. 2018; Naitam and Kaushik 2021).

6.4 The Diversity of Archaea in the Rhizosphere Region

Soil particularly rhizosphere region represents an attractive area for microbes of all domains, whether they are harmful or beneficial to the plant. The rhizosphere is a sink of microorganisms which are diverse in nature. It also comprised nitrifying, denitrifying, nitrogen-fixing, and ammonia-oxidizing organisms involved in N-cycle. The dynamic and complicated regions are full of interactions that are most often in favor of the plant, where beneficial microbes predominate. The presence, dominance, and entrance to the plant body are all controlled by genetic factors as well as the surrounding environment conditions (Moissl-Eichinger et al. 2018). Therefore, the soil microbial community looks like a solution waiting for more investigation to be discovered, characterized, and optimally used in agriculture to bridge the gap between the food production and consumption that is increasing along with the world population (Meena et al. 2017). In both the aboveground and belowground phytobiomes, Archaea represent a significant component of plant-associated ecosystems. Archaea are recently found to occupy plant ectorhizosphere, rhizoplane as well as endo-rhizosphere regions (Knief et al. 2012; Oburger and Schmidt 2016). Euryarchaeota was reported to inhabit the rhizosphere of *Oryza sativa* (Knief et al. 2012), *Coffea arabica* (Oliveira et al. 2013), *Olea europaea* (Müller et al. 2015), *Eruca sativa* (Taffner et al. 2018), *Jatropha curcas* (Dubey et al. 2016), and *Zea maize* (Chelius and Triplett 2001). Furthermore, metagenomic studies revealed the presence of Crenarchaeota in association with *Lycopersicum esculentus, Phragmites australis, Olea europaea, Eruca sativa, Jatropha curcas,* and *Erica andevalensis* (Simon et al. 2005; Mendes et al. 2013; Oliveira et al. 2013; Liu et al. 2015; Müller et al. 2015; Dubey et al. 2016; Taffner et al. 2018). There occurrence was found to be beneficial to hosts due to the presence of different beneficial traits including IAA and siderophores production, phosphate solubilization, CO_2 fixation, and sulfur reduction (Knief et al. 2012) like plant growth-promoting bacteria (PGPB) (Mowafy et al. 2022).

Crenarchaeota and Euryarchaeota were found in the rhizosphere of *Jatropha curcas*, which adapted to flourish under high-temperature and salt stress. Although the exact mechanisms are yet to be known, Crenarchaeota and Euryarchaeota may aid in the adaptation of Jatropha curcas to salt stress and high temperatures (Dubey et al. 2016). Archaea have been implicated in increasing plant immune responses, such as generating induced systemic resistance (ISR) to pathogenic bacteria in *Arabidopsis* (Song et al. 2019). Away from the rhizosphere, a recent report shows the signature of Archaea as a part of microbial community inhabiting the seeds of wild alpine trees indicating the integrated role in supporting survival and seed germination along with other microbes (Wassermann et al. 2019). Table 6.2 shows some Archaea phyla that have been reported to be associated with different crop plants.

6.5 Future Prospective

It became evident that the use of microbes as an alternative or at least to limit the use of chemicals will undoubtedly achieve the goals of sustainable agriculture to increase plant production in an environmentally friendly way with economic benefits. Beneficial microorganisms in particular archaea, with their potential to survive in harsh conditions and support plant life under abiotic stresses, make them important candidates. However, more is yet to reveal about the diversity of Archaea and their mechanisms to support plant growth and how could we employ them to improve crop productivity. Archaea may show less functional adaptation to agricultural plants compared to their wild relatives due to changes in nutrient and energy levels in the soil because of excessive fertilizer applications (Taffner et al. 2018; Qi et al. 2021). Such problems need huge effort to formulate Archaeal applications in sustainable agricultural practices.

Table 6.2 Some archaea phyla that and their association with crop plants

Archaea phylum	Crop	References
Euryarchaeota	*Oryza sativa*	Großkopf et al. (1998)
Euryarchaeota, Crenarchaeota	*Cherries of Coffea arabica*	Oliveira et al. (2013)
Crenarchaeota	*Lycopersicum esculentus*	Simon et al. (2005)
Methanogens	*Oryza sativa*	Pump et al. (2015)
Crenarchaeota, Euryarchaeota	*Jatropha curcas*	Dubey et al. 2016)
Thaumarchaeota, Crenarchaeota, Euryarchaeota	*Olea europaea L.*	Müller et al. (2015)
Crenarchaeota, Euryarchaeota	*Zea mays*	Chelius and Triplett (2001)
Nitrosopumilus, Nitrososphaera	*Halocnemum strobilaceum*	He et al. (2017)
Nitrosocosmicusoleophilus MY3	*Arabidopsis thaliana*	Song et al. (2019)
Crenarchaeota, and Euryarchaeota	*Erica andevalensis*	Alori et al. (2020)
Methanobacterium, Methanoregula, Methanospirillum, Methanomethylovorans, Methanosarcina, Methanosaeta, Crenarchaeota	*Phragmites australis*	Liu et al. (2015)
Thaumarchaeota, Euryarchaeota, Crenarchaeota, Methanosarcina	*Eruca sativa Mill.*	Dubey et al. (2016)
Methanocellales, Methanosaetaceae, Thaumarchaeota	*Oryza sativa*	Knief et al. (2012)
Halobacteria, Methanobacteria, Methanomicrobia, Thermoprotei	*Rhizophora mangle*	Pires et al. (2012)
Methanobacteriales, Methanosarcinales, Methanocellales	*Oryza sativa*	Moissl-Eichinger et al. (2018)

6.6 Conclusions

Safer biosimilars have become a priority of modern agriculture. We could not rely only on synthetic fertilizers for their negative effect on the environment resources. More attention is thus being given to Archaea as plant growth promotors as given to bacteria and fungi. The ability of Archaea to thrive under extreme conditions such as moisture-deficient saline regimes along with their ability to affect plant growth by solubilizing phosphate and phytohormones like IAA and chelate iron by siderophores makes them good potential biotechnological candidates for plant growth promotion particularly for their adaptability for various ecosystems. It has been proven that there are many strains of Archaea in the rhizosphere of different plants. In addition to recent studies on its presence in the microbial community of seeds to stimulate growth and support immunity, the studies on supporting Archaea in sustainable agriculture are still emerging day after day.

Acknowledgement Not applicable.

Conflicts of Interest The author(s) declares(declare) no conflicts of interest.

References

Alori ET, Emmanuel OC, Glick BR, Babalola OO (2020) Plant–archaea relationships: a potential means to improve crop production in arid and semi-arid regions. World J Microbiol Biotechnol 36(9):1–10

Bae H-S, Morrison E, Chanton JP, Ogram A (2018) Methanogens are major contributors to nitrogen fixation in soils of the Florida Everglades. Appl Environ Microbiol 84(7):e02222–e02217

Bloom AJ (2015) The increasing importance of distinguishing among plant nitrogen sources. Curr Opin Plant Biol 25:10–16

Borrel G, Brugère JF, Gribaldo S, Schmitz RA, Moissl-Eichinger C (2020) The host-associated archaeome. Nat Rev Microbiol 18(11):622–636

Boyd E, Anbar A, Miller S, Hamilton T, Lavin M, Peters J (2011) A late methanogen origin for molybdenum-dependent nitrogenase. Geobiology 9(3):221–232

Brockwell J, Bottomley PJ (1995) Recent advances in inoculant technology and prospects for the future. Soil Biol Biochem 27(4):683–697

Cabello P, Roldan MD, Moreno-Vivian C (2004) Nitrate reduction and the nitrogen cycle in archaea. Microbiology 150(11):3527–3546

Chelius M, Triplett E (2001) The diversity of archaea and bacteria in association with the roots of *Zea mays* L. Microb Ecol 41(3):252–263

Dave BP, Anshuman K, Hajela P (2006) Siderophores of halophilic archaea and their chemical characterization. Int J Exp Biol 44(4):340–344

De Bruijn FJ (2015) Biological nitrogen fixation. In: Lugtenberg B (ed) Principles of plant-microbe interactions. Springer, Cham

Dent D, Cocking E (2017) Establishing symbiotic nitrogen fixation in cereals and other non-legume crops: the greener nitrogen revolution. Agric Food Sec 6(1):7

Dhakephalkar PK, Prakash O, Lanjekar VB, Tukdeo MP, Ranade DR (2019) Methanogens for human welfare: more boon than bane. In: Satyanarayana T, Das S, Johri B (eds) Microbial diversity in ecosystem sustainability and biotechnological applications. Springer, Singapore, pp 565–591

Dubey G, Kollah B, Gour VK, Shukla AK, Mohanty SR (2016) Diversity of bacteria and archaea in the rhizosphere of bioenergy crop *Jatropha curcas*. 3 Biotech 6(2):1–10

Fleischman D, Kramer D (1998) Photosynthetic rhizobia. Biochim Biophys Acta Bioenerg 1364(1): 17–36

Galloway JN, Townsend AR, Erisman JW, Bekunda M, Cai Z, Freney JR, Martinelli LA, Seitzinger SP, Sutton MA (2008) Transformation of the nitrogen cycle: recent trends, questions, and potential solutions. Science 320(5878):889–892

Großkopf R, Stubner S, Liesack W (1998) Novel euryarchaeotal lineages detected on rice roots and in the anoxic bulk soil of flooded rice microcosms. Appl Environ Microbiol 64(12):4983–4989

Hecht U, Mohr H (1990) Factors controlling nitrate and ammonium accumulation in mustard (*Sinapis alba*) seedlings. Physiol Plant 78(3):379–387

He Y, Hu W, Ma D, Lan H, Yang Y, Gao Y (2017) Abundance and diversity of ammonia-oxidizing archaea and bacteria in the rhizosphere soil of three plants in the Ebinur Lake wetland. Canadian J Microbiol 63(7):573–582

Hemerly A (2016) Genetic controls of biomass increase in sugarcane by association with beneficial nitrogen-fixing bacteria. In: Plant and animal genome XXIV conference. Plant and animal genome.

Herridge DF, Peoples MB, Boddey RM (2008) Global inputs of biological nitrogen fixation in agricultural systems. Plant Soil 311(1–2):1–18

Im YJ, Ji M, Lee A, Killens R, Grunden AM, Boss WF (2009) Expression of *Pyrococcus furiosus* superoxide reductase in *Arabidopsis* enhances heat tolerance. Plant Physiol 151(2):893–904

Kaschuk G, Hungria M (2017) Diversity and importance of diazotrophic bacteria to agricultural sustainability in the tropics. In: De Azevedo JL, Quecine MC (eds) Diversity and benefits of microorganisms from the tropics. Springer, Cham, pp 269–292

Kaur B, Kaur G, Asthir B (2017) Biochemical aspects of nitrogen use efficiency: an overview. J Plant Nutr 40(4):506–523

Knief C, Delmotte N, Chaffron S, Stark M, Innerebner G, Wassmann R, Von Mering C, Vorholt JA (2012) Metaproteogenomic analysis of microbial communities in the phyllosphere and rhizosphere of rice. ISME J 6(7):1378–1390

Leigh JA (2000) Nitrogen fixation in methanogens: the archaeal perspective. Curr Issues Mol Biol 2(4):125–131

Liu Y, Li H, Liu QF, Li YH (2015) Archaeal communities associated with roots of the common reed (*Phragmites australis*) in Beijing Cuihu wetland. World J Microbiol Biotechnol 31(5):823–832

Ma M, Du H, Sun T, An S, Yang G, Wang D (2019) Characteristics of archaea and bacteria in rice rhizosphere along a mercury gradient. Sci Total Environ 650:1640–1651

MacLeod F, Kindler GS, Wong HL, Chen R, Burns BP (2019) Asgard archaea: diversity, function, and evolutionary implications in a range of microbiomes. AIMS Microbiol 5(1):48

Meena VS, Meena SK, Verma JP, Kumar A, Aeron A, Mishra PK, Bisht JK, Pattanayak A, Naveed M, Dotaniya M (2017) Plant beneficial rhizospheric microorganism (PBRM) strategies to improve nutrients use efficiency: a review. Ecol Eng 107:8–32

Mehta MP, Butterfield DA, Baross JA (2003) Phylogenetic diversity of nitrogenase (nifH) genes in deep-sea and hydrothermal vent environments of the Juan de Fuca Ridge. Appl Environ Microbiol 69(2):960–970

Mendes R, Garbeva P, Raaijmakers JM (2013) The rhizosphere microbiome: significance of plant beneficial, plant pathogenic, and human pathogenic microorganisms. FEMS Microbiol Rev 37(5):634–663

Miller A, Cramer M (2005) Root nitrogen acquisition and assimilation. Plant Soil 274(1):1–36

Mitsui A, Kumazawa S, Takahashi A, Ikemoto H, Cao S, Arai T (1986) Strategy by which nitrogen-fixing unicellular cyanobacteria grow photoautotrophically. Nature 323(6090):720

Moissl-Eichinger C, Pausan M, Taffner J, Berg G, Bang C, Schmitz RA (2018) Archaea are interactive components of complex microbiomes. Trend Microbiol 26(1):70–85

Mowafy MA, Agha MS, Haroun SA, Abbas MA, Elbalkini M (2022) Insights in nodule-inhabiting plant growth promoting bacteria and their ability to stimulate *Vicia faba* growth. Egypt J Basic Appl Sci 9(1):51–64

Müller H, Berg C, Landa BB, Auerbach A, Moissl-Eichinger C, Berg G (2015) Plant genotype-specific archaeal and bacterial endophytes but similar *Bacillus* antagonists colonize Mediterranean olive trees. Front Microbiol 6:138

Naitam MG, Kaushik R (2021) Archaea: an agro-ecological perspective. Curr Microbiol 78(7): 2510–2521

Oburger E, Schmidt H (2016) New methods to unravel rhizosphere processes. Trends Plant Sci 21(3):243–255

Oliveira MN, Santos TM, Vale HM, Delvaux JC, Cordero AP, Ferreira AB, Miguel PS, Tótola MR, Costa MD, Moraes CA (2013) Endophytic microbial diversity in coffee cherries of *Coffea arabica* from southeastern Brazil. Can J Microbiol 59(4):221–230

Pandey A, Nigam P, Soccol CR, Soccol VT, Singh D, Mohan R (2000) Advances in microbial amylases. Biotechnol Appl Biochem 31(2):135–152

Patil J, Suryawanshi P, Bajekal S (2016) Siderophores of haloalkaliphilic archaea from Lonar lake, Maharashtra, India. Cur Sci 111(4):621–623

Pires AC, Cleary DF, Almeida A, Cunha Â, Dealtry S, Mendonça-Hagler LC, Smalla K, Gomes NC (2012) Denaturing gradient gel electrophoresis and barcoded pyrosequencing reveal

unprecedented archaeal diversity in mangrove sediment and rhizosphere samples. Appl Environ Microbiol 78(16):5520–5528
Peoples M, Freney J, Mosier A (1995a) Minimizing gaseous losses of nitrogen. In: Bacon PE (ed) Nitrogen fertilization in the environment. Marcel Dekker, New York, pp 565–602
Peoples M, Herridge D, Ladha JK (1995b) Biological nitrogen fixation: an efficient source of nitrogen for sustainable agricultural production? Plant Soil 174(1-2):3–28
Prudence SM, Worsley S, Balis L, Murrell JC, Lehtovirta-Morley L, Hutchings ML (2019) Root-associated archaea: investigating the niche occupied by ammonia oxidising archaea within the wheat root microbiome. Access Microbiol 1(1A):253
Pump J, Pratscher J, Conrad R (2015) Colonization of rice roots with methanogenic archaea controls photosynthesis-derived methane emission. Environ Microbiol 17(7):2254–2260
Qi L, Ma Z, Chang SX, Zhou P, Huang R, Wang Y, Wang Z, Gao M (2021) Biochar decreases methanogenic archaea abundance and methane emissions in a flooded paddy soil. Sci Total Environ 752:141958
Rosswall T (1983) The nitrogen cycle. In: Bolin B, Cook RB (eds) The major biogeochemical cycles and their interactions. Wiley, New York, pp 21–46–50
Simon HM, Jahn CE, Bergerud LT, Sliwinski MK, Weimer PJ, Willis DK, Goodman RM (2005) Cultivation of mesophilic soil crenarchaeotes in enrichment cultures from plant roots. Appl Environ Microbiol 71(8):4751–4760
Song GC, Im H, Jung J, Lee S, Jung MY, Rhee SK, Ryu CM (2019) Plant growth-promoting archaea trigger induced systemic resistance in *Arabidopsis thaliana* against *Pectobacterium carotovorum* and *Pseudomonas syringae*. Environ Microbiol 21(3):940–948
Sun W, Shahrajabian MH, Cheng Q (2021) Nitrogen fixation and diazotrophs—a review. Rom Biotechnol Lett 26:2834–2845
Taffner J, Erlacher A, Bragina A, Berg C, Moissl-Eichinger C, Berg G (2018) What is the role of Archaea in plants? New insights from the vegetation of alpine bogs. MSphere 3(3):e00122-00118
Taffner J, Cernava T, Erlacher A, Berg G (2019) Novel insights into plant-associated archaea and their functioning in arugula (*Eruca sativa* Mill.). J Adv Res 19:39–48
Turk KA, Rees AP, Zehr JP, Pereira N, Swift P, Shelley R, Lohan M, Woodward EMS, Gilbert J (2011) Nitrogen fixation and nitrogenase (nifH) expression in tropical waters of the eastern North Atlantic. ISME J 5(7):1201
Wang S, Zheng Z, Zou H, Li N, Wu M (2019) Characterization of the secondary metabolite biosynthetic gene clusters in archaea. Comput Biol Chem 78:165–169
Wassermann B, Cernava T, Müller H, Berg C, Berg G (2019) Seeds of native alpine plants host unique microbial communities embedded in cross-kingdom networks. Microbiome 7(1):1–12
Wenli S, Shahrajabian MH, Cheng Q (2021) Archaea, bacteria and termite, nitrogen fixation and sustainable plants production. Not Bot Horti Agrobot Cluj Napoca 49(2):12172–12172
White RH (1987) Indole-3-acetic acid and 2-(indol-3-ylmethyl) indol-3-yl acetic acid in the thermophilic archaebacterium *Sulfolobus acidocaldarius*. J Bacteriol 169(12):5859–5860
Woese CR, Fox GE (1977) Phylogenetic structure of the prokaryotic domain: the primary kingdoms. Proc Natl Acad Sci U S A 74(11):5088–5090
Wu F, Speth DR, Philosof A, Crémière A, Narayanan A, Barco RA, Connon SA, Amend JP, Antoshechkin IA, Orphan VJ (2022) Unique mobile elements and scalable gene flow at the prokaryote–eukaryote boundary revealed by circularized *Asgard archaea* genomes. Nat Microbiol 7(2):200–212
Yadav AN, Sharma D, Gulati S, Singh S, Dey R, Pal KK, Kaushik R, Saxena AK (2015) Haloarchaea endowed with phosphorus solubilization attribute implicated in phosphorus cycle. Sci Rep 5(1):1–10
Yadav AN, Verma P, Kaushik R, Dhaliwal H, Saxena A (2017) Archaea endowed with plant growth promoting attributes. EC Microbiol 8(6):294–298
Yadav AN, Gulati S, Sharma D, Singh RN, Rajawat MVS, Kumar R, Dey R, Pal KK, Kaushik R, Saxena AK (2019) Seasonal variations in culturable archaea and their plant growth promoting attributes to predict their role in establishment of vegetation in Rann of Kutch. Biologia 74(8): 1031–1043

Part II
Plant Growth Promotion: Exploring Benefits

Chapter 7
Root Nodule Bacteria-Rhizobia: Exploring the Beneficial Effects on Non-legume Plant Growth

Magdalena Knežević, Aneta Buntić, Dušica Delić, and Olivera Stajković-Srbinović

Abstract For the improvement of crop productivity in a sustainable manner, inoculation of legume crops by root nodule endophytic bacteria is an inevitable approach. Recently, the research regarding the use of rhizobial bacteria to improve growth, yield, nutrient composition and quality of non-legume plants has been augmented. Plant growth promoting characteristics of rhizobia (production of siderophores, indole-3-acetic acid, 1-aminocyclopropane-1-carboxylic acid deaminase and the ability to solubilize inorganic phosphate) have been characterized as beneficial in the production of both legume and non-legume plants. In addition, the ability of rhizobia to promote non-legume growth under unfavourable environmental conditions, as well as its biocontrol potential (antifungal and nematicidal), also qualified these bacteria for the application in the contemporary production of non-legumes. The efficiency of inoculum and harmlessness of native rhizobia to the environment are some of the major benefits of using rhizobial inoculants to promote growth and increase the yield of non-legumes. This chapter will summarize recent research regarding the beneficial effects of rhizobia belonging to *Rhizobium*, *Bradyrhizobium*, *Mesorhizobium* and *Ensifer* (*Sinorhizobium*) to non-legume plants, by highlighting rhizobial mechanisms responsible for the promotion of plant growth, biocontrol characteristics and interactions between rhizobia and non-legumes.

Keywords Rhizobia · Root nodule bacteria · Non-legumes · Plant growth promotion · Biocontrol

M. Knežević (✉) · A. Buntić · D. Delić · O. Stajković-Srbinović
Institute of Soil Science, Belgrade, Serbia
e-mail: magdalena.knezevic@soilinst.rs

© The Author(s), under exclusive license to Springer Nature Singapore Pte Ltd. 2022

D. K. Maheshwari et al. (eds.), *Nitrogen Fixing Bacteria: Sustainable Growth of Non-legumes*, Microorganisms for Sustainability 36,
https://doi.org/10.1007/978-981-19-4906-7_7

7.1 Introduction

Soil bacteria having the ability to induce highly specialized structures on the roots of leguminous plants (nodules), where the fixation of atmospheric nitrogen (N_2) is performed, are designated as rhizobia, the symbiotic nitrogen–fixing bacteria. Rhizobia are facultative microsymbionts that live freely in the soil (and at that stage do not fix N) or enter into symbiosis with the roots of specific legumes, whereby they fix N from the air (Garrity et al. 2004; Lindström and Mousavi 2019). Establishing a symbiosis between rhizobia and legumes is a highly specific reaction, where only certain types of rhizobia can effectively nodulate specific legumes. Nodules provide a favourable environment for bacteria where they are protected from adverse environmental conditions (van Rhijn and Vanderleyden 1995). Plants supply bacteria with nutrients (predominantly dicarboxylic acids), while in return they receive N from the bacteria (Lodwig and Poole 2003; O'Gara and Shanmugam 1976). Rhizobia (microsymbionts) fix N only in symbiosis with plants (macrosymbionts). The symbiotic community of legumes and rhizobia participates with 20% of the total biological N fixation (BNF). Besides symbiotic bacteria in the root nodules, a variety of non-rhizobial taxa have been isolated from different legumes, including strains from genera: *Agrobacterium*, *Bacillus*, *Curtobacterium*, *Enterobacter*, *Erwinia*, *Mycobacterium*, *Paenibacillus*, *Pseudomonas*, *Phyllobacterium*, *Ochrobactrum*, *Sphingomonas* and others.

In addition to N fixation, rhizobial species have been shown to associate with the roots of non-leguminous plants without nodule formation and improve their growth by one or more direct or indirect mechanisms (Mehboob et al. 2009). Therefore, the rhizobia can be considered like other plant growth-promoting rhizobacteria (PGPR). Some of the direct mechanisms by which rhizobia promote the plant growth are the production of plant hormones (auxin, cytokinin, ethylene) or other chemical substances (lipo-chito-oligosaccharides or lumichrome), phosphate solubilization, improving the uptake of plant nutrients by altering root morphology, siderophore production, hydrogen cyanide (HCN) production, etc. (Antoun and Prevost 2005; Etesami 2022; Katiyar et al. 2021; Vessey 2003). Indirect mechanisms include improving the plant growth through biocontrol, competition with pathogens, or induction of resistance mechanisms in plants (Agarwal et al. 2017; Gopalakrishnan et al. 2015; Maheshwari et al. 2019).

The application of rhizobia-based inoculants in the promotion of legumes growth, as well as the overall crop health and composition has been extensively used (Buntić et al. 2019a; Knežević et al. 2022; Stajković-Srbinović et al. 2020). Recently, the attempts to increase seed germination, yield, and quality, as well as to promote plant growth of legume and non-legume plants have been made by the application of non-rhizobial PGP bacteria (Abulfaraj and Jalal 2021; Buntić et al. 2019b; Knežević et al. 2021b; Li et al. 2021; Rakić et al. 2021). The main limitation of using PGPR in promoting the plant growth is regarded to its capability to associate with targeted plants (Mayer et al. 2019). However, as PGP rhizobia can also colonize roots of non-legumes, their application is lately being directed to promoting the growth of

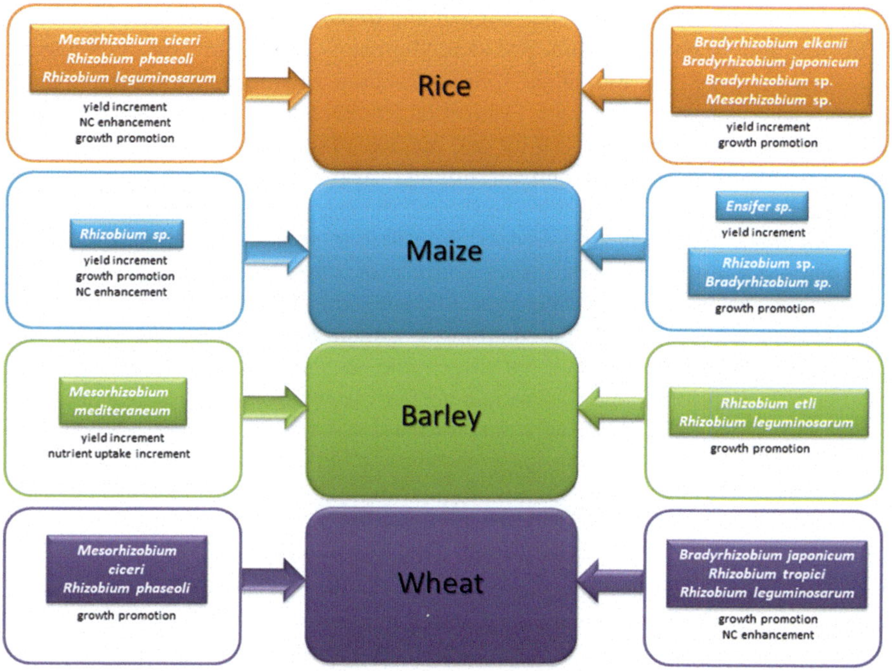

Fig. 7.1 Most common rhizobial species used for growth promotion of rice, maize, barley and wheat. *NC:* nutrient composition

cultivars such as pepper, mustard, lettuce and tomato (Flores-Félix et al. 2013). In addition, rice, maize, barley and wheat are currently one of the most represented crops for determining the beneficial effects of rhizobial inoculation on non-legume plants (Fig. 7.1).

Some of the major benefits of using rhizobial inoculants to promote growth and yield of non-legumes are reflected in the efficiency of inoculum preparation (as rhizobia have been extensively used in legume crops production), as well as in the harmlessness of native rhizobia to the environment (Antoun et al. 1998). In addition, the harmlessness of rhizobia is also important as most of the non-legumes treated with these PGPR are being consumed raw, and treatment with potentially infectious bacteria could cause infections in humans and animals. Regarding the consumption of vegetables, it is of great importance to use exclusive bacterial inoculums consisted of non-pathogenic bacteria, which do not pose any adverse effect on human health. As rhizobial biofertilizers have been widely used for legume inoculation, their safety has been comprehensively studied and proven (García-Fraile et al. 2012). Similar precautions need to be taken in the case of vegetables which are generally being consumed fresh, such as lettuce, spinach, carrots, tomatoes, peppers or radishes (Flores-Félix et al. 2013; García-Fraile et al. 2012; Jiménez-Gómez et al. 2018; Menéndez et al. 2020; Verma et al. 2020).

To secure healthy food and environment and to reduce/avoid the use of chemical fertilizers and plant protection products, humanity needs to conserve natural resources for a healthy biological solution in crop production, considering both legume and non-legume plants. Rhizobial inoculants, used as plant growth enhancers, influence the soil fertility and plant productivity, etc. in a healthy way by improving crop nutrition and physiology, crop fungal protection, and stress tolerance.

This chapter will summarize recent and relevant research regarding the use of rhizobia such as *Rhizobium*, *Bradyrhizobium*, *Mesorhizobium* and *Ensifer* (*Sinorhizobium*) genera for promoting the non-legumes growth (both under optimal and unfavourable environmental conditions), their biocontrol potential against various phytopathogenes, mechanisms underlying in their activity, as well as the strategies for improving plant growth and development.

7.2 Plant Growth Promotion Rhizobial Traits

Root nodulating bacteria were widely studied for their symbiotic association with legumes and BNF capacity. However, some members of *Rhizobium*, *Bradyrhizobium*, *Mesorhizobium* and *Sinorhizobium* genera are capable of inducing plant growth and development through biochemical or physiological mechanisms other than N fixation (Boiero et al. 2007). They are capable of regulating plant growth through the production and release of various metabolites such as phytohormone, siderophores, ACC deaminase, antibiotics, HCN and others (Ahmad et al. 2019; Boiero et al. 2007; Chandra et al. 2007; Saghafi et al. 2019; Serova et al. 2019; Vargas et al. 2009).

It is considered that the most important direct PGP mechanism, besides BNF, is the synthesis of phytohormones (plant growth-regulating compounds) (Zahir et al. 2004). The production of auxins, cytokinins, gibberellins (GA), and abscisic acid (ABA) is a relatively common characteristic of rhizobia (Bhattacharjee et al. 2012; Boiero et al. 2007; Ferreira et al. 2020; Ullah et al. 2017b). It is known that hormone-producing rhizobial strains can improve the growth of plants by inoculating seeds or roots (Chandra et al. 2007; Humphry et al. 2007; Mishra et al. 2006; Yanni et al. 2001). The ability of root nodulating bacteria to produce plant hormones is different, and the type and amount of hormone vary with the type of genera (Boiero et al. 2007; Humphry et al. 2007). One strain can be capable to produce one or more different phytohormones (indole-3-acetic acid (IAA), GA_3, zeatin, ABA, salicylic acid (SA), jasmonic acid (JA)) as stated by Boiero et al. (2007). Masciarelli et al. (2014) showed for the first time that *Bradyrhizobium japonicum* was capable to produce five different phytohormones simultaneously: indole-3-acetic acid (IAA), GA_3, ABA, salicylic acid (SA) and jasmonic acid (JA).

Besides other bacteria, algae and fungi, rhizobial species are evidenced as one of the major IAA producers (the most important auxin-type plant hormones) (Boiero et al. 2007; Chi et al. 2005; Pandey and Maheshwari 2007). Indole-3-acetic acid (a by-product of L-tryptophan metabolism) is involved in the control of cell division, tissue differentiation, while such rhizobia can induce an increase of amino acids within the invaded plant tissues (Jaiswal et al. 2021). Among various root nodulating bacteria, *Sinorhizobium meliloti* showed the high level of IAA production (612 μg/ml), when adding 0.1% tryptophan to the growth medium (Kanaan and Al-Barhawee 2021). In addition, *Bradyrhizobium* and *Rhizobium* strains have also been shown to be efficient IAA producers. *Bradyrhizobium japonicum* produced about 41 $\mu g\ ml^{-1}$ and 61.25 $\mu g\ ml^{-1}$ of IAA, respectively, under standard assay condition after 48 h and 72 h in presence of yeast extract as N source (Kiruthika and Arunkumar 2021). With increasing L-tryptophan concentrations, *B. japonicum* reached the maximum production of 78.33 of IAA $\mu g\ ml^{-1}$. Further, Datta and Basu (2000) found that *Rhizobium* sp. isolated from the root nodules of pigeon pea (*Cajanus cajan*), produced high amounts of IAA (99.7 $\mu g\ ml^{-1}$) during growth in basal medium supplemented with L-tryptophan.

Gibberellins (GA) are a type of phytohormones produced by bacteria and fungi which have a role in plant development and can manipulate the host plant. Although the GA biosynthetic pathway in bacteria remains elusive, the presence of putative GA biosynthetic gene clusters/operons has been documented in rhizobial species (Nett 2017). The first report on gibberellin synthesis in rhizobial species was given by Atzorn et al. (1988) in *Rhizobium phaseoli*. The manipulation of legume plant host by the expression of GA operons reflects mostly during the symbiosis, by increasing the size, number and development of root nodules and by bifurcation of nodule meristem, while in the case of non-legumes the GA production allows rhizobia a selective advantage over other bacteria (McAdam et al. 2018; Nett 2017; Serova et al. 2019). Additionally, it has been observed that the application of gibberellins in combination with *Rhizobium* inoculant has a synergistic effect on the growth, yield and nutrient content of chickpea (Rafique et al. 2021).

Another significant metabolite produced by rhizobia is the enzyme 1-aminociclopropane-1-carboxylase (ACC) deaminase. This enzyme reduces the levels of ethylene (which is elevated in plant tissues during stress conditions) in higher plants by converting a precursor of ethylene (ACC) into ammonia and α-ketobutyrate (Orozco-Mosqueda et al. 2019). The first report documenting the presence of ACC deaminase in *Rhizobium* strains (*R. leguminosarum* bv. *viciae* and *R. hedysari*) was given by Ma et al. (2003). In addition, ACC deaminase producers were found among the strains of *Ensifer* (*Sinorhizobium*), *Mesorhizobium* and *Bradyrhizobium* (Ma et al. 2003; Piromyou et al. 2017; Saghafi et al. 2019). Similarly, the production of higher levels of rhizobial trehalose (non-reducing disaccharide) has been linked with higher rates of survival under unfavourable environmental conditions, such as drought or saline stress (Sarapat et al. 2020). In addition, rhizobia also produce bacterial metabolites which can act as inducers of plant growth (lumichrome and riboflavin) or as a barrier against pathogens (thiamine, ethylene, brassinosteroid, biotin, niacin and ascorbic acid) (Jaiswal et al. 2021; Nouwen et al. 2021).

Iron is important in the process of photosynthesis, but in general, the amount of available Fe^{3+} ions is low in an aerobic environment, both at low and high soil pH

(Masalha et al. 2000). Siderophores are known as small, chelating agents of high affinity for iron, which, secreted by microorganisms, form soluble complexes with Fe^{3+} that can be introduced into cells by active transport (Neilands 1993). For example, the proportion of root nodule siderophore-producing strains appears to increase in iron-stressed soils, while the total population of root nodule bacteria remains unchanged. Root nodulating bacteria are capable to produce a number of siderophores: carboxylates (rhizobactin, citrate and anthranilate), catechols, dihydroxamates (rhizobactin 1021 and vicibactin) and trihydroxamates (Carson et al. 2000; Datta and Chakrabartty 2014; Storey et al. 2006). The production of catechols or carboxylates is sporadic and may be strain-specific. Dihydroxamate siderophores are mainly a characteristic of *Sinorhizobium* species, while trihydroxamates of *Rhizobium* species (Carson et al. 2000). The first report about the capability of *Rhizobium* species to produce a dihydroxamate type siderophore was given by Storey et al. (2006). They showed that *R. leguminosarum* is capable to produce schizokinen which has a similar structure with rhizobactin 1021. 'Rhizobactin' is the first chemically characterized siderophore produced by a strain of *Sinorhizobium meliloti* (Storey et al. 2006). In addition, Carson et al. (2000) observed that bradyrhizobia could not produce any types of these siderophores.

Phosphorus (P) is one of the main macronutrients necessary for plant growth and development. However, in the soil, it is often found in an insoluble form, as $CaPO_4$ in alkaline soils, or as phosphates of Al and Fe in acid soils. The ability to solubilize phosphates (organic and inorganic) to a form available to plants is an important trait of many PGP bacteria, including root nodulating bacteria (Abd-Alla 1994; Halder and Chakrabarty 1993; Peix et al. 2001; Rodríguez et al. 2006). In both plants and microorganisms, the main mechanism of P solubilization is the excretion of hydrogen ions, the production of organic acids, and the biosynthesis of acid phosphatase (Arcand and Schneider 2006; Bais et al. 2006). The production of inorganic and organic acids lowers the pH of the soil solution, whereby phosphorus is released from $CaPO_4$. Enzymatic solubilization of phosphates takes place during the mineralization of organic phosphates. The main advantage of using root-nodulating bacteria as a phosphate-solubilizing microorganism is their beneficial nutritional effect, resulting both phosphate mobilization and N fixation (Sridevi and Mallaiah 2009). For instance, *Mesorhizobium mediterraneum* and *Mesorhizobium ciceri* are known for their high phosphate-solubilizing efficiency (Rivas et al. 2006). Sridevi and Mallaiah (2009) reported that *Rhizobium* isolated from *Cassia absus* reached maximum solubilization of 620 μg ml^{-1}.

The production of various secondary metabolites by rhizobia which are toxic to other microorganisms (antibiosis) has been documented. The production of antibiotics (bacteriocines) (Bardin et al. 2004; Chandra et al. 2007; Deshwal et al. 2003b), hydrogen cyanide (HCN) (Arfaoui et al. 2006; Chandra et al. 2007), siderophores which inhibit fungal pathogens (Deshwal et al. 2003a; Chandra et al. 2007) and mycolytic enzymes (chitinases and β-1,3-glucanases involved in fungal cell wall lysis) (Kumar et al. 2011) were also recorded.

Induced systemic resistance (ISR) is a state of plant stimulated to elicit its defence response, through the production of plant defence enzymes, phenolics, flavonoids, or other phytoalexins (Das et al. 2017). Phenolic compounds may act as structural barriers, activators of plant defence genes, and modulators of pathogenicity (Ramos et al. 1997). Numerous rhizobial species were able to induce systemic resistance by producing bio-stimulatory agents (Arfaoui et al. 2006; Rabie 1998). It was demonstrated that individual cellular components of the rhizobia induce ISR: lipopolysaccharides, flagella, cyclic lipopeptides, homoserine lactones, acetoin and butanediol (Lugtenberg and Kamilova 2009).

Moreover, rhizobial bacteria can be useful in the process of improvement in crop production on marginal soils (Naseer et al. 2019). These are some examples that bioremediation with the usage of rhizobial strains, tolerant to heavy metals and with some PGP traits, can be a cost-effective substitution of seed fertilizer used in conventional remediation. Stan et al. (2011) suggested the reason for the survival of *Rhizobium* strains in a heavy metal-polluted area is the alteration of different genes, including those involved in the symbiotic process of N_2 fixation.

7.3 Potential of Rhizobia to Promote the Growth of Different Non-legumes

Plant growth-promoting mechanisms of species belonging to *Rhizobium*, *Bradyrhizobrium*, *Mesorhizobium* and *Ensifer* genus have been considered by various authors (Gopalakrishnan et al. 2015). Some of the studies covering the use of rhizobia in a growth promotion of different kinds of non-legume crops are presented in the Table 7.1.

7.3.1 Rhizobia: Rice, Wheat and Barley

There is a growing number of rhizobial species that are being discovered and introduced into the microbiome of rice (Nadarajah 2017). Inoculation of rice with *R. leguminosarum* (isolated from lentil), *M. ciceri* (isolated from chickpea) and *R. phaseoli* (isolated from mung bean) enhanced the yield (paddy yield, plant biomass and straw dry biomass), growth parameters (number of tiller and grains panicle and plant height) and nutrient composition (increase in nitrogen, potassium and phosphorus) of rice, in the potted soil (Husssain et al. 2009). Furthermore, promotion of rice seedling emergence and establishment of early seedling development has been achieved by inoculation with rhizobial strains (Baset Mia et al. 2012). Hahn et al. (2016) found that inoculation of rice by *Mesohizobium* (or *Mesorhizobium* in combination with *Azospirillum*) promotes the growth of wetland rice. Inoculation of rice by ACC-producing *Bradyrhizobium* sp. can reduce ethylene synthesis, indicating its potential use in enhancing the rice growth and grain yield (Sarapat et al. 2020). Recently, Padukkage et al. (2021) demonstrated that

Table 7.1 Recent relevant research regarding the use of rhizobia in the cultivation of non-legumes

Rhizobia	Non-legume species	Reference
Rhizobium leguminosarum	Wheat	Mehboob et al. (2011)
	Maize	Ali (2017)
	Rice	Bhattacharjee et al. (2012)
	Wheat	Roque et al. (2021)
	Barley	Santillana et al. (2012)
	Wheat	Ullah et al. (2017a)
Rhizobium selenitireducens	*Arabidopsis thaliana*	Mayer et al. (2019)
	Lettuce	
	Basil	
	Bok choy	
Rhizobium spp.	Cotton	Qureshi et al. (2019)
	Tomato (var. "cherry")	García-Fraile et al. (2012)
	Canola	Saghafi et al. (2018)
	Pepper (var. "verde italiano")	García-Fraile et al. (2012)
	Carrot	Flores-Félix et al. (2013)
Rhizobium spp.	Maize	Qureshi et al. (2013)
	Maize	Cavalcanti et al. (2020)
	Sunflower	Ullah et al. (2017b)
	Sunflower	Ferreira et al. (2020)
	Potato	Naqqash et al. (2016)
Rhizobia	Oat	Silva et al. (2020)
	Maize	Silva et al. (2020)
	Wheat (var. "Siran")	Adnan et al. (2014)
Rhizobium phaseoli	Maize	Hussain et al. (2014)
	Rice	Husssain et al. (2009)
	Maize	Hussain et al. (2016a, b)
	Wheat	Mehboob et al. (2011)
Rhizobium etli	Barley	Santillana et al. (2012)
	Potato	Gervasio et al. (2019)
Rhizobium pusense	Onion	Afify et al. (2019)
Rhizobium rubi	Broccoli	Yildirim et al. (2011)
Rhizobium symbiovar *viciae*	Wheat	Bartoli et al. (2020)
Rhizobium tropici	Wheat	Roque et al. (2021)
Rhizobium radiobacter syn. *Agrobacterium tumefaciens*	Lettuce	Verma et al. (2020)
Rhizobium laguerreae	Spinach	Jiménez-Gómez et al. (2018)
Rhizobium alamii	Rapeseed	Tulumello et al. (2021)
Mesorhizobium ciceri	Rice	Husssain et al. (2009)
	Wheat	Ullah et al. (2017a)
	Maize	Hussain et al. (2014)
	Tomato	Menéndez et al. (2020)
	Wheat	Hussain et al. (2018)
	Maize	Ali (2017)

(continued)

Table 7.1 (continued)

Rhizobia	Non-legume species	Reference
	Wheat	Mehboob et al. (2011)
Mesorhizobium tamadayense	Tomato	Menéndez et al. (2020)
Mesorhizobium loti	*Arabidopsis*	Poitout et al. (2017)
Bradyrhizobium japonicum	Wheat	Roque et al. (2021)
	Rice	Padukkage et al. (2021)
Bradyrhizobium elkani	Potato	Gervasio et al. (2019)
	Rice	Padukkage et al. (2021)
Bradyrhizobium sp.	Rice	Sarapat et al. (2020)
	Maize	Cavalcanti et al. (2020)
Ensifer meliloti	Maize	Ibarra-Galeana et al. (2017)
Ensifer spp.	Maize	Chen et al. (2020a)

B. japonicum had the best PGP effect on rice growth, under the laboratory conditions, while *B. elkanii* was most efficient in the field and increased overall rice yield. Moreover, IAA and ACC deaminase producing *R. leguminosarum* bv. *trifolii* (isolated from root nodules of *Trifolium alexandrinum* L.) promoted the growth of rice, in terms of biomass, root branching and N content (Bhattacharjee et al. 2012).

On the other hand, Roque et al. (2021) showed that co-inoculation of wheat by rhizobia, such as *B. japonicum*, *R. tropici* and *R. leguminosarum* (isolated from legumes), can improve growth and composition of wheat grain. Strains of rhizobia (determined as *R. phaseoli*, *M. ciceri* and *R. leguminosarum*) isolated from root nodules of mung bean, chickpea and lentil, respectively, with the ability to produce siderophores, IAA, chitinases, EPS and to solubilize inorganic P, showed the potential to promote the growth of wheat (Mehboob et al. 2011). Furthermore, Mehboob et al. (2011) showed that although these rhizobial treatments increased wheat growth parameters (plant height, number of tillers, straw yield, grain yield, root length, as well as the content of P and K), some of the applied strains achieved values which were not statistically different from the uninoculated control. Adnan et al. (2014) reported that inoculation by rhizobial strains (isolated from lentil, pea and chickpea) significantly increased the number of tillers and grains, plant height and grain weight of wheat. In addition, authors indicated that rhizobia isolated from peas had better PGP effect on wheat (except for plant height), in comparison to the other inoculum, while isolates from chickpea showed increase in the overall plant parameters (Adnan et al. 2014). Bearing in mind that the entrance and interaction of rhizobia with the non-legumes is often conditioned by plant species and environmental factors, this may be due to the occurrence of better interactions of different rhizobial species with different non-legumes (Nadarajah 2017). In addition, Bartoli et al. (2020) indicated that the level of wheat growth promotion induced by *R. leguminosarum* symbiovar *viciae* is conditioned by the level of root colonization by these rhizobia.

Earlier, Peix et al. (2001) examined that *M. mediterraneum* enhance the growth (increased dry matter) and nutrient uptake (increased nitrogen, potassium, calcium and magnesium content) in barley (*Hordeum vulgare*) plants. Treatment of barley with this rhizobial strain increased phosphorus content up to 125% in soil enriched with insoluble phosphates, indicating its effectiveness in mobilization and solubilization of phosphorus (Peix et al. 2001). Furthermore, the growth of barley was also improved by *R. leguminosarum* and *R. etli* under greenhouse conditions (dry matter increase up to 37%), while these strains also showed the ability to antagonize phytopathogenic fungi *Alternaria solani* and *Fusarium* sp. in vitro (Santi et al. 2013) causing diseases in barley.

7.3.2 Maize

Application of rhizobia in growth promotion of maize has been studied in different researches (Ali 2017; Hussain et al. 2014, 2016a, b; Qureshi et al. 2013; Silva et al. 2020). Silva et al. (2020) showed that inoculation of maize and oat (*Avena sativa*) by native rhizobia isolated from *Desmodium incanum* can improve the growth of both plant species. They suggested that native phosphate solubilizing and IAA-producing rhizobia could be used for increasing the yield of these cultures, especially as native rhizobia may express a better adaptability under disrupted soil conditions (Silva et al. 2020). Qureshi et al. (2013) revealed that inoculation of maize by IAA-producing *Rhizobium* sp. isolated from chickpea, berseem and lentil can improve the growth, nitrogen, phosphorus and chlorophyll content, as well as the transpiration and photosynthetic rate in maize. Cavalcanti et al. (2020) showed that IAA and siderophore producing and calcium phosphate solubilizing rhizobia, such as *Bradyrhizobium sp.* and *Rhizobium sp.*, have potential to promote the growth of maize at the same level as the commercial strain of *Azospirillum brasilense*. Chen et al. (2020a) showed that inoculation of PGP *Ensifer* sp. (isolated from the maize rhizosphere) significantly increased grain yield of maize (by 29%, in comparison to the untreated control).

7.3.3 Lettuce and Spinach

Few leafy vegetables also showed beneficial effect of rhizobia applied as biofertilizer. Verma et al. (2020) showed that *R. radiobacter* syn. *Agrobacterium tumefaciens* (isolated from root nodules of *Leucaena leucocephala*) with N-fixing, Zn and P solubilization activity, IAA, siderophores and EPS producing ability, improved nutritional content and growth parameters of lettuce (*Lactuca sativa*) in saline soil. The parameters such as antioxidant activity, total carbohydrate, chlorophyll, flavonoid, nitrate, total protein, phosphate and zinc content were improved after treatment of lettuce with *R. radiobacter*, in comparison to the chemical fertilizer, vermicompost and farmyard manure, under field conditions (Verma et al. 2020). Earlier, Chabot et al. (1996) suggested that P solubilization activity of

R. leguminosarum bv. *phaseoli* is responsible for the promotion of lettuce and maize growth in moderately or very fertile soils. Inoculation of lettuce and canola (*Brassica campestris*) with *R. leguminosarum* induced significant promotion of growth of both plant species. The production of IAA and cytokinin by these rhizobia could be responsible for the direct growth promotion of leafy vegetable and oil seed crops, as bacterial tryptophan and adenosine mutants did not promote the plant growth, in comparison to the wild-type rhizobial strain (Noel et al. 1996).

Inoculation by *R. laguerreae* significantly increased yield and growth parameters (leaf number, size and weight, and chlorophyll and N content) of spinach (*Spinacia oleracea*), suggesting that its PGP mechanisms (phosphate solubilization activity, IAA and siderophores production, as well as the production of cellulase and the ability to form biofilm) could be the cause of its beneficial effect on spinach (Jiménez-Gómez et al. 2018).

7.3.4 Vegetables

Rhizobial strains with various direct and indirect PGP attributes, such as siderophores, ACC deaminase and IAA production, showed the ability to promote the growth of pepper and tomato by increasing the seedling biomass, number of flowers and fruits, as well as the weight of fruits in the time of harvest (an increase of pepper fruit fresh weight of 30%, in comparison to the uninoculated control) (García-Fraile et al. 2012). The authors also suggested that IAA and siderophores production ability of these rhizobial strains could indicate their potential to assist in the iron acquisition. Menéndez et al. (2020) observed that *M. ciceri* and *M. tamadayense* (with the ability to produce IAA, siderophores, cellulases, to solubilize phosphate and potassium and to form biofilm) had positive impact of tomato seedling growth and development.

Gervasio et al. (2019) concluded that *R. etli* and *B. elkani* positively influenced the growth parameters of potato (height of plant, leaf surface and the number of stems, leaves and stolons) in semi-controlled conditions. In addition, it has been shown that *Bradyrhizobium* sp. isolated from sweet potato actively expresses nitrogenase activity in planta, suggesting its potential to increase N content in this crop (Terakado-Tonooka et al. 2012). Nitrogen-fixing and IAA-producing *Rhizobium* sp. showed a good potential to colonize and to maintain its density in the potato rhizosphere, which implies its potential use as potato biofertilizer (Naqqash et al. 2016).

Previously, Antoun et al. (1998) suggested that specific bradyrhizobia could potentially be used for improving the growth of non-legumes, as they discovered that *B. japonicum* significantly increased the growth of radish in growth cabinets. Flores-Félix et al. (2013) demonstrated that rhizobia can colonize roots and promote the plant growth of carrot (*Daucus carota*) and lettuce, by increasing the dry weight of shoots and roots. Authors also suggested that the ability of these rhizobia to produce siderophores, IAA and to solubilize inorganic phosphates makes them suitable for biofertilizer formulation specified for non-legumes (Flores-Félix et al. 2013). The beneficial effect of *R. rubi* on the growth of broccoli was also recorded, with observed increase in yield (20.2%), chlorophyll content (14.0%) and macro and micro-nutrient composition, in comparison to the control (Yildirim et al. 2011). An increase in onion growth parameters (number of leaves and dry weight of plant) was observed 120 days after treatment by *R. pusence*, in comparison to the recommended dose of chemical fertilizer (Afify et al. 2019).

7.3.5 Other Non-legumes

Alami et al. (2000) showed that inoculation of sunflower by rhizobia isolated from dystric cambisol soil can cause a significant increase in shoot and root dry mass, both under normal and disrupted soil conditions (water stress). In addition, the authors observed that the used *Rhizobium* was a major producer of exopolysaccharides (EPS), and that the inoculation affected the soil macropore volume (Alami et al. 2000). Different *Rhizobium* sp. isolated from legume plants, such as lentil, mung bean, vegetable pea, berseem and chickpea, significantly increased growth of sunflower (Ullah et al. 2017b). These applied rhizobia showed the ability to produce IAA, while the most effective strain was isolated from mung bean, which also significantly increased the content of chlorophyll a, N and P, in comparison to the untreated plants. Ferreira et al. (2020) also showed that inoculation of sunflower by *Rhizobium* sp. positively affected the growth and sprouting of sugarcane under greenhouse conditions, probably due to the production of gibberellins, suggesting that this strain could be used as sugarcane inoculant.

Species belonging to *Rhizobiales,* capable to colonize *Arabidopsis thaliana* roots without negative effect on the plant growth, indicated the ability of rhizobia to actively interact with non-legume roots and to potentially affect its host niche (Garrido-Oter et al. 2018). Furthermore, *M. loti* induced promotion of shoot growth, stimulation of root hair elongation and inhibited primary root growth of *Arabidopsis* in vitro, thus the authors also concluded that auxin transport and signalling pathway is necessary for inhibition of primary root growth, while it has no effect on root hair elongation (Poitout et al. 2017).

Application of different strain of *Rhizobium* sp. to the cotton plants improved the plant physiological parameters, such as the number and weight of bolls, chlorophyll content, photosynthetic, transpiration and photo-active radiation rate, and promoted the plant growth (plant height) (Qureshi et al. 2019). *Rhizobium selenitireducens* improved the total length of roots of basil, bok choy and lettuce, grown in aquaponic conditions (Mayer et al. 2019).

7.3.6 Promotion of Non-legumes Growth Under Unfavourable Conditions

It has been shown that the N-fixing and plant growth-promoting rhizobia could enhance the growth of plants under different unfavourable environmental conditions (Granada et al. 2014; Knežević et al. 2021a). Several studies revealed that strains from genera *Rhizobium*, *Ensifer* and *Mesorhizobium* could tolerate low soil pH (Brígido and Oliveira 2012; Correa et al. 1999; Knežević et al. 2022; Stajković-Srbinović et al. 2015a, b). As the availability of essential nutrients required for plant growth is influenced by the physicochemical properties of the soil, mainly by the soil pH (Dinić et al. 2019; Maksimović et al. 2021), application of rhizobia that can improve nutrient composition under acid soil conditions is crucial. By secretion of rhizobial metabolites into the rhizosphere, microorganisms reduce the mobility and bioavailability of metals that become biodegradable and less toxic and benefit plants by sequestering heavy metals during the phytoremediation process (Delić et al. 2022) Also, heavy metal-tolerant *Bradyrhizobium*, *Rhizobium* and *Mesorhizobium* strains promoted the plant growth in soils with heavy metal elevated concentrations (Stajković-Srbinović et al. 2020; Sujkowska-Rybkowska et al. 2020).

Soil salinity is also a major factor in the sustainable plant growth and management. Worldwide, approximately 7% of the soils on Earth and 20% of the total arable areas are affected by salinity (Rizwan et al. 2015). Inoculation of different non-legumes with rhizobial strains had a positive effect on plant growth, especially on the plant height and shoot dry weight indices under salinity stress (Saghafi et al. 2018). Tulumello et al. (2021) showed that *Rhizobium alamii* GBV030 had a PGP effect on rapeseed growth, increasing its tolerance to water stress, probably involving its capacity to produce EPS, and other PGP traits. The inoculation of rice by *Bradyrhizobium* strain containing ACC deaminase, improved leaf relative water content, survival, recovery rates, and improved the crop yield in field conditions (Sarapat et al. 2020). The inoculation of seeds or roots of non-legumes with ACC deaminase-producing rhizobial strains has emerged as a new approach for diminishing stress-induced physiological changes in plants, as this rhizobial enzyme lowers the level of ethylene in developing seedling and plants (Katiyar et al. 2021; Saghafi et al. 2019).

7.3.7 Crop Rotation

Crop rotation of legumes and non-legumes can have positive effect on the soil quality (Tokhetova et al. 2021; Nurymova et al. 2020). By enriching the soil microbiome by rotation of inoculated legumes and non-legumes, the establishment of a beneficial rhizobial microbiome occurs, while the amount of soil N elevates due to the presence of N-fixing bacteria. Several studies indicated that besides direct inoculation of non-legume crops by rhizobia, the positive effect on their growth can also be observed as a consequence of subsequent planting after inoculated legume crops or vice versa (Delić et al. 2013). On the other hand, Piromyou et al. (2017) stated that stubbles of rice inoculated by *Bradyrhizobium* sp. can serve as inoculum in the rice-legume crop rotation system. Positive correlation was found between yield and N content in barley, wheat and canola, grown after peas inoculated by rhizobia (Lupwayi et al. 2004). Dubova et al. (2017) recorded an increase in onion leaf yield, when onions were grown as a subsequent crop after beans (inoculated by rhizobia in the combination with mycorrhiza fungi), in a pot experiment. Inoculants based on *R. leguminosarum* isolated from clover root nodules improved the growth of rice shoots and roots, as well as the seedling vigour, in the Egyptian Nile delta, where rice has been successfully rotated with clover (Yanni et al. 2001). Planting of broccoli after faba bean (inoculated by *R. laguerreae*) and common bean (inoculated by *R. tropici*) produced a similar yield of broccoli, in comparison to the conventional practices during the first year (Karavidas et al. 2020). Similarly, planting of wheat and barley after faba bean (inoculated by *Rhizobium*) improved soil fertility, lowered the requirements for N fertilization and increased the yield up to 66% and 42% for wheat and barley, respectively (Alemayehu 2020). Galindo et al. (2021) recorded improvements in growth parameters of wheat, when it was grown as a subsequent crop of cowpea (inoculated by *Bradyrhizobium* sp. and *Bradyrhizobium* sp. in co-inoculation with *Azospirillum brasilense*). In addition, Kumar et al. (2021) showed that inoculation with *B. japonicum* improves the yield and N content of wheat and soybean grown in rotation, but effects on the soybean were lower in comparison to the direct inoculation.

7.4 Role of Rhizobia in the Biocontrol of Non-legume Phytopathogens

In the last few years, there has been a trend in the world to give preference to the use of biological control agents to control plant diseases, instead of using of resistant cultivars or pesticides (Das et al. 2017; Leila and El-Hafid 2020). The application of synthetic pesticides (fungicides and nematicides) contributes to groundwater and soil pollution, poses a serious risk for environment and human health, and also leads to eradication of non-target beneficial microorganisms in the soil. It is highly desirable that the control of fungi and nematodes pursuits for an alternative non-chemical and environment friendly strategy (Das et al. 2017; Mhatre et al. 2019). Out of different environment friendly approaches, PGP strains may act as an efficient fungi and nematode biocontrol, and as agents for plant growth and yield increment, at the same time (Agarwal et al. 2017; Knežević et al. 2021a; Maheshwari et al. 2019; Mhatre et al. 2019). Studies on the potential of *Rhizobium* as a biological control agent are far rarer, compared to the other bacteria, such as species of the genus *Bacillus*, *Pseudomonas*, *Paenibacillus*, *Streptomyces*, *Serratia*, *Klebsiella*, *Actinomycetes* and *Azotobacter* (Charpe 2019; Das et al. 2017; Knežević et al.

2021a, b; Mhatre et al. 2019; Pundir and Jain 2015; Singh et al. 2019). A few of them are even commercialized in a form of bacterial inoculant, and most of them are successful in combating plant pathogens (Das et al. 2017; Singh et al. 2019).

Many studies indicated the importance of the presence of rhizobia in the rhizosphere of non-legumes, as they can inhibit the growth of root pathogens. By protecting plant roots from different diseases caused by fungi or pests, rhizobia indirectly affect the plant growth (Mehboob et al. 2009; Zaim et al. 2017). Control of plant diseases with rhizobia can be achieved by various mechanisms, depending on the pathogen type. The suppression of fungal pathogens by rhizobia includes: (1) mycoparasitism (parasitism on hyphal tips, inhibition of reproductive structures like sclerotia or zoospores, or the production of enzymes like chitinase, which usually degrade the cell wall of the plant pathogenic fungi); (2) production of secondary metabolites (antibiotics, hydrogen cyanide (HCN)); (3) competition between pathogens and rhizobia for iron (siderophore production), nutrients and niches; (4) reduction of susceptibility to pathoges by inducting plant defence mechanisms and stimulating plant growth (Charpe 2019; Das et al. 2017; Gopalakrishnan et al. 2015; Kenawy et al. 2019; Leila and El-Hafid 2020; Mehboob et al. 2012). The mechanisms of nematode suppression by rhizobia can be categorized as direct antagonism and indirect effect. Direct antagonism includes antibiosis (production of low molecular weight organic compounds), production of lytic enzymes and inducting systemic resistance, while indirect effects include phytohormone production, N fixation, phosphate and potassium solubilization, siderophores and ammonia production (Mhatre et al. 2019). In addition, the literature suggests that the combination of rhizobial strains (two or more), or in combination with other PGPR strains (which have diverse modes of antagonistic activity against phytopathogens) is more effective than single-strain inoculum (Das et al. 2017) for the biocontrol of non-legume phytopathogens.

7.4.1 Production of Secondary Metabolites and Lytic Enzymes

The production of one or more antifungal and nematicidal secondary metabolites (antibiotics and HCN) and lytic enzymes by rhizobia can provide direct biocontrol of plant diseases through inhibition, restriction and elimination of phytopathogens (Das et al. 2017; Deshwal et al. 2003b). These bacteria play an important role in the biocontrol of several fungi and pests by means of competition and parasitism, and present the most powerful biocontrol mechanism for combating phytopathogens (Martínez-Viveros et al. 2010; Mhatre et al. 2019). There are several reports on the production of antibiotics or lytic enzymes by various rhizobia. Bradyrhizobial strains (*B. japonicum*, *B. elkanii* and *Bradyrhizobium* sp.) had the capability to produce rhizobitoxine (Deshwal et al. 2003b; Yuhashi et al. 2000), while rhizobial strains (*R. leguminosarum* bv. *trifolii* and *Rhizobium* spp.) were capable to produce trifolitoxin (Breil et al. 1996; Deshwal et al. 2003b) and rhizobitoxine (Siddiqui et al. 2007). *Rhizobium leguminosarum* has been reported to produce increased levels of phytoalexins (4-hydroxy-2,3,9-trimethoxy pterocarpan) and reduced the nematode population (Siddiqui et al. 2007). Hydrogen cyanide (HCN) is volatile antimicrobial secondary metabolites that can also be produced by rhizobial strains, which improves their efficiency for the suppression of plant diseases (Ahmad et al. 2019). Although the production of HCN is generally rare among rhizobial species, some isolates showed the ability to produce HCN. *Rhizobium* isolates RR-1, GNR-1 and SFR-2 showed antifungal potential against both *R. solani* and *S. rolfsii* (in vitro) and exhibited strong, moderate and weak HCN production, respectively (Manasa et al. 2017).

In the case of lytic enzymes, it has been determined that some cell-wall lysing enzymes (chitinase, lyase and cellulase) cause the destruction of pathogens (Sindhu et al. 2010; Volpiano et al. 2019). Two *Ensifer* (*Sinorhizobium*) strains and one *Rhizobium* strain isolated from nodules of fenugreek showed inhibition of *F. oxysporum* (in vitro), due to the loss of structural integrity of the mycelium, hyphal perforation, lysis, fragmentation and degradation. *Ensifer* (*Sinorhizobium*) *meliloti* strains were capable of producing chitinase, while *R. leguminosarum* (TR2) showed β-1,3-glucanase activity (Kumar et al. 2011). Chitinase activity was confirmed among bradyrhizobial strains isolated from root nodules of *Vigna mungo*. *Bradyrhizobium* sp. strains were able to inhibit *M. phaseolina* mycelial growth (in vitro) and showed chitinase activity (Dubey et al. 2012).

7.4.2 Siderophores Production

Few rhizobia evolved a specific mechanism to chelate insoluble forms of iron by siderophores. Siderophores production by rhizobia normally occurs in neutral to alkaline pH soils, due to low iron solubility at elevated pH, as a response to iron deficiency. These compounds could increase rhizobial competition ability under iron-deficient conditions, consequentially limiting the availability of iron for pathogenic fungi and nematodes, and at the same time promoting plant growth (Mhatre et al. 2019; Volpiano et al. 2019). The role of siderophore production in the suppression of plant pathogens is not completely clear, thus the antagonism often occurs in siderophore-producing rhizobia.

Among the studied bradyrhizobial strains isolated from *Arachis hypogaea* (peanut), *Bradyrhizobium* sp. showed inhibition of *M. phaseolina* radial growth in vitro and simultaneous production of siderophores (Deshwal et al. 2003a). The variation in antagonistic potential among siderophores-producing rhizobia could be due to the differences in the type of produced siderophores. Among ten *R. leguminosarum* bv. *trifolii* isolates, which showed some antifungal activity against *Verticillium* sp., two of them had the strongest antagonistic activity (CXS-12, AGR-3 and ELD-15) and two (IRG-17 and SBO-3) displayed less pronounced antagonistic effect as siderophore producers (Vargas et al. 2009).

7.4.3 Indole Acetic Acid (IAA) Production

Among the phytohormones produced by rhizobia (auxins, cytokinins, gibberellic acid, abscisic acid, ethylene, polyamines, brassinosteroids, jasmonates, salicylic acid, strigolactones), indole acetic acid (IAA) is the most common phytohormone (Mhatre et al. 2019). The phytostimulatory effects of IAA produced by rhizobial species could be helpful in biocontrol of various pathogens. Moreover, rhizobia could also directly affect the growth of plant pathogens by IAA production (Volpiano et al. 2019). *Rhizobium* strains isolated from common bean (*Phaseolus vulgaris* L.) proved to be strong antagonists of *Sclerotium rolfsii* growth. During screening, out of 33 antagonistic *Rhizobium* strains, 16 were able to inhibit *S. rolfsii* mycelial growth (for more than 84%). Volpiano et al. (2018) verified a direct correlation between IAA production and inhibition of mycelial growth, with antagonistic strains producing up to 36.5 μg mL^{-1} of IAA. Furthermore, *Rhizobium* isolate GNR-1 showed the high level of IAA production among all tested *Rhizobium* spp. isolates (24.12 μg mL^{-1}) and strong inhibitory effect against *S. rolfsii* (Manasa et al. 2017).

7.4.4 Suppression of Fungal Pathogens of Non-legumes by Rhizobia

Like legumes, non-legume crops are susceptible to many fungal pathogens (most of them belong to the Ascomycetes and the Basidiomycetes) (Das et al. 2017; Glick 2015). Yield losses due to these diseases can vary each year, depending on whether it is a severe epidemic or a regular feature (Das et al. 2017). Some of the fungal phytopathogens of non-legumes belong to the following genera: *Fusarium*, *Rhizoctonia*, *Macrophomina*, *Sclerotinia*, *Ascochyta*, *Alternaria*, *Aspergillus* and *Xanthomonas* (Behera et al. 2020; Das et al. 2017; Dubey and Maheshwari 2011; Fatima et al. 2009; Mehboob et al. 2009; Pundir and Jain 2015; Singh and Pathak 2015; Zaim et al. 2017). Different species of rhizobia belonging to genera *Ensifer* (*Sinorhizobium*), *Bradyrhizobium*, *Rhizobium* and *Mesorhizobium* have been used in the literature to suppress fungal pathogens of non-legumes (Table 7.2).

Strains of the *Ensifer* have shown good potential as biocontrol agents for different fungal disease of non-legumes plants. *Ensifer meliloti* KUMH 139 and KUMH 555 have been used as soil drench cover in controlling the soil-borne root-infecting fungi (*Macrophomina phaseolina*, *Fusarium* spp., and *Rhizoctonia solani*) on non-leguminous plants like okra (*Abelmoschus esculentus* L.) and sunflower (*Helianthus annuus* L.). These strains inhibited *M. phaseolina* and *Fusarium* infection for more than 50% in both tested plants. The infection caused by *Rhizoctonia solani* was reduced for more than a 50% on okra, due to application of *E. meliloti*, while the infection was reduced by more than 50% on sunflower only with KUMH 139 (Ehteshamul-Haque and Ghaffar 1993). *Ensifer meliloti* R5 isolated from sweet

Table 7.2 The use of rhizobia in the biocontrol of fungi on non-legumes (literature review)

Rhizobia	Non-legumes (host)	Fungi	Reference
Ensifer meliloti	Okra Sunflower Chili	*Macrophomina phaseolina* *Fusarium* spp. *Rhizoctonia solani* *F. oxysporum* *F. solani*	Ehteshamul-Haque and Ghaffar (1993) Sheikh et al. (2006) Parveen et al. (2020)
Ensifer sahelens	Chili	*F. oxysporum* *F. solani* *R. solani* *M. phaseolina*	Parveen et al. (2020)
Bradyrhizobium japonicum	Tomato Sunflower Okra	*M. phaseolina* *F. solani* *R. solani* *Fusarium* spp.	Siddiqui and Shaukat (2002) Siddiqui et al. (1998) Ehteshamul-Haque and Ghaffar (1993) Omar and Abd-Alla (1998)
Bradyrhizobium sp.	Okra Sunflower Chili	*F. solani* *M. phaseolina* *R. solani* *F. oxysporum*	Omar and Abd-Alla (1998) Parveen et al. (2020)
Rhizobium leguminosarum bv. *viceae*	Sugar beet	*Pythium* sp.	Bardin et al. (2004)
Rhizobium leguminosamm bv. *phaseoli*	Okra Sunflower Rice	*M. phaseolina* *F. solani*, *R. solani*	Omar and Abd-Alla (1998)
Rhizobium leguminosarum bv. *trifolii*	Rice	*R. solani*	Mishra et al. (2006)
Rhizobium leguminosarum	Okra Sunflower	*R. solani* *Fusarium* spp.	Ehteshamul-Haque and Ghaffar (1993)
Rhizobium trifolii	Sunflower	*M. phaseolina* *F. solani*, *R. solani*	Siddiqui et al. (1998)
Rhizobium sp.	Micropropagated Garnem rootstock Sunflower Chili	*R. solani* *Sclerotium rolfsii* *M. phaseolina* *F. oxysporum* *F. solani*	Jemai et al. (2021) Jatoi et al. (2018) Parveen et al. (2020)
Mesorhizobium loti	Indian mustard	*Sclerotinia sclerotiorum*	Chandra et al. (2007)

clover (*Melilotus indica* L.) exhibited improved plant growth with a significant decrease in infection by the root-infecting fungi on okra, when applied with and without locally available nursery fertilizers (flourish, frutan, NPK, urea and fish meal) as seed dressing and soil drenching. No infection of *M. phaseolina*, *Fusarium* spp., and *R. solani* on okra was observed when fishmeal+*E. meliloti* R5 treatment was used as soil drench (Sheikh et al. 2006).

Two *E. meliloti* strains (NFB-28 and NFB-29) and three *E. sahelens* strains (NFB-30, NFB-31 and NFB-32), applied individually and in combination with *Pseudomonas aeruginosa* strains (PGPR-6, PGPR-11 and PGPR-37), were used by Parveen et al. (2020) to control infection of *F. oxysporum*, *F. solani*, *R. solani* and *M. phaseolina* in chili (*Capsicum annuum* L). Rhizobial strains NFB-30, NFB-31 and NFB-32 completely prevented the *F. solani* infection on chili roots under the greenhouse conditions. NFB-28 also reduced *F. solani* infection in comparison to other rhizobial strains. The lowest biocontrol potential was exhibited by NFB-29 strain. Infection (%) of *M. phaseolina* and *F. oxysporum* were reduced by all tested rhizobial strains: 6.2% (NFB-31 and NFB-32), 25% (NFB-29, NFB-30) and 12.5% (NFB-28). In the field experiments, stains NFB-28 and NFB-30 completely prevented the *F. solani* and *R. solani* infection after 75 and 45 days, respectively. Inoculation with dual inoculant NFB-28+NFB-30 controlled the infection of *F. oxysporum*, *M. phaseolina* and *R. solani* significantly after 45, 75 and 45 days, respectively. In dual inoculation in field experiments with *Pseudomonas* strains, two rhizobial strains were used (NFB-28 and NBF30). *Fusarium solani* infection was completely controlled by PGPR-37+NFB-28 and PGPR-11+NFB-28 after 45 and 75 days, respectively and infection was significantly reduced after 45 days by PGPR-37+NFB-30 and PGPR-11+NFB-30. Interestingly, *F. oxysporum* infection was reduced by PGPR-11+NFB-30 and was completely controlled by PGPR-37+NFB-30, PGPR-11+NFB-28, NFB-28+PGPR-6, NFB-30+PGPR-6. In addition, infection of *M. phaseolina* was suppressed by NFB-28+PGPR-6, NFB-30+PGPR-6 and *R. solani* infection was completely controlled by PGPR-37+NFB-30, NFB-28 +PGPR-6 and NFB-30+ PGPR-6 (Parveen et al. 2020).

Similar to *Ensifer* and other fast-growing rhizobia, bradyrhizobial strains have also been used successfully as biocontrol agents for charcoal rot caused by *M. phaseolina*, *F. solani* and *R. solani* on plants such as tomato, okra, sunflower and chili. Siddiqui and Shaukat (2002) used *B. japonicum* in the glasshouse and field experiments to reduce the growth of tomato (*Solanum lycopersicum*) pathogens (*M. phaseolina*, *F. solani* and *R. solani*). The highest efficiency of inoculation with this strain was recorded against *M. phaseolina*, where the fungal infection was reduced for about 80% in tomato (Siddiqui and Shaukat 2002). On the other hand, *B. japonicum* was successfully used as soil drench cover in controlling *R. solani* infection on sunflower, as well as *Fusarium* spp. and *M. phaseolina* infection on okra and sunflower (Ehteshamul-Haque and Ghaffar 1993). In both cases, the applied bradyrhizobial strain reduced infection by more than 50%. Omar and Abd-Alla (1998) used two bradyrhizobial inoculums (*B. japonicurn* and *Bradyrhizobium* sp.) for inoculation of soil during okra and sunflower seeds planting. These treatments exhibited different relative efficiency against *F. solani*, *M. phaseolina* and *R. solani* on okra and sunflower seedlings. Both strains significantly reduced root rot severity caused by these pathogenic fungi (Omar and Abd-Alla 1998). *Bradyrhizobium japonicum* strain 811 (chickpea isolate) and

B. japonicum strain KUCC-823 also showed potential in the control of root infecting fungi (*M. phaseolina, R. solani, F. solani*) on sunflower (Siddiqui et al. 1998). The maximum reduction of *F. solani* and *R. solani* infection occurred by strain KUCC-843. The KUCC-811 strain showed greater efficacy in reducing *M. phaseolina* growth on sunflower, while *Bradyrhizobium* sp. NFB-1 showed decrease in infection of *M. phaseolina, F.* solani and *F. oxysporum* in chili (*Capsicum annuum* L), in greenhouse experiment. Infection (%) of *M. phaseolina* and *F. oxysporum* were controlled by strain NFB-1 (25%), while *F. solani* by 62.5% (Parveen et al. 2020).

In addition to the single inoculant, bradyrhizobial strains were used in the dual inoculant to control fungal diseases. Using *B. japonicum* strain 569Sm^r in combination with *P. fluorescens* strain CHA0 or *P. aeruginosa* strain IE-6S$^+$ caused significant suppression of multiple tomato pathogens (*M. phaseolina, F. solani* and *R. solani*) in the field. The application of dual inoculant containing strains 569Sm^r and CHA0 provided complete protection of tomato roots against *M. phaseolina*. Efficiency in controlling *R. solani* remained similar for both dual inoculants (17% of infection), while slightly better effects of IE-6S$^+$+569Sm^r were recorded for the reduction of *F. solani* infection (Siddiqui and Shaukat 2002).

Among the species of the genus *Rhizobium, R. leguminosarum* is the most studied for control of fungal diseases. *R. leguminosarum* bv. *viceae* strains (isolated from pea and lentil root nodules) showed high efficiency as biological control agents for controlling the sugar beet (*Beta vulgaris* L) damping-off caused by *Pythium* (*Pythium* sp. "group G" strain), as observed by Bardin et al. (2004). Similarly, *R. leguminosarum* bv. *phaseoli* significantly reduced root rot severity caused by *M. phaseolina, F. solani* and *R. solani* in okra and sunflower, when used as soil treatment during the seed planting (Omar and Abd-Alla 1998). In addition, these strains were effective as the fungicide treatment for protection of sugar beet seedlings against *Pythium* damping-off, in the field experiment (Bardin et al. 2004). Inoculation of rice (*Oryza sativa* L.) with *R. leguminosarum* bv. *phaseoli* or *R. leguminosarum* bv. *trifolii* mediated in the induction of systemic resistance against *R. solani*. These strains were effective in inducing resistance in rice by a rapid accumulation of phenolics (gallic, ferulic, tannic, and cinnamic acids) in the plant. The accumulation of phenolic compounds was especially enhanced in the presence of pathogen *R. solani* (Mishra et al. 2006). *Rhizobium leguminosarum* was also successfully used as soil drench cover in controlling *R. solani* infection in sunflower or okra, as well as *Fusarium* spp. infections in okra and sunflower (Ehteshamul-Haque and Ghaffar 1993). *Rhizobium trifolii* (berseem clover isolate) showed a good potential in the control of root infecting fungi (*M. phaseolina, F. solani* and *R. solani*) in sunflower. The highest infection reduction was achieved for *R. solani*, followed by *M. phaseolina* and *F. solani*, respectively (Siddiqui et al. 1998). Recently, Jemai et al. (2021) demonstrated the potential of four *Rhizobium* strains to inhibit *R. solani* affecting micropropagated Garnem (*Prunus amygdalus* × *Prunus persica*) rootstock. Incorporation of inoculums consisted of these strains in sterilized peat (one week prior to inoculation with *R. solani*) allowed the relative protection of Garnem micropropagated plantlets during acclimatization. PP6 (pea isolate) and HaD4002 (bean isolate) conferred the highest survival rate (20–21%), followed by PP29 (pea isolate) (10%) and Pch Kass, respectively. All plants grew normally in the absence of fungus and rhizobia, while no plants survived in the presence of fungus alone (Jemai et al. 2021). Ten *Rhizobium* isolates (S1–S10), isolated from cluster bean (*Cyamopsis tetragonoloba* L.), were used to suppress collar rot of sunflower caused by *S. rolfsii*. The *Rhizobium* strains S7 and S6 were highly efficient in the reduction of *S. rolfsii* growth in soil in pot experiment (Jatoi et al. 2018). More recently, Parveen et al. (2020) used *Rhizobium* sp. NFB-2 to decrease the infection caused by *M. phaseolina*, *F. oxysporum* and *F. solani* in chili (*Capsicum annuum* L.), in greenhouse experiment. Infection (%) of *M. phaseolina* and *F. oxysporum* were significantly controlled by strain NFB-2 and it was 25%, while the infection (%) of *F. solani* was 62.5% (Parveen et al. 2020).

Earlier, *M. loti* (isolated from root nodules of *Mimosa pudica*) inhibited the growth of *S. sclerotiorum* (by 75%) which caused the white rot in Indian mustard *(Brassica campestris)* (Chandra et al. 2007). HCN and hydroxamate type siderophores production probably played a significant role in the inhibition of fungi.

7.4.5 Suppression of Nematodes in Non-legumes by Rhizobia

Most of the studies focused on two major groups of plant parasitic nematodes: cyst (*Heterodera* and *Globodera* sp.) and root knot nematodes (*Meloidogyne* sp.). During their life cycle, they have six stages: the egg, juvenile stages (J1, J2, J3 and J4) and the adult stage (males or females), while the second juvenile stage (J2) is regarded as an invasive stage when the infection of plant roots occurs (Maheshwari et al. 2012; Mhatre et al. 2019; Oro et al. 2020). Among root-nodulating rhizobia, strains from three genera, *Bradyrhizobium*, *Rhizobium* and *Ensifer* were used for controling nematodes in non-legumes plants (such as tomato, chilli and potato) (Table 7.2).

Parveen et al. (2019) used *Bradyrhizobium* spp. NFB-1 for biocontrol of root knot nematode (*Meloidogyne javanica*) on chili roots. The NFB-1 strain showed a significant reduction of galls and number of nematode penetrations in the roots, compared to the uninoculated control. Other bradyrhizobial strain, *B. japonicum* 569m^r, alone or in combination with *P. fluorescens* strain CHA0 or *P. aeruginosa* strain IE-6S$^+$, caused a substantial mortality of *M. javanica* juveniles in tomato rhizosphere (Siddiqui and Shaukat 2002). In addition, in both iron-deficient and iron-sufficient soils, strain 569Sm^r significantly suppressed *Meloidogyne incognita*. Used as a soil drench, these rhizobia not only suppressed root knot nematodes, but also enhanced the growth of tomato plants, both under glasshouse and field conditions. Inoculation with dual inoculums, IE-6S$^+$+569Sm^r induced higher reduction in galls (about 42%) than CHA0+569Sm^r (about 32%), compared to the control. The strongest effect on the decrement in gall number (about 53%) was achieved by using the inoculant that consisted of all three strains (Siddiqui and Shaukat 2002).

Phytoparasitic nematodes have generally been controlled by nematicides. The application of *Rhizobium* strains, *R. etli* G12 and *Rhizobium* spp. NFB-2 controlled

M. incognita and *M. javanica*, respectively (Hallmann et al. 2001; Parveen et al. 2020). The number of galls formed by *M. incognita* on potato was significantly decreased when *R. etli* G12 treatment was applied. The reduction in the number of galls was 34 and 39% lower for the treatment with G12 and G12 (pGT-trp), respectively than on root treated with nematode alone (Hallmann et al. 2001). *Rhizobium* spp. NFB-2 did not show a significant reduction of galls on chili roots, but the reduction in the number of nematode penetrations in the roots compared to the uninoculated control was significant (Parveen et al. 2020). In addition, *Rhizobium etli* G12, alone or in the combination with *Glomus intraradices,* was applied as the biocontrol agent of *M. incognita* on tomato. The application of *Rhizobium etli* G12 alone resulted in a significant reduction of galling up to 39% and a significant reduction in the number of egg masses, but combining the two microorganisms led to a significant reduction in the numbers of galls (up to 61%) and egg masses (up to 54%), compared to the control (Reimann et al. 2008). Reitz et al. (2000) examined the ability of *R. etli* G12 to control cyst nematode *Globodera pallida*. They showed that both living and heat-killed cells of *R. etli* strain G12 can induce systemic resistance in potato roots to *G. pallida* nematode infection. The pretreatment of potato roots, with living G12 cells or different concentrations of LPS extract (from *R. etli* G12), resulted in a significant reduction in *G. pallida* infection in potato roots. The greatest reduction in *G. pallida* infection was achieved using 1 mg ml^{-1} LPS extract (up to 44%), while treatment with living cells of *R. etli* reduced the nematode infection by 34% (Reitz et al. 2000).

Ensifer meliloti NFB-28, *E. meliloti* NFB-29, *E. sahelens* NFB-30, *E. sahelens* NFB-31 and *E. sahelens* NFB-32, individually or in combination with *P. aeruginosa* strains (PGPR-6, PGPR-11, PGPR-37), showed biocontrol potential against *M. javanica* on chili roots (Parveen et al. 2020). All *Ensifer* strains caused inhibitory effect on root knot nematode by reducing the number of galls and nematode penetrations in the roots, under greenhouse conditions. Significant reduction of galls was achieved by strains NFB-28, NFB-29, NGB-30, but NFB-29 was the most effective. The number of nematodal penetration in roots was significantly reduced by all strains, in comparison to the control. Strains NFB-30 and NBB-28 showed the highest reduction in juvenile root penetration. In the field experiment, NFB-28, NFB-30 and inoculation with dual inoculum of these strains decreased the number of galls and nematode penetration in chili roots after 45 days. In addition, the reduction of nematodes was also achieved by dual inoculation with rhizobium strains (NFB-28 and NFB-30) and *P. aeruginosa* strains (PGPR-6, PGPR-11, PGPR-37) (Parveen et al. 2020).

7.5 Interactions Between Rhizobia and Non-legume Plants

Rhizobium, *Mesorhizobium*, *Bradyrhizobium* and *Ensifer* species are conventionally known as symbionts of legume crops. The main advantage of the formation of symbiotic systems inside the root nodules of leguminous plants is regarded as the protection of bacteria from the competition with other microorganisms in the rhizosphere (including other bacteria or fungi), as well as in the possibility of bidirectional metabolites exchange between the legume plant and symbiotic bacterium (Webster et al. 1997). In addition, root nodules enable lower energy consumption, both for host plant and symbiotic bacteria, as well as the protection from N fixation inhibitors, such as elevated concentration of oxygen (Dent and Cocking 2017). Nodule formation is induced by the secretion of flavonoid compounds by legume plants into the rhizosphere, where they activate the transcription of bacterial *nod* genes which produce a lipochitooligosaccharidic signal necessary for the development of symbiosomes (Oldroyd et al. 2011). The recognition between flavonoides and Nod proteins is responsible for defining the specificity of the plant-rhizobia symbiosis (Wang et al. 2018). However, as *nod* genes are generally located on plasmids, they can be transferred by horizontal gene transfer by conjugation between different species in the rhizosphere or within the root nodules/symbiosomes, which can consequently alter the bacterial host-range specificity (Bañuelos-Vazquez et al. 2020).

The discovery of an effective nodulation of *Parasponia andersonii* (Cannabaceae) induced by *Bradyrhizobium* opened a new perspective in the research of rhizobial nodulation, especially regarding the non-legume plants (Trinick 1979). It has been shown that *Parasponia* can induce the transcription of rhizobial *nod* genes, indicating that this plant species produces *nod* geneinducing compounds (fluorescent substances with flavonoid-like properties) (Reddy et al. 2007). Although the mechanism responsible for the induction of non-legume plants nodulation is not completely determined, it can be concluded that the control mechanisms over rhizobial bacteria are not exclusive to the legumes (Dupin et al. 2020).

Despite being unable to induce nodulation, rhizobial species can enter the root system of different non-legumes at the place of lateral roots emergence. After entering into the plant root system by a 'crack entry', rhizobia penetrate into the cortex cells of young emerging lateral roots, which induces thicker and shorter roots (Cocking et al. 1995). The production of specific rhizobial lythic enzymes such as cellulase, amylase, and pectinase could underlie in the effectiveness of its penetration in the roots cortex system. Besides colonization of the cortical cells, rhizobia are also able to colonize xylem vessels of the non-legume roots (Reddy et al. 2007). Thus, as invasion of rhizobia into the tissues of non-legume plants bypasses the plant defence mechanisms, it induces no harm for the colonized plant tissues. The mechanisms between rhizobia and non-legumes required for successful colonization are achieved by the exchange of different signal molecules (Santi et al. 2013). Regarded to the induction of rhizobial *nod* genes transcription, it has been shown that different types of non-legumes can secrete phenolic compounds such as vanillin and isovanillin, which can consequently induce transcription of *nod* genes (Le Strange et al. 1990; Reddy et al. 2000, 2007). Although exogenous application of specific flavonoids can promote colonization of non-legumes roots by rhizobia, there is insufficient data whether the colonization of non-legume roots by rhizobia is dependent from *nod* genes activation. The complete nature and significance of *nod* genes inducing compounds produced by non-legume plants is yet to be determined.

The production of biofilm by rhizobia has been extensively studied (Janczarek et al. 2015; Robledo et al. 2012). The fast-growing rhizobia, *E. meliloti* and *R. leguminosarum* formed biofilm on abiotic surfaces, where the level of EPS production and the presence of flagella were one of the major factors for the biofilm formation performance (Fujishige et al. 2006). The production of EPS is highly significant in the associative interactions of rhizobia with non-legume plants roots, such as tomato (Vershinina et al. 2021). The ability of *E. meliloti* to form a 3-dimensional biofilm is induced by the expression of *nod* genes (Fujishige et al. 2008). Although the transcription of *nod* genes can be induced by plant-derived substances, the presence of flavonoids is not essential for the maturation of rhizobial biofilm, suggesting that their role in biofilm formation is probably an ancestral function (Fujishige et al. 2008). However, different plant-derived substances are likely to induce biofilm formation. As mentioned before, rhizobia enter the non-legumes root system at the place of lateral roots emergence. The formation of biofilm at the 'crack entry' could be explained by the higher concentrations of plant exudates found around actively growing tissues (Rinaudi and Giordano 2010). The mechanisms of flavonoid-independent biofilm production are yet to be comprehensively studied, as it may underlie in the pattern of non-legume roots colonization.

It has been shown that the attachment of *R. leguminosarum* to tomato and pepper seedling gradually increases during the first 9 days of inoculation, after which bacteria became firmly attached to seedling root surfaces (García-Fraile et al. 2012). In the same research, the initiation of *Rhizobium* biofilm formation was also observed in intercellular spaces of seedlings roots (García-Fraile et al. 2012). Similarly, Bhattacharjee et al. (2012) showed that *R. leguminosarum* bv. *trifolii* has the ability to colonize the root surface of rice, with a strong chemotaxis response found between rhizobia and the rice seed leachates and root exudates. In the rice seedlings, rhizobia were observed at the lateral root junctions and at the main root tip where it formed an infection thread-like structure towards the base of the rice root hair (Perrine-Walker et al. 2007). Chen and Zhu (2013) demonstrated that infection of rice roots by rhizobia is a process which is independent from the common symbiosis *Sym* genes. Furthermore, transcriptomic analysis of rice shoots showed that differentially-expressed genes (DEGs) were up-regulated by rhizobia even before its ascendance from the roots to the shoots, suggesting that rhizobia may be involved in the promotion of rice shoot growth by long-distance signalling (Wu et al. 2018). Glaeser et al. (2015) showed that conglomerates of rhizobial cells were particularly present at lateral barley root protrusions and that they probably serve as entry sites. In addition, the expression of non-expressor of pathogenesis-related genes 1 (NPR1) has been identified as a requirement for colonization of barley by *R. radiobacter*, which allows the establishment of typical spatiotemporal colonization pattern and bacterial multiplication (Kumar et al. 2020). Several *Rhizobium* or *Bradyrhizobium* species have been observed in the roots of non-legumes such as wheat, banana and maize (Bartoli et al. 2020; Martínez et al. 2003; Rosenblueth and Martínez-Romero 2004; Yoneyama et al. 2019). The formation of nodule-like structures consisted of scattered centrally located cells enclosed by several layers of cells, with rhizobia mainly localized in spaces between the cell layers, was also observed on rice seedling roots (Dent and Cocking 2017). Similarly, the formation of these structures was also observed in the oilseed roots after the enzymatic treatment, thus rhizobia invaded only dead root cells, while significant nitrogenase activity was not observed both in oilseed and rice roots (Dent and Cocking 2017). Regarding the colonization of sugarcane (*Poaceae*), bradyrhizobia is found to express N fixation genes in planta (Thaweenut et al. 2010). A metagenomic approach based on the improved amplification of *nifH* genes showed the presence of *Bradyrhizobium* in the root compartment of sugarcane, indicated the importance of beneficial rhizobia associated with this plant species (Gaby et al. 2017). In addition, metagenomic analysis of bacterial communities associated with seeds of grasses such as *Festuca rubra* (red fescue), *Lolium arundinacea* (tall fescue) and *Lolium perenne* (perennial ryegrass) showed that the greater diversity and density of bacteria (including Rhizobiaceae) can improve seedling growth in stress conditions, while the extremely high concentration of these bacteria can interfere with the seedling development by competing for the nutrients from rhizosphere (Chen et al. 2020b). Also, it has been shown that there is a strong influence of soil microbiome on the structure of the root microbiome of barley (Yang et al. 2017).

7.5.1 Rhizobia and Non-legumes Association Development

The necessity of achieving quality and sustainable crop production with higher yield arises due to the increment of worldwide consumption. Recently, research regarding the improvement of N content in economically significant plants by reducing the use of chemical fertilizers and by creating an artificial symbiosis between PGP rhizobia and non-legumes has been gaining the wider perspective (Santi et al. 2013). The understanding of complex interactions between N-fixing rhizobia and non-legumes offers a new insight into the improvement of sustainable agriculture. Although there are some findings that rhizobial nitrogenase genes can be expressed *in planta* (determined by culture-independent methods), the level of N obtained in this way remains insufficient for an effective plant growth promotion, in comparison to the chemical fertilizers (Rosenblueth et al. 2018). Besides a conventional approach, where the potent PGP rhizobia is being applied to non-legume plants in order to promote the growth of selected plants by expressing the aforementioned PGP traits, a novel research approach based on the methods of genetic engineering is being scrutinized. As molecular biology methods evolve and become more accessible, the possibilities of their application to improve the properties that would contribute to the greater efficiency of rhizobia and non-legume interactions are frequently used.

Genetic engineering of rhizobia is commonly referred to as the enhancement of N fixation or colonization patterns regarded to efficient recognition, chemotaxis and root invasion (Priyadarshini et al. 2021; Rosenblueth et al. 2018). Dispensability of different rhizobial genes provides flexibility for expanding nitrogenase activity by methods of genetic engineering (Goyal et al. 2021). Bloch et al. (2020) showed that signalling pathways involved in the formation of symbiosomes (such as signalling for nodules formation induced by plant hormones) are common for all plants, thus these signalling cascades may be blocked in the non-legume plants. As the specific signalling pathways in cereals are analogue to the ones present in the legume crops in effective nodule organogenesis, engineering a nodule-like symbiosis between rhizobia and cereal crops offers a promising alternative (Bloch et al. 2020). In addition, the expression of transcription factors that regulate the induction of flavonoid biosynthetic pathway could be engineered to promote *nod* gene induction in rhizobia (Reddy et al. 2007).

Elevating the competitive potential of rhizobia is another target for engineering, especially when considering unfavourable environmental conditions, which generally suppress rhizobial activity and viability. Regarding stressors coupled with water deficiency (salinity or drought stress), modification of bacterial chaperones showed potential for increasing the stress tolerance and adaptability of rhizobia (Goyal et al. 2021). This kind of modification is of great importance as interactions between non-legumes and rhizobia can be altered or even inhibited by unfavourable environment conditions (Mayer et al. 2019).

7.6 Future Prospects of Rhizobial Inoculant Formulation

Rhizobial inoculants can replace mineral N fertilizers in sustainable agriculture or can be a fertilizer supplement in the conventional agricultural production due to their ability to increase plant yield, provide high-quality protein food and improve N fixation potential of the soil. These can also be used as a plant probiotic for some non-legumes (cereals, fruits, vegetables, etc.), with the aim to enhance the plant growth (Jiménez-Gómez et al. 2018). Besides being used as single inoculums, the potential of using synergistic rhizobia with other soil microorganisms in sustainable agriculture has been elucidated with examples, followed by their future plant growth potential and quality prospects (Naseer et al. 2019; Stajković-Srbinović et al. 2021; Siddiqui and Shaukat 2002). Rhizobial inoculants with PGP rhizobacteria (binary or polyvalent) are the future in the production of biofertilizers, with a potential use in the production of leguminous and non-leguminous crops. Recently, co-inoculation of different strains of rhizobia or with other PGP rhizobacteria which have a complementary or synergistic effect on improving plant growth has been intensively used in agricultural production (Delić et al. 2012). The results of some authors showed the ability of rhizobia to promote the growth of non-legumes in co-inoculation with particular non-rhizobial bacteria (Knežević et al. 2021a; Rosenblueth et al. 2018; Stajković-Srbinović et al. 2015b). In that way, an effective rhizobial strain, as active agent of bivalent or polyvalent bio-inoculant, can be intended for inoculation of plant mixtures, which involve legumes as well. In addition, rhizobial inoculation can reduce mineral N application in the production of non-legumes, which can be used as cost-effective way of crop production in the soils with limited fertility. Seed inoculation of non-legumes can enhance quantity and quality of not only non-legumes but also legumes grown as subsequent crops in crop rotation (Alemayehu 2020; Galindo et al. 2021; Kumar et al. 2021). In order to obtain modified strains for various purposes, different genetic methods should be used (Das et al. 2017).

Quality of active agents of bio-inoculant is another important task in agricultural biotechnology. One of the most important steps in selecting the proper active agents is its competitivness to native rhizobacteria. Highly efficient strains which simultaneously have more than two PGP traits should be obtained during the selection process. Moreover, carrier quality and improved formulations are equally important for the success of inoculants under field conditions. Liquid inoculant represents a complex biological formulation consisted from active agents (bacteria) and its metabolites secreted to the used growth medium (Buntić et al. 2021; Boiero et al. 2007). In that regard, during the inoculant preparation, the attention should also be given to the physiological traits (production of different metabolites) of bacteria in selected medium, which can have potentially negative secondary effects. Boiero et al. (2007) demonstrated that *B. japonicum* can produce ethylene in a medium enriched by of L-methionine. The authors suggested that the evaluation of all the inoculant components for each rhizobial strain is necessary in order to obtain an accurate quality control. Biofilm-based formulations showed a good ability to protect microorganisms and keep their survival under stressful environmental conditions (Swarnalakshmi et al. 2013; Triveni et al. 2012; Das et al. 2017). Polymer-based formulations allow the optimal storage life of the bio-inoculant (Tittabutr et al. 2007). Water-in-oil emulsion technology for developing liquid formulations is beneficial for bacterial inoculants (Vandergheynst et al. 2006). Das et al. (2017) pointed out that the application of nanotechnology as new class of bacterial inoculants should provide suitable carrier for bacteria due to nanostructures which improve stability and high surface area. Suitable nano-formulations may contribute to a stability of bacterial inoculants under the influence of high temperature, UV influence or desiccation.

7.7 Concluding Remarks

The application of rhizobia for the growth promotion of non-legumes, as well as for protection against different kinds of phytopathogenes has been gaining more attention. Currently, the use of non-rhizobial inoculants based on species belonging to *Pseudomonas*, *Bacillus* and *Serratia* is being potentiated. However, there are several concerns regarding the safety of these inoculants. On the contrary, crops treated by rhizobial inoculants have been regarded as safe, both for human and animal consumption. Although there is a variety of literature regarding the beneficial effect of rhizobia to legumes, more research on the potential use of rhizobia for improving the growth of non-legumes is required. Besides studying the mechanisms responsible for the promotion of plant growth, identification of mechanisms necessary for the successful interaction between rhizobia and host non-legume plant is also of great importance. As there are many commercially available rhizobia-based products used for promotion of legume growth, their effect should also be tested on the non-legumes. By using the same rhizobial fertilizer for different crops (including legumes and non-legumes), the use of chemical fertilizers could be drastically lowered, production could be facilitated, and costs could be lower for the end-users. Inoculums containing rhizobia with multiple beneficial properties would have many advantages, as they would enable a more complete diet of both legumes and non-leguminous plants and improve the general fertility of the soil. In addition, the selection of effective and competitive rhizobia to be used as inoculants for growing non-legumes represents an emerging solution for improving the growth and quality of significant crops. In conclusion, enriching the rhizosphere microbiome of non-legumes with competitive and effective PGP rhizobia, and engineering the plants to provide better conditions for symbiosis should be the focus of further research (Priyadarshini et al. 2021).

References

Abd-Alla MH (1994) Use of organic phosphorus by *Rhizobium leguminosarum* biovar viceae phosphatases. Biol Fertil Soils 18:216–218. https://doi.org/10.1007/BF00647669

Abulfaraj A, Jalal R (2021) Use of plant growth-promoting bacteria to enhance salinity stress in soybean (*Glycine max* L.) plants. Saudi Biol Sci 28(7):3823–3834. https://doi.org/10.1016/j.sjbs.2021.03.053

Adnan M, Shah Z, Khan SM, Khan G, Ali A, Khan N, Saleem N, Nawaz S, Akbar S, Samreen S, Zaib K (2014) Integrated effects of rhizobial inoculum and inorganic fertilizers on wheat yield and yield components. Am J Plant Sci 05(13):2066–2073. https://doi.org/10.4236/ajps.2014.513222

Afify A, Hauka F, Elsawah A (2019) Plant growth-promoting rhizobacteria enhance onion (*Allium cepa* L.) productivity and minimize requisite chemical fertilization. Environ Biodivers Soil Secur 2:9

Agarwal M, Dheeman S, Dubey RC, Kumar P, Maheshwari DK, Bajpai VK (2017) Differential antagonistic responses of *Bacillus pumilus* MSUA3 against *Rhizoctonia solani* and *Fusarium oxysporum* causing fungal diseases in *Fagopyrum esculentum* Moench. Microbiol Res 205:40–47. https://doi.org/10.1016/j.micres.2017.08.012

Ahmad M, Nadeem SM, Zahir ZA (2019) Plant-microbiome interactions in agroecosystem: an application. In: Kumar V et al (eds) Microbiome in plant health and disease. Springer, Singapore, pp 251–291

Alami Y, Achouak W, Marol C, Heulin T (2000) Rhizosphere soil aggregation and plant growth promotion of sunflowers by an exopolysaccharide-producing *Rhizobium* sp. strain isolated from sunflower roots. Appl Environ Microbiol 66(8):3393–3398. https://doi.org/10.1128/aem.66.8.3393-3398.2000

Alemayehu N (2020) The use of biofertilizer by smallholder farmers and its impact on productivity of pulse cereal cropping system in Arsi Zone, Oromia regional state, Southeastern Ethiopia. East Afr J Sci 14(1):1–12

Ali Q (2017) Inoculation with rhizobial consortium for improving the growth, yield and quality of maize under salt-stressed conditions. Pak J Agric Sci 54(01):97–105. https://doi.org/10.21162/PAKJAS/17.5388

Antoun H, Prevost D (2005) Ecology of plant growth promoting rhizobacteria. In: Siddiqui ZA (ed) PGPR: biocontrol and biofertilization. Springer, New York, NY, pp 1–38

Antoun H, Beauchamp CJ, Goussard N, Chabot R, Lalande R (1998) Potential of *Rhizobium* and *Bradyrhizobium* species as plant growth promoting rhizobacteria on non-legumes: effect on radishes (*Raphanus sativus* L.). Plant Soil 204:57–67. https://doi.org/10.1023/A:1004326910584

Arcand MM, Schneider KD (2006) Plant- and microbial-based mechanisms to improve the agronomic effectiveness of phosphate rock: a review. An Acad Bras Cienc 78:791–807. https://doi.org/10.1590/s0001-37652006000400013

Arfaoui A, Sifi B, Boudabous A, Hadrami IE, Chérif M (2006) Identification of *Rhizobium* isolates possessing antagonistic activity against *Fusarium oxysporum* f. sp. ciceris, the causal agent of *Fusarium* wilt of chickpea. J Plant Pathol 88:67–75. https://doi.org/10.4454/JPP.V88I1.832

Atzorn R, Crozier A, Wheeler CT, Sandberg G (1988) Production of gibberellins and indol-3-acetic acid by *Rhizobium phaseoli* in relation to nodulation of *Phaseolus vulgaris* roots. Planta 175: 532–538. https://doi.org/10.1007/BF00393076

Bais HP, Weir TL, Perry LG, Gilroy S, Vivanco JM (2006) The role of root exudates in rhizosphere interactions with plants and other organisms. Annu Rev Plant Biol 57:233–266. https://doi.org/10.1146/annurev.arplant.57.032905.105159

Bañuelos-Vazquez L, Cazares D, Rodríguez S, Cervantes-De la Luz L, Sánchez-López R, Castellani L, Tejerizo G, Brom S (2020) Transfer of the symbiotic plasmid of *Rhizobium etli* CFN42 to endophytic bacteria inside nodules. Front Microbiol 11:1752. https://doi.org/10.3389/fmicb.2020.01752

Bardin SD, Huang HC, Pinto J, Amundsen EJ, Erickson RS (2004) Biological control of Pythium damping-off of pea and sugar beet by *Rhizobium leguminosarum* bv. *viceae*. Can J Bot 82(3):291–296. https://doi.org/10.1139/b04-003

Bartoli C, Boivin S, Marchetti M et al (2020) *Rhizobium leguminosarum* symbiovar *viciae* strains are natural wheat endophytes and can stimulate root development and colonization by arbuscular mycorrhizal fungi. BioRxiv. https://doi.org/10.1101/2020.08.07.241844

Baset Mia MA, Shamsuddin ZH, Mahmood M (2012) Effects of rhizobia and plant growth promoting bacteria inoculation on germination and seedling vigor of lowland rice. Afr J Biotechnol 11(16):1337. https://doi.org/10.5897/AJB09.1337

Behera B, Das TK, Raj R, Ghosh S, Raza MB, Sen S (2020) Microbial consortia for sustaining productivity of non-legume crops: prospects and challenges. Agric Res 2020:1–14. https://doi.org/10.1007/s40003-020-00482-3

Bhattacharjee RB, Jourand P, Chaintreuil C, Dreyfus B, Singh A, Mukhopadhyay SN (2012) Indole acetic acid and ACC deaminase-producing *Rhizobium leguminosarum* bv. *trifolii* SN10 promote rice growth, and in the process undergo colonization and chemotaxis. Biol Fertil Soils 48(2):173–182. https://doi.org/10.1007/s00374-011-0614-9

Bloch S, Ryu M, Ozaydin B, Broglie R (2020) Harnessing atmospheric nitrogen for cereal crop production. Curr Opin Biotechnol 62:181–188. https://doi.org/10.1016/j.copbio.2019.09.024

Boiero L, Perrig D, Masciarelli O, Penna C, Cassan F, Luna V (2007) Phytohormone production by three strains of *Bradyrhizobium japonicum* and possible physiological and technological implications. Appl Microbiol Biotechnol 74:874–880. https://doi.org/10.1007/s00253-006-0731-9

Breil I, Koch T, Heller A, Schlotzer E, Grunert A, van Ackern K, Neuhof H (1996) Alteration of n-3 fatty acid composition in lung tissue after short-term infusion of fish oil emulsion attenuates inflammatory vascular reaction. Crit Care Med 24(11):1893–1902. https://doi.org/10.1097/00003246-199611000-00021

Brígido C, Oliveira S (2012) Most acid-tolerant chickpea mesorhizobia show induction of major chaperone genes upon acid shock. Microb Ecol 65(1):145–153. https://doi.org/10.1007/s00248-012-0098-7

Buntić A, Stajković-Srbinović O, Knežević M, Kuzmanović Đ, Rasulić N, Delić D (2019a) Development of liquid rhizobial inoculants and pre-inoculation of alfalfa seeds. Arch Biol Sci 71(2):379–387. https://doi.org/10.2298/ABS181008062B

Buntić A, Stajković-Srbinović O, Knežević M, Kuzmanović Đ, Rasulić N, Kuzmanović Đ, Dimitrijević-Branković S, Delić D (2019b) The effect of bacterial isolates from rhizosphere soils on wheat and barley seed germination. Zemlj Biljka 68(2):1–12. https://doi.org/10.5937/ZemBilj1902001B

Buntić A, Stajković-Srbinović O, Knežević M, Rasulić N, Ugrenović V, Kuzmanović Đ, Delić D (2021) Efficiency of alfalfa seed priming and rhizobial-based liquid inoculants in an Eutric Cambisol. Zemdirbyste 108(4):339–346. https://doi.org/10.13080/z-a.2021.108.043

Carson KC, Meyer JM, Dillworth MJ (2000) Hydroxamate siderophores of root nodule bacteria. Soil Biol Biochem 32:11–21. https://doi.org/10.1016/S0038-0717(99)00107-8

Cavalcanti MIP, de Carvalho NR, Rodrigues DR et al (2020) Maize growth and yield promoting endophytes isolated into a legume root nodule by a cross-over approach. Rhizosphere 2020: 100211. https://doi.org/10.1016/j.rhisph.2020.100211

Chabot R, Antoun H, Cescas MP (1996) Growth promotion of maize and lettuce by phosphate-solubilizing *Rhizobium leguminosarum* biovar. *phaseoli*. Plant Soil 184:311–321. https://doi.org/10.1007/BF00010460

Chandra S, Choure K, Dubey RC, Maheshwari DK (2007) Rhizosphere competent *Mesorhizobium loti* MP6 induces root hair curling, inhibits *Sclerotinia sclerotiorum* and enhances growth of Indian mustard (*Brassica campestris*). Braz J Microbiol 38:124–130. https://doi.org/10.1590/S1517-83822007000100026

Charpe AM (2019) Free-living PGPRs in biotic stress management. In: Sayyed RZ (ed) Plant growth promoting rhizobacteria for sustainable stress management. Springer, Singapore, pp 275–324

Chen C, Zhu H (2013) Are common symbiosis genes required for endophytic rice-rhizobial interactions? Plant Signal Behav 8(9):e25453. https://doi.org/10.4161/psb.25453

Chen L, Hao Z, Li K et al (2020a) Effects of growth-promoting rhizobacteria on maize growth and rhizosphere microbial community under conservation tillage in Northeast China. Microb Biotechnol. https://doi.org/10.1111/1751-7915.13693

Chen Q, Meyer W, Zhang Q, White J (2020b) 16S rRNA metagenomic analysis of the bacterial community associated with turf grass seeds from low moisture and high moisture climates. PeerJ 8:e8417. https://doi.org/10.7717/peerj.8417

Chi F, Shen SH, Cheng HP, Jing YX, Yanni YG, Dazzo FB (2005) Ascending migration of endophytic rhizobia, from roots to leaves, inside rice plants and assessment of benefits to rice growth physiology. Appl Environ Microbiol 71:7271–7278. https://doi.org/10.1128/AEM.71.11.7271-7278.2005

Cocking EC, Kothari SL, Batchelor CA et al (1995) Interaction of rhizobia with non-legume crops for symbiotic nitrogen fixation nodulation. In: Fendrik I, del Gallo M, Vanderleyden J, de Zamaroczy M (eds) *Azospirillum* VI and related microorganisms, vol 37. Springer, Berlin. https://doi.org/10.1007/978-3-642-79906-8_20

Correa OS, Rivas EA, Barneix AJ (1999) Cellular envelopes and tolerance to acid pH in *Mesorhizobium loti*. Curr Microbiol 38(6):329–334. https://doi.org/10.1007/PL00006812

Das K, Prasanna R, Saxena AK (2017) Rhizobia: a potential biocontrol agent for soilborne fungal pathogens. Folia Microbiol 62(5):425–435. https://doi.org/10.1007/s12223-017-0513-z

Datta C, Basu PS (2000) Indole acetic acid production by a *Rhizobium* species from root nodules of a leguminous shrub, *Cajanus cajan*. Microbiol Res 155(2):123–127. https://doi.org/10.1016/S0944-5013(00)80047-6

Datta B, Chakrabartty PK (2014) Siderophore biosynthesis genes of *Rhizobium* sp. isolated from *Cicer arietinum* L. 3 Biotech 4:391–401. https://doi.org/10.1007/s13205-013-0164-y

Delić D, Stajković-Srbinović O, Živković S, Protić N, Rasulić N, Kuzmanović Đ, Simić A (2012) Growth promotion of Italian ryegrass (*Loliummultiflorum lam.*) by application of plant growth promoting rhizobacteria. Zaštita Bilja 63(2):93–99

Delić D, Stajković-Srbinović O, Kuzmanović Đ, Rasulić N, Maksimović S, Radović J, Simić A (2013) Influence of plant growth promoting rhizobacteria on alfalfa, *Medicago sativa* L. yield by inoculation of a preceding Italian ryegrass, *Lolium multiflorum* Lam. In: Barth S, Milbourne D (eds) Breeding strategies for sustainable forage and turf grass improvement. Springer, Dordrecht

Delić D, Stajković-Srbinović O, Buntić A (2022) Hazards and usability of coal fly ash. In: Saljnikov E, Mueller L, Lavrishchev A, Eulenstein F (eds) Advances in understanding soil degradation. Innovations in landscape research. Springer, Cham

Dent D, Cocking E (2017) Establishing symbiotic nitrogen fixation in cereals and other non-legume crops: the greener nitrogen revolution. Agric Food Secur 6(1):84. https://doi.org/10.1186/s40066-016-0084-2

Deshwal VK, Dubey RC, Maheshwari DK (2003a) Isolation of plant growth-promoting strains of *Bradyrhizobium* (Arachis) sp. with biocontrol potential against *Macrophomina phaseolina* causing charcoal rot of peanut. Curr Sci 84:443–448

Deshwal VK, Pandey P, Kang SC, Maheshwari DK (2003b) Rhizobia as a biological control agent against soil borne plant pathogenic fungi. Indian J Exp Biol 41(10):1160–1164

Dinić Z, Maksimović J, Stanojković-Sebić A, Pivić R (2019) Prediction models for bioavailability of Mn, Cu, Zn, Ni and Pb in soils of Republic of Serbia. Agronomy 9(12):856. https://doi.org/10.3390/agronomy9120856

Dubey RC, Maheshwari DK (2011) Role of PGPR in integrated nutrient management of oil seed crops. In: Maheshwari DK (ed) Bacteria in agrobiology: plant nutrient management. Springer, Berlin, pp 1–15

Dubey P, Gupta GP, Dubey RC (2012) Culture filtrates of plant growth promoting *Bradyrhizobium* sp. (*Vigna*) strains VR1 and VR2 inhibit growth and sclerotia germination of *Macrophomina phaseolina in vitro*. N Y Sci J 5:1–9

Dubova L, Šenberga A, Alsiņa I (2017) Inoculated broad beans (*Vicia faba*) as a precrop for spring onions (*Allium cepa*). Res Rural Dev 2:33–39. https://doi.org/10.22616/RRD.23.2017.046

Dupin S, Geurts R, Kiers E (2020) The non-legume *Parasponia andersonii* mediates the fitness of nitrogen-fixing rhizobial symbionts under high nitrogen conditions. Front Plant Sci 10. https://doi.org/10.3389/fpls.2019.01779

Ehteshamul-Haque S, Ghaffar A (1993) Use of rhizobia in the control of root rot diseases of sunflower, okra, soybean and mungbean. J Phytopathol 138(2):157–163. https://doi.org/10.1111/j.1439-0434.1993.tb01372.x

Etesami H (2022) Root nodules of legumes: a suitable ecological niche for isolating non-rhizobial bacteria with biotechnological potential in agriculture. Curr Res Biotechnol 4:78–86. https://doi.org/10.1016/j.crbiot.2022.01.003

Fatima Z, Saleemi M, Zia M, Sultan T, Aslam M, Rehman R, Chaudhary MF (2009) Antifungal activity of plant growth-promoting rhizobacteria isolates against *Rhizoctonia solani* in wheat. Afr J Biotechnol 8(2). https://doi.org/10.5897/AJB2009.000-9040

Ferreira NS, Matos GF, Meneses CH et al (2020) Interaction of phytohormone-producing rhizobia with sugarcane mini-setts and their effect on plant development. Plant Soil 451(1):221–238. https://doi.org/10.1007/s11104-019-04388-0

Flores-Félix J, Menéndez E, Rivera L et al (2013) Use of *Rhizobium leguminosarum* as a potential biofertilizer for *Lactuca sativa* and *Daucus carota* crops. J Plant Nutr Soil Sci 176(6):876–882. https://doi.org/10.1002/jpln.201300116

Fujishige N, Kapadia N, De Hoff P, Hirsch A (2006) Investigations of *Rhizobium* biofilm formation. FEMS Microbiol Ecol 56(2):195–206. https://doi.org/10.1111/j.1574-6941.2005.00044.x

Fujishige N, Lum M, De Hoff P, Whitelegge J, Faull K, Hirsch A (2008) *Rhizobium* common nod genes are required for biofilm formation. Mol Microbiol 67(3):504–515. https://doi.org/10.1111/j.1365-2958.2007.06064.x

Gaby JC, Rishishwar L, Valderrama-Aguirre LC, Green SJ, Valderrama-Aguirre A, Jordan IK, Kostka JE (2017) Diazotroph community characterization via a high-throughput *nif*h amplicon sequencing and analysis pipeline. Appl Environ Microbiol 84(4):e01512. https://doi.org/10.1128/AEM.01512-17

Galindo FS, da Silva EC, Pagliari PH et al (2021) Nitrogen recovery from fertilizer and use efficiency response to *Bradyrhizobium* sp. and *Azospirillum brasilense* combined with N rates in cowpea-wheat crop sequence. Appl Soil Ecol 157:103764. https://doi.org/10.1016/j.apsoil.2020.103764

García-Fraile P, Carro L, Robledo M et al (2012) *Rhizobium* promotes non-legumes growth and quality in several production steps: towards a biofertilization of edible raw vegetables healthy for humans. PLoS One 7(5):e38122. https://doi.org/10.1371/journal.pone.0038122

Garrido-Oter R, Nakano R, Dombrowski N, Ma K, McHardy A, Schulze-Lefert P (2018) Modular traits of the *Rhizobiales* root microbiota and their evolutionary relationship with symbiotic rhizobia. Cell Host Microbe 24(1):155–167. https://doi.org/10.1016/j.chom.2018.06.006

Garrity GM, Bell JA, Lilburn TG (2004) Taxonomic outline of the prokaryotes. In: Bergey's manual of systematic bacteriology, 2nd edn. Bergey's Manual Trust, Athens. https://doi.org/10.1007/bergeysoutline

Gervasio G, Jerez-Mompie E, Belquis-Morales M, Caridad-Nápoles M (2019) Selección de una rizobacteria promotora del crecimiento en papa (*Solanum tuberosum* L.). Cult Trop 40(2):e07

Glaeser SP, Imani J, Alabid I et al (2015) Non-pathogenic *Rhizobium radiobacter* F4 deploys plant beneficial activity independent of its host *Piriformospora indica*. ISME J 10(4):871–884. https://doi.org/10.1038/ismej.2015.163

Glick BR (2015) Beneficial plant-bacterial interactions. Springer, Heidelberg, pp 1–28

Gopalakrishnan S, Sathya A, Vijayabharathi R, Varshney RK, Gowda CL, Krishnamurthy L (2015) Plant growth promoting rhizobia: challenges and opportunities. 3 Biotech 5(4):355–377. https://doi.org/10.1007/s13205-014-0241-x

Goyal RK, Schmidt MA, Hynes MF (2021) Molecular biology in the improvement of biological nitrogen fixation by rhizobia and extending the scope to cereals. Microorganisms 9:125. https://doi.org/10.3390/microorganisms9010125

Granada CE, Strochein M, Vargas LK, Bruxel M, de Sá EL, Passaglia LM (2014) Genetic diversity and symbiotic compatibility among rhizobial strains and *Desmodium incanum* and *Lotus* spp. plants. Genet Mol Biol 37(2):396–405. https://doi.org/10.1590/s1415-47572014000300012

Hahn L, de Sá ELS, Osório Filho BD, Machado RG, Damasceno RG, Giongo A (2016) Rhizobial inoculation, alone or coinoculated with *Azospirillum brasilense*, promotes growth of wetland rice. Rev Bras Ciênc Solo 40:6. https://doi.org/10.1590/18069657rbcs20160006

Halder AK, Chakrabarty PK (1993) Solubilization of inorganic phosphate by *Rhizobium*. Folia Microbiol 38:325–330. https://doi.org/10.1007/BF02898602

Hallmann J, Quadt-Hallmann A, Miller WG, Sikora RA, Lindow SE (2001) Endophytic colonization of plants by the biocontrol agent *Rhizobium etli* G12 in relation to *Meloidogyne incognita* infection. Phytopathology 91(4):415–422. https://doi.org/10.1094/PHYTO.2001.91.4.415

Humphry DR, Andrews M, Santos SR et al (2007) Phylogenetic assignment and mechanism of action of a crop growth promoting *Rhizobium radiobacter* strain used as a biofertilizer on graminaceous crops in Russia. Antonie Van Leeuwenhoek 91:105–113. https://doi.org/10.1007/s10482-006-9100-z

Hussain M, Zahir Z, Asghar H, Mahmood S (2014) Scrutinizing rhizobia to rescue maize growth under reduced water conditions. Soil Sci Soc Am J 78(2):538–545. https://doi.org/10.2136/sssaj2013.07.0315

Hussain M, Mehmood S, Zahir Z et al (2016a) Modulating nutrition, physiology and production of maize through *Rhizobium phaseoli* and *Mesorhizobium ciceri* inoculation under drought stress conditions. In: 7th international conference on water resources and the arid environments (ICWRAE 7), p 549–559

Hussain M, Zahir Z, Asghar H, Mubaraka R, Naveed M (2016b) Efficacy of rhizobia for improving photosynthesis, productivity, and mineral nutrition of maize. Clean Soil Air Water 44(11):1564–1571. https://doi.org/10.1002/clen.201500912

Hussain MB, Zahir ZA, Mehboob I et al (2018) *Mesorhizobium ciceri*-CR-39 inoculation to wheat for drought tolerance at critical growth stages. Pak J Bot 51(1):26. https://doi.org/10.30848/PJB2019-1(26)

Husssain MB, Mehboob I, Zahir ZA, Naveed M, Asghar HN (2009) Potential of *Rhizobium* spp. for improving growth and yield of rice (*Oryza sativa* L.). Soil Environ 28(1):49–55

Ibarra-Galeana JA, Castro-Martínez C, Fierro-Coronado RA, Armenta-Bojórquez AD, Maldonado-Mendoza IE (2017) Characterization of phosphate-solubilizing bacteria exhibiting the potential for growth promotion and phosphorus nutrition improvement in maize (*Zea mays* L.) in calcareous soils of Sinaloa, Mexico. Ann Microbiol 67:801–811. https://doi.org/10.1007/s13213-017-1308-9

Jaiswal S, Mohammed M, Ibny F, Dakora F (2021) Rhizobia as a source of plant growth-promoting molecules: potential applications and possible operational mechanisms. Front Sustain Food Syst 4:619676. https://doi.org/10.3389/fsufs.2020.619676

Janczarek M, Rachwał K, Cieśla J, Ginalska G, Bieganowski A (2015) Production of exopolysaccharide by *Rhizobium leguminosarum* bv. *trifolii* and its role in bacterial attachment and surface properties. Plant Soil 388:211–227. https://doi.org/10.1007/s11104-014-2320-5

Jatoi AR, Ujjan AA, Ramzan N, Naureen N, Memon RA, Shahzad MSS (2018) Antagonistic activity of *Rhizobium* isolates to fungus *Sclerotium rolfsii* Sacc. causing collar rot of sunflower (*Helianthus annuus* L.). Int J Biol Biotechnol 15(4):677–682

Jemai N, Gargouri S, Hemissi I, Mahmoud KB, Ksouri MF, Jemmali A (2021) *Rhizoctonia solani* affecting micropropagated Garnem (*Prunus amygdalus* x *Prunus persica*) rootstock-characterization and biocontrol with Rhizobia. J Plant Pathol 103(1):207–215. https://doi.org/10.1007/s42161-020-00712-1

Jiménez-Gómez A, Flores-Félix J, García-Fraile P, Mateos P, Menéndez E, Velázquez E, Rivas R (2018) Probiotic activities of *Rhizobium laguerreae* on growth and quality of spinach. Sci Rep 8(1):18632. https://doi.org/10.1038/s41598-017-18632-z

Kanaan II, Al-Barhawee NIK (2021) Isolation, characterization and optimization of wild type sinorhizobium meliloti to produce high concentrations of indole acetic acid. Iraqi J Sci 2021: 1415–1422. https://doi.org/10.24996/ijs.2021.62.5.3

Karavidas I, Ntatsi G, Ntanasi T, Vlachos I, Tampakaki A, Iannetta PP, Savvas D (2020) Comparative assessment of different crop rotation schemes for organic common bean production. Agronomy 10(9):1269. https://doi.org/10.3390/agronomy10091269

Katiyar P, Dubey RC, Maheshwari DK (2021) ACC deaminase-producing *Ensifer adhaerens* KS23 enhances proximate nutrient of *Pisum sativum* L. cultivated in high altitude. Arch Microbiol 203(5):2689–2698. https://doi.org/10.1007/s00203-021-02250-5

Kenawy A, Dailin DJ, Abo-Zaid GA et al (2019) Biosynthesis of antibiotics by PGPR and their roles in biocontrol of plant diseases. In: Sayyed R (ed) Plant growth promoting rhizobacteria for sustainable stress management, vol 13. Springer, Singapore, pp 1–35

Kiruthika S, Arunkumar M (2021) A comprehensive study on IAA production by *Bradyrhizobium japonicum* and *Bacillus subtilis* and its effect on *Vigna radiata* plant growth. Indian J Agric Res 55(5):570–576

Knežević M, Berić T, Buntić A, Delić D, Nikolić I, Stanković S, Stajković-Srbinović O (2021a) Potential of root nodule nonrhizobial endophytic bacteria for growth promotion of *Lotus corniculatus* L. and *Dactylis glomerata* L. J Appl Microbiol 131(6):2929–2940. https://doi.org/10.1111/jam.15152

Knežević M, Stajković-Srbinović O, Assel M, Milić M, Mihajlovski K, Delić D, Buntić A (2021b) The ability of a new strain of *Bacillus pseudomycoides* to improve the germination of alfalfa seeds in the presence of fungal infection or chromium. Rhizosphere 18:100353. https://doi.org/10.1016/j.rhisph.2021.100353

Knežević M, Berić T, Buntić A et al (2022) Native mesorhizobium strains improve yield and nutrient composition of the common bird's-foot trefoil grown in an acid soil. Rhizosphere 21: 100487. https://doi.org/10.1016/j.rhisph.2022.100487

Kumar H, Dubey RC, Maheshwari DK (2011) Effect of plant growth promoting rhizobia on seed germination, growth promotion and suppression of *Fusarium* wilt of fenugreek (*Trigonella foenum-graecum* L.). Crop Prot 30(11):1396–1403. https://doi.org/10.1016/j.cropro.2011.05.001

Kumar N, Galli M, Dempsey D, Imani J, Moebus A, Kogel K (2020) NPR1 is required for root colonization and the establishment of a mutualistic symbiosis between the beneficial bacterium *Rhizobium radiobacter* and barley. Environ Microbiol 23(4):2102–2115. https://doi.org/10.1111/1462-2920.15356

Kumar V, Rawat AK, Rao DLN (2021) Improving the performance of *Bradyrhizobium japonicum* by double inoculation in non-fertilized and fertilized wheat–soybean rotation. Agric Res. https://doi.org/10.1007/s40003-021-00600-9

Le Strange KK, Bender GL, Djordjevic MA, Rolfe BG, Redmond JW (1990) The *Rhizobium* strain NGR234 *nodD1* gene product responds to activation by simple phenolic compounds vanillin and isovanillin present in wheat seedling extracts. Mol Plant-Microbe Interact 3:214–220. https://doi.org/10.1094/MPMI-3-214

Leila B, El-Hafid N (2020) Biofertilizers and biopesticides: microbes for sustainable agriculture. In: Yadav AN, Rastegari AA, Yadav N, Kour D (eds) Advances in plant microbiome and sustainable agriculture. Springer, Singapore, pp 257–279

Li Q, Xing Y, Fu X et al (2021) Biochemical mechanisms of rhizospheric *Bacillus subtilis*-facilitated phytoextraction by alfalfa under cadmium stress – microbial diversity and metabolomics analyses. Ecotoxicol Environ Saf 212:112016. https://doi.org/10.1016/j.ecoenv.2021.112016

Lindström K, Mousavi S (2019) Effectiveness of nitrogen fixation in rhizobia. Microb Biotechnol 13(5):1314–1335. https://doi.org/10.1111/1751-7915.13517

Lodwig EM, Poole PS (2003) Metabolism of *Rhizobium* bacteroids. Crit Rev Plant Sci 22:37–38. https://doi.org/10.1080/713610850

Lugtenberg B, Kamilova F (2009) Plant growth promoting rhizobacteria. Annu Rev Microbiol 63: 541–556. https://doi.org/10.1146/annurev.micro.62.081307.162918

Lupwayi NZ, Clayton GW, Hanson KG, Rice WA, Biederbeck VO (2004) Endophytic rhizobia in barley, wheat and canola roots. Can J Plant Sci 84(1):37–45. https://doi.org/10.4141/p03-087

Ma W, Sebestianova SB, Sebestian J, Burd GI, Guinel FC, Glick BR (2003) Prevalence of 1-aminocyclopropane-1-carboxylate deaminase in *Rhizobium* spp. Antonie Van Leeuwenhoek 83(3):285–291. https://doi.org/10.1023/A:1023360919140

Maheshwari DK, Shukla S, Aeron A et al (2012) Rhizobacteria for management of nematode disease in plants. In: Maheshwari DK (ed) Bacteria in agrobiology: disease management. Springer, Berlin, pp 379–404. https://doi.org/10.1007/978-3-642-33639-3_14

Maheshwari DK, Saraf M, Dheeman S (2019) Plant growth-promoting rhizobacteria (PGPR) as protagonists of ever-sustained agriculture: an introduction. In: Maheshwari DK, Dheeman S (eds) Field crops: sustainable management by PGPR, sustainable development and biodiversity, vol 23. Springer, Cham, pp 1–10. https://doi.org/10.1007/978-3-030-30926-8_1

Maksimović J, Pivić R, Stanojković-Sebić A, Jovković M, Jaramaz D, Dinić Z (2021) Influence of soil type on the reliability of the prediction model for bioavailability of Mn, Zn, Pb, Ni and Cu in the soils of the Republic of Serbia. Agronomy 11(1):141. https://doi.org/10.3390/agronomy11010141

Manasa K, Reddy S, Triveni S (2017) Characterization of potential PGPR and antagonistic activities of *Rhizobium i*solates from different rhizosphere soils. J Pharmacogn Phytochem 6(3):51–54

Martínez L, Caballero-Mellado J, Orozco J, Martínez-Romero E (2003) Diazotrophic bacteria associated with banana (*Musa* spp.). Plant Soil 257(1):35–47. https://doi.org/10.1023/A:1026283311770

Martínez-Viveros O, Jorquera MA, Crowley DE, Gajardo GM, Mora ML (2010) Mechanisms and practical considerations involved in plant growth promotion by rhizobacteria. J Soil Sci Plant Nutr 10(3):293–319. https://doi.org/10.4067/S0718-95162010000100006

Masalha J, Kosegarten H, Elmaci O, Mengel K (2000) The central role of microbial activity for iron acquisition in maize and sunflower. Biol Fertil Soils 30:433–439. https://doi.org/10.1007/s003740050021

Masciarelli O, Llanes A, Luna V (2014) A new PGPR co-inoculated with *Bradyrhizobium japonicum* enhances soybean nodulation. Microbiol Res 169(7-8):609–615. https://doi.org/10.1016/j.micres.2013.10.001

Mayer E, Dörr de Quadros P, Fulthorpe R (2019) *Plantibacter flavus*, *Curtobacterium herbarum*, *Paenibacillus taichungensis*, and *Rhizobium selenitireducens* endophytes provide host-specific growth promotion of *Arabidopsis thaliana*, basil, lettuce, and bok choy plants. Appl Environ Microbiol 85:e00383. https://doi.org/10.1128/AEM.00383-19

McAdam EL, Reid JB, Foo E (2018) Gibberellins promote nodule organogenesis but inhibit the infection stages of nodulation. J Exp Bot 69(8):2117–2130. https://doi.org/10.1093/jxb/ery046

Mehboob I, Naveed M, Zahir ZA (2009) Rhizobial association with non-legumes: mechanisms and applications. Crit Rev Plant Sci 28(6):432–456. https://doi.org/10.1080/07352680903187753

Mehboob I, Zahir ZA, Arshad M, Tanveer A, Azam FE (2011) Growth promoting activities of different *Rhizobium* spp., in wheat. Pak J Bot 43(3):1643–1650

Mehboob I, Naveed M, Zahir ZA, Ashraf M (2012) Potential of rhizobia for sustainable production of non-legumes. In: Ashraf M, Öztürk M, Ahmad M, Aksoy A (eds) Crop production for agricultural improvement. Springer, Dordrecht, pp 659–704

Menéndez E, Pérez-Yépez J, Hernández M, Rodríguez-Pérez A, Velázquez E, León-Barrios M (2020) Plant growth promotion abilities of phylogenetically diverse *Mesorhizobium* strains: effect in the root colonization and development of tomato seedlings. Microorganisms 8(3):412. https://doi.org/10.3390/microorganisms8030412

Mhatre PH, Karthik C, Kadirvelu K et al (2019) Plant growth promoting rhizobacteria (PGPR): a potential alternative tool for nematodes bio-control. Biocatal Agric Biotechnol 17:119–128. https://doi.org/10.1016/j.bcab.2018.11.009

Mishra RPN, Singh RK, Jaiswal HK, Kumar V, Maurya S (2006) *Rhizobium*-mediated induction of phenolics and plant growth promotion in rice (*Oryza sativa* L.). Curr Microbiol 52:383–389. https://doi.org/10.1007/s00284-005-0296-3

Nadarajah KK (2017) *Rhizobium* in rice yield and growth enhancement. In: Alexander P, Hansen AP, Choudhary DK, Agrawal PK, Varma A (eds) *Rhizobium* biology and biotechnology. Springer, Cham, pp 83–103

Naqqash T, Hameed IA, Hanif MK, Majeed A, van Elsas JD (2016) Differential response of potato toward inoculation with taxonomically diverse plant growth promoting rhizobacteria. Front Plant Sci 7:144. https://doi.org/10.3389/fpls.2016.00144

Naseer I, Ahmad M, Nadeem SM, Ahmad I, Zahir ZA (2019) Rhizobial inoculants for sustainable agriculture: prospects and applications. In: Biofertilizers for sustainable agriculture and environment. Springer, Cham, pp 245–283

Neilands JB (1993) Siderophores. Arch Biochem Biophys 302:1–3. https://doi.org/10.1006/abbi.1993.1172

Nett RS (2017) Gibberellin biosynthesis by bacteria and its effect on the rhizobia-legume symbiosis. Graduate Theses and Dissertations. Iowa State University 16288. https://lib.dr.iastate.edu/etd/16288

Noel T, Sheng C, Yost C, Pharis R, Hynes M (1996) *Rhizobium leguminosarum* as a plant growth-promoting rhizobacterium: direct growth promotion of canola and lettuce. Can J Microbiol 42(3):279–283. https://doi.org/10.1139/m96-040

Nouwen N, Arrighi JF, Gully D, Giraud E (2021) RibBX of *Bradyrhizobium* ORS285 plays an important role in intracellular persistence in various *Aeschynomene* host plants. Mol Plant-Microbe Interact 34(1):88–99. https://doi.org/10.1094/MPMI-07-20-0209-R

Nurymova R, Tokhetova L, Baizhanova B, Daldabaueva G (2020) Influence of barley seeding rate and fertilizer dose on the yield of melilot in the subcover sowing in the rice crop rotation. Zemlj Biljka 69(1):65–73. https://doi.org/10.5937/ZemBilj2001065N

O'Gara F, Shanmugam KT (1976) Regulation of nitrogen fixation by *Rhizobia*. Export of fixed N_2 as NH_4^+. Biochim Biophys Acta 437:313–321. https://doi.org/10.1016/0304-4165(76)90001-5

Oldroyd G, Murray J, Poole P, Downie J (2011) The rules of engagement in the legume-rhizobial symbiosis. Annu Rev Genet 45(1):119–144. https://doi.org/10.1146/annurev-genet-110410-132549

Omar SA, Abd-Alla MH (1998) Biocontrol of fungal root rot diseases of crop plants by the use of rhizobia and bradyrhizobia. Folia Microbiol 43(4):431–437. https://doi.org/10.1007/BF02818587

Oro V, Knežević M, Dinić Z, Delić D (2020) Bacterial microbiota isolated from cysts of *Globodera rostochiensis* (Nematoda: Heteroderidae). Plan Theory 9(9):1146. https://doi.org/10.3390/plants9091146

Orozco-Mosqueda M, Duan J, DiBernardo M, Zetter E, Campos-García J, Glick B, Santoyo G (2019) The production of ACC deaminase and trehalose by the plant growth promoting bacterium *Pseudomonas* sp. UW4 synergistically protect tomato plants against salt stress. Front Microbiol 10:1392. https://doi.org/10.3389/fmicb.2019.01392

Padukkage D, Geekiyanage S, Reparaz JM, Bezus R, Balatti PA, Degrassi G (2021) *Bradyrhizobium japonicum*, *B. elkanii* and *B. diazoefficiens* interact with rice (*Oryza sativa*), promote growth and increase yield. Curr Microbiol 78:417–428. https://doi.org/10.1007/s00284-020-02249-z

Pandey P, Maheshwari DK (2007) Two-species microbial consortium for growth promotion of *Cajanus cajan*. Curr Sci 92:1137–1141

Parveen G, Noreen R, Shafique HA, Sultana V, Ehteshamul-Haque S, Athar M (2019) Role of rhizobia in suppressing the root diseases of soybean under soil amendment. Planta Daninha 37: 38. https://doi.org/10.1590/S0100-83582019370100038

Parveen G, Urooj F, Shafique HA, Rahman A, Ehteshamul-Haque S (2020) Role of rhizobia in suppressing the root rot and root knot disease of chili used alone or with *Pseudomonas aeruginosa*. Pak J Bot 52(3):1097–1104

Peix A, Rivas-Boyero A, Mateos P, Rodriguez-Barrueco C, Martínez-Molina E, Velazquez E (2001) Growth promotion of chickpea and barley by a phosphate solubilizing strain of *Mesorhizobium mediterraneum* under growth chamber conditions. Soil Biol Biochem 33(1):103–110. https://doi.org/10.1016/S0038-0717(00)00120-6

Perrine-Walker FM, Prayitno J, Rolfe BG, Weinman JJ, Hocart CH (2007) Infection process and the interaction of rice roots with rhizobia. J Exp Bot 58(12):3343–3350. https://doi.org/10.1093/jxb/erm181

Piromyou P, Greetatorn T, Teamtisong K, Tittabutr P, Boonkerd N, Teaumroong N (2017) Potential of rice stubble as a reservoir of bradyrhizobial inoculum in rice-legume crop rotation. Appl Environ Microbiol 83(22):e01488. https://doi.org/10.1128/AEM.01488-17

Poitout A, Martinière A, Kucharczyk B, Queruel N, Silva-Andia J, Mashkoor S et al (2017) Local signalling pathways regulate the *Arabidopsis* root developmental response to *Mesorhizobium loti* inoculation. J Exp Bot 68(5):1199–1211. https://doi.org/10.1093/jxb/erw502

Priyadarshini P, Choudhury S, Tilgam J, Bharati A, Sreeshma N (2021) Nitrogen fixing cereal: a rising hero towards meeting food security. Plant Physiol Biochem 167:912–920. https://doi.org/10.1016/j.plaphy.2021.09.012

Pundir RK, Jain P (2015) Mechanism of prevention and control of medicinal plant-associated diseases. In: Egamberdieva D, Shrivastava S, Varma A (eds) Plant-growth-promoting rhizobacteria (PGPR) and medicinal plants. Springer, Cham, pp 231–246

Qureshi MA, Shahzad H, Imran Z, Mushtaq M, Akhtar N, Ali MA, Mujeeb F (2013) Potential of *Rhizobium* species to enhance growth and fodder yield of maize in the presence and absence of L-tryptophan. J Anim Plant Sci 23(5):1448–1454

Qureshi M, Shahzad H, Saeed M, Ullah S, Ali M, Mujeeb F, Anjum M (2019) Relative potential of *Rhizobium* species to enhance the growth and yield attributes of cotton (*Gossypium hirsutum* L.). Eur J Soil Sci 8(2):159–166. https://doi.org/10.18393/ejss.544747

Rabie GH (1998) Induction of fungal disease resistance in *Vicia faba* by dual inoculation with *Rhizobium leguminosarum* and vesiculararbuscular mycorrhizal fungi. Mycopathologia 141: 159–166. https://doi.org/10.1023/a:1006937821777

Rafique M, Naveed M, Mustafa A et al (2021) The combined effects of gibberellic acid and *Rhizobium* on growth, yield and nutritional status in chickpea (*Cicer arietinum* L.). Agronomy 11:105. https://doi.org/10.3390/agronomy11010105

Rakić T, Pešić M, Kostić N et al (2021) Rhizobacteria associated with *Miscanthus* × *giganteus* improve metal accumulation and plant growth in the flotation tailings. Plant Soil 462(1-2):349–363. https://doi.org/10.1007/s11104-021-04865-5

Ramos T, Bellaj ME, Idrissi-Tourane AE, Daayf F, Hadrami IE (1997) Les phénolamides des rachis de palmes, composants de la réaction de défense du palmier dattier vis-à-vis de *Fusarium oxysporum* f.sp. albedinis, agent causal du bayoud. J Phytopathol 145:487–493. https://doi.org/10.1111/j.1439-0434.1997.tb00355.x

Reddy PM, Hernandez-Oane RJ, Kouchi H, Stacey G, Ladha JK (2000) Exploring the genetic potential of rice for forming symbiotic associations with rhizobia. In: Pedrosa FO, Hungria M, Yates MG, Newton WE (eds) Nitrogen fixation: from molecules to crop productivity. Kluwer Academic Publishers, Dordrecht, pp 331–332

Reddy PM, Rendón-Anaya M, Soto del Río MD, Khandual S (2007) Flavonoids as signaling molecules and regulators of root nodule development. Dyn Soil Dyn Plant 1(2):83–94

Reimann S, Hauschild R, Hildebrandt U, Sikora RA (2008) Interrelationships between *Rhizobium etli* G12 and *Glomus intraradices* and multitrophic effects in the biological control of the root-knot nematode *Meloidogyne incognita* on tomato. J Plant Dis Prot 115(3):108–113. https://doi.org/10.1007/BF03356249

Reitz M, Rudolph K, Schroder I, Hoffmann-Hergarten S, Hallmann J, Sikora R (2000) Lipopolysaccharides of *Rhizobium etli* strain G12 act in potato roots as an inducing agent of systemic resistance to infection by the cyst nematode *Globodera pallida*. Appl Environ Microbiol 66(8):3515–3518. https://doi.org/10.1128/AEM.66.8.3515-3518.2000

Rinaudi LV, Giordano W (2010) An integrated view of biofilm formation in rhizobia. FEMS Microbiol Lett 304(1):1–11. https://doi.org/10.1111/j.1574-6968.2009.01840.x

Rivas R, Peix A, Mateos PF, Trujillo ME, Martínez-Molina E, Velázquez E (2006) Biodiversity of populations of phosphate solubilizing rhizobia that nodulates chickpea in different Spanish soils. Plant Soil 287(1):23–33. https://doi.org/10.1007/s11104-006-9062-y

Rizwan M, Ali S, Ibrahim M et al (2015) Mechanisms of silicon-mediated alleviation of drought and salt stress in plants: a review. Environ Sci Pollut Res Int 22(20):15416–15431. https://doi.org/10.1007/s11356-015-5305-x

Robledo M, Rivera L, Jiménez-Zurdo JI et al (2012) Role of *Rhizobium* endoglucanase CelC2 in cellulose biosynthesis and biofilm formation on plant roots and abiotic surfaces. Microb Cell Factories 11:125. https://doi.org/10.1186/1475-2859-11-125

Rodríguez H, Fraga R, Gonzalez T, Bashan Y (2006) Genetics of phosphate solubilization and its potential applications for improving plant growth promoting bacteria. Plant Soil 287:15–21. https://doi.org/10.1007/978-1-4020-5765-6_2

Roque W, Guimarães S, Bonfim-Silva E (2021) Development of wheat plants coinoculated with rhizobium strains. Rev Bras Eng Agric Ambient 25(11):758–763. https://doi.org/10.1590/1807-1929/agriambi.v25n11p758-763

Rosenblueth M, Martínez-Romero E (2004) *Rhizobium etli* maize populations and their competitiveness for root colonization. Arch Microbiol 181:337–344. https://doi.org/10.1007/s00203-004-0661-9

Rosenblueth M, Ormeño-Orrillo E, López-López A et al (2018) Nitrogen fixation in cereals. Front Microbiol 9:e01794. https://doi.org/10.3389/fmicb.2018.01794

Saghafi D, Ghorbanpour M, Lajayer B (2018) Efficiency of *Rhizobium* strains as plant growth promoting rhizobacteria on morpho-physiological properties of *Brassica napus* L. under salinity stress. J Soil Sci Plant Nutr. https://doi.org/10.4067/S0718-95162018005000903

Saghafi D, Ghorbanpour M, Ajirloo HS, Lajayer BA (2019) Enhancement of growth and salt tolerance in *Brassica napus* L. seedlings by halotolerant *Rhizobium* strains containing ACC-deaminase activity. Plant Physiol Rep 24(2):225–235. https://doi.org/10.1007/s40502-019-00444-0

Santi C, Bogusz D, Franche C (2013) Biological nitrogen fixation in non-legume plants. Ann Bot 111(5):743–767. https://doi.org/10.1093/aob/mct048

Santillana N, Zúñiga D, Arellano C (2012) Growth promoting capacity in barley (*Hordeum vulgare*) and antagonistic potential of *Rhizobium leguminosarum* and *Rhizobium etli*. Agrocienc Urug 16(2):11–17. https://doi.org/10.31285/AGRO.17.534

Sarapat S, Longtonglang A, Umnajkitikorn K, Girdthai T, Boonkerd N, Tittabutr P, Teaumroong N (2020) Application of rice endophytic *Bradyrhizobium strain* SUTN9-2 containing modified ACC deaminase to rice cultivation under water deficit conditions. J Plant Interact 15(1):322–334. https://doi.org/10.1080/17429145.2020.1824028

Serova T, Tsyganova A, Tikhonovich I, Tsyganov V (2019) Gibberellins inhibit nodule senescence and stimulate nodule meristem bifurcation in pea (*Pisum sativum L.*). Front Plant Sci 10:285. https://doi.org/10.3389/fpls.2019.00285

Sheikh LI, Dawar S, Zaki MJ, Ghaffar A (2006) Efficacy of *Bacillus thuringiensis* and *Rhizobium meliloti* with nursery fertilizers in the control of root infecting fungi on mung bean and okra plants. Pak J Bot 38(2):465

Siddiqui IA, Shaukat SS (2002) Mixtures of plant disease suppressive bacteria enhance biological control of multiple tomato pathogens. Biol Fertil Soils 36(4):260–268. https://doi.org/10.1007/s00374-002-0509-x

Siddiqui IA, Ehteshamul-Haque S, Ghaffar A (1998) Effect of *Rhizobia* and fungal antagonists in the control of root infecting fungi on sun flower and chickpea. Pak J Bot 30:279–286

Siddiqui ZA, Baghel G, Akhtar MS (2007) Biocontrol of *Meloidogyne javanica* by *Rhizobium* and plant growth-promoting rhizobacteria on lentil. World J Microbiol Biotechnol 23(3):435–441. https://doi.org/10.1007/s11274-006-9244-z

Silva F, Winck B, Borges C et al (2020) Native rhizobia from southern Brazilian grassland promote the growth of grasses. Rhizosphere 16:100240. https://doi.org/10.1016/j.rhisph.2020.100240

Sindhu SS, Dua S, Verma MK, Khandelwal A (2010) Growth promotion of legumes by inoculation of rhizosphere bacteria. In: Khan MS, Zaidi A, Musarrat J (eds) Microbes for legume improvement. Springer, Vienna, pp 195–235

Singh SK, Pathak R (2015) Ecological manipulations of *Rhizobacteria* for curbing medicinal plant diseases. In: Egamberdieva D, Shrivastava S, Varma A (eds) Plant-growth-promoting rhizobacteria (PGPR) and medicinal plants, Soil biology, vol 42. Springer, Cham, pp 217–230

Singh A, Bhardwaj R, Singh IK (2019) Biocontrol agents: potential of biopesticides for integrated pest management. In: Giri B, Prasad R, Wu QS, Varma A (eds) Biofertilizers for sustainable agriculture and environment, Soil biology, vol 55. Springer, Cham, pp 413–433

Sridevi M, Mallaiah KV (2009) Phosphate solubilization by *Rhizobium* strains. Indian J Microbiol 49(1):98–102. https://doi.org/10.1007/s12088-009-0005-1

Stajković-Srbinović O, Delić D, Rasulić N, Buntić A, Kuzmanović D, Kostić-Kravljanac L, Sikirić B (2015a) Alfalfa growth on acid soil as influenced by calcification and *Ensifer* strains inoculation. Zemlj Biljka 64(2):1–8

Stajković-Srbinović O, Delić D, Rasulić N, Čakmak D, Kuzmanović Đ, Sikirić B (2015b) Mineral composition of red clover under *Rhizobium* inoculation and lime application in acid soil. Not Bot Hortic Agrobot Cluj Napoca 43(2):554–560. https://doi.org/10.15835/nbha43210007

Stajković-Srbinović O, De Meyer S, Kuzmanović Đ, Dinić Z, Delić D, Willems A (2020) Soybean seed chemical composition as influenced by *Bradyrhizobium* inoculation in soils with elevated nickel concentrations. Appl Soil Ecol 153:103576. https://doi.org/10.1016/j.apsoil.2020.103576

Stajković-Srbinović O, Kuzmanović Đ, Rasulić N, Knežević M, Ugrenović V, Buntić A, Delić D (2021) Effect of inoculation with *Bradyrhizobium* and phosphate solubilizing bacteria on soybean seed yield and composition. Zemlj Biljka 70(2):56–66. https://doi.org/10.5937/ZemBilj2102056S

Stan V, Gament E, Cornea CP, Voaideş C, Mirela D, Plopeanu G (2011) Effects of heavy metal from polluted soils on the rhizobium diversity. Not Bot Hortic Agrobot Cluj 39(1):88–95

Storey EP, Boghozian R, Little JL, Lowman DW, Chakraborty R (2006) Characterization of 'Schizokinen'; a dihydroxamate-type siderophore produced by *Rhizobium leguminosarum* IARI 917. Biometals 19(6):637–649. https://doi.org/10.1007/s10534-006-9001-7

Sujkowska-Rybkowska M, Kasowska D, Gediga K, Banasiewicz J, Stępkowski T (2020) *Lotus corniculatus*-rhizobia symbiosis under Ni, Co and Cr stress on ultramafic soil. Plant Soil 451(1-2):459–484. https://doi.org/10.1007/s11104-020-04546-9

Swarnalakshmi K, Prasanna R, Kumar A et al (2013) Evaluating the influence of novel cyanobacterial biofilmed biofertilizers on soil fertility and plant nutrition in wheat. Eur J Soil Biol 55:107–116. https://doi.org/10.1016/j.ejsobi.2012.12.008

Terakado-Tonooka J, Fujihara S, Ohwaki Y (2012) Possible contribution of *Bradyrhizobium* on nitrogen fixation in sweet potatoes. Plant Soil 367(1-2):639–650. https://doi.org/10.1007/s11104-012-1495-x

Thaweenut N, Hachisuka Y, Ando S, Yanagisawa S, Yoneyama T (2010) Two seasons' study on *nifH* gene expression and nitrogen fixation by diazotrophic endophytes in sugarcane (*Saccharum* spp. hybrids): expression of *nif*H genes similar to those of rhizobia. Plant Soil 338(1-2):435–449. https://doi.org/10.1007/s11104-010-0557-1

Tittabutr P, Payakapong W, Teaumroong N, Singleton P, Boonkerd N (2007) Growth, survival and field performance of bradyrhizobial liquid inoculant formulations with polymeric additives. Sci Asia 33(1):69. https://doi.org/10.2306/scienceasia1513-1874.2007.33.069

Tokhetova L, Baizhanova B, Baykenzhieva A, Kultasov B, Predić T (2021) Perspectives for cultivation of diversified crops in a rice (*Oryza sativa* L.)-based crop rotation in the Kyzylorda region, Kazakhstan. Zemlj Biljka 70(1):68–85. https://doi.org/10.5937/ZemBilj2101068T

Trinick MJ (1979) Structures of nitrogen fixing root nodules formed on *Parasponia andersonii* Planch. Can J Microbiol 25:565–578. https://doi.org/10.1139/m79-082

Triveni S, Prasanna R, Saxena AK (2012) Optimization of conditions for in vitro development of *Trichoderma viride* based biofilms as potential inoculants. Folia Microbiol 57:431–437. https://doi.org/10.1007/s12223-012-0154-1

Tulumello J, Chabert N, Rodriguez J, Long J, Nalin R, Achouak W, Heulin T (2021) *Rhizobium alamii* improves water stress tolerance in a non-legume. Sci Total Environ 797:148895. https://doi.org/10.1016/j.scitotenv.2021.148895

Ullah S, Khan MY, Asghar HN, Zahir ZA (2017a) Differential response of single and co-inoculation of *Rhizobium leguminosarum* and *Mesorhizobium ciceri* for inducing water deficit stress tolerance in wheat. Ann Microbiol 67:739–749. https://doi.org/10.1007/s13213-017-1302-2

Ullah S, Qureshi M, Ali M, Mujeeb F, Yasin S (2017b) Comparative potential of *Rhizobium* species for the growth promotion of sunflower (*Helianthus annuus* L.). Eur J Soil Sci 6(3):189–189. https://doi.org/10.18393/ejss.286694

van Rhijn P, Vanderleyden J (1995) The rhizobium–plant symbiosis. Microbiol Rev 59:124–142. https://doi.org/10.1128/mr.59.1.124-142.1995

Vandergheynst J, Scher H, Guo HY, Schultz D (2006) Water-in-oil emulsions that improve the storage and delivery of the biolarvacide *Lagenidium giganteum*. BioControl 52(2):207–229. https://doi.org/10.1007/s10526-006-9021-9

Vargas LK, Lisboa BB, Schlindwein G, Granada CE, Giongo A, Beneduzi A, Passaglia LMP (2009) Occurrence of plant growth-promoting traits in clover-nodulating rhizobia strains isolated from different soils in Rio Grande do Sul state. Rev Bras Ciênc Solo 33:1227–1235. https://doi.org/10.1590/S0100-06832009000500016

Verma M, Singh A, Dwivedi D, Arora NK (2020) Zinc and phosphate solubilizing *Rhizobium radiobacter* (LB2) for enhancing quality and yield of loose leaf lettuce in saline soil. Environ Sustain 3(2):209–218. https://doi.org/10.1007/s42398-020-00110-4

Vershinina Z, Chubukova O, Nikonorov Y et al (2021) Effect of *rosR* gene overexpression on biofilm formation by *Rhizobium leguminosarum*. Microbiology 90(2):198–209. https://doi.org/10.1134/S0026261721020144

Vessey JK (2003) Plant growth promoting rhizobacteria as biofertilizers. Plant Soil 255:571–586. https://doi.org/10.1023/A:1026037216893

Volpiano CG, Lisboa BB, São José JFB, de Oliveira AMR, Beneduzi A, Passaglia LMP, Vargas LK (2018) *Rhizobium* strains in the biological control of the phytopathogenic fungi *Sclerotium* (Athelia) *rolfsii* on the common bean. Plant Soil 432(1):229–243. https://doi.org/10.1007/s11104-018-3799-y

Volpiano CG, Lisbia BB, Granada CE et al (2019) Rhizobia for biological control of plant diseases. In: Kumar V, Prasad R, Kumar M, Choudhary D (eds) Microbiome in plant health and disease. Springer, Singapore, pp 315–336

Wang Q, Liu J, Zhu H (2018) Genetic and molecular mechanisms underlying symbiotic specificity in legume-rhizobium interactions. Front Plant Sci 9:e00313. https://doi.org/10.3389/fpls.2018.00313

Webster G, Gough C, Vasse J et al (1997) Interactions of rhizobia with rice and wheat. Plant Soil 194:115–122. https://doi.org/10.1023/A:1004283819084

Wu Q, Peng X, Yang M et al (2018) Rhizobia promote the growth of rice shoots by targeting cell signaling, division and expansion. Plant Mol Biol 97:507–523. https://doi.org/10.1007/s11103-018-0756-3

Yang L, Danzberger J, Schöler A, Schröder P, Schloter M, Radl V (2017) Dominant groups of potentially active bacteria shared by barley seeds become less abundant in root associated microbiome. Front Plant Sci 8:1005. https://doi.org/10.3389/fpls.2017.01005

Yanni YG, Rizk RY, El-Fattah FKA et al (2001) The beneficial plant growth-promoting association *of Rhizobium leguminosarum* bv. *trifolii* with rice roots. Aust J Plant Physiol 28(9):845. https://doi.org/10.1071/pp01069

Yildirim E, Karlidag H, Turan M, Dursun A, Goktepe F (2011) Growth, nutrient uptake, and yield promotion of broccoli by plant growth promoting rhizobacteria with manure. Hortic Sci 46(6):932–936. https://doi.org/10.21273/HORTSCI.46.6.932

Yoneyama T, Terakado-Tonooka Y, Bao Z, Minamisawa K (2019) molecular analyses of the distribution and function of diazotrophic *Rhizobia* and methanotrophs in the tissues and rhizosphere of non-leguminous plants. Plan Theory 8(10):408. https://doi.org/10.3390/plants8100408

Yuhashi KI, Ichikawa N, Ezura H et al (2000) Rhizobitoxine production by *Bradyrhizobium elkanii* enhances nodulation and competitiveness on *Macroptilium atropurpureum*. Appl Environ Microbiol 66(6):2658–2663. https://doi.org/10.1128/aem.66.6.2658-2663.2000

Zahir ZA, Arshad M, Frankenberger WT (2004) Plant growth promoting rhizobacteria: applications and perspectives in agriculture. Adv Agron 81:97–168. https://doi.org/10.1016/S0065-2113(03)81003-9

Zaim S, Bekkar AA, Belabid L (2017) *Rhizobium* as a crop enhancer and biofertilizer for increased non-legume production. In: Hansen A, Choudhary D, Agrawal P, Varma A (eds) *Rhizobium* biology and biotechnology. Soil biology, vol 50. Springer, Cham, pp 25–37

Chapter 8
Interactions of Nitrogen-Fixing Bacteria and Cereal Crops: An Important Dimension

Prashant Katiyar, Sandeep Kumar, and Naveen Kumar Arora

Abstract Cereals have been a crucial part of diets of humans as well as animals since the advent of agriculture. These crops along with many other crops on earth require a large amount of nitrogen (N) for their growth and enhanced productivity. This can be achieved through biological nitrogen fixation (BNF) by utilizing diazotrophs for converting atmospheric nitrogen to plant available form (NH_4^+). It has been established that rhizobia are the symbiotic nitrogen-fixers of legumes forming nodules in their roots, thus promoting the growth and health of legumes. However, rhizobial interactions with non-legumes have also been explored by scientists for their applied benefits. During the last 30 years or so, the studies on interactions of rhizobia in non-legumes are gaining interest, as it has been found that rhizobia are capable of associating with the roots of non-legumes, although without forming true root nodules. There are several mechanisms for beneficial interactions between rhizobia and non-legumes. Some of the direct mechanisms are phytohormones, signal molecules such as lipo-polysaccharides, lumichrome, siderophores, 1-aminocyclopropane-1-carboxylic acid (ACC) deaminase activity and solubilization of minerals such as phosphates for uptake by plants. Indirect methods for growth promotion of plants by rhizobia are differential breeding, gene editing, alteration of root morphology, inducing systemic resistance, increase in exopolysaccharide production, etc. These mechanisms will be discussed in this chapter to explain the importance of rhizobial interactions for sustainable production of non-leguminous crops.

Keywords Biological nitrogen fixation · Cereals · Phytohormones · Rhizobia

P. Katiyar · S. Kumar
Department of Botany and Microbiology, Gurukula Kangri (Deemed to be University), Haridwar, Uttarakhand, India

N. K. Arora (✉)
Department of Environmental Science, School of Earth and Environmental Sciences, Babasaheb Bhimrao Ambedkar University, Lucknow, Uttar Pradesh, India

© The Author(s), under exclusive license to Springer Nature Singapore Pte Ltd. 2022

D. K. Maheshwari et al. (eds.), *Nitrogen Fixing Bacteria: Sustainable Growth of Non-legumes*, Microorganisms for Sustainability 36,
https://doi.org/10.1007/978-981-19-4906-7_8

8.1 Introduction

Long-term viability of an agricultural system depends on the efficient use of internal resources. Microbial inoculation of plant growth-promoting microorganisms (PGPM) is a cost-effective and eco-friendly way to improve yields and quality of agri-produce. The targets of sustainable agriculture can only be achieved by the involvement of beneficial soil microorganisms (BSMs). Rhizobia, a group of very important BSMs are non-spore forming, Gram negative rods belonging to the family Rhizobiaceae comprising of genera *Rhizobium*, *Mesorhizobium*, *Ensifer* (formerly *Sinorhizobium*), *Bradyrhizobium*, *Phyllobacterium*, *Microvirga*, *Azorhizobium*, *Methylobacterium*, *Ochrobactrum*, *Devosia*, *Shinella* (α-proteobacteria), *Burkholderia*, *Cupriavidus* (previously *Ralstonia*) (β-proteobacteria) with some of the γ-proteobacteria (*Pseudomonadales*) (https://lpsn.dsmz.de/family/rhizobiaceae). Most of the members of this family have the extraordinary ability to form mutualistic association with plants and form nodules in the roots, which are the sight of biological nitrogen fixation (BNF) and make it (nitrogen) available to their symbiotic partners in a very efficient manner (De Lajudie and Young 2020). Rhizobia, on the other hand, can also form non-specific associative interactions with the roots of non-legumes without forming true nodules (Reyes and Schmidt 1979; Mehboob et al. 2009; Ullah et al. 2017). These implicit relationships between plant roots and beneficial bacteria stimulate growth and are of great significance due to the fact that many crops have shown boost in growth and yield production after rhizobial inoculation (Höflich et al. 1994; Yanni et al. 1997; Qureshi et al. 2019; Ali 2021).

Rhizobia are now being reported to enhance the growth and productivity of a wide range of cereals, grasses and other non-legumes (Dent and Cocking 2017; Silva et al. 2020). Rhizobia can directly affect non-legume plant development by producing phytohormones and vitamins, restricting plant ethylene production, improving nutrient absorption, increasing stress tolerance, and solubilizing minerals such as of phosphate, potassium and zinc. In addition, by communicating with other beneficial microorganisms, rhizobia can indirectly facilitate the growth of non-leguminous plants (Fig. 8.1).

Rhizobia are also capable of facilitating growth of crops under stressful conditions. For example, *R. leguminosarum* KS09 and *R. phaseoli* KR16 isolated from pea nodules under stress conditions promoted growth of *Pisum sativum* var. *arkel* under heavy metal stress conditions (Katiyar et al. 2021). Prior to that, Antoun et al. (1998) and Antoun and Prévost (2005) reported that specific rhizobial strains can be used as PGPR on non-legumes as well.

This chapter is aimed to discuss different mechanisms of action used by diazotrophs and non-legume for their interactions. Further, the importance of nitrogen fixation in crop improvement will be described.

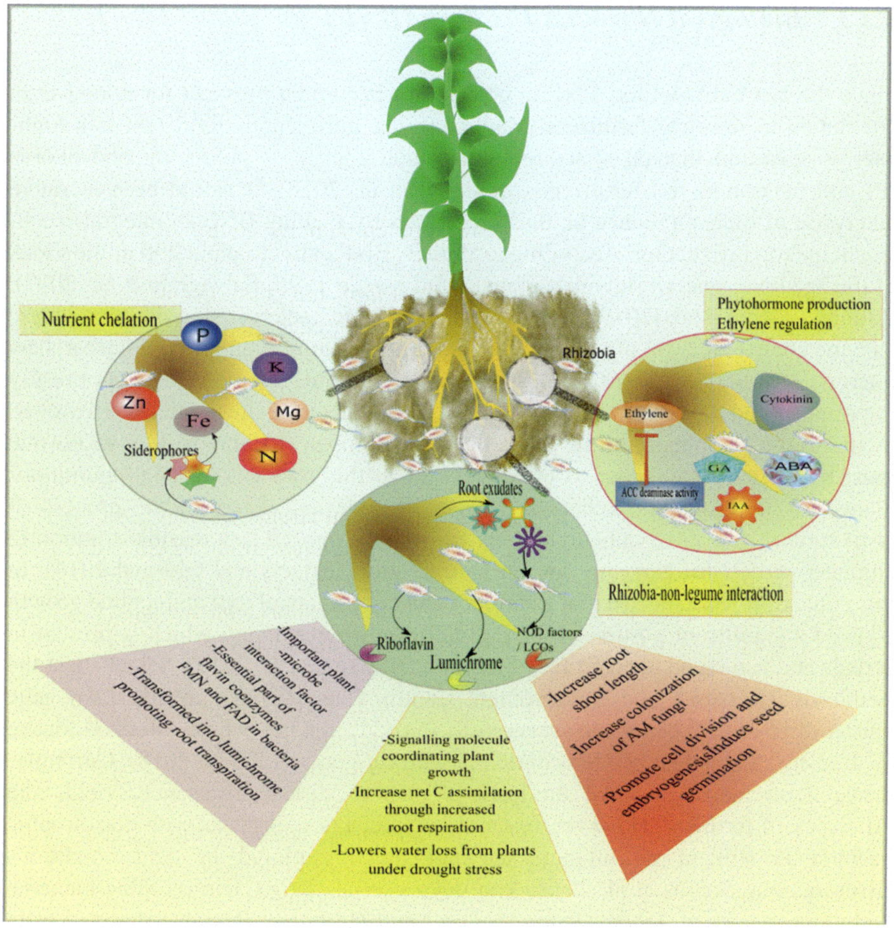

Fig. 8.1 Role of rhizobia in promoting growth of non-leguminous plants through various mechanisms and interactions

8.2 Plant Growth: Direct Mechanisms

Direct mechanisms include such processes which impact the plant growth directly by providing nutrients, releasing growth regulators, solubilizing nutrients, etc. These mechanisms induce the metabolic activity of plants, resulting in improvement in their adaptation and growth (Govindasamy et al. 2011; Glick 2014; Verma et al. 2019).

8.2.1 Biological Nitrogen Fixation (BNF)

Since the last half century, BNF in cereals has become a prospect for a long-term alternative to nitrogen fertilizers. Manufacturing nitrogenous fertilizers is a high-energy operation that takes six times as much energy as producing phosphorus (P) and potassium (K) fertilizers (Montalvo et al. 2016). It would be a lucrative enterprise to reduce reliance on these fertilizers by making BNF an integral mechanism in crop production. According to studies, BNF extracts about 200 million tons of nitrogen per year worldwide (Graham and Vance 1995; Salvagiotti et al. 2009). Ramírez-Puebla et al. (2019) found that diazotrophic bacteria using the nitrogenase enzyme contribute 50% of field N. As a result, BNF attracts agricultural researchers interest, encouraging them to introduce this trait into staple and economically important food crops as evidenced by rhizobia-legume interactions. However, owing to the complexity of the BNF, there are number of technical limitations that must be overcome in order to effectively solve this trait among non-leguminous plants. As a result, researchers are yet to decode and establish the process of N-fixation in cereals. Synthetic biology has been proposed as a feasible solution to engineer cereals for carrying out nitrogen fixation (Rogers and Oldroyd 2014). In this context, it was reported that a similar pathway for lateral root and nodule growth showed a significant portion of the nodule-forming machinery which is present in cereals and can be engineered for N-fixation (Katarína et al. 2019). Scientists at the Lethbridge Research and Development Centre, Canada, have created a *nif* gene cluster of 16 critical genes and successfully introduced it into wheat mitochondria, so as to restore the wheat plantlets enriched with *nif* genes (Li et al. 2018). Currently, research on maize is based on features such as disease resistance, yield, and efficiency of fertilizer. However, research on interaction of maize with diazotrophic bacteria is still in its infancy and not much explored in field conditions (Brusamarello-Santos et al. 2017; Van deynze et al. 2018). Diazotrophic bacteria, including species of *Azospirillum*, *Azoarcus* and *Herbaspirillum*, are reported from intercellular plant tissues as endophytes. This opens up the possibility of studying the importance of endophytes in maize cultivation and breeding programmes so that dependence on nitrogen-based chemical fertilizers gets reduced. Diazotrophs also aid plant growth by facilitating pathways such as vitamin and phytohormone synthesis, nutrient uptake, reduction of ethylene, and pathogen tolerance (Kemper et al. 1998; Karoney et al. 2020). As a result, under the right circumstances, a diazotrophic relationship may also support plant growth or wellbeing (Chelius and Triplett 2000; Teixeira et al. 2006). Taking all of this into account, combining the BNF mechanism with maize breeding techniques would aid in achieving long-term agricultural development (Yang and Zhu 2013; Puri et al. 2018). Since maize cannot fix nitrogen, therefore, there is no experimental evidence to quantify its effect on maize yields (Breedt et al. 2017). An examination of the energy cost of legumes such as soybean can provide a rough estimate of the energy cost of N-fixation in other crops. According to the study, soybeans absorb 19% of photosynthate, but this is offset by N-fixation, nitrate assimilation, and other pathways such as maintenance of carbon sink, which induces photosynthesis and boosts the health and the growth of plant (Ladha and Reddy 2000; Rodrigues et al. 2008).

8.2.2 Phytohormones Production

Phytohormones are the natural plant products that influence flowering, ageing, root development, bud, stem, and other component distortion and decay, as well as stem elongation suppression or promotion, fruit colour enhancement, leafing and/or leaf senescence prevention, and many other factors. The well-known plant hormones are auxins, cytokinins, gibberellins, abscisic acid and ethylene (Khalid et al. 2004). Production of phytohormones is an important feature of rhizobia (Taller and Sturtevant 1991; Matiru and Dakora 2004) and is considered as one of the most feasible approach to influence plant growth. The rhizobial genera have the ability to develop various forms of plant hormones. Various rhizobial species can secret cytokinins (Phillips and Torrey 1970; Miri et al. 2016), abscisic acid (ABA) (Figueiredo et al. 2008), gibberellic acid (Rafique et al. 2021), and indole acetic acid (IAA) (Etesami and Maheshwari 2018). Production of phytohormones helps plants in a variety of ways and influences a range of biological processes in the host, such as root hair formation, defense against pathogens, acquisition of nutrients, etc. (Dupuy et al. 2018; Khare et al. 2018).

8.2.3 Lumichrome

Lumichrome is a signal compound isolated from *Sinorhizobium meliloti* cell culture filtrates that has the ability to stimulate plant growth (Kanu et al. 2007; Dakora et al. 2015) by improving carbon assimilation through increased root respiration (Phillips et al. 1999). Lumichrome in addition to traditional plant hormones is reported to coordinate plant growth. It promotes growth at a nanomolar concentration (5 nm), while higher dosage (50 nm) inhibits root growth in non-legumes. Besides, rhizobacterial inoculation of plants has been proposed to mitigate the effects of drought conditions by producing lumichrome, which lowers leaf transpiration and reduces water loss through evaporative cooling in the leaves (Phillips et al. 1999).

8.2.4 Riboflavin

Root-colonizing bacteria are commonly known to release riboflavin and this biomolecule may have evolved due to plant-bacteria interactions and plays a very important role in the mutualism (García-Angulo 2017). Riboflavin is an essential part of the flavin coenzymes in bacteria and is reported to play important role in plant-bacteria interactions (Dakora et al. 2015; Lopez et al. 2019). Rhizobia release riboflavin, a vitamin that is quickly transformed via photo-proteolysis into lumichrome, which promotes plant growth by inducing root transpiration (Phillips et al. 1999; Dakora et al. 2015; Wheatley et al. 2020). This biomolecule can also play a crucial role in rhizobia and non-legume interactions.

8.2.5 *Ethylene Regulation*

Ethylene is a ripening hormone but also required in seed germination to lower the seed dormancy by promoting adventitious root and root hair creation (Ma et al. 2002; Arora et al. 2012; Maheshwari et al. 2015). Root elongation is impaired if the ethylene concentration stays elevated after germination (Le et al. 2001). Plant growth promotion by PGPR is due to the lowering of ethylene levels in plants by the synthesis of enzymc 1-aminocyclopropane-1-carboxylate (ACC) deaminase enzymes (Glick 2005), that breaks down ACC which is the immediate biosynthetic precursor to ethylene in plants. ACC-deaminase-producing PGPR protects plants from abiotic stresses supporting better growth under water stress conditions (Penrose and Glick 2003) and better ability to withstand the inhibitory effects of ethylene stress due to heavy metals (Arshad et al. 2007; Kotoky et al. 2019), salinity (Bharti and Barnawal 2019), drought (Ali et al. 2014; Saikia et al. 2018) and other abiotic stresses. It has been reported that inoculating plants with ACC-deaminase producing bacteria can lower the ACC and ethylene levels by two- to fourfold (Singha et al. 2018). *Pseudomonas* sp. strain ACP (Honma 1993) producing ACC-deaminase enzyme is used in a diverse variety of crop plants to alleviate stresses (Kaneko et al. 2002; Deshwal et al. 2003a, b; Ma et al. 2003; Conforte et al. 2010; Checcucci et al. 2017) and in biological control of plant pathogens. Many researchers are involved in studying ACC deaminase producing rhizobial strains to support growth of non-legumes under biotic and abiotic stresses (Singh et al. 2010).

8.2.6 *Siderophore Production*

Iron is a very important micro nutrient and is required for a number of biochemical activities including chlorophyll synthesis, and as co-factor of several enzymes. Its deficiency can result in diseases and makes plants susceptible to infection. Its overuse also results in negative effects such as the leaves turning bronze and developing tiny brown spots (Pourbabaee et al. 2020). In an aerobic climate, iron is insoluble at biological pH, where it occurs as oxyhydroxide in the trivalent state (Schwyn and Neilands 1987; Arora and Verma 2017), and the free Fe^{3+} supply at low soil pH (e.g. pH 4) is insufficient to satisfy plant demand. The concentration of free Fe^{3+} is too low in calcareous soils with high pH, to support optimum plant growth (Sharma and Johri 2003). Fe^{3+} chelation is the most effective pathway for plant roots to absorb iron under deficient conditions (Reid et al. 1984). PGPR are known to produce siderophores, which are low-molecular-weight, iron chelating agents produced accumulating ferric irons from the surroundings. Catecholates and hydroxamates are the two major types of siderophores. Bacterial siderophore as well as other chelating metabolites are thought to be important for the chelation of iron to plants, and a variety of plants may use bacterial siderophore complexes to meet out the needs of iron (Alaylar et al. 2020). Moreover, the use of microorganisms that develop chelating compounds under iron-deficient conditions can improve plant Fe acquisition (Crowley et al. 1992). Diverse species of the family *Rhizobiaceae* are reported to produce siderophores. *Rhizobium meliloti* (Schwyn and Neilands 1987; Arora et al. 2001), *Sinorhizobium meliloti*, *R. leguminosarum* bv. *viciae*, *R. leguminosarum* bv. *trifolii*, *R.leguminosarum* bv. *phaseoli*, *Rhizobium tropici* (Carson et al. 2000), *Rhizobium* sp. (Roy and Chakrabartty 2000; Ahemad and Khan 2012), *Bradyrhizobium* (Abd-Alla 1998), etc. are commonly reported to produce siderophores under iron-deficient conditions (Arora et al. 2001; Sullivan et al. 2006; Boiteau et al. 2016) both in the rhizosphere of legumes and non-legume plants.

8.2.7 Stress Tolerance

Plant growth-promoting bacteria producing ACC-deaminase have shown to shield plants from the negative effects of a variety of environmental stresses, including floods (Grichko and Glick 2001), phytopathogens (Chernin and Glick 2012; Li et al. 2013), heavy-metals (Yadav 2010; Etesami 2018; Katiyar et al. 2021) and drought (Vurukonda et al. 2016; Saikia et al. 2018). Inoculation of non-leguminous plants by stress-tolerant rhizobial strain roots has been proposed to mitigate water stress (Farwell et al. 2007) by modifying leaf stomatal conductance, transpiration, photo-respiration potential (Chi et al. 2005), and root development, resulting in increased nutrient and water usage efficiency and drought resistance. Abscisic acid or lumichrome (Phillips et al. 1999; Dakora et al. 2015) are rhizobacterial compounds that minimize leaf stomatal conductance and thus water loss through transpiration. So far, experiments have revealed rhizobia's growth-promoting effects on plant growth under drought stress through the development of various metabolites in cells and tissues of leguminous plants, but no significant correlation to the mechanism has been established. Halo and metal-tolerant rhizobia are also reported to protect non-legumes from high soil salinity and heavy metal contamination (Arora 2014).

Apart from above, rhizobia are now well known to carry out phosphate, zinc and potassium solubilization (Alikhani et al. 2006; Kamran et al. 2017; Menéndez et al. 2020). Recently, Verma et al. (2020) reported a zinc and phosphate solubilizing strain of *Rhizobium radiobacter* enhancing growth of lettuce. Similarly, Menéndez et al. (2020) reported phosphate and potassium solubilizing strains of *Mesorhizobium* species able to enhance growth of tomatoes. Antoun et al. (1998) reported phosphate-solubilizing strains of *Rhizobium* and *Bradyrhizobium* with ability to enhance growth of raddish. However, more research is required to use the symbiotic nitrogen fixers as solubilizers of other minerals. The use of such rhizobial strains can be beneficial not only for their symbiotic legume partners but also with non-legumes.

8.3 Plant Growth: Indirect Mechanisms

The indirect mechanisms involve biological control through which rhizobacteria promote plant growth by minimizing the effect of diseases, such as microbial infections, by providing systemic resistance, competence for resources and altering the root adherence.

8.3.1 Biological Control

Biological control is a process through which microorganisms, including different genera of N_2 fixing bacteria facilitate plant growth by limiting pathogen growth through the secretion of secondary metabolites such as antibiotics (Arguelles-Arias et al. 2009; Deshwal et al. 2003a, b), hydrogen cyanide production (HCN) (Chandra et al. 2007), anti-fungal enzymes (Kumar et al. 2009), and other antimicrobial processes (Maheshwari et al. 2015). Siderophore production by rhizobia gives them an advantage, resulting in pathogen exclusion (Arora et al. 2001). Rhizobia can fight soil-borne pathogens in a variety of ways such as parasitism, competition, etc. Pathogen displacement may occur as a result of nutrient competition between biocontrol bacteria and pathogens. The iron competition is a well-known example of microbial competition. *R. meliloti* was reported to secrete high Fe^{3+} affinity siderophores, sequestering iron and simultaneously inhibiting phytopathogen *Macrophomina phaseolina* (Arora et al. 2001). It is the non-availability of iron required by the phytopathogen that makes it weak and incompetent.

Rhizobia are also known to produce antimicrobial compounds that suppress or destroy phytopathogens. *R. leguminosarum* bv. *trifolii* has been found to produce antibiotic peptide trifolitoxin (Breil et al. 1993), that shows broad range of activity against bacteria (Breil et al. 1993; Scupham and Triplett 2006). Parasitism is also checked by PGPR by producing chitinolytic enzymes that break down the cell walls of pathogenic fungi. *R. leguminosarum*, *S. meliloti*, and *Bradyrhizobium japonicum* are rhizobial species that have been used extensively against the biocontrol of several phytopathogenic fungi of diverse crops (Ehteshamul Haque and Ghaffar 1993; Hemissi et al. 2011). Rhizobia are reported to prevent the development of *Fusarium oxysporum*, which causes root rot in tomato and sunflower (Perveen et al. 1994). *R. leguminosarum* bv. *viciae* is reported to control deleterious phytopathogens such as species of *Pythium*, which cause sugar beet damping-off (Huang and Erickson 2007). *Mesorhizobium loti* and *B. japonicum* are reported to control a variety of phytopathogens (*Phytophthora megasperma*, *Pythium ultimum*, *Fusarium oxysporum*) causing diseases in mustard and sunflower (Chandra et al. 2007). *R. meliloti* strains are also known to control root knot caused by *Meloidogyne incognita* in various crop plants (Dawar et al. 2008; Kumar et al. 2021).

8.3.2 Disease Resistance

Disease suppression may be achieved by microbial induction of tolerance in plants (Rabie 1998), and this process known as induced systemic resistance (ISR) occurs when PGPR strains trigger immune response by the plants against the pathogens. Rhizobia are known to activate plant defence mechanisms to control pathogens (Dutta et al. 2008), such as suppression of fungal pathogens of *Helianthus annus* and *Glycine max* (Alami et al. 1999; Dean et al. 2009). Rhizobia help non-leguminous plants flourish by developing a variety of biostimulants (Van Oosten et al. 2017; Backer et al. 2018). Lipopolysaccharides (LPS) formed by *Rhizobium etli* result in ISR in potato against infection by the cyst nematode *Globodera pallida* through a signal transduction (Reitz et al. 2002). Furthermore, rice plants inoculated with *R. leguminosarum* bv. *phaseoli* and *R. leguminosarum* bv. *trifolii* produce higher levels of phenolic compounds, which can result in ISR resulting in protection from phytopathogens (Hussain et al. 2009).

8.3.3 Interaction of Rhizobia with Other PGPR

Co-inoculation/dual inoculation of rhizobia with other PGPR can result in even better performance and yield of legumes as well as non-legumes, and there are success stories reported for diverse crops (Kumar et al. 2009; Widawati and Suliasih 2018) such as barley (Mirshekari et al. 2012; Baris et al. 2014), rice (Ashrafuzzaman et al. 2009), and maize (Gholami et al. 2009). Furthermore, a mixed inoculation with N_2 fixing and phosphate solubilizing bacteria is more efficient than a single inoculation for providing a more balanced nutrition to plants. Co-inoculating sugar beets with nitrogen-fixing and P-solubilizing bacteria increased sugar content and yield (Şahin et al. 2004). Co-inoculation of *R. meliloti* and *Bacillus thuringiensis* strains improved seed germination along with the fresh and dry weight of okra plants and also resulted in a reduction in infection by root infecting fungi. Consequently, using a balanced mixture of *Rhizobium*, and PGPR strains increase plant production in nutrient-deficient and degraded environments (Nadeem et al. 2014). For instance, under salinity stress, the co-inoculation effect of *Serratia* sp. and *Bradyrhizobium* in soybean significantly increased fresh and dry weight (Han and Lee 2005). However, further research is required to prepare and use consortia-based inoculants involving diazotrophs and other PGPR for non-leguminous plants in order to increase productivity in a sustainable manner. For this compatibility check and other ecological aspects involving biochemical and molecular level studies are required.

8.3.4 Altering Root Adherence

Root adhering soil (RAS) is very important because it is the immediate environment of the plant root and site of several activities crucial for growth and development of the plant. The uptake of nutrients, minerals and water along with development of mutualistic relationships (due to root exudation) depend on RAS zone (Guyonnet et al. 2018). As a result, around the root system, soil composition and aggregate stabilization are much more important (Gill et al. 2007). The secretion biopolymers known as exopolysaccharides (EPS) by PGPR are well known to improve aggregation of rhizosphere soil. This was confirmed by experimental findings showing that amendment of soil with EPS from PGPR resulted in better soil qualities and aggregation (Ashraf et al. 2006; Fatima and Arora 2020). The synthesis of significant quantities of EPS is a common feature of many rhizobia (Niehaus et al. 1993; Ghosh and Maiti 2016). Production of EPS helps microbes and the plants in many ways and is particularly important under abiotic stress conditions. Being hygroscopic in nature, EPS plays a very important role in the management and uptake of water, particularly in drought conditions, resulting in improved root dry mass, root structure and reduced fertilizer use by better nutrient uptake through improved RAS (Diaz et al. 1989). EPS produced by rhizobia is known to play a crucial role in *Rhizobium*-legume symbiosis. The role of EPS in establishing successful interactions of rhizobia with non-legumes is yet to be elucidated. However, it is certain that EPS is an important metabolite and has a significant role in formation of biofilm in the soil and on root surface, and helps in tiding over the biotic and abiotic stresses.

8.4 Role of Rhizobia in Improving Productivity of Cereals

8.4.1 Wheat

Various genera of nitrogen-fixing bacteria have a role to play in growth and development of cereals and other crops and are known to promote root growth and improve yield. Webster et al. (1997) investigated the entry of *Azorhizobium caulinodans* strain ORS571 into the lateral roots of wheat, reporting that rhizobia are present inside the cracks associated with emerging lateral roots after inoculation with ORS571 carrying a lacZ reporter gene. These researchers reported that the flavanone naringenin facilitated the colonization of lateral root cracks and intercellular colonization of cells at concentrations of 10^4 and 10^5 M. Sabry et al. (1997) studied lateral roots of wheat plants inoculated with *A. caulinodans* to measure its endophytic potential. Authors found that *A. caulinodans* cells were present between the cortical cells, within the xylem, and in the root meristem. In addition, as compared to uninoculated plants, they found substantial improvements in dry weight and nitrogen content of seeds of wheat crop. Höflich (2000) reported that *R. leguminosarum* bv. *trifolii* R39 resulted in better shoot formation and growth of wheat in experiments conducted under greenhouse conditions. Hilali et al. (2001) reported *R. leguminosarum* bv. *trifolii* from the roots of wheat grown in rotation with clover from two different Moroccan soils. They also reported that the rhizobial strain promoted wheat growth in greenhouse conditions. Some isolates showed their ability to increase shoots' fresh or dry matter yield. Strain IAT168 acted like a PGPR, exhibiting a 24% increase in shoot dry mass and grain yields in the loamy sand. Although, no PGPR activity was found in the silty clay Merchouch, and few of the isolates had a noticeable negative impact on yields. These researchers proposed that choosing an *R. leguminosarum* bv. *trifolii* strain was successful with clover and exhibited its beneficial effect on wheat in a crop rotation scheme. *Rhizobium* has been reported as a wheat growth booster in field trials. On the other hand, Anyia et al. (2009) found that the effect of *A. caulinodans* inoculation on wheat enhanced grain yield and total biomass by 34 and 49% respectively. The inoculated plants developed more tillers and had a larger leaf area in comparison to that of non-inoculated plants. Supplementation of bacterial inoculum gave the maximum grain yield, with a difference of 106% over the control. *Rhizobium* is also reported to improve the uptake of minerals in wheat resulting in its biofortification. By inoculating wheat plantlets with *Rhizobium* KYGT207, Kaci et al. (2005) were able to report a significant increase in shoot and root dry mass, root adhering soil (RAS) dry mass (dm) per root dm (RAS/RT), and RAS moisture stabilization. *Rhizobium* strain KYGT207 produced EPS in sandy soils under water stress, and its populations contributed to soil aggregation, in the rhizosphere (Afzal and Bano 2008).

8.4.2 Maize

Hossain et al. (2007) tested rhizobial strains on six different non-leguminous plants. Rhizobial strain PAR-401 was found to be most effective for *Zea mays* and resulted in more than double increase in shoot and root dry weight in comparison to control. Prévost et al. (2000) conducted an experiment to investigate the growth and mineral absorption of maize inoculated with *B. japonicum* strains. Strains 532C and USDA 136 enhanced root dry weight of corn by 8.5 and 6.7% respectively, while strain 532C enhanced shoot dry weight by 8.55%. In another experiment, rhizobial strains significantly improved the shoot dry weight of maize in comparison to uninoculated plants. Strains of *B. japonicum* may have an impact on maize growth and mineral nutrition. Souleimanov et al. (2002) conducted greenhouse tests in hydroponics solutions with four concentrations of Nod factor so as to determine the influence of Nod factor Nod Bj-v of *B. japonicum* on corn growth. They observed an increase in maize biomass of up to 11 and 7% at two different Nod factors, with 12% rise in total root length. Thus it was hypothesized that the 'hormone-like' actions of Nod factor were responsible for the growth enhancement of maize. According to Höflich (2000), *R. leguminosarum* bv. *trifolii* strain R39 boosted maize shoot development in greenhouse trials, however, in field trials on two different textured soils, inoculation of strain R39 also accelerated maize growth. In the field trials, author found a considerable increase in the absorption of nitrogen, phosphorus and potassium in maize and showed improved shoot and root growth. In a pure culture based metabolic experiment rhizobial strain R39 synthesized cytokinin, and auxin (Mehboob et al. 2009). *R. phaseoli* strains enhanced root length, shoot length, and seedling biomass of maize by 49, 21, and 35%, respectively, over the uninoculated control. *R. leguminosarum* increased biomass, root length, and shoot length of maize. The potential of these rhizobial strains to enhance the development of non-legumes has also been demonstrated in field trial experiments. Inoculating maize with strain R39 of *R. leguminosarum* bv. *trifolii* resulted in a considerable increase in shoot dry matter (Chabot et al. 1996; Egamberdiyeva 2007). Mandimba (1995) conducted studies on maize intercropped with common bean (*Phaseolus vulgaris*) inoculated with rhizobial strain. Results showed increase in growth and yields. Authors stated that maize benefitted from the nitrogen fixed by the rhizobia-legume association or bean consumed less nitrogen from the soil making it available to maize; nonetheless, the increment of maize yield showed that the *Rhizobium* strains were involved in some PGPR activity. Similarly, Chabot et al. (1996) investigated the effects of *R. leguminosarum* bv. *phaseoli* in a field inoculation trial for forage maize. The rhizobial strain showed a correlation in performance with simultaneous P-fertilization. When the prescribed quantity of P fertilizer was added, rhizobial inoculation showed best result in improving the maize growth in comparison to un-inoculated control.

8.4.3 Rice

Rhizobia can improve rice growth and yield production by acting as natural elicitors. Many researchers have reported rice growth enhancement due to *Rhizobium* inoculation (Biswas et al. 2000; Hussain et al. 2009; Yanni and Dazzo 2010). According to Singh et al. (2009), gfp-tagged rhizobia inoculated with rice generated significantly more root and shoot biomass, as well as other growth, biochemical, and physiological parameters. Several researchers have looked at how to improve rice growth and production in a laboratory or greenhouse setting. Singh et al. (2009) employed four rhizobia isolates for rice in lab and greenhouse experiments. The plant growth was improved by all the four isolates in terms of vegetative parameters and dry weight. In an experiment, Mishra et al. (2006) observed significant increase in biomass and grain production of rice plants inoculated with rhizobial strains. In a series of tests, Yanni and Dazzo (2010) proved the ability of *R. leguminosarum* bv. *trifolii* to colonize rice roots and its impact on rice crop. In addition, under field conditions strains *R. leguminosarum* bv. *trifolii* showed improvement in shoot and root development and grain yield of rice plants. Infact, *R. leguminosarum* bv. *trifolii* strains were found to be inner root colonizers with the ability to boost rice growth and production in both laboratory and field circumstances. Yanni et al. (2001) looked at the effects of rhizobial inoculation on rice development in the lab and in the greenhouse. They found that the rhizobia-rice relationship increased root and shoot development, subsequently improved seedling vigour, resulted in a considerable increase in grain output at maturity. Peng et al. (2002) did experiments to assess the effect of rhizobial inoculation on the growth and photosynthesis by rice plants. Three rhizobial strains were inoculated into the rice seed and pot soil, along with N fertilizer. In all three tests, rhizobial inoculation resulted in a considerable increase in photosynthetic rate and a significant rise in grain yield, suggesting that rhizobial strains may stimulate rice growth and production via mechanisms that boosted net photosynthetic rate. Mishra et al. (2006) used reverse phase-high performance liquid chromatography to estimate phenolic compounds from different regions of rice plants inoculated with rhizobial strains infected with *Rhizoctonia solani*. They found that rhizobium-inoculated rice plants produced more phenolic compounds than uninoculated control plants. Furthermore, it was shown that when rice plants were infected with the rhizobial strain RRE6, the phenolic content was more than when plants were treated with another rhizobial strain ANU843. In the case of both rhizobial strains, phenolic acids mediated ISR resulted in protection of plants from the attack of pathogens, which in turn boosted plant growth and production.

8.4.4 Barley

Peix et al. (2001) and Beatty et al. (2010) evaluated the effectiveness of a *Mesorhizobium mediterraneum* strain in enhancing barley growth in a plant growth chamber experiment. The strain PECA21 significantly improved the dry matter, N, P, Ca, and Mg contents of barley plants. Humphry et al. (2007) investigated the mode of action of *R. radiobacter* strain 204 in crop growth fostering due to involvement of GA production. They also found that the strain 204 culture and culture supernatant accelerated the metabolization of barley seed and facilitated shoot growth. Further, strain 204 stimulated barley growth by producing plant growth stimulating compounds, and GA-like activities play a key role. Thus, the strain was found to be instrumental in inducing early vegetative growth in barley.

Apart from these crops, nitrogen fixers (including both symbiotic and asymbiotic) are now being reported to enhance growth through nitrogen fixation and other mechanisms in non-legumes (Table 8.1). However, association of symbiotic diazotrophs, largely including rhizobia, needs to be explored more because of their efficiency and endophytic nature.

Table 8.1 Association of nitrogen-fixing bacteria with non-legume agricultural crops

Bacterial strains	Non-legume crops	Mechanism of action	Reference
Burkholderia sp. *Glucanoacetobacter diazotrophicus* *Serratia marcescens* *Pantoea agglomerans*	*Oryza sativa* L. Common name: rice	Biological nitrogen fixation and phytohormone production	Baldani et al. (2000) Muthukumarasamy et al. (2005, 2007) Gyaneshwar et al. (2001) Feng et al. (2006)
Burkholderia silvatlantica *Azospirillum brasilense* *Pseudomonas* sp. *A. brasilense*	*Zea mays* L. Common name: maize	Biological nitrogen fixation	Estrada et al. (2005) Riggs et al. (2001) Shaharoona et al. (2006) Dobbelaere et al. (2002)
Herbaspirillum seropedicae *H. rubrisubalbicans* *Gluconacetobacter diazotrophicus* *Klebsiella* sp. GR9 *Enterobacter* sp.	*Saccharum officinarum* L. Common name: sugarcane	Biological nitrogen fixation	Oliveira et al. (2002) Suman et al. (2005) Govindarajan et al. (2007) Mirza et al. (2001)
Rhizobium trifolii *Cellulomonas* sp.	*Triticum aestivum* L. Common name: wheat	Biological nitrogen fixation Increased uptake of N, P and K	Hilali et al. (2001) Egamberdiyeva and Höflich (2002)
Azospirillum brasilense *Achromobacter insolitus* *Zoogloea ramigera*	*Triticum aestivum hard* L. Common name: wheat	Biological nitrogen fixation, promoted root shoot length, chlorophyll content enhanced, IAA production increased	da Silveira et al. (2016)
Paenibacillus beijingensis BJ-18	*Triticum aestivum* L., *Zea mays* L., *Cucumis sativus* L. Common name: wheat, maize, and cucumber respectively	Enhancement of activity of glutamine synthetase (GS) and nitrate reductase (NR) in plants; upregulation of expression levels of N uptake and N metabolism genes: *AMT* (ammonium transporter), *NRT* (nitrate transporter), *NiR* (nitrite reductase), *NR*, *GS* and *GOGAT* (glutamate synthase)	Li et al. (2019)
Azotobacter sp. strain Avi2	*Oryza sativa* L. Common name: rice	Nitrogen fixation and enhancement of vegetative and reproductive growth in plant	Banik et al. (2019)
Azospirillum brasilense and *Pseudomonas fluorescens*	*Oryza sativa* L. Common name: rice	Co-inoculation accelerated N transformations and improved the N-supplying capacity of the rhizosphere	Zhang et al. (2018)

(continued)

Table 8.1 (continued)

Bacterial strains	Non-legume crops	Mechanism of action	Reference
		soil, and increased rice biomass	
Bacillus megaterium *Bacillus mycoides*	*Saccharum officinarum* L. Common name: sugarcane	Nitrogenase activity, and disease resistance	Singh et al. (2020)
Klebsiella sp. Br1 *Klebsiella pneumoniae* Fr1 *Bacillus pumilus* S1r1 *Acinetobacter* sp.	*Zea mays* L. Common name: maize	Nitrogen fixation and nitrogen remobilization Phosphate solubilization and Auxin production	Kuan et al. (2016)
Paenibacillus polymyxa P2b-2R	*Brassica napus* L. Common name: canola	Biological nitrogen fixation	Puri et al. (2016)
Lysinibacillus sphaericus (L1) *Klebsiella pneumoniae* (S2) *Bacillus cereus* (R2)	*Oryza sativa* L. Common name: rice	Nitrogen fixation, production of phytohormones, ACC deaminase activity, Biocontrol activity	Shabanamol et al. (2018)
Burkholderia kururiensis *Burkholderia tropica* *Herbaspirillum seropedicae*	*Sorghum bicolor* L. Moench Common name: sorghum	Biological nitrogen fixation, improve grain yield, fertilizer-N recovery	dos Santos et al. (2017)
Bacillus pumilus	*Solanum lycopersicum* L. Common name: tomato	Biological nitrogen fixation	Masood et al. (2020)
Brevundimonas naejangsanensis HWG-A15 *Brevundimonas terrae* KSL-145 *Brevundimonas* sp. X60 *Brevundimonas* sp. MM68May	*Solanum tuberosum* L. Common name: potato	Biological nitrogen fixation, phosphate solubilization	Naqqash et al. (2020)

8.5 Conclusion and Future Prospects

Researchers are now interested to raise the yields of cereals crops by application of N-fixing bacteria. Cereal crops are nitrogen-exhaustive crops, requiring tonnes of nitrogen for high yield, while a high dose of N with a poor nitrogen usage efficiency (20–30%) causes a slew of negative side effects on the soil, water, and climate. As a result, transferring N-fixation traits to cereals are needed to address the above issues and ensure agro-ecosystem sustainability. However, so far it has not been possible to engineer the cereal crops for BNF and is still a challenge for the researchers around the globe.

Scientific analysis and modification of diazotrophs and plants can be possible because of technological advancement in molecular techniques involving next-generation sequencing (NGS), genome editing, along with bioinformatics tools. Functional genomics, transcriptomics, and metabolomics can help in elucidating underlying mechanisms of legume-rhizobia symbiosis for successful implementation in non-legumes. Transcriptomics, proteomics, and metabolomics may produce large datasets by monitoring, identifying, and quantifying the gene expression of hosts and diazotrophs, and assessing the requirements or success of symbiosis-specific metabolites and genes. A combination of high-throughput and advanced technologies would allow the discovery of novel genes, monitoring of their expression at various stages of mutualistic interactions, and elucidation of genetic elements needed for effective coexistence. Non-leguminous model systems must be studied in order to harness them by breeding or engineering to improve N-fixation rates. Significant omics datasets for integrative analysis and mining of symbiotic genes can be created, allowing for the identification of genotypes for use in breeding programmes. Metagenomic research is expected to uncover microbiomes in the rhizosphere and other areas surrounding plants, both of which may be active in BNF. To make BNF a possibility in non-leguminous plants, targeted investment is needed to unearth the housekeeping and molecular-ecosystem-based mechanisms. Incorporating crop and microbial activity with genetic modification would make it easier to find short- and long-term options for increasing food productivity and utilizing the biodiversity in a proper way.

Acknowledgement Not applicable

Conflict of Interest Author(s) declares no conflict of interest.

References

Abd-Alla MH (1998) Growth and siderophore production in vitro of *Bradyrhizobium* (Lupin) strains under iron limitation. Eur J Soil Biol 34(2):99–104

Afzal A, Bano A (2008) *Rhizobium* and phosphate solubilizing bacteria improve the yield and phosphorus uptake in wheat (*Triticum aestivum*). Int J Agric Biol 10(1):85–88

Ahemad M, Khan MS (2012) Ecological assessment of biotoxicity of pesticides towards plant growth promoting activities of pea (*Pisum sativum*)-specific *Rhizobium* sp. strain MRP1. Emir J Food Agric 24:334–343

Alami Y, Champolivier L, Merrien A, Heulin T (1999) The role of *Rhizobium* sp., a rhizobacterium that produces exopolysaccharide in the aggregation of the rhizospherical soil of the sunflower: effects on plant growth and resistance to hydric constraints. Oleagi C Gras Lip 6:524–528

Alaylar B, Egamberdieva D, Gulluce M, Karadayi M, Arora NK (2020) Integration of molecular tools in microbial phosphate solubilization research in agriculture perspective. World J Microbiol Biotechnol 36:93

Ali Q (2021) Rhizobial consortium improves germination attributes and salinity tolerance in maize (*Zea mays* L.). Plant Environ 1(2):130–137

Ali SZ, Sandhya V, Rao LV (2014) Isolation and characterization of drought-tolerant ACC deaminase and exopolysaccharide-producing fluorescent *Pseudomonas* sp. Ann Microbiol 64(2):493–502

Alikhani HA, Saleh-Rastin N, Antoun H (2006) Phosphate solubilization activity of rhizobia native to Iranian soils. Plant Soil 287:35

Antoun H, Prévost D (2005) Ecology of plant growth promoting rhizobacteria. In: Siddiqui ZA (ed) PGPR: biocontrol and biofertilization. Springer, Dordrecht, pp 1–38

Antoun H, Beauchamp CJ, Goussard N, Chabot R, Lalande R (1998) Potential of *Rhizobium* and *Bradyrhizobium* species as plant growth promoting rhizobacteria on non-legumes: effect on radishes (*Raphanus sativus* L.). In: Hardarson G, Broughton WJ (eds) Molecular microbial ecology of the soil. Developments in plant and soil sciences, vol 83. Springer, Dordrecht, pp 57–67

Anyia AO, Archambault DJ, Bécquer CJ, Slaski JJ (2009) Plant growth-promoting diazotrophs and productivity of wheat on the Canadian prairies. In: Khan M, Zaidi A, Musarrat J (eds) Microbial strategies for crop improvement. Springer, Berlin, pp 287–300

Arguelles-Arias A, Ongena M, Halimi B, Lara Y, Brans A, Joris B, Fickers P (2009) *Bacillus amyloliquefaciens* GA1 as a source of potent antibiotics and other secondary metabolites for biocontrol of plant pathogens. Microb Cell Factories 8(1):1–12

Arora NK (2014) Plant microbe symbiosis: applied facets. Springer, New Delhi

Arora NK, Verma M (2017) Modified microplate method for rapid and efficient estimation of siderophore produced by bacteria. 3 Biotech 7(6):381

Arora NK, Kang SC, Maheshwari DK (2001) Isolation of siderophore-producing strains of *Rhizobium meliloti* and their biocontrol potential against *Macrophomina phaseolina* that causes charcoal rot of groundnut. Curr Sci 81:673–677

Arora NK, Tewari S, Singh S, Lal N, Maheshwari DK (2012) PGPR for protection of plant health under saline conditions. In: Maheshwari DK (ed) Bacteria in agrobiology: stress management. Springer, Heidelberg, pp 239–258

Arshad M, Saleem M, Hussain S (2007) Perspectives of bacterial ACC deaminase in phytoremediation. Trends Biotechnol 25(8):356–362

Ashraf M, Hasnain S, Berge O (2006) Effect of exo-polysaccharides producing bacterial inoculation on growth of roots of wheat (*Triticum aestivum* L.) plants grown in a salt-affected soil. Int J Environ Sci Technol 3(1):43–51

Ashrafuzzaman M, Hossen FA, Ismail MR, Hoque A, Islam MZ, Shahidullah SM, Meon S (2009) Efficiency of plant growth-promoting rhizobacteria (PGPR) for the enhancement of rice growth. Afr J Biotechnol 8(7):1247–1252

Backer R, Rokem JS, Ilangumaran G, Lamont J, Praslickova D, Ricci E, Smith DL (2018) Plant growth-promoting rhizobacteria: context, mechanisms of action, and roadmap to commercialization of biostimulants for sustainable agriculture. Front Plant Sci 9:1473

Baldani VLD, Baldani JI, Döbereiner J (2000) Inoculation of rice plants with the endophytic diazotrophs *Herbaspirillum seropedicae* and *Burkholderia* spp. Biol Fertil Soils 30:485–491

Banik A, Dash GK, Swain P, Kumar U, Mukhopadhyay SK, Dangar TK (2019) Application of rice (*Oryza sativa* L.) root endophytic diazotrophic *Azotobacter* sp. strain Avi2 (MCC 3432) can increase rice yield under greenhouse and field condition. Microbiol Res 219:56–65

Baris O, Sahin F, Turan M, Orhan F, Gulluce M (2014) Use of plant-growth-promoting rhizobacteria (PGPR) seed inoculation as alternative fertilizer inputs in wheat and barley production. Commun Soil Sci Plant Anal 45(18):2457–2467

Beatty PH, Anbessa Y, Juskiw P, Carroll RT, Wang J, Good AG (2010) Nitrogen use efficiencies of spring barley grown under varying nitrogen conditions in the field and growth chamber. Ann Bot 105(7):1171–1182

Bharti N, Barnawal D (2019) Amelioration of salinity stress by PGPR: ACC deaminase and ROS scavenging enzymes activity. In: Singh AK, Kumar A, Singh PK (eds) PGPR amelioration in sustainable agriculture. Woodhead Publishing, Sawston, pp 85–106

Biswas JC, Ladha JK, Dazzo FB (2000) Rhizobia inoculation improves nutrient uptake and growth of lowland rice. Soil Sci Soc Am J 64(5):1644–1650

Boiteau RM, Mende DR, Hawco NJ, McIlvin MR, Fitzsimmons JN, Saito MA, Repeta DJ (2016) Siderophore-based microbial adaptations to iron scarcity across the eastern Pacific Ocean. Proc Natl Acad Sci 113(50):14237–14242

Breedt G, Labuschagne N, Coutinho TA (2017) Seed treatment with selected plant growth-promoting rhizobacteria increases maize yield in the field. Ann Appl Biol 171(2):229–236

Breil BT, Ludden PW, Triplett EW (1993) DNA sequence and mutational analysis of genes involved in the production and resistance of the antibiotic peptide trifolitoxin. J Bacteriol 175(12):3693–3702

Brusamarello-Santos LC, Gilard F, Brulé L, Quilleré I, Gourion B, Ratet P, Hirel B (2017) Metabolic profiling of two maize (*Zea mays* L.) inbred lines inoculated with the nitrogen fixing plant-interacting bacteria *Herbaspirillum seropedicae* and *Azospirillum brasilense*. PLoS One 12(3):e0174576

Carson KC, Meyer JM, Dilworth MJ (2000) Hydroxamate siderophores of root nodule bacteria. Soil Biol Biochem 32(1):11–21

Chabot R, Antoun H, Cescas MP (1996) Growth promotion of maize and lettuce by phosphate-solubilizing *Rhizobium leguminosarum* biovar. *phaseoli*. Plant Soil 184(2):311–321

Chandra S, Choure K, Dubey RC, Maheshwari DK (2007) Rhizosphere competent *Mesorhizobium loti* MP6 induces root hair curling, inhibits *Sclerotinia sclerotiorum* and enhances growth of Indian mustard (*Brassica campestris*). Braz J Microbiol 38(1):124–130

Checcucci A, Azzarello E, Bazzicalupo M, De Carlo A, Emiliani G, Mancuso S, Mengoni A (2017) Role and regulation of ACC deaminase gene in *Sinorhizobium meliloti*: Is it a symbiotic, rhizospheric or endophytic gene. Front Genet 8:6

Chelius MK, Triplett EW (2000) Immunolocalization of dinitrogenase reductase produced by *Klebsiella pneumoniae* in association with *Zea mays* L. Appl Environ Microbiol 66(2):783–787

Chernin L, Glick BR (2012) The use of ACC deaminase to increase the tolerance of plants to various phytopathogens. In: Maheshwari DK (ed) Bacteria in agrobiology: stress management. Springer, Heidelberg, pp 279–299

Chi LS, Liu RS, Lee BJ (2005) Synthesis of Y2O3: Eu, Bi red phosphors by homogeneous coprecipitation and their photoluminescence behaviors. J Electrochem Soc 152(8):93

Conforte VP, Echeverria M, Sánchez C, Ugalde RA, Menéndez AB, Lepek VC (2010) Engineered ACC deaminase-expressing free-living cells of *Mesorhizobium loti* show increased nodulation efficiency and competitiveness on *Lotus* spp. J Gen Appl Microbiol 56(4):331–338

Crowley DE, Römheld V, Marschner H, Szaniszlo PJ (1992) Root-microbial effects on plant iron uptake from siderophores and phytosiderophores. Plant Soil 142(1):1–7

da Silveira APD, Sala VMR, Cardoso EJBN, Labanca EG, Cipriano MAP (2016) Nitrogen metabolism and growth of wheat plant under diazotrophic endophytic bacteria inoculation. Appl Soil Ecol 107:313–319

Dakora FD, Matiru V, Kanu AS (2015) Rhizosphere ecology of lumichrome and riboflavin, two bacterial signal molecules eliciting developmental changes in plants. Front Plant Sci 6:700

Dawar S, Sattar A, Zaki MJ (2008) Seed dressing with biocontrol agents and nematicides for the control of root knot nematode on sunflower and okra. Pak J Bot 40(6):2683–2691

De Lajudie P, Young JPW (2020) International committee on systematics of prokaryotes subcommittee on the taxonomy of rhizobia and agrobacteria. Minutes of the closed meeting by video conference

Dean JM, Mescher MC, De Moraes CM (2009) Plant–rhizobia mutualism influences aphid abundance on soybean. Plant Soil 323(1):187–196
Dent D, Cocking E (2017) Establishing symbiotic nitrogen fixation in cereals and other non-legume crops: the greener nitrogen revolution. Agric Food Secur 6(1):1–9
Deshwal VK, Dubey RC, Maheshwari DK (2003a) Isolation of plant growth-promoting strains of *Bradyrhizobium* (*Arachis*) sp. with biocontrol potential against *Macrophomina phaseolina* causing charcoal rot of peanut. Curr Sci 84:443–448
Deshwal VK, Pandey P, Kang SC, Maheshwari DK (2003b) Rhizobia as a biological control agent against soil borne plant pathogenic fungi. Ind J Exp Bot 4:1160–1164
Diaz CL, Melchers LS, Hooykaas PJ, Lugtenberg BJ, Kijne JW (1989) Root lectin as a determinant of host–plant specificity in the *Rhizobium*–legume symbiosis. Nature 338(6216):579–581
Dobbelaere S, Croonenborghs A, Thys A, Ptacek D, Okon Y, Vanderleyden J (2002) Effect of inoculation with wild type *Azospirillum brasilense* and *A. irakense* strains on development and nitrogen uptake of spring wheat and grain maize. Biol Fertil Soils 36(4):284–297
dos Santos CLR, Alves GC, de Matos Macedo AV, Giori FG, Pereira W, Urquiaga S, Reis VM (2017) Contribution of a mixed inoculant containing strains of *Burkholderia* spp. and *Herbaspirillum* spp. to the growth of three sorghum genotypes under increased nitrogen fertilization levels. Appl Soil Ecol 113:96–106
Dupuy LX, Mimault M, Patko D, Ladmiral V, Ameduri B, MacDonald MP, Ptashnyk M (2018) Micromechanics of root development in soil. Curr Opin Genet Dev 51:18–25
Dutta S, Mishra AK, Kumar BD (2008) Induction of systemic resistance against fusarial wilt in pigeon pea through interaction of plant growth promoting rhizobacteria and rhizobia. Soil Biol Biochem 40(2):452–461
Egamberdiyeva D (2007) The effect of plant growth promoting bacteria on growth and nutrient uptake of maize in two different soils. Appl Soil Ecol 36(2-3):184–189
Egamberdiyeva D, Höflich G (2002) Root colonization and growth promotion of winter wheat and pea by *Cellulomonas* spp. at different temperatures. Plant Growth Regul 38(3):219–224
Ehteshamul Haque S, Ghaffar A (1993) Use of rhizobia in the control of root rot diseases of sunflower, okra, soybean and mungbean. J Phytopathol 138(2):157–163
Estrada P, Mavingui P, Cournoyer B, Fontaine F, Balandreau J, Caballero-Mellado J (2005) A N_2-fixing endophytic *Burkholderia* sp. associated with maize plants cultivated in Mexico. Int J Syst Evol Microbiol 55:1233–1237
Etesami H (2018) Bacterial mediated alleviation of heavy metal stress and decreased accumulation of metals in plant tissues: mechanisms and future prospects. Ecotoxicol Environ Saf 147:175–191
Etesami H, Maheshwari DK (2018) Use of plant growth promoting rhizobacteria (PGPRs) with multiple plant growth promoting traits in stress agriculture: action mechanisms and future prospects. Ecotoxicol Environ Saf 156:225–246
Farwell AJ, Vesely S, Nero V, Rodriguez H, McCormack K, Shah S, Glick BR (2007) Tolerance of transgenic canola plants (*Brassica napus*) amended with plant growth-promoting bacteria to flooding stress at a metal-contaminated field site. Environ Pollut 147(3):540–545
Fatima T, Arora NK (2020) *Pseudomonas entomophila* PE3 and its exopolysaccharides as biostimulants for enhancing growth, yield and tolerance responses of sunflower under saline conditions. Microbiol Res 244:126671
Feng Y, Shen D, Song W (2006) Rice endophyte *Pantoea agglomerans* YS19 promotes host plant growth and affects allocations of host photosynthates. J Appl Microbiol 100(5):938–945
Figueiredo MV, Burity HA, Martinez CR, Chanway CP (2008) Alleviation of drought stress in the common bean (*Phaseolus vulgaris* L.) by co-inoculation with *Paenibacillus polymyxa* and *Rhizobium tropici*. Appl Soil Ecol 40(1):182–188
García-Angulo VA (2017) Overlapping riboflavin supply pathways in bacteria. Crit Rev Microbiol 43(2):196–209

Gholami A, Shahsavani S, Nezarat S (2009) The effect of plant growth promoting rhizobacteria (PGPR) on germination, seedling growth and yield of maize. World Acad Sci Eng Technol 49: 19–24

Ghosh PK, Maiti TK (2016) Structure of extracellular polysaccharides (EPS) produced by rhizobia and their functions in legume–bacteria symbiosis: a review. Ach Life Sci 10(2):136–143

Gill S, Abid M, Azam F (2007) Ammonium and nitrate fertilization effects on biomass yield of maize (*Zea mays* L.) with special emphasis on rhizodeposition. Pak J Agric Sci 44(3):443–449

Glick BR (2005) Modulation of plant ethylene levels by the bacterial enzyme ACC deaminase. FEMS Microbiol Lett 251(1):1–7

Glick BR (2014) Bacteria with ACC deaminase can promote plant growth and help to feed the world. Microbiol Res 169:30–39

Govindarajan M, Kwon SW, Weon HY (2007) Isolation, molecular characterization and growth-promoting activities of endophytic sugarcane diazotroph *Klebsiella* sp. GR9. World J Microbiol Biotechnol 23(7):997–1006

Govindasamy V, Senthilkumar M, Magheshwaran V, Kumar U, Bose P, Sharma V, Annapurna K (2011) *Bacillus* and *Paenibacillus* spp.: potential PGPR for sustainable agriculture. In: Maheshwari DK (ed) Microbiology monograph: plant growth and health promoting bacteria. Springer, Heidelberg, pp 333–364

Graham PH, Vance CP (1995) Nitrogen fixation in agriculture: application and perspectives. In: Tikhonovich IA, Provorov NA, Romanov VI, Newton WE (eds) Nitrogen fixation: fundamentals and applications. Current plant science and biotechnology in agriculture, vol 27. Springer, Dordrecht, pp 77–86

Grichko VP, Glick BR (2001) Amelioration of flooding stress by ACC deaminase-containing plant growth-promoting bacteria. Plant Physiol Biochem 39(1):11–17

Guyonnet JP, Cantarel AA, Simon L, Haichar FEZ (2018) Root exudation rate as functional trait involved in plant nutrient-use strategy classification. Ecol Evol 8(16):8573–8581

Gyaneshwar P, James EK, Mathan N, Reddy PM, Reinhold-Hurek B, Ladha JK (2001) Endophytic colonization of rice by a diazotrophic strain of Serretia marcessens. J Bacteriol 183:2634–2635

Han HS, Lee KD (2005) Physiological responses of soybean-inoculation of *Bradyrhizobium japonicum* with PGPR in saline soil conditions. Res J Agric Biol Sci 1(3):216–221

Hemissi I, Mabrouk Y, Neila A, Bouraoui M, Saidi M, Bouaziz S (2011) Effects of some *Rhizobium* strains on chickpea growth and biological control of *Rhizoctonia solani*. Afr J Microbiol Res 5(24):4080–4090

Hilali A, Prévost D, Broughton WJ, Antoun H (2001) Effets de l'inoculation avec des souches de *Rhizobium leguminosarum* biovar *trifolii* sur la croissance du blé dans deux sols du Maroc. Can J Microbiol 47(6):590–593

Höflich G (2000) Colonization and growth promotion of non-legumes by *Rhizobium* bacteria. In: Bell CR, Brylinsky M, Johnson-Green P (eds) Microbial biosystems: new Frontiers: proceedings of the 8th international symposium on microbial ecology. Atlantic Canada Society for Microbial Ecology, Halifax, pp 827–830

Höflich G, Wiehe W, Kühn G (1994) Plant growth stimulation by inoculation with symbiotic and associative rhizosphere microorganisms. Experientia 50(10):897–905

Honma M (1993) Stereospecific reaction of 1-aminocyclopropane-1-carboxylate deaminase. In: Pech JC, Latché A, Balagué C (eds) Cellular and molecular aspects of the plant hormone ethylene. Current plant science and biotechnology in agriculture, vol 16. Springer, Dordrecht, pp 111–116

Hossain F, Boddupalli PM, Sharma RK, Kumar P, Singh BB (2007) Evaluation of quality protein maize genotypes for resistance to stored grain weevil *Sitophilus oryzae* (*Coleoptera: Curculionidae*). Int J Tropic Ins Sci 27(2):114–121

Huang HC, Erickson RS (2007) Effect of seed treatment with *Rhizobium leguminosarum* on Pythium damping-off, seedling height, root nodulation, root biomass, shoot biomass, and seed yield of pea and lentil. J Phytopathol 155(1):31–37

Humphry DR, Andrews M, Santos SR, James EK, Vinogradova LV, Perin L, Cummings SP (2007) Phylogenetic assignment and mechanism of action of a crop growth promoting *Rhizobium radiobacter* strain used as a biofertiliser on graminaceous crops in Russia. Antonie Van Leeuwenhoek 91(2):105–113

Hussain MB, Mehboob I, Zahir ZA, Naveed M, Asghar HN (2009) Potential of *Rhizobium* spp. for improving growth and yield of rice (*Oryza sativa* L.). Soil Environ 28(1):49–55

Kaci Y, Heyraud A, Barakat M, Heulin T (2005) Isolation and identification of an EPS-producing *Rhizobium* strain from arid soil (Algeria): characterization of its EPS and the effect of inoculation on wheat rhizosphere soil structure. Res Microbiol 156(4):522–531

Kamran S, Shahid I, Baig DN, Rizwan M, Malik KA, Mehnaz S (2017) Contribution of zinc solubilizing bacteria in growth promotion and zinc content of wheat. Front Microbiol 8:2593

Kaneko T, Nakamura Y, Sato S, Minamisawa K, Uchiumi T, Sasamoto S, Tabata S (2002) Complete genomic sequence of nitrogen-fixing symbiotic bacterium *Bradyrhizobium japonicum* USDA110. DNA Res 9(6):189–197

Kanu S, Matiru VN, Dakora FD (2007) Strain and species differences in rhizobial secretion of lumichrome and riboflavin, measured using thin-layer chromatography. Symbiosis 43:37–43

Karoney EM, Ochieno DM, Baraza DL, Muge EK, Nyaboga EN, Naluyange V (2020) *Rhizobium* improves nutritive suitability and tolerance of *Phaseolus vulgaris* to *Colletotrichum lindemuthianum* by boosting organic nitrogen content. Appl Soil Ecol 149:103534

Katarína K, Elena M, Josef J (2019) Plant responses to stress induced by toxic metals and their nanoforms. In: Pessarakli M (ed) Handbook of plant and crop stress, 4th edn. CRC Press, Boca Raton, pp 479–522

Katiyar P, Dubey RC, Maheshwari DK (2021) ACC deaminase-producing *Ensifer adhaerens* KS23 enhances proximate nutrient of *Pisum sativum* L. cultivated in high altitude. Arch Microbiol 203:2689–2698

Kemper WD, Alberts EE, Foy CD, Clark RB, Ritchie JC, Zobel RW (1998) Aerenchyma, acid tolerance, and associative N fixation enhance carbon sequestration in soil. In: Management of carbon sequestration in soil. CRC Press, Boca Raton, pp 221–234

Khalid A, Arshad M, Zahir ZA (2004) Screening plant growth-promoting rhizobacteria for improving growth and yield of wheat. J Appl Microbiol 96(3):473–480

Khare E, Mishra J, Arora NK (2018) Multifaceted interactions between endophytes and plant: developments and prospects. Front Microbiol 9:2732

Kotoky R, Nath S, Maheshwari DK, Pandey P (2019) Cadmium resistant plant growth promoting rhizobacteria *Serratia marcescens* S2I7 associated with the growth promotion of rice plant. Environ Sustain 2(2):135–144

Kuan KB, Othman R, Abdul Rahim K, Shamsuddin ZH (2016) Plant growth-promoting rhizobacteria inoculation to enhance vegetative growth, nitrogen fixation and nitrogen remobilisation of maize under greenhouse conditions. PLoS One 11(3):e0152478

Kumar S, Pandey P, Maheshwari DK (2009) Reduction in dose of chemical fertilizers and growth enhancement of sesame (*Sesamum indicum* L.) with application of rhizospheric competent *Pseudomonas aeruginosa* LES4. Eur J Soil Biol 45(4):334–340

Kumar P, Kumar S, Dubey RC (2021) Biocontrol of *Macrophomina phaseolina* (tassi) Goid causing charcoal rot disease in *Lycopersicon esculentum* L. by using multi species bacterial consortia. Environ Conserv J 10(3):77–79

Ladha JK, Reddy PM (2000) The quest for nitrogen fixation in rice. International Rice Research Institute, Los Banos, p 354

Le J, Vandenbussche F, Van Der Straeten D, Verbelen JP (2001) In the early response of *Arabidopsis* roots to ethylene, cell elongation is up-and down-regulated and uncoupled from differentiation. Plant Physiol 125(2):519–522

Li N, Kong L, Zhou W, Zhang X, Wei S, Ding X, Chu Z (2013) Overexpression of Os_2H_{16} enhances resistance to phytopathogens and tolerance to drought stress in rice. PCTOC 115(3): 429–441

Li L, Niu Y, Ruan Y, DePauw RM, Singh AK, Gan Y (2018) Agronomic advancement in tillage, crop rotation, soil health, and genetic gain in durum wheat cultivation: a 17-year Canadian story. Agronomy 8(9):193

Li Y, Li Y, Zhang H, Wang M, Chen S (2019) Diazotrophic *Paenibacillus beijingensis* BJ-18 provides nitrogen for plant and promotes plant growth, nitrogen uptake and metabolism. Front Microbiol 10:1119

Lopez BR, Palacios OA, Bashan Y, Hernández-Sandoval FE, de Bashan LE (2019) Riboflavin and lumichrome exuded by the bacterium *Azospirillum brasilense* promote growth and changes in metabolites in *Chlorella sorokiniana* under autotrophic conditions. Algal Res 44:101696

Ma W, Penrose DM, Glick BR (2002) Strategies used by rhizobia to lower plant ethylene levels and increase nodulation. Can J Microbiol 48(11):947–954

Ma W, Guinel FC, Glick BR (2003) *Rhizobium leguminosarum* biovar viciae 1-aminocyclopropane-1-carboxylate deaminase promotes nodulation of pea plants. Appl Environ Microbiol 69(8):4396–4402

Maheshwari DK, Dheeman S, Agarwal M (2015) Phytohormone-producing PGPR for sustainable agriculture. In: Maheshwari DK (ed) Bacterial metabolites in sustainable agroecosystem. Springer, Cham, pp 159–182

Mandimba GR (1995) Contribution of nodulated legumes on the growth of *Zea mays* L. under various cropping systems. Symbiosis 19:213–222

Masood S, Zhao XQ, Shen RF (2020) *Bacillus pumilus* promotes the growth and nitrogen uptake of tomato plants under nitrogen fertilization. Sci Hortic 272:109581

Matiru VN, Dakora FD (2004) Potential use of rhizobial bacteria as promoters of plant growth for increased yield in landraces of African cereal crops. Afr J Biotechnol 3(1):1–7

Mehboob I, Naveed M, Zahir ZA (2009) Rhizobial association with non-legumes: mechanisms and applications. Crit Rev Plant Sci 28(6):432–456

Menéndez E, Pérez-Yépez J, Hernández M, Rodríguez-Pérez A, Velázquez E, León-Barrios M (2020) Plant growth promotion abilities of phylogenetically diverse *Mesorhizobium* strains: effect in the root colonization and development of tomato seedlings. Microorganisms 8:412

Miri M, Janakirama P, Held M, Ross L, Szczyglowski K (2016) Into the root: how cytokinin controls rhizobial infection. Trends Plant Sci 21(3):178–186

Mirshekari B, Hokmalipour SS, Sharifi RS, Farahvash F, Ebadi-Khazine-Gadim A (2012) Effect of seed biopriming with plant growth promoting rhizobacteria (PGPR) on yield and dry matter accumulation of spring barley (*Hordeum vulgare* L.) at various levels of nitrogen and phosphorus fertilizers. J Food Agric Environ 10(3/4):314–320

Mirza SM, Ahmad W, Latif F, Haurat J, Bally R, Normand P, Malik KA (2001) Isolation, partial characterization, and the effect of plant growth-promoting bacteria (PGPB) on micro-propagated sugarcane in vitro. Plant Soil 237(1):47–54

Mishra RP, Singh RK, Jaiswal HK, Kumar V, Maurya S (2006) Rhizobium-mediated induction of phenolics and plant growth promotion in rice (*Oryza sativa* L.). Curr Microbiol 52(5):383–389

Montalvo D, Degryse F, Da Silva RC, Baird R, McLaughlin MJ (2016) Agronomic effectiveness of zinc sources as micronutrient fertilizer. Adv Agron 139:215–267

Muthukumarasamy R, Cleenwerck I, Revathi G, Vadivelu M, Janssens D, Hoste B, Ui Gum K, Park K, Son CY, Sa T, CaballeroMellado J (2005) Natural association of *Gluconacetobacter diazotrophicus* and diazotrophic *Acetobacter peroxydans* with wetland rice. Syst Appl Microbiol 28:277–286

Muthukumarasamy R, Kang UG, Park KD, Jeon W-T, Park CY, Cho YS, Kwon S-W, Song J, Roh D-H, Revathi G (2007) Enumeration, isolation and identification of diazotrophs from Korean wetland rice varieties grown with long-term application of N and compost and their short-term inoculation effect on rice plants. J Appl Microbiol 102:981–991

Nadeem SM, Ahmad M, Zahir ZA, Javaid A, Ashraf M (2014) The role of mycorrhizae and plant growth promoting rhizobacteria (PGPR) in improving crop productivity under stressful environments. Biotechnol Adv 32(2):429–448

Naqqash T, Imran A, Hameed S, Shahid M, Majeed A, Iqbal J, Malik KA (2020) First report of diazotrophic *Brevundimonas* spp. as growth enhancer and root colonizer of potato. Sci Rep 10(1):1–14

Niehaus K, Kapp D, Pühler A (1993) Plant defence and delayed infection of alfalfa pseudonodules induced by an exopolysaccharide (EPS I)-deficient *Rhizobium meliloti* mutant. Planta 190(3): 415–425

Oliveira ALM, Urquiaga S, Döbereiner J, Baldani JI (2002) The effect of inoculating endophytic N_2-fixing bacteria on micropropagated sugarcane plants. Plant Soil 242:205–215

Peix A, Rivas-Boyero AA, Mateos PF, Rodriguez-Barrueco C, Martınez-Molina E, Velazquez E (2001) Growth promotion of chickpea and barley by a phosphate solubilizing strain of *Mesorhizobium mediterraneum* under growth chamber conditions. Soil Biol Biochem 33(1): 103–110

Peng S, Biswas JC, Ladha JK, Gyaneshwar P, Chen Y (2002) Influence of rhizobial inoculation on photosynthesis and grain yield of rice. Agronomy 94(4):925–929

Penrose DM, Glick BR (2003) Methods for isolating and characterizing ACC deaminase-containing plant growth-promoting rhizobacteria. Physiol Plant 118(1):10–15

Perveen S, Ehteshamul-Haque S, Ghaffar A (1994) Biological control of soilborne root infecting fungi in tomato and okra. Pak J Bot 26:181–181

Phillips DA, Torrey JG (1970) Cytokinin production by *Rhizobium japonicum*. Physiol Plant 23(6): 1057–1063

Phillips DA, Joseph CM, Yang GP, Martínez-Romero E, Sanborn JR, Volpin H (1999) Identification of lumichrome as a *Sinorhizobium* enhancer of alfalfa root respiration and shoot growth. Proc Natl Acad Sci 96(22):12275–12280

Pourbabaee AA, Shoaibi AA, Alikhani AA, Emami S (2020) Evaluation of iron content in calcareous soil treated with bacteria producing siderophore under drought stress. Iran J Soil Water Res 51(5):1305–1315

Prévost D, Saddiki S, Antoun H (2000) Growth and mineral nutrition of corn inoculated with effective strains of *Bradyrhizobium japonicum*. In: Proceedings of the 5th international PGPR workshop. Villa Carlos Paz, Córdoba

Puri A, Padda KP, Chanway CP (2016) Evidence of nitrogen fixation and growth promotion in canola (*Brassica napus* L.) by an endophytic diazotroph *Paenibacillus polymyxa* P2b-2R. Biol Fertil Soils 52(1):119–125

Puri A, Padda KP, Chanway CP (2018) Nitrogen-fixation by endophytic bacteria in agricultural crops: recent advances. In: Amanullah, Fahad S (eds) Nitrogen in agriculture. IntechOpen, London, pp 73–94

Qureshi MA, Shahzad H, Saeed MS, Ullah S, Ali MA, Mujeeb F, Anjum MA (2019) Relative potential of *Rhizobium* species to enhance the growth and yield attributes of cotton (*Gossypium hirsutum* L.). Eur J Soil Sci 8(2):159–166

Rabie GH (1998) Induction of fungal disease resistance in *Vicia faba* by dual inoculation with *Rhizobium leguminosarum* and vesicular-arbuscular mycorrhizal fungi. Mycopathology 141(3): 159–166

Rafique M, Naveed M, Mustafa A, Akhtar S, Munawar M, Kaukab S, Salem MZ (2021) The combined effects of gibberellic acid and *Rhizobium* on growth, yield and nutritional status in chickpea (*Cicer arietinum* L.). Agronomy 11(1):105

Ramírez-Puebla ST, Hernández MAR, Ruiz GG, Ormeño-Orrillo E, Martinez-Romero JC, Servín-Garcidueñas LE, Martínez-Romero E (2019) Nodule bacteria from the cultured legume *Phaseolus dumosus* (belonging to the *Phaseolus vulgaris* cross-inoculation group) with common tropici phenotypic characteristics and symbiovar but distinctive phylogenomic position and chromid. Syst Appl Microbiol 42(3):373–382

Reid CPP, Crowley DE, Kim HJ, Powell PE, Szaniszlo PJ (1984) Utilization of iron by oat when supplied as ferrated synthetic chelate or as ferrated hydroxamate siderophore. J Plant Nutr 7(1-5):437–447

Reitz M, Oger P, Niehaus K, Farrand SK, Hallmann J, Meyer A, Sikora RA (2002) Importance of the O-antigen, core-region and lipid A of rhizobial lipopolysaccharides for the induction of systemic resistance in potato to *Globodera pallida*. Nematology 4(1):73–79

Reyes VG, Schmidt EL (1979) Population densities of *Rhizobium japonicum* strain 123 estimated directly in soil and rhizospheres. Appl Environ Microbiol 37(5):854–858

Riggs PJ, Chelius MK, Iniguez AL, Kaeppler SM, Triplett EW (2001) Enhanced maize productivity by inoculation with diazotrophic bacteria. Aust J Plant Physiol 28:829–836

Rodrigues EP, Rodrigues LS, de Oliveira ALM, Baldani VLD, dos Santos Teixeira KR, Urquiaga S, Reis VM (2008) *Azospirillum amazonense* inoculation: effects on growth, yield and N_2 fixation of rice (*Oryza sativa* L.). Plant Soil 302(1):249–261

Rogers C, Oldroyd GE (2014) Synthetic biology approaches to engineering the nitrogen symbiosis in cereals. J Exp Bot 65(8):1939–1946

Roy N, Chakrabartty PK (2000) Effect of aluminum on the production of siderophore by *Rhizobium* sp. (*Cicer arietinum*). Curr Microbiol 41(1):5–10

Sabry SR, Saleh SA, Batchelor CA, Jones J, Jotham J, Webster G, Cocking EC (1997) Endophytic establishment of *Azorhizobium caulinodans* in wheat. Proc R Soc Lond B Biol Sci 264(1380): 341–346

Şahin F, Çakmakçi R, Kantar F (2004) Sugar beet and barley yields in relation to inoculation with N_2-fixing and phosphate solubilizing bacteria. Plant Soil 265(1):123–129

Saikia J, Sarma RK, Dhandia R, Yadav A, Bharali R, Gupta VK, Saikia R (2018) Alleviation of drought stress in pulse crops with ACC deaminase producing rhizobacteria isolated from acidic soil of Northeast India. Sci Rep 8(1):1–16

Salvagiotti F, Specht JE, Cassman KG, Walters DT, Weiss A, Dobermann A (2009) Growth and nitrogen fixation in high-yielding soybean: impact of nitrogen fertilization. Agronomy 101(4): 958–970

Schwyn B, Neilands JB (1987) Universal chemical assay for the detection and determination of siderophores. Anal Biochem 160(1):47–56

Scupham AJ, Triplett EW (2006) Determination of the amino acid residues required for the activity of the anti-rhizobial peptide antibiotic trifolitoxin. J Appl Microbiol 100(3):500–507

Shabanamol S, Divya K, George TK, Rishad KS, Sreekumar TS, Jisha MS (2018) Characterization and in planta nitrogen fixation of plant growth promoting endophytic diazotrophic *Lysinibacillus sphaericus* isolated from rice (*Oryza sativa*). Physiol Mol Plant Pathol 102:46–54

Shaharoona B, Arshad M, Zahir ZA, Khalid A (2006) Performance of *Pseudomonas* spp. containing ACC-deaminase for improving growth and yield of maize (*Zea mays* L.) in the presence of nitrogenous fertilizer. Soil Biol Biochem 38(9):2971–2975

Sharma A, Johri BN (2003) Growth promoting influence of siderophore-producing *Pseudomonas* strains GRP3A and PRS9 in maize (*Zea mays* L.) under iron limiting conditions. Microbiol Res 158(3):243–248

Silva FB, Winck B, Borges CS, Santos FL, Bataiolli RD, Backes T, Sá ELS (2020) Native rhizobia from southern Brazilian grassland promote the growth of grasses. Rhizosphere 16:100240

Singh MK, Kushwaha C, Singh RK (2009) Studies on endophytic colonization ability of two upland rice endophytes, *Rhizobium* sp. and *Burkholderia* sp., using green fluorescent protein reporter. Curr Microbiol 59(3):240–243

Singh N, Kumar S, Bajpai VK, Dubey RC, Maheshwari DK, Kang SC (2010) Biological control of *Macrophomina phaseolina* by chemo-tactic fluorescent *Pseudomonas aeruginosa* PN1 and its plant growth promontory activity in Chir-pine. Crop Prot 29(10):1142–1147

Singh RK, Singh P, Li HB, Song QQ, Guo DJ, Solanki MK, Li YR (2020) Diversity of nitrogen-fixing rhizobacteria associated with sugarcane: a comprehensive study of plant-microbe interactions for growth enhancement in *Saccharum* spp. BMC Plant Biol 20(1):1–21

Singha LP, Sinha N, Pandey P (2018) Rhizoremediation prospects of polyaromatic hydrocarbon degrading rhizobacteria, that facilitate glutathione and glutathione-S-transferase mediated stress response, and enhance growth of rice plants in pyrene contaminated soil. Ecotoxicol Environ Saf 164:579–588

Souleimanov A, Prithiviraj B, Smith DL (2002) The major Nod factor of *Bradyrhizobium japonicum* promotes early growth of soybean and corn. J Exp Bot 53(376):1929–1934

Sullivan JT, Jeffery EF, Shannon JD, Ramakrishnan G (2006) Characterization of the siderophore of *Francisella tularensis* and role of fslA in siderophore production. J Bacteriol 188(11): 3785–3795

Suman A, Gaur A, Shrivastava AK, Yadav RL (2005) Improving sugarcane growth and nutrient uptake by inoculating *Gluconacetobacter diazotrophicus*. Plant Growth Regul 47:155–162

Taller BJ, Sturtevant DB (1991) Cytokinin production by rhizobia. In: Hennecke H, Verma DPS (eds) Advances in molecular genetics of plant-microbe interactions. Current plant science and biotechnology in agriculture, vol 10. Springer, Dordrecht, pp 215–221

Teixeira FC, Reinert F, Rumjanek NG, Boddey RM (2006) Quantification of the contribution of biological nitrogen fixation to *Cratylia mollis* using the ^{15}N natural abundance technique in the semi-arid Caatinga region of Brazil. Soil Biol Biochem 38(7):1989–1993

Ullah S, Qureshi MA, Ali MA, Mujeeb F, Yasin S (2017) Comparative potential of *Rhizobium* species for the growth promotion of sunflower (*Helianthus annuus* L.). Eur J Soil Sci 6(3): 189–196

Van Deynze A, Zamora P, Delaux PM, Heitmann C, Jayaraman D, Rajasekar S, Bennett AB (2018) Nitrogen fixation in a landrace of maize is supported by a mucilage-associated diazotrophic microbiota. PLoS Biol 16(8):e2006352

Van Oosten MJ, Pepe O, De Pascale S, Silletti S, Maggio A (2017) The role of biostimulants and bioeffectors as alleviators of abiotic stress in crop plants. Chem Biol Technol Agric 4(1):1–12

Verma M, Mishra J, Arora NK (2019) Plant growth-promoting rhizobacteria: diversity and applications. In: Sobti R, Arora NK, Kothari R (eds) Environmental biotechnology: for sustainable future. Springer, Singapore, pp 129–173

Verma M, Singh A, Dwivedi DH, Arora NK (2020) Zinc and phosphate solubilizing *Rhizobium radiobacter* (LB2) for enhancing quality and yield of loose leaf lettuce in saline soil. Environ Sustain 3:209–218

Vurukonda SS, Vardharajula S, Shrivastava M, SkZ A (2016) Enhancement of drought stress tolerance in crops by plant growth promoting rhizobacteria. Microbiol Res 184:13–24

Webster G, Gough C, Vasse J, Batchelor CA, O'callaghan KJ, Kothari SL, Cocking EC (1997) Interactions of rhizobia with rice and wheat. In: Ladha JK, de Bruijn FJ, Malik KA (eds) Opportunities for biological nitrogen fixation in rice and other non-legumes. Developments in plant and soil sciences, vol 75. Springer, Dordrecht, pp 115–122

Wheatley RM, Ford BL, Li L, Aroney ST, Knights HE, Ledermann R, Poole PS (2020) Lifestyle adaptations of *Rhizobium* from rhizosphere to symbiosis. Proc Natl Acad Sci 117(38): 23823–23834

Widawati S, Suliasih (2018) The effect of plant growth promoting rhizobacteria (PGPR) on germination and seedling growth of *Sorghum bicolor* L. Moench. IOP Conf Ser Earth Environ Sci 166:012022

Yadav SK (2010) Heavy metals toxicity in plants: an overview on the role of glutathione and phytochelatins in heavy metal stress tolerance of plants. S Afr J Bot 76(2):167–179

Yang DT, Zhu X (2013) Modernization of agriculture and long-term growth. J Monet Econ 60(3): 367–382

Yanni YG, Dazzo FB (2010) Enhancement of rice production using endophytic strains of *Rhizobium leguminosarum bv. trifolii* in extensive field inoculation trials within the Egypt Nile delta. Plant Soil 336(1):129–142

Yanni YG, Rizk RY, Corich V, Squartini A, Ninke K, Philip-Hollingsworth S, Dazzo FB (1997) Natural endophytic association between *Rhizobium leguminosarum* bv. *trifolii* and rice roots and assessment of its potential to promote rice growth. In: Ladha JK, de Bruijn FJ, Malik KA (eds) Opportunities for biological nitrogen fixation in rice and other non-legumes. Developments in plant and soil sciences, vol 75. Springer, Dordrecht, pp 99–114

Yanni YG, Rizk RY, Abd El-Fattah FK, Squartini A, Corich V, Giacomini A, Dazzo FB (2001) The beneficial plant growth-promoting association of *Rhizobium leguminosarum bv. trifolii* with rice roots. Funct Plant Biol 28(9):845–870

Zhang J, Hussain S, Zhao F, Zhu L, Cao X, Yu S, Jin Q (2018) Effects of *Azospirillum brasilense* and *Pseudomonas fluorescens* on nitrogen transformation and enzyme activity in the rice rhizosphere. J Soils Sediments 18(4):1453–1465

Chapter 9
Microbiome to the Rescue: Nitrogen Cycling and Fixation in Non-legumes

Papri Nag and Sampa Das

Abstract Nitrogen fertilizers are essential for producing higher crop yields and are used extensively by farmers. However, nitrogenous fertilizers are also highly polluting sources in the agroecosystem. A pollution-free alternative to synthetic fertilizer is the process by which prokaryotes (diazotroph) can fix nitrogen from the atmosphere and release it into the rhizosphere in plant utilizable forms. This process is known as biological nitrogen fixation (BNF). The process of BNF has been studied and exploited in leguminous crops, but the potential of BNF has not been exploited in non-leguminous crops like rice, wheat or maize. The successful establishment of BNF into the rhizosphere depends not only on the diazotroph but also in its interaction with microbes already present in the soil. The presence of nitrifying, denitrifying, ammonia-oxidizing or nitrate/nitrite reducing bacteria in the rhizosphere microbiome determines the availability of nitrogen for plant growth. Hence, understanding the processes controlling N transformation is important for creating microbial consortia with the ability to the supplement nitrogen requirement of the crop plant and to reduce dependence on synthetic nitrogenous fertilizers.

Keywords Microbial consortia · N Cycle · Nitrogen fixation · Wheat · Non-legume · Diazotrophs

9.1 Introduction

Nitrogen fertilizer is one of the major causes of pollution from agricultural systems (Martínez-Dalmau et al. 2021). It was estimated that approximately 52% of the applied N is lost to the environment as ammonia, nitrate, and nitrogen oxides causing soil, groundwater, and air pollution (Ladha et al. 2016). Among cereals, rice has one of the lowest nitrogen use efficiency (NUE), biomass yield kg^{-1} N applied) (Norton et al. 2015). NUE is controlled by the genome of the host plant; however, part of

P. Nag (✉) · S. Das
Division of Plant Biology, Bose Institute, Kolkata, West Bengal, India

© The Author(s), under exclusive license to Springer Nature Singapore Pte Ltd. 2022

D. K. Maheshwari et al. (eds.), *Nitrogen Fixing Bacteria: Sustainable Growth of Non-legumes*, Microorganisms for Sustainability 36,
https://doi.org/10.1007/978-981-19-4906-7_9

NUE is governed by the ability of the plant to harbour beneficial bacteria in its rhizosphere (Wang et al. 2020). Specialized microbes expressing the enzyme nitrogenase can convert nitrogen (N_2) from the atmosphere to plant utilizable forms like NH_3. Unlike legumes, which can form specialized structures like root nodules, providing a niche for host-specific nitrogen-fixing microbes, cereals can benefit from associative and endophytic diazotrophs which accumulate in the rhizosphere as a result of chemotaxis towards the root exudates (Lugtenberg and Kamilova 2009). The rhizosphere is an eclectic mix of microorganisms, consisting of nitrifying, denitrifying, ammonia-oxidizing and nitrogen-fixing bacteria. Thus, the NUE of plants depends on the presence or absence of other microbes in the rhizosphere, which in turn determine the microbial transformation of the fixed nitrogen. Hence, for augmenting the contribution of BNF in non-legumes, understanding the microbial nitrogen cycle in the rhizosphere is essential.

The plant along with its microbial community in the phyllosphere, rhizosphere and endosphere is known as a 'holobiont', and the combined microbial genome is called the 'microbiome'. Each plant species contains a 'core microbiome' which does not change with the environmental or soil conditions (Zilber-Rosenberg and Rosenberg 2008). Although according to the 'hologenome theory of evolution' variations in host and microbiome may lead to evolutionary change in the 'core microbiome' (Zilber-Rosenberg and Rosenberg 2008). Plants recruit beneficial bacteria by providing carbon in the form of sugar, amino acids, organic acids, fatty acids and growth factors (Mavrodi et al. 2021). It may be predicted that the abundance of microbial taxa in the 'core microbiome' will be affected by any change in host plant exudates. However, the role of 'core microbiome' directly responsible for recruitment of the microbial network has not been demonstrated till now (Coskun et al. 2017a, b). The functional genomic approach of defining a 'core microbiome' may solve this question in the future (Lemanceau et al. 2017). The microbiome helps the host plant by providing plant growth-promoting metabolites (Spence et al. 2014), bio-protection against biotic and abiotic stresses (Naylor et al. 2017). One of the most important benefits host plants derive from the microbiome is the acquisition of nutrients by recruiting phosphate-solubilizing, iron-acquiring, or nitrogen-fixing bacteria.

9.2 Nitrogen Cycling in the Rhizosphere of Non-legumes

Nitrogen cycling in an agroecosystem can be divided into three steps: input into the system, retention or immobilization within the system, and loss from the system. The input components may be from naturally occurring sources (rain, decomposition of biomaterials like plants, microbes, small eukaryotes, or manures), or from synthetic fertilizers. Immobilization into the ecosystem can occur when the bio-available N is utilized by non-diazotrophic microbes and released upon their decomposition or by charges on the soil particles to be later absorbed by the plants. The third step is the loss occurring from the agro-ecosystem through leaching, volatilization and

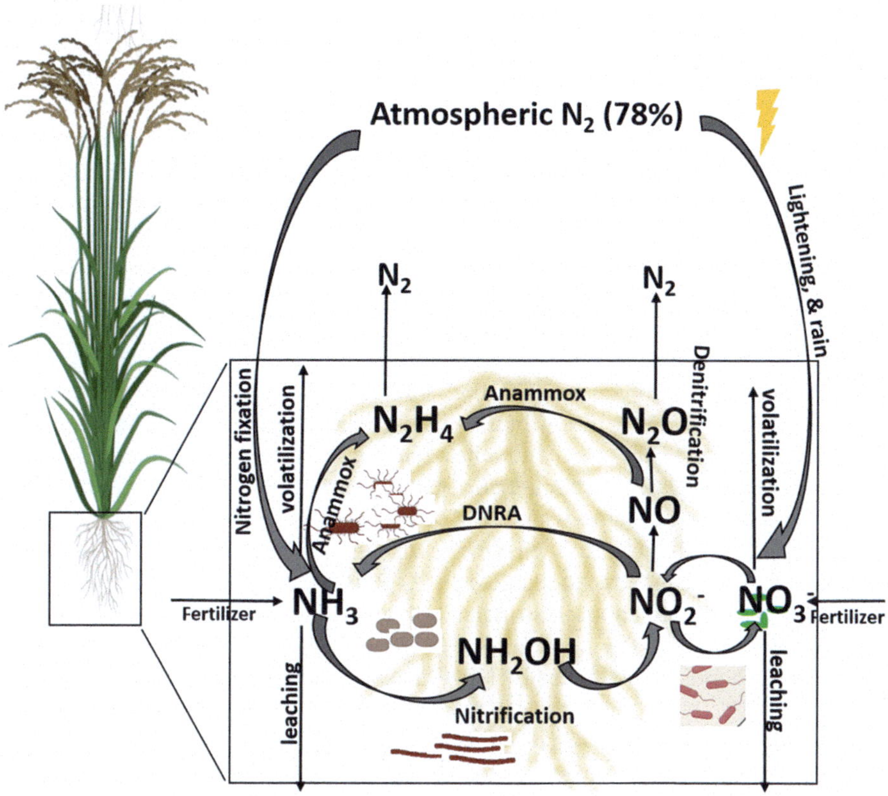

Fig. 9.1 Schematic diagram of rhizosphere microbial N-cycle. Part of the figure was created using BioRender (https://biorender.com/)

microbial processes. Microbial processes like biological nitrogen fixation (BNF), nitrification and denitrification play essential roles in all the three steps of N-cycling. Ammonification and dissimilatory nitrification also play important role in microbial N-cycling (Fig. 9.1).

9.2.1 Biological Nitrogen Fixation

The microbial N-cycle commences with the recruitment of prokaryotes capable of fixing nitrogen into the system. The diazotrophs can fix nitrogen under asymbiotic conditions as free-living forms in the rhizosphere, associative nitrogen fixers on the root surface, endophytes inside plant roots and as symbionts inside root nodules (reviewed in Santi et al. 2013). Free-living nitrogen fixers like *Azotobacter*, *Pseudomonas* and others can fix nitrogen in the absence of any host, while, associative diazotrophs like *Azospirillum*, *Burkholderia*, and *Klebsiella* can fix nitrogen more efficiently in association with plant roots. Many plant endophytic diazotrophs have been demonstrated in cereals: *Pantoea agglomerans* (Quecine et al. 2012) and *Gluconacetobacter diazotrophicus* (Sevilla et al. 2001) in sugarcane; *Azospirillum* spp. (Boddey and Dobereiner 1982), *Azoarcus* spp. (Reinhold-Hurek and Hurek 1998), *Herbaspirillum* (Baldani et al. 1986) and *Pseudomonas stutzeri* (Desnoues et al. 2003) in rice; and *Burkholderia* spp. (Caballero-Mellado et al. 2004) in maize. Endophytic diazotrophs can fix nitrogen inside plant roots and do not survive well in the soil, while symbiotic diazotrophs can fix nitrogen only inside the micro-aerobic conditions of the root nodules (Carvalho et al. 2014).

The process of biological nitrogen fixation is exclusive to prokaryotes expressing the enzyme nitrogenase. Three types of nitrogenases have been discovered in prokaryotes: Fe-Molybdenum nitrogenase, Fe-Vanadium containing nitrogenase and only Fe-containing nitrogenase. The best-studied and most commonly occurring nitrogenase is the Fe-Mo nitrogenase which is discussed in this article. Fe-Mo nitrogenase enzyme has two subunits: the Fe subunit, also known as the dinitrogenase reductase, is a homomeric subunit encoded by *nifH*, and the Mo-Fe subunit, also known as dinitrogenase, is a heteromeric subunit encoded by *nifD* and *nifK*. Nitrogen fixation by this enzyme is energy consuming for the cell, each molecule of dinitrogen reduced requires 16 ATP molecules (Table 9.1). Hence, the process of nitrogen fixation is stringently regulated at multiple levels (Santos-Medellín et al. 2017). The structural genes, *nifH*, *nifD* and *nifK*, coding for the nitrogenase subunits are regulated by NifA, an enhancer binding protein (EBP), and its inhibitor NifL. NifL/NifA regulates the transcription of *nif-H*, *D*, *K* and other *nif* genes required for maturation of nitrogenase depending on the energy status, nitrogen availability and oxygen status of the cell (Dixon et al. 1997). Another level of control at the transcriptional level is exerted by NtrC and NtrB and the global nitrogen PII-like proteins (GlnK, GlnB, GlnZ) (Zhang et al. 2001a, b). At the post-translational level, DraT/DraG can deactivate/reactivate Fe protein, respectively, in response to N availability and at the maturation level ClpX, an ATP-dependent protease, controls the formation of Fe protein (Martínez-Noël et al. 2011).

9.2.2 *Nitrification*

The process of formation of nitrate from ammonium is called nitrification. NH_3 released during BNF or released as a result of decomposition is converted to NH_4^+ which in turn gets oxidized to NO_2^- and ultimately to NO_3^-. Ammonia-oxidizing archaea (AOA) and ammonia-oxidizing bacteria (AOB) are responsible for the transformation of NH_4^+ to NO_2^-. Based on DGGE profiles, Jia and Conrad (2009) concluded that AOB is more abundant in agricultural soil. Similar results were obtained by Banning et al. (2015) based on qPCR analysis of 16S rRNA and amoA gene abundance. *Nitrosospira* and *Nitrosomonas* are the most abundant AOB in agro-ecosystems (Hendriks et al. 2000), *Nitrososphaera viennensis* is the only AOA isolated so far (Tourna et al. 2011). Ammonia is converted to

Table 9.1 Enzymes involved in microbial N-cycle. PDB (https://www.rcsb.org/) entries of the enzymes used for illustration are given in parenthesis

Process	Enzyme		Agriculturally important organisms	References
Biological nitrogen fixation	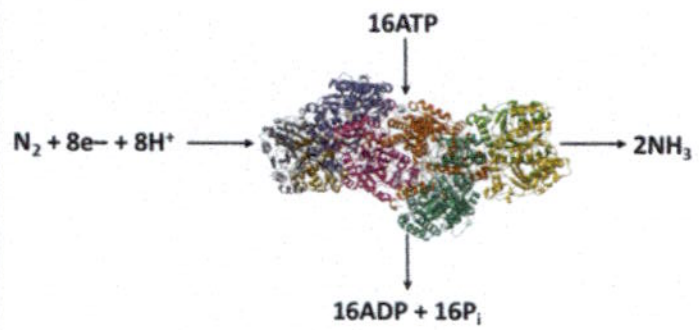	Nitrogenase (1N2C)	*Azotobacter vinelandii,Pantoea agglomerans, and Gluconacetobacter diazotrophicus; Azospirillum* spp., *Azoarcus* spp., *Herbaspirillum and Pseudomonas stutzeri and Burkholderia* spp.	Quecine et al. (2012) Sevilla et al. (2001) Boddey and Dobereiner (1982) Reinhold-Hurek and Hurek (1998) Baldani et al. (1986) Desnoues et al. (2003) Caballero-Mellado et al. (2004)
Dissimilatory nitrate reduction	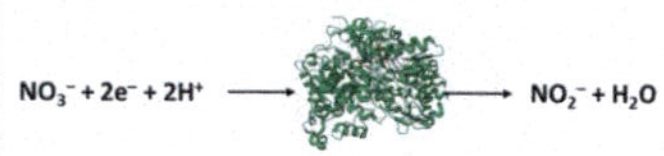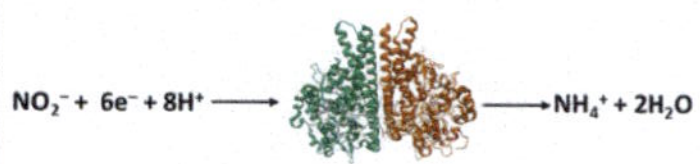	Periplasmic nitrate reductase (NAP) (2NYA) Cytochrome c Nitrite Reductase (Nrf) (1OAH) Apomolybdo-NarGHI (1SIW) NADH-dependent nitrite reductase small subunit (4AIV)	*Thiobacillus denitrificans, Desulfovibrio desulfuricans, Desulfobacterium* spp *Pseudomonas and Bacillus* sp.	Dalsgaard and Bak (1994) Brunet and Garcia-Gil (1996) Seitz and Cypionka (1986) Samuelsson (1985) Caskey and Tiedje (1979)

(continued)

Table 9.1 (continued)

Process	Enzyme		Agriculturally important organisms	References
	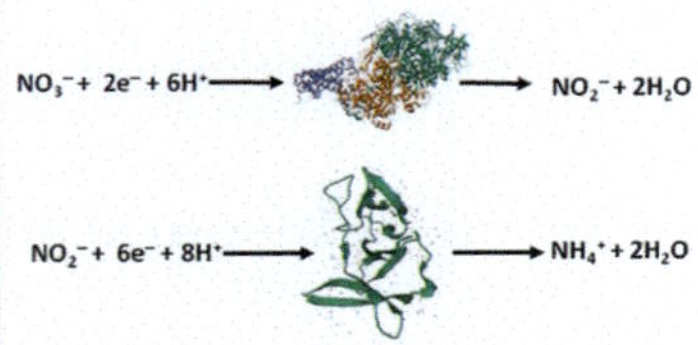			
Anaerobic Ammonium oxidation	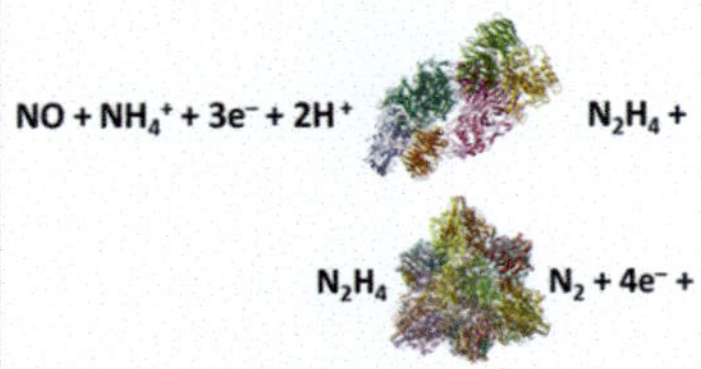	Hydrazine Synthase (5C2V) Hydrazine dehydrogenase (6HIF)	*Kuenenia stuttgartiensis* *Planctomycetes*	Kartal et al. (2011) Yang et al. (2015)
Aerobic ammonium oxidation	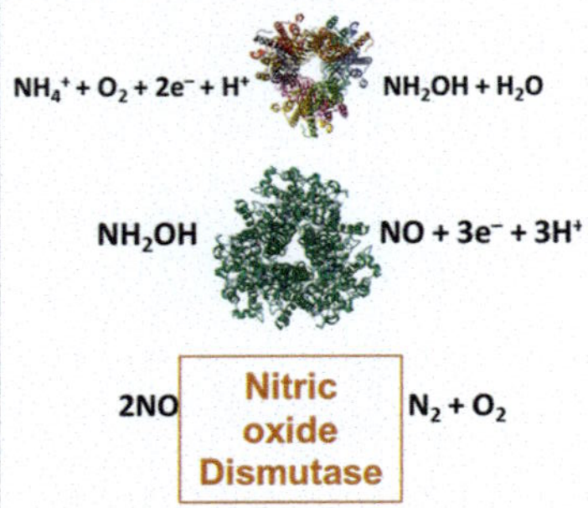	Methane monooxygenase (7EV9) Hydroxylamine oxidoreductase (1FGJ)	*Nitrosospira*, *Nitrosomonas Nitrososphaera viennensis*	Mendum et al. (1999) Tourna et al. (2011)
Denitrification		Nitrate reductase (apomolybdo-NarGHI)	*Alcaligenes*, *Pseudomonas and Bacillus*	Yoon et al. (2015)

	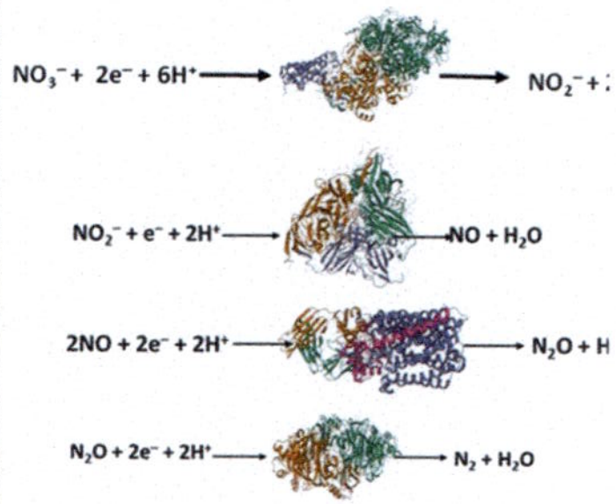	(1SIW) Nitrite reductase (1AQ8) Nitric oxide reductase (3O0R) Nitrous oxide reductase (3SBR)		

hydroxylamine using the enzyme ammonia monooxygenase (AMO) in AOA and AOB. All AOB contain the enzyme hydroxylamine oxidoreductase (HAO), which can oxidize hydroxylamine to nitrate. The oxidation of hydroxylamine supplies electrons to both the AMO and a typical electron transport chain composed of cytochrome c proteins (Walker et al. 2010). Both NO_3^- and NO_2^- can be taken up by plants and other microbes in the rhizosphere or retained in the soil.

Anaerobic ammonia oxidation (anammox) is coupled to the reduction of nitrite and the production of hydrazine (N_2H_4). Kartal et al. (2011) demonstrated the molecular mechanism of anaerobic oxidation in great detail in *Kuenenia stuttgartiensis*. The process of anammox occurs inside a membrane-bound organelle inside the cell called 'anammoxosome'. All the enzymes hydrazine synthase (HZS), hydrazine dehydrogenase (HDH), nitrite reductase (NiR) and hydroxylamine oxidoreductase (HOA) have been demonstrated to be localized inside this organelle (de Almeida et al. 2016). NiR reduces NO_2^- to NO to provide a substrate for the first step of anammox. The first step in anammox is the production of hydrazine (N_2H_4) by the enzyme hydrazine synthase (HZS). HZS reduces NO to hydroxylamine (NH_2OH) followed by the formation of N–N bond between NH_4^+ and NH_2OH to form N_2H_4. Finally, hydrazine is oxidized to N_2 by hydrazine dehydrogenase (HDH) to complete the process.

HZS is encoded by two genes, *hzsA* and *hzsB*, and can be used as a phylogenetic marker for anammox (Dietl et al. 2015). HAO and HZO in anammox may act as backup systems that efficiently trap hydrazine (Kartal et al. 2011). The significant presence of Planctomycetes in rice rhizosphere led to the speculation that anammox may play a role in N-transformation in paddy fields. Yang et al. (2015) demonstrated the presence of anammox bacteria in paddy fields by the presence of *hzsB*.

9.2.3 Denitrification

Microbial denitrification is a respiratory process of reducing nitrate to dinitrogen in several steps by four enzymes ($NO_3 \rightarrow NO_2 \rightarrow NO \rightarrow N_2O \rightarrow N_2$) (Table 9.1). Nitrate reductase, the first enzyme in the denitrification process, can convert NO_3^- to NO_2^- and is present in many bacteria. All nitrate reductases (NRs) are molybdopterin proteins. Based on the cellular location and function, nitrate reductases can be divided into three distinct types: respiratory NR (Nar), assimilatory NR (Nas) and periplasmic NR (Nap) (Moreno-Vivián et al. 1999). Nar and Nap are membrane-associated and generate ATP as a result of NO_3^- reduction, the assimilatory NR (Nas) will be discussed later under assimilatory nitrate reduction.

The second step in the denitrification process is the reduction of NO_2 to NO by nitrite reductase (NiR). NiR exists as two iso-functional periplasmic enzymes encoded by *nirS* and *nirK* (Giles et al. 2012). NiR is responsible for a crucial step in the denitrification process, which decides the direction of electron flow towards the denitrification pathway, the dissimilatory pathway or the assimilatory process (Ding et al. 2019). The NiR encoded by *nirK* contains copper as a co-factor (Cu-Nir), while the *nirS* encoded enzyme contains heme c and heme d1 (cd1-Nir). The third enzyme, Nitric oxide reductase (NO reduction to N_2O), is of two types- cNor, which is encoded by *norC* and *norB* and uses ac-type cytochrome as an electron donor; and qNor, which accepts electrons from quinols and may be encoded by *norB* (Hendriks et al. 2000). The last step in the denitrification process is the reduction of N_2O to N_2 by nitrous oxide reductase (NOS). NOS is encoded by *nosZ* forming a homodimer with a bi-nuclear copper centre (Moura and Moura 2001). The last steps in the denitrification process are oxygen sensitive and lacking in some of the bacterial species, resulting in the production of the harmful N_2O. The most common types of denitrifying bacteria in an agroecosystem are organotrophs, which use organic carbon compounds as a source of energy *Alcaligenes*, *Pseudomonas* and *Bacillus* species as well as Actinobacteria (especially *Streptomycetes*) (Pandey et al. 2020).

9.2.4 Assimilatory Nitrate Reduction

The anabolic process of incorporation of nitrate for growth and development by converting NO_3^- to NH_3 occurs in almost all organisms including bacteria, fungi and eukaryotes. Nitrate transported into the cell is reduced by Nas NRs and NiRs. In *Klebsiella*, the nasFEDCBA operon is responsible for assimilatory nitrate reduction (Lin et al. 1994). NasA, the NR and NasD, the assimilatory NiR which converts nitrite into NH_3. In *Azotobacter vinelandii* two assimilatory NiRs exist in one operon *nasA* and *nasB* (Ramos et al. 1993). NasC and NirA have been identified as the assimilatory NR and NiR, respectively, in *Bradyrhizobium japonicum* (Ruiz et al. 2019). The cytoplasmic location of the assimilatory NRs and NiRs necessitates the presence of nitrate and nitrite transporters. Several nitrate and nitrite transporters are reported in bacteria; the genes for these transporters mostly exist in the same operon as the NRs and NiRs.

9.2.5 Dissimilatory Nitrate Reduction to Ammonia (DNRA)

Dissimilatory reduction of nitrate to ammonium is coupled with the release of NH_4^+ into the environment and is also known as the short circuit to nitrogen fixation (Table 9.1). Two systems of DNRA occur in microbes: Nap/Nrf and Nar/Nir (Wang and Gunsalus 2000; Wang et al. 2019). NO_3^- reduction to NO_2^- by Nap or Nar NRs marks the beginning of the DNRA process. Nitrite reducers like the periplasmic NrfA (Stewart et al. 2002) or cytoplasmic NirB (Wang et al. 2019) further reduce NO_2^- to NH_3. Like denitrification, the first step of reduction in DNRA can occur in respiratory conditions and is coupled to energy production (Simon and Klotz 2012). The second step may occur in either respiratory or fermentative conditions. NirB connects NH_4^+ production with NADH oxidation and NADH dehydrogenase, which utilizes the ETC to transfer electrons to menaquinone (Wang et al. 2019). Under high

NO_3^- both nitrate reductase (NarGHI) and nitrite reductase (NirB) are expressed. NapAGHBFLD (periplasmic nitrate reductase) and NrfABCD (periplasmic nitrite reductase) are expressed under low NO_3^- concentration (Yoon et al. 2015). NirB utilizes the fermentative pathway to reduce NO_2^- to NH_4^+, whereas, NrfA utilizes the respiratory pathway. NrfA is considered to be the marker gene for the presence of DNRA in microbes (Pandey et al. 2020). Respiratory form of DNRA may be more prevalent in agricultural soil e.g. *Thiobacillus denitrificans*, *Desulfovibrio desulfuricans*, *Desulfobacterium* spp *Pseudomonas* and *Bacillus* (Pandey et al. 2020). However, DNRA is an important N-retention factor in flooded soils and is dependent on the C/NO_3^- ratio of the soil (Kraft et al. 2011; van den Berg et al. 2017; Yoon et al. 2015).

9.3 Effect of Soil Fertilization on Microbiota

Rice plants are known to prefer NH_4^+ to NO_3^- (Sasakawa and Yamamoto 1978) while Wheat and *Arabidopsis* plants prefer NO_3^- to NH_4^+ (Bloom et al. 2010). However, in elevated atmospheric CO_2 conditions, Wheat and Arabidopsis plants prefer NH_4^+ to NO_3^- (Bloom et al. 2010). Hence, the type of N fertilizer applied affects yield, instigating farmers to apply excessive fertilizers. N-fertilizers are also potent sources for the release of nitrous oxide (N_2O) (Winiwarter et al. 2017). An estimate by NOAA shows that the amount of N_2O has increased from 300 ppb in 1980 to 332 ppb in 2020 (Udvardi et al. 2021). It was estimated that out of the total N-fertilizer applied to the crops, only 48% was utilized by the crops and the rest was lost to the environment, causing pollution (Ladha et al. 2016). This in turn may leave a permanent damage to the soil in its ability to support microbial growth. To fully assess the effect of N-fertilization on the soil microbiota, several comparative studies have been done.

Cross-biome studies for soil N-cycling show that soil moisture, C/N ratio and microbial biomass had strong correlation to N-cycling (Yang et al. 2017). One of the significant observations in cross-biomes studies was the contribution of DNRA to N-cycling across all soil types (Yang et al. 2017). Studies across N-gradients in soils show that N-rich soil encourages the growth of copiotrophic bacteria, and long-term fertilization may lead to a shift in soil microbiota from oligotrophic to copiotrophic bacteria (Fierer et al. 2012). Short-term fertilization studies with inorganic and organic fertilizers show that temporal variability in microbial communities correlates with N-availability (Norton et al. 2015). It was also observed that the diversity of microbiota was significantly higher after compost application (Ouyang and Norton 2020). It is expected that microbial activity is higher in plant rhizosphere (Barea et al. 2005; Billings et al. 2004;); however, modern high throughput technologies like next generations sequencing does not support this hypothesis (Trivedi et al. 2020). Differences in the microbiota in the rhizosphere of different plant genotype (Edwards et al. 2015; Shenton et al. 2016), depending on different growth stages (Edwards et al. 2018; Imchen et al. 2019), and presence of different abiotic and biotic stresses (Antoniou et al. 2017; Berendsen et al. 2012; Santos-Medellín et al. 2017) were quite significant.

The microbiota normally present in rice rhizosphere varies with the genotype and geographical location (Edwards et al. 2015). However, some of the microbes like *Burkholderia*, *Bradyrhizobium* and *Methylosinus* were found to be predominant in rice rhizosphere under low nitrogen conditions (Ikeda et al. 2014). Contrastingly, fertilized wheat rhizosphere contained *Paraburkholderia*, *Rhizobium*, *Pseudomonas* and *Rhodanobacter* (Reid et al. 2021). In paddy fields, prolonged (32 years) effect of fertilization with chemical fertilizer (CF) and CF combined with farmyard manure (CFM) show increased total carbon and total nitrogen content. Application of CF and CFM also reduced the nitrate-N and ammonium-N content, showing increased activity of ammonia-oxidizers and nitrate oxidizers in the soil compared to no fertilizer application (Gu et al. 2017). Yi et al. (2018) showed that the addition of urea had a positive effect on urease activity, available potassium, AOB, and production of NO_3-N. Ammonium addition to the soil has a stimulatory effect on nitrification when the carbon content is high (Yi et al. 2018). The moisture content of soil also has a significant effect on the soil nitrate content and N_2O release. Low-land rice has a unique water requirement pattern. Paddy grows under waterlogged conditions from transplantation to panicle maturation stage. Waterlogged conditions create anaerobic condition, and the oxygen leakage from aerenchym tissues create sub-oxic levels near rice roots (Ding et al. 2019; Wang et al. 2015b). With waterlogging, the AOA communities increased with *Candidatus nitrosocaldus* being the most dominant (Imchen et al. 2019). AOB *Nitrosospira* is not affected by the waterlogged conditions and increased in abundance with plant growth (Imchen et al. 2019). Nitrate oxidizing bacteria (NOB), *Nitrospira*, favour paddy waterlogged conditions and is the predominant NOB (Wang et al. 2015c). AOA, AOB and NOB bacteria increased in abundance after urea application (Ding et al. 2019); however, the interaction between these communities remain to be studied. Hussain et al. (2011) observed that ammonia-oxidizers dominated over nitrate-oxidizers (Hussain et al. 2011). Low nitrogen supported the growth of β-proteobacteria in roots compared to high nitrogen conditions in paddy rhizosphere. Studies show that *Indica* varieties of rice have higher NUE efficiency compared to *Japonica* attributed to the nrt1.1b gene sequence variation (Rakotoson et al. 2021). Using nrt1.1b mutants, it was demonstrated that the nitrate sensor and transporter are responsible for recruiting diverse microbiota in the rhizosphere (Zhang et al. 2019); however, the exact reason for such recruitment is yet to be demonstrated. Evidence of carbon content available to the plant is reflected in its root exudate (Yuan et al. 2019); similar studies for nitrogen and its role in recruitment of microbiome needs to be investigated further.

9.4 Root Exudates as Modulators of Microbes in Rhizosphere

The first step in plant-microbe interaction is the recognition of plant exudates by soil microbes, chemotaxis towards the roots, and ultimately adhesion or entry into the root. Sugar released by the roots is one of the most important components as it is used as an energy source for microbial growth. It is thought that 5–30% of the carbon fixed by the plants are secreted in the form of root exudates (Dakora and Phillips 2002; Kuzyakov et al. 2000, 2003; Nguyen 2003). Using ^{14}C isotopes Yuan et al. (2019) concluded that within 6 hours of carbon assimilation by plants, the rhizosphere microbial communities can utilize the fixed carbon. The carbon assimilated by the plants reaches the roots via the phloem. Wang et al. (2015a) have shown that elevated sucrose loading in the phloem of *Oryza* and *Arabidopsis* led to more sucrose in the roots. Weisskopf et al. (2006) showed that organic acids released by roots also act as chemo-attractants for root microbiota.

In wild maize, it has recently been demonstrated that the mucilage produced by brace roots could provide a rich nutrient for growth of microbes including the nitrogen-fixing *Burkholderia* (Amicucci et al. 2019; Van Deynze et al. 2018). Studies of wheat microbiome across multiple soil types also showed that *Burkholderiaceae* and *Pseudomonadaceae* were recruited by the wheat root exudates (Prudence et al. 2021). In addition to sugars, secondary metabolites like phenolics and anthocyanins in root exudates influence the rhizosphere microbiome by encouraging growth of beneficial bacteria, while at the same time, discouraging the growth of harmful microbes. Maize plants secrete benzoxazinoids like 2,4-dihydroxy-7-methoxy-2H-1,4-benzoxazin-3(4H)-one (DIMBOA-Glc) during early growth stages. The breakdown product 6-methoxy-benzoxazolin-2-one (MBOA) is converted to phenoxazinones by microbes. Phenoxazinones can repel *Bacillus* and *Fusarium*; but at the same time it attracts *Pseudomonas putida* (Hu et al. 2018; Kudjordjie et al. 2019). A 'Red Root' phenotypic mutant of sorghum show higher secretion of phenolics and anthocyanin, and the rhizosphere microbiota is enriched for Actinobacteria (Balasubramanian et al. 2021).

Many perennial grasses, rice and sorghum roots can inhibit nitrifying bacteria in the rhizosphere (Subbarao et al. 2021; Zhou et al. 2020). This biological inhibition of nitrification (BNI) has been attributed to the presence of a metabolite, sorgoleone, in the root exudate of sorghum (Einhellig and Souza 1992). Sorgoleone inhibits the first step of nitrification by inhibiting ammonium oxidizing archaea (AOA) (Sarr et al. 2020). Another metabolite, methyl. 3-(4-hydroxyphenyl) propionate (MHPP), also present in the root exudate of Sorghum, has the same effect on BNI (Zakir et al. 2008). In rice, 1,9-decanediol, was identified to be responsible for BNI by inhibiting the ammonia monooxygenase (AMO) (Sun et al. 2016).

9.5 Strategies for Incorporating Microbiomes with Net N Benefit in Cereals

The first step for increasing N absorption by plants is to introduce diazotrophs capable of efficient N-fixation into the rhizosphere. Several strategies have been used for introducing one or combinations of microbes into the rhizosphere. The creation of a synthetic consortium would require in-depth studies of the effect of one community on other members of the consortium, especially when complex bio-geochemical processes like N-cycle are considered. Several studies have been done to determine the effect of one community on the other. Using a 62-strain synthetic community isolated from the *A. thaliana* phyllosphere, drop-out and late introduction experiments were conducted to show that keystone species have the greatest effect on community structures (Carlström et al. 2019). Rhizosphere microbes isolated from plant phosphate starvation response demonstrated that microbes can be classified into functional bacterial blocks and that a subset selected from within these microbes can predict the outcome of the whole consortia (Herrera Paredes et al. 2018). Statistical approaches like co-occurrence network (Huang et al. 2021) metabolic utilization network (Muller et al. 2017; McClure et al. 2020; Roume et al. 2015; and Permutational multivariate analysis of variance (PERMANOVA) (Anderson et al. 2017) can help gain insights into the interdependencies and antagonisms between microbes in a consortium.

The classical approach is to increase the adherence efficiency of the diazotroph to the host root by seed treatment or seedling dip treatment before transplantation, ensuring maximum benefit to the host plant (Hata et al. 2021; Khan et al. 2021). Elevating the adherence capacity would ensure survival of the diazotroph on the root surface as well as benefit for the host plant. Introducing a single microbe (Young et al. 2006) or consortia (Bergero et al. 2017) encapsulated into alginate beads with root adhering properties can also benefit the host plant (Bashan et al. 2014). The second strategy would be to manipulate the host plant to release root exudates for inhibiting denitrification and nitrification (Coskun et al. 2017a, b). It is thought that the domestication of plants followed by high fertilizer application regimes to sustain the high yields has resulted in the loss of root microbial diversity (Shenton et al. 2016). Thus, isolating root microbes from wild relatives and creating synthetic consortiums can help in restoration of microbes capable of absorbing nutrients at a higher rate, therefore, minimizing loss from soil. It has been hypothesised that the functional core microbiome comprises similar microbes, it remains to be understood how the rest of the microbiome is recruited in different soil conditions. Evidence regarding heritability of microbiomes is slowly being revealed (Hoecker et al. 2006; Paschold et al. 2010; Rakotoson et al. 2021); hence, plant breeding strategies to transfer the traits from wild relatives to commonly cultivated varieties are gaining popularity.

9.6 Conclusions

Manipulating the rhizosphere microbiome for elevating NUE requires a thorough understanding of the microbial N cycle, N transformation capacity among the microbes and N uptake by the host plant. Successful creation and introduction of a synthetic microbial consortium for net N benefit into the rhizosphere will depend on the ability of the microbes to secrete, retain and release plant available forms of nitrogen. Unfortunately, introduction of microbial consortium into the rhizosphere is not a one-size-fits-all process and depends on the genotype of the host plant, the interactions with indigenous microbes and environmental conditions like temperature, soil moisture, pH, etc. However, the most important among all the microbial processes would be augmenting BNF in synchronization with the N requirement of the host plant.

References

Amicucci MJ, Galermo AG, Guerrero A, Treves G, Nandita E, Kailemia MJ, Higdon SM et al (2019) Strategy for structural elucidation of polysaccharides: elucidation of a maize mucilage that harbors diazotrophic bacteria. Anal Chem 91(11):7254–7265. https://doi.org/10.1021/acs.analchem.9b00789

Anderson MJ, Walsh DC, Robert Clarke K, Gorley RN, Guerra-Castro E (2017) Some solutions to the multivariate Behrens–Fisher problem for dissimilarity-based analyses. Aus N Z J Stat 59(1): 57–79

Antoniou A, Tsolakidou MD, Stringlis IA, Pantelides IS (2017) Rhizosphere microbiome recruited from a suppressive compost improves plant fitness and increases protection against vascular wilt pathogens of Tomato. Front Plant Sci 8:2022. https://doi.org/10.3389/fpls.2017.02022

Balasubramanian VK, Dampanaboina L, Cobos CJ, Yuan N, Xin Z, Mendu V (2021) Induced secretion system mutation alters rhizosphere bacterial composition in *Sorghum bicolor* (L.) Moench. Planta 253(2):33. https://doi.org/10.1007/s00425-021-03569-5

Baldani JI, Baldani VLD, Seldin L, Döbereiner J (1986) Characterization of *Herbaspirillum seropedicae* gen. nov., sp. nov. a root-associated nitrogen-fixing bacterium. Int J Syst Bacteriol 36:86–93

Banning NC, Maccarone LD, Fisk LM, Murphy DV (2015) Ammonia-oxidising bacteria not archaea dominate nitrification activity in semi-arid agricultural soil. Sci Rep 5:11146. https://doi.org/10.1038/srep11146

Barea JM, Pozo MJ, Azcón R, Azcón-Aguilar C (2005) Microbial co-operation in the rhizosphere. J Exp Bot 56(417):1761–1778. https://doi.org/10.1093/jxb/eri197

Bashan Y, de Bashan LE, Prabhu SR, Hernandez JP (2014) Advances in plant growth-promoting bacterial inoculant technology: formulations and practical perspectives (1998–2013). Plant Soil 378(1):1–33

Berendsen RL, Pieterse CM, Bakker PA (2012) The rhizosphere microbiome and plant health. Trends Plant Sci 17(8):478–486. https://doi.org/10.1016/j.tplants.2012.04.001

Bergero M, Liffourrena A, Opizzo B, Fochesatto AS, Lucchesi G (2017) Immobilization of a microbial consortium on Ca-alginate enhances degradation of cationic surfactants in flasks and bioreactor. Int Biodeterior Biodegradation 117:18. https://doi.org/10.1016/j.ibiod.2016.11.018

Billings SA, Schaeffer SM, Evans RD (2004) Soil microbial activity and N availability with elevated CO2 in Mojave desert soils. Global Biogeochem Cycles 18:11

Bloom AJ, Burger M, Asensio JS, Cousins AB (2010) Carbon dioxide enrichment inhibits nitrate assimilation in wheat and Arabidopsis. Science 328(5980):899–903

Boddey RM, Dobereiner J (1982) Association of *Azospirillum* and other diazotrophs with tropical Gramineae in non-symbiotic nitrogen fixation and organic matter in the tropics. 12th Intl. Congress of Soil Sci, New Delhi, India, p 28–48

Brunet RC, Garcia-Gil LJ (1996) Sulfide-induced dissimilatory nitrate reduction to ammonia in anaerobic freshwater sediments. FEMS Microbiol Ecol 21:131–138

Caballero-Mellado J, Martínez-Aguilar L, Paredes-Valdez G, Santos PE (2004) *Burkholderia unamae* sp. nov., an N2-fixing rhizospheric and endophytic species. Int J Syst Evol Microbiol 54(4):1165–1172. https://doi.org/10.1099/ijs.0.02951-0

Carlström CI, Field CM, Bortfeld-Miller M, Müller B, Sunagawa S, Vorholt JA (2019) Synthetic microbiota reveal priority effects and keystone strains in the *Arabidopsis* phyllosphere. Nat Ecol Evol 3(10):1445–1454. https://doi.org/10.1038/s41559-019-0994-z

Carvalho TL, Balsemão-Pires E, Saraiva RM, Ferreira PC, Hemerly AS (2014) Nitrogen signalling in plant interactions with associative and endophytic diazotrophic bacteria. J Exp Bot 65(19): 5631–5642. https://doi.org/10.1093/jxb/eru319

Caskey WH, Tiedje JM (1979) Evidence for Clostridia as agents of dissimilatory reduction of nitrate to ammonium in soils. Soil Sci Soc Am J 43:931–936

Coskun D, Britto DT, Shi W, Kronzucker HJ (2017a) How plant root exudates shape the nitrogen cycle. Trends Plant Sci 22(8):661–673. https://doi.org/10.1016/j.tplants.2017.05.004

Coskun D, Britto DT, Shi W, Kronzucker HJ (2017b) Nitrogen transformations in modern agriculture and the role of biological nitrification inhibition. Nat Plants 3:17074. https://doi.org/10.1038/nplants.2017.74

Dakora F, Phillips D (2002) Root exudates as mediators of mineral acquisition in low-nutrient environments. Plant Soil 245:35–47. https://doi.org/10.1023/A:1020809400075

Dalsgaard T, Bak F (1994) Nitrate reduction in a sulfate-reducing bacterium, Desulfovibrio desulfuricans, isolated from rice paddy soil: sulfide inhibition, kinetics, and regulation. Appl Environ Microbiol 60:291–297

de Almeida NM, Wessels HJ, de Graaf RM, Ferousi C, Jetten MS, Keltjens JT, Kartal B (2016) Membrane-bound electron transport systems of an anammox bacterium: a complexome analysis. Biochim Biophys Acta 1857(10):1694–1704. https://doi.org/10.1016/j.bbabio.2016.07.006

Desnoues N, Lin M, Guo X, Ma L, Carreño-Lopez R, Elmerich C (2003) Nitrogen fixation genetics and regulation in a *Pseudomonas stutzeri* strain associated with rice. Microbiology 149(8): 2251–2262. https://doi.org/10.1099/mic.0.26270-0

Dietl A, Ferousi C, Maalcke WJ, Menzel A, de Vries S, Keltjens JT, Jetten MS, Kartal B, Barends TR (2015) The inner workings of the hydrazine synthase multiprotein complex. Nature 527(7578):394–397. https://doi.org/10.1038/nature15517

Ding LJ, Cui HL, Nie SA, Long XE, Duan GL, Zhu YG (2019) Microbiomes inhabiting rice roots and rhizosphere. FEMS Microbiol Ecol 95(5):40. https://doi.org/10.1093/femsec/fiz040

Dixon R, Cheng Q, Shen GF, Day A, Dowson-Day M (1997) *nif* gene transfer and expression in chloroplasts: prospects and problems. Plant Soil 194:193–203. https://doi.org/10.1023/A:1004296703638

Edwards J, Johnson C, Santos-Medellín C, Lurie E, Podishetty NK, Bhatnagar S, Eisen JA et al (2015) Structure, variation, and assembly of the root-associated microbiomes of rice. Proc Natl Acad Sci U S A 112(8):E911–E920. https://doi.org/10.1073/pnas.1414592112

Edwards JA, Santos-Medellín CM, Liechty ZS, Nguyen B, Lurie E, Eason S, Phillips G et al (2018) Compositional shifts in root-associated bacterial and archaeal microbiota track the plant life cycle in field-grown rice. PLoS Biol 16(2):e2003862. https://doi.org/10.1371/journal.pbio.2003862

Einhellig FA, Souza IF (1992) Phytotoxicity of sorgoleone found in grain Sorghum root exudates. J Chem Ecol 18(1):1–11. https://doi.org/10.1007/BF00997160

Fierer N, Lauber CL, Ramirez KS, Zaneveld J, Bradford MA, Knight R (2012) Comparative metagenomic, phylogenetic and physiological analyses of soil microbial communities across nitrogen gradients. ISME J 6(5):1007–1017. https://doi.org/10.1038/ismej.2011.159

Giles M, Morley N, Baggs EM, Daniell TJ (2012) Soil nitrate reducing processes - drivers, mechanisms for spatial variation, and significance for nitrous oxide production. Front Microbiol 3:407. https://doi.org/10.3389/fmicb.2012.00407

Gu Y, Wang Y, Lu S, Xiang Q, Yu X, Zhao K, Zou L et al (2017) Long-term fertilization structures bacterial and archaeal communities along soil depth gradient in a paddy soil. Front Microbiol 8: 1516. https://doi.org/10.3389/fmicb.2017.01516

Hata EM, Yusof MT, Zulperi D (2021) Induction of systemic resistance against bacterial leaf streak disease and growth promotion in rice plant by *Streptomyces shenzhenesis* TKSC3 and *Streptomyces* sp. SS8. Plant Pathol J 37(2):173–181. https://doi.org/10.5423/PPJ.OA.05.2020.0083

Hendriks J, Oubrie A, Castresana J, Urbani A, Gemeinhardt S, Saraste M (2000) Nitric oxide reductases in bacteria. Biochim Biophys Acta 1459(2-3):266–273. https://doi.org/10.1016/s0005-2728(00)00161-4

Herrera Paredes S, Gao T, Law TF, Finkel OM, Mucyn T, Teixeira PJPL, , Salas González I et al (2018) Design of synthetic bacterial communities for predictable plant phenotypes. PLoS Biol 16(2): e2003962. DOI: https://doi.org/10.1371/journal.pbio.2003962

Hoecker N, Keller B, Piepho HP, Hochholdinger F (2006) Manifestation of heterosis during early maize (*Zea mays* L.) root development. Theor Appl Genet 112(3):421–429. https://doi.org/10.1007/s00122-005-0139-4

Hu L, Robert CAM, Cadot S, Zhang X, Ye M, Li B, Manzo D et al (2018) Root exudate metabolites drive plant-soil feedbacks on growth and defense by shaping the rhizosphere microbiota. Nat Commun 9(1):2738. https://doi.org/10.1038/s41467-018-05122-7

Huang L, Zhang G, Bai J, Xia Z, Wang W, Jia J, Wang X, Liu X, Cui B (2021) Desalinization via freshwater restoration highly improved microbial diversity, co-occurrence patterns and functions in coastal wetland soils. Sci Total Environ 765:142769

Hussain Q, Liu Y, Zhang A, Pan G, Li L, Zhang X, Song X et al (2011) Variation of bacterial and fungal community structures in the rhizosphere of hybrid and standard rice cultivars and linkage to CO2 flux. FEMS Microbiol Ecol 78(1):116–128. https://doi.org/10.1111/j.1574-6941.2011.01128.x

Ikeda S, Sasaki K, Okubo T, Yamashita A, Terasawa K, Bao Z, Liu D et al (2014) Low nitrogen fertilization adapts rice root microbiome to low nutrient environment by changing biogeochemical functions. Microbes Environ 29(1):50–59. https://doi.org/10.1264/jsme2.me13110

Imchen M, Kumavath R, Vaz ABM, Góes-Neto A, Barh D, Ghosh P, Kozyrovska N et al (2019) 16S rRNA gene amplicon based metagenomic signatures of rhizobiome community in rice field during various growth stages. Front Microbiol 10:2103. https://doi.org/10.3389/fmicb.2019.02103

Jia Z, Conrad (2009) Bacteria rather than Archaea dominate microbial ammonia oxidation in an agricultural soil. Environ Microbiol 11(7):1658–1671. https://doi.org/10.1111/j.1462-2920.2009.01891.x

Kartal B, Maalcke WJ, de Almeida NM, Cirpus I, Gloerich J, Geerts W, Op den Camp HJ et al (2011) Molecular mechanism of anaerobic ammonium oxidation. Nature 479(7371):127–130. https://doi.org/10.1038/nature10453

Khan MR, Haque Z, Ahamad F, Zaidi B (2021) Biomanagement of rice root-knot nematode *Meloidogyne graminicola* using five indigenous microbial isolates under pot and field trials. J Appl Microbiol 130(2):424–438. https://doi.org/10.1111/jam.14788

Kraft B, Strous M, Tegetmeyer HE (2011) Microbial nitrate respiration–genes, enzymes and environmental distribution. J Biotechnol 155(1):104–117. https://doi.org/10.1016/j.jbiotec.2010.12.025

Kudjordjie EN, Sapkota R, Steffensen SK, Fomsgaard IS, Nicolaisen M (2019) Maize synthesized benzoxazinoids affect the host associated microbiome. Microbiome 7(1):59. https://doi.org/10.1186/s40168-019-0677-7

Kuzyakov A, Yakov A, Domanski A, Grzegorz A (2000) Carbon input by plants into the soil. Review. J Plant Nutr Soil Sci 163:421–431

Kuzyakov Y, Raskatov A, Kaupenjohann M (2003) Turnover and distribution of root exudates of *Zea mays*. Plant Soil 254:317–327. https://doi.org/10.1023/A:1025515708093

Ladha JK, Tirol-Padre A, Reddy CK, Cassman KG, Verma S, Powlson DS, van Kessel C et al (2016) Global nitrogen budgets in cereals: a 50-year assessment for maize, rice, and wheat production systems. Sci Rep 6:19355. https://doi.org/10.1038/srep19355

Lemanceau P, Blouin M, Muller D, Moënne-Loccoz Y (2017) Let the core microbiota be functional. Trends Plant Sci 22:583–595. https://doi.org/10.1016/j.tplants.2017.04.008

Lin JT, Goldman BS, Stewart V (1994) The nasFEDCBA operon for nitrate and nitrite assimilation in *Klebsiella pneumoniae* M5al. J Bacteriol 176(9):2551–2559. https://doi.org/10.1128/jb.176.9.2551-2559.1994

Lugtenberg B, Kamilova F (2009) Plant-growth-promoting rhizobacteria. Annu Rev Microbiol 63: 541–556. https://doi.org/10.1146/annurev.micro.62.081307.162918

Martínez-Dalmau J, Berbel J, Ordóñez-Fernández R (2021) Nitrogen fertilization. A review of the risks associated with the inefficiency of its use and policy responses. Sustain For 13:5625. https://doi.org/10.3390/su13105625

Martínez-Noël G, Curatti L, Hernandez JA, Rubio LM (2011) NifB and NifEN protein levels are regulated by ClpX2 under nitrogen fixation conditions in *Azotobacter vinelandii*. Mol Microbiol 79:1182–1193. https://doi.org/10.1111/j.1365-2958.2011.07540.x

Mavrodi OV, McWilliams JR, Peter JO, Berim A, Hassan KA, Elbourne LDH, LeTourneau MK et al (2021) Root exudates alter the expression of diverse metabolic, transport, regulatory, and stress response genes in rhizosphere *Pseudomonas*. Front Microbiol 12:651282. https://doi.org/10.3389/fmicb.2021.651282

McClure RS, Lee JY, Chowdhury TR, Bottos EM, White RA 3rd, Kim YM, Nicora CD et al (2020) Integrated network modeling approach defines key metabolic responses of soil microbiomes to perturbations. Sci Rep 10(1):10882. https://doi.org/10.1038/s41598-020-67878-7

Mendum TA, Sockett RE, Hirsch PR (1999) Use of molecular and isotopic techniques to monitor the response of autotrophic ammonia-oxidizing populations of the beta subdivision of the class proteobacteria in arable soils to nitrogen fertilizer. Appl Environ Microbiol 65(9):4155–4162. https://doi.org/10.1128/AEM.65.9.4155-4162.1999

Moreno-Vivián C, Cabello P, Martínez-Luque M, Blasco R, Castillo F (1999) Prokaryotic nitrate reduction: molecular properties and functional distinction among bacterial nitrate reductases. J Bacteriol 181(21):6573–6584. https://doi.org/10.1128/JB.181.21.6573-6584.1999

Moura I, Moura JJG (2001) Structural aspects of denitrifying enzymes. Curr Opin Chem Biol 5: 168–175

Muller E, Faust K, Widder S, Herold M, Martinez Arbas S, Wilmes P (2017) Using metabolic networks to resolve ecological properties of microbiomes. Curr Opin Syst Biol 8:4. https://doi.org/10.1016/j.coisb.2017.12.004

Naylor D, DeGraaf S, Purdom E, Coleman-Derr D (2017) Drought and host selection influence bacterial community dynamics in the grass root microbiome. ISME J 11:2691–2704

Nguyen C (2003) Rhizodeposition of organic C by plants: mechanisms and controls. Agronomie 23:375–396

Norton R, Davidson E, Roberts T (2015) Position paper - nitrogen use efficiency and nutrient performance indicators. Global Partnership on Nutrient Management (GPNM), Nairobi

Ouyang Y, Norton JM (2020) Short-term nitrogen fertilization affects microbial community composition and nitrogen mineralization functions in an agricultural soil. Appl Environ Microbiol 86(5):e02278. https://doi.org/10.1128/AEM.02278-19

Pandey CB, Kumar U, Kaviraj M, Minick KJ, Mishra AK, Singh JS (2020) DNRA: a short-circuit in biological N-cycling to conserve nitrogen in terrestrial ecosystems. Sci Total Environ 738: 139710. https://doi.org/10.1016/j.scitotenv.2020.139710

Paschold A, Marcon C, Hoecker N, Hochholdinger F (2010) Molecular dissection of heterosis manifestation during early maize root development. Theor Appl Genet 120(2):383–388. https://doi.org/10.1007/s00122-009-1082-6

Prudence SM, Newitt JT, Worsley SF, Macey MC, Murrell JC, Lehtovirta-Morley LE, Hutchings MI (2021) Soil, senescence and exudate utilisation: characterisation of the Paragon var. spring bread wheat root microbiome. Environ Microb 16(1):12. https://doi.org/10.1186/s40793-021-00381-2

Quecine MC, Araújo WL, Rossetto PB, Ferreira A, Tsui S, Lacava PT, Mondin M et al (2012) Sugarcane growth promotion by the endophytic bacterium *Pantoea agglomerans*. Appl Environ Microbiol 78(21):7511–7518. https://doi.org/10.1128/AEM.00836-12

Rakotoson T, Dusserre J, Letourmy P, Frouin J, Ratsimiala IR, Rakotoarisoa NV, Vom Brocke K, Ramanantsoanirina A, Ahmadi N, Raboin LM (2021) Genome-wide association study of nitrogen use efficiency and agronomic traits in upland rice. Rice Sci 28(4):379–390

Ramos F, Blanco G, Gutiérrez JC, Luque F, Tortolero M (1993) Identification of an operon involved in the assimilatory nitrate-reducing system of *Azotobacter vinelandii*. Mol Microbiol 8(6):1145–1153. https://doi.org/10.1111/j.1365-2958.1993.tb01659.x

Reid TE, Kavamura VN, Abadie M, Torres-Ballesteros A, Pawlett M, Clark IM, Harris J, Mauchline TH (2021) Inorganic chemical fertilizer application to wheat reduces the abundance of putative plant growth-promoting rhizobacteria. Front Microbiol 2021:458

Reinhold-Hurek B, Hurek T (1998) Life in grasses: diazotrophic endophytes. Trends Microbiol 6(4):139–144. https://doi.org/10.1016/s0966-842x(98)01229-3

Roume H, Heintz-Buschart A, Muller EEL, May P, Satagopam VP, Laczny CC, Narayanasamy S et al (2015) Comparative integrated omics: identification of key functionalities in microbial community-wide metabolic networks. NPJ Biofilms Microb 1:15007. https://doi.org/10.1038/npjbiofilms.2015.7

Ruiz B, Le Scornet A, Sauviac L, Rémy A, Bruand C, Meilhoc E (2019) The nitrate assimilatory pathway in *Sinorhizobium meliloti*: Contribution to NO production. Front Microbiol 10:1526. https://doi.org/10.3389/fmicb.2019.01526

Samuelsson MO (1985) Dissimilatory nitrate reduction to nitrite, nitrous oxide and ammonium by *Pseudomonas putrefaciens*. Appl Environ Microbiol 50:812–815

Santi C, Bogusz D, Franche C (2013) Biological nitrogen fixation in non-legume plants. Ann Bot 111:743–767. https://doi.org/10.1093/aob/mct048

Santos-Medellín C, Edwards J, Liechty Z, Nguyen B, Sundaresan V (2017) Drought stress results in a compartment-specific restructuring of the rice root-associated microbiomes. MBio 8(4): e00764. https://doi.org/10.1128/mBio.00764-17

Sarr PS, Ando Y, Nakamura S, Deshpande S, Subbarao GV (2020) Sorgoleone release from sorghum roots shapes the composition of nitrifying populations, total bacteria, and archaea and determines the level of nitrification. Biol Fertil Soils 56:145–166

Sasakawa H, Yamamoto Y (1978) Comparison of the uptake of nitrate and ammonium by rice seedlings: influences of light, temperature, oxygen concentration, exogenous sucrose, and metabolic inhibitors. Plant Physiol 62(4):665–669. https://doi.org/10.1104/pp.62.4.665

Seitz HJ, Cypionka H (1986) Chemolithotrophic growth of *Desulfovibrio desulfuricans* with hydrogen coupled to ammonification of nitrate or nitrite. Arch Microbiol 146:63–67

Sevilla M, Burris RH, Gunapala N, Kennedy C (2001) Comparison of benefit to sugarcane plant growth and 15N2 incorporation following inoculation of sterile plants with *Acetobacter diazotrophicus* wild-type and Nif-mutants strains. Mol Plant-Microbe Interact 14(3):358–366. https://doi.org/10.1094/MPMI.2001.14.3.358

Shenton M, Iwamoto C, Kurata N, Ikeo K (2016) Effect of wild and cultivated rice genotypes on rhizosphere bacterial community composition. Rice 9(1):42. https://doi.org/10.1186/s12284-016-0111-8

Simon J, Klotz M (2012) Diversity and evolution of bioenergetic systems involved in microbial nitrogen compound transformations. Biochim Biophys Acta 2012:1827. https://doi.org/10.1016/j.bbabio.2012.07.005

Spence C, Alff E, Johnson C, Ramos C, Donofrio N, Sundaresan V, Bais H (2014) Natural rice rhizospheric microbes suppress rice blast infections. BMC Plant Biol 14:130. https://doi.org/10.1186/1471-2229-14-130

Stewart V, Lu Y, Darwin AJ (2002) Periplasmic nitrate reductase (NapABC enzyme) supports anaerobic respiration by *Escherichia coli* K-12. J Bacteriol 184(5):1314–1323. https://doi.org/10.1128/JB.184.5.1314-1323.2002

Subbarao GV, Kishii M, Bozal-Leorri A, Ortiz-Monasterio I, Gao X, Ibba MI, Karwat H et al (2021) Enlisting wild grass genes to combat nitrification in wheat farming: a nature-based solution. Proc Natl Acad Sci U S A 118(35):e2106595118. https://doi.org/10.1073/pnas.2106595118

Sun L, Lu Y, Yu F, Kronzucker HJ, Shi W (2016) Biological nitrification inhibition by rice root exudates and its relationship with nitrogen-use efficiency. New Phytol 212(3):646–656. https://doi.org/10.1111/nph.14057

Tourna M, Stieglmeier M, Spang A, Könneke M, Schintlmeister A, Urich T, Engel M et al (2011) *Nitrososphaera viennensis*, an ammonia oxidizing archaeon from soil. Proc Natl Acad Sci U S A 108(20):8420–8425. https://doi.org/10.1073/pnas.1013488108

Trivedi P, Leach JE, Tringe SG, Sa T, Singh BK (2020) Plant-microbiome interactions: from community assembly to plant health. Nat Rev Microbiol 18(11):607–621. https://doi.org/10.1038/s41579-020-0412-1

Udvardi M, Below FE, Castellano MJ, Eagle AJ, Giller KE, Ladha JK, Liu X et al (2021) A research road map for responsible use of agricultural nitrogen. Front Sustain Food Syst 5:660155. https://doi.org/10.3389/fsufs.2021.660155

van den Berg EM, Elisário MP, Kuenen JG, Kleerebezem R, van Loosdrecht MCM (2017) Fermentative bacteria influence the competition between denitrifiers and DNRA bacteria. Front Microbiol 8:1684. https://doi.org/10.3389/fmicb.2017.01684

Van Deynze A, Zamora P, Delaux PM, Heitmann C, Jayaraman D, Rajasekar S, Graham D et al (2018) Nitrogen fixation in a landrace of maize is supported by a mucilage-associated diazotrophic microbiota. PLoS Biol 16(8):e2006352. https://doi.org/10.1371/journal.pbio.2006352

Walker CB, de la Torre JR, Klotz MG, Urakawa H, Pinel N, Arp DJ, Brochier-Armanet C et al (2010) *Nitrosopumilus maritimus* genome reveals unique mechanisms for nitrification and autotrophy in globally distributed marine crenarchaea. Proc Natl Acad Sci U S A 107(19):8818–8823. https://doi.org/10.1073/pnas.0913533107

Wang H, Gunsalus RP (2000) The nrfA and nirB nitrite reductase operons in *Escherichia coli* are expressed differently in response to nitrate than to nitrite. J Bacteriol 182(20):5813–5822. https://doi.org/10.1128/JB.182.20.5813-5822.2000

Wang L, Lu Q, Wen X, Lu C (2015a) Enhanced sucrose loading improves rice yield by increasing grain size. Plant Physiol 169(4):2848–2862. https://doi.org/10.1104/pp.15.01170

Wang N, Ding LJ, Xu HJ, Li HB, Su JQ, Zhu YG (2015b) Variability in responses of bacterial communities and nitrogen oxide emission to urea fertilization among various flooded paddy soils. FEMS Microbiol Ecol 91(3):13. https://doi.org/10.1093/femsec/fiv013

Wang B, Zhao J, Guo Z, Ma J, Xu H, Jia Z (2015c) Differential contributions of ammonia oxidizers and nitrite oxidizers to nitrification in four paddy soils. ISME J 9(5):1062–1075. https://doi.org/10.1038/ismej.2014.194

Wang X, Tamiev D, Alagurajan J, DiSpirito AA, Phillips GJ, Hargrove MS (2019) The role of the NADH-dependent nitrite reductase, Nir, from *Escherichia coli* in fermentative ammonification. Arch Microbiol 201(4):519–530. https://doi.org/10.1007/s00203-018-1590-3

Wang W, Hu B, Li A, Chu C (2020) NRT1.1s in plants: functions beyond nitrate transport. J Exp Bot 71(15):4373–4379. https://doi.org/10.1093/jxb/erz554

Weisskopf L, Abou-Mansour E, Fromin N, Tomasi N, Santelia D, Edelkott I, Neumann G et al (2006) White lupin has developed a complex strategy to limit microbial degradation of secreted citrate required for phosphate acquisition. Plant Cell Environ 29(5):919–927. https://doi.org/10.1111/j.1365-3040.2005.01473.x

Winiwarter W, Höglund-Isaksson L, Klimont Z, Schoepp W, Amann M (2017) Technical opportunities to reduce global anthropogenic emissions of nitrous oxide. Environ Res Lett. https://doi.org/10.1088/1748-9326/aa9ec9

Yang XR, Li H, Nie SA, Su JQ, Weng BS, Zhu GB, Yao HY et al (2015) Potential contribution of anammox to nitrogen loss from paddy soils in Southern China. Appl Environ Microbiol 81(3): 938–947. https://doi.org/10.1128/AEM.02664-14

Yang W, Ryals R, Cusack D, Silver W (2017) Cross-biome assessment of gross soil nitrogen cycling in California ecosystems. Soil Biol Biochem 107:144–155. https://doi.org/10.1016/j.soilbio.2017.01.004

Yi X, Yuan J, Zhu Y, Yi X, Zhao Q, Fang K, Cao L (2018) Comparison of the abundance and community structure of n-cycling bacteria in paddy rhizosphere soil under different rice cultivation patterns. Int J Mol Sci 19(12):3772. https://doi.org/10.3390/ijms19123772

Yoon S, Cruz-García C, Sanford R, Ritalahti KM, Löffler FE (2015) Denitrification versus respiratory ammonification: environmental controls of two competing dissimilatory NO_3^-/NO_2^- reduction pathways in *Shewanella loihica* strain PV-4. ISME J 9(5):1093–1104. https://doi.org/10.1038/ismej.2014.201

Young CC, Rekha PD, Lai WA, Arun AB (2006) Encapsulation of plant growth-promoting bacteria in alginate beads enriched with humic acid. Biotechnol Bioeng 95(1):76–83. https://doi.org/10.1002/bit.20957

Yuan C-L, Zhang L, Wang J, Hu H, Shen J-P, He J-Z (2019) Distributions and environmental drivers of archaea and bacteria in paddy soils. J Soils Sediments 19:1997. https://doi.org/10.1007/s11368-018-1997-0

Zakir HAKM, Subbarao GV, Pearse SJ, Gopalakrishnan S, Ito O, Ishikawa T, Kawano N et al (2008) Detection, isolation and characterization of a root-exuded compound, methyl 3-(4-hydroxyphenyl) propionate, responsible for biological nitrification inhibition by sorghum (*Sorghum bicolor*). New Phytol 180(2):442–451. https://doi.org/10.1111/j.1469-8137.2008.02576.x

Zhang Y, Pohlmann EL, Halbleib CM, Ludden PW, Roberts GP (2001a) Effect of P(II) and its homolog GlnK on reversible ADP-ribosylation of dinitrogenase reductase by heterologous expression of the *Rhodospirillum rubrum* dinitrogenase reductase ADP-ribosyl transferase-dinitrogenase reductase-activating glycohydrolase regulatory system in *Klebsiella pneumoniae*. J Bacteriol 183(5):1610–1620. https://doi.org/10.1128/JB.183.5.1610-1620.2001

Zhang Y, Pohlmann EL, Ludden PW, Roberts GP (2001b) Functional characterization of three GlnB homologs in the photosynthetic bacterium *Rhodospirillum rubrum*: roles in sensing ammonium and energy status. J Bacteriol 183:6159–6168

Zhang J, Liu YX, Zhang N, Hu B, Jin T, Xu H, Qin Y et al (2019) NRT1.1B is associated with root microbiota composition and nitrogen use in field-grown rice. Nat Biotechnol 37(6):676–684. https://doi.org/10.1038/s41587-019-0104-4

Zhou Y, Lambrides CJ, Li J, Xu Q, Toh R, Tian S, Yang P et al (2020) Nitrifying microbes in the rhizosphere of perennial grasses are modified by biological nitrification inhibition. Microorganisms 8(11):1687. https://doi.org/10.3390/microorganisms8111687

Zilber-Rosenberg I, Rosenberg E (2008) Role of microorganisms in the evolution of animals and plants: the hologenome theory of evolution. FEMS Microbiol Rev 32(5):723–735

Chapter 10
Biological Nitrogen Fixation in the Rhizosphere of Cacao (*Theobroma cacao* L.) and Coffee (*Coffea* spp.) and its Role in Sustainable Agriculture

Angelbert D. Cortes and Shamdee Nahar-Cortes

Abstract Agricultural farming systems for cacao and coffee crops usually depend on chemical fertilization; however, the excessive use of fertilizers poses environmental threats. Alternative farming inputs are exploited to maintain soil fertility using renewable, environmentally friendly, and cost-effective farming practices. Biological nitrogen fixation is an essential alternative mechanism to chemical fertilization to achieve sustainable agricultural production and healthy ecosystem functioning. Nitrogen fixation is facilitated by diazotrophic rhizobacteria that establish mutual relationships with plants as either associative, endophytic, or symbiotic. These plant growth-promoting rhizobacteria are isolated and characterized to assess their potential use as a sustainable alternative to agrochemical inputs. Despite the numerous studies that demonstrated the promising nitrogen-fixation capabilities of cacao- and coffee-associated microbes, the present review found the lack of studies on the direct application of diazotrophic bacteria on cacao and coffee plants. Meanwhile, intercropping of N_2-fixing trees with cacao and coffee in agroforestry ecosystems is extensively studied due to the beneficial effects of N_2-fixing trees for the perennial crops in providing available N sources, shade, and protection from biological agents. Hence, this minireview emphasized the two mechanisms by which cacao and coffee plants can acquire N sources from the environment, which are through: (i) intercropping N_2-fixing trees and (ii) inoculation of diazotrophic rhizobacteria.

A. D. Cortes (✉)
Department of Biological Sciences, College of Arts and Sciences, Cavite State University, Cavite, Philippines
e-mail: angelbert.cortes@cvsu.edu.ph

S. Nahar-Cortes
Graduate School, University of the Philippines Los Baños, College, Laguna, Philippines
e-mail: snnahar@up.edu.ph

© The Author(s), under exclusive license to Springer Nature Singapore Pte Ltd. 2022
D. K. Maheshwari et al. (eds.), *Nitrogen Fixing Bacteria: Sustainable Growth of Non-legumes*, Microorganisms for Sustainability 36,
https://doi.org/10.1007/978-981-19-4906-7_10

Keywords Biological nitrogen fixation · Plant growth-promoting rhizobacteria · Cacao · Coffee

10.1 Introduction

Plant-microbe association exemplifies a complex and multi-organ system with varying degrees of intimacy and mutual dependence (De Mandal et al. 2021). Beneficial microorganisms in the rhizosphere form unique symbiotic and non-symbiotic relationships with plants by maintaining nutrient recycling, hormone production, preventing microbial infections, and improving tolerance towards potentially hazardous compounds (dos Santos et al. 2020; De Mandal et al. 2021). Meanwhile, plants facilitate root exudation at the root tips by producing primary metabolites (sugars, amino acids, and organic acids) useable for many rhizosphere-dwelling microbes (Canarini et al. 2019). These beneficial microbes are considered as plant growth-promoting rhizobacteria that may provide a biological alternative to fix and immobilize nutrients, like nitrogen, stimulating crop growth and production (Backer et al. 2018).

Nitrogen (N) is one of the major nutrients limiting plant growth in agro ecosystems (Mahmud et al. 2020). Nitrogen precursors play important roles in many cellular processes in plants such as energy metabolism (Foyer et al. 2011), photosynthesis (Zhang et al. 2020), signal transduction (Forman et al. 2008), and synthesis of macromolecules (Kaur et al. 2021). Biological nitrogen fixation (BNF) is an essential N supply route for many terrestrial plants and is an essential mechanism for sustainable agricultural production and healthy ecosystem functioning (Cooper and Scherer 2012; Mahmud et al. 2020). BNF is commonly driven by nitrogenase-possessing prokaryotes known as 'diazotrophs' (Cooper and Scherer 2012; Cortes et al. 2020). Diazotrophic symbionts use oxygen-sensitive nitrogenase to reduce N_2 gas to ammonia (NH_3) which is performed at optimum conditions (Cooper and Scherer 2012). This biological process takes place when excess amounts of mineral nitrogen compounds in the environment are absent (Volkogon et al. 2021). Moreover, BNF as biofertilizers is shown to be directly proportional to agricultural sustainability because they have the capacity to reduce the use of nitrogen fertilizers to approximately 0.160 billion tons per year (Soumare et al. 2020). Thus, BNF is an alternative option in sustainable agriculture, particularly in cacao and coffee production.

Cacao (*Theobroma cacao* L.) and Coffee (*Coffea* spp.) are important perennial crops belonging to the family of Malvaceae (de Souza et al. 2018) and Rubiaceae (Ferreira et al. 2019), respectively. Cacao beans are mainly used for the production of chocolate and other cocoa products (de Souza et al. 2018), whereas coffee beans are mainly used to produce coffee beverages with unique taste and aroma (Ferreira et al. 2019). In 2020/21, a global record production of 5.024 million tons of cacao products is forecasted, with Africa contributing the largest cocoa output (77%), followed by the Americas (17%) and Asia-Oceania (6%) regions (International Cocoa Organization 2021). Coffee consumption is increasing at an annual growth rate of 3.6%, and the global production of green beans reached 20.7 MT in 2019. The total production of coffee for the year 2020/21 has revamped marginally (0.4% increase) to 169.99 million 60-kg bags from 169.00 million bags from the previous year (International Coffee Organization 2021). In agroforestry systems, these crops are vulnerable to environmental threats due to climate change, thus there is an urgency for land use planning to define the best areas and growing systems for the production of coffee and cacao (de Sousa et al. 2019). In addition, excessive use of chemical fertilization in cacao and coffee crops tends to cause environmental consequences, thus biological nitrogen fixation is being exploited to increase agronomic efficiency while reducing production costs and environmental pollution caused by chemical fertilizers (Souza et al. 2015).

There are several reports about biological nitrogen fixation in legumes (Masson-Boivin et al. 2009) and limited in case of non-legume plants (Santi et al. 2013). However, this minireview will highlight the contribution of BNF in non-legume perennial crops, namely, *T. cacao* and *Coffea* spp., and its role in sustainable agriculture. We summarize here the mechanisms of how the non-fixing perennial crops acquire N sources from the environment to sustain their growth and biological functions.

10.2 Biological Nitrogen Fixation

Biological nitrogen fixation through conversion of atmospheric N_2 to NH_3 is performed by the small subset of taxonomically diverse organisms within the prokaryotes, which can be symbiotic, endophytic, and associative or free-living in relation to the host plants (Cooper and Scherer 2012; Souza et al. 2015). Diazotrophic microorganisms perform BNF through nitrogenase, a highly conserved and oxygen-sensitive enzyme, consisting of two metalloproteins—MoFe and Fe proteins—which contain different metal clusters (Dixon and Kahn 2004). Symbiotic nitrogen fixation is largely limited to legumes in agricultural systems, but there are groups of diazotrophs that inhabit the rhizosphere of other crop plants, like non-legumes, and have shown to enhance plant growth and development (Mus et al. 2016). This activity is highly abundant in terrestrial ecosystems as it helps maintain the biogeochemical recycling of nitrogen molecules in the environment. Several studies have reported the plant growth promotion traits of many plant growth-promoting rhizobacteria (PGPR), such as biological nitrogen fixation (Igiehon and Babalola 2018), zinc and phosphate solubilization (Mehta et al. 2015), S-oxidation stress alleviation through ACC deaminase activity (Gupta and Pandey 2019; Duan et al. 2021), production of siderophores and phytohormones (Sayyed et al. 2013; Maheshwari et al. 2015: Baliyan et al. 2021), and secretion of secondary metabolites acting as biocontrol agents (Compant et al. 2005).

Besides, the degree of intimacy and interdependency between plants and microbes are categorized into three associations: intracellular symbiotic associations, intercellular endophytic associations, and loose associations with associative or free-living nitrogen fixers (Mus et al. 2016). These types of associations based on the relationship between plants and nitrogen-fixing bacteria vary in terms of their strategy and biological processes.

10.2.1 Endosymbiotic Nitrogen-Fixation

Nitrogen-fixing symbiosis between legumes and rhizobia is the most extensively studied and exploited plant-bacteria association (Fig. 10.1). In relationship, legumes provide rhizobia with reduced C and a protected, anaerobic environment required for nitrogenase activity, while rhizobia provide the legumes with biologically available N sources (Backer et al. 2018). All rhizobia elicit the formation of root nodules, the plant organs dedicated to the fixation and assimilation of nitrogen (Masson-Boivin et al. 2009). Within root nodules, the symbiotic association of rhizobial bacteria with leguminous plants generates symbiosomes, the structures that are formed when differentiated bacteria are enclosed in an intracellular plant-derived compartment (Oldryod et al. 2011). Another common plant-microbe symbiosis involving the role of nitrogen fixation is the association of actinorhizal plants with Gram-positive actinobacterial species belonging to the genus *Frankia*. The strategies and rate of nitrogen fixation in actinorhizal nodules are comparable with legume symbioses (Sellstedt and Richau 2013; Mus et al. 2016).

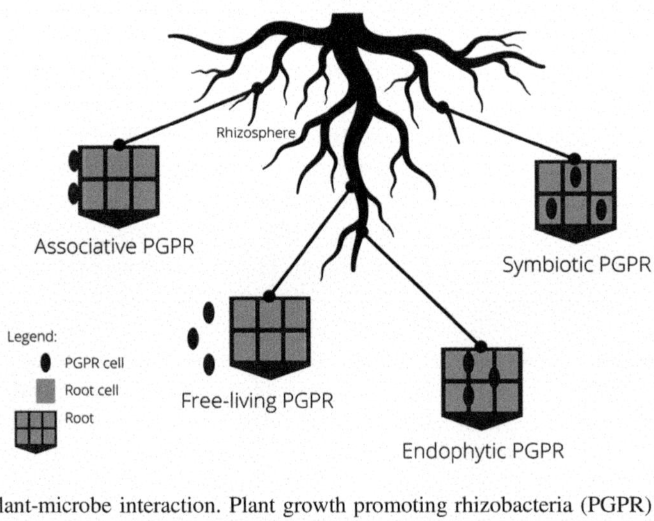

Fig. 10.1 Plant-microbe interaction. Plant growth promoting rhizobacteria (PGPR) can establish relationship with the host plant through several mechanisms: associative and free-living, endophytic, and symbiotic. The plant growth-promoting traits of these rhizobacteria improve the growth and development of cacao and coffee crops. (The visual concept was derived from Souza et al. (2015))

10.2.2 Endophytic Nitrogen-Fixation

Endophytic PGPR (Fig. 10.1) are good inoculant candidates because they colonize the roots (within apoplastic spaces) and create a favorable environment for plant development and function (Souza et al. 2015). These microbes establish a tight association with the host plant by spreading and multiplying within plant tissues without causing damage and eliciting significant defense reactions (Mus et al. 2016). The utilization of endophytic bacteria in agricultural systems provides a potential contribution to sustainable agricultural practices by alleviating several biotic and abiotic stresses of plants (Prasad et al. 2020). They secrete metabolites (e.g., flavonoids, peptides, quinones, alkaloids, steroids, etc.) that inhibit the growth and development of biotic stressors through antibiosis (Kumar and Dara 2021). Nitrogen-fixing endophytes also respond to plant exudates by regulating the expression of several genes, such as those associated with exopolysaccharide biosynthesis and biofilm formation (Meneses et al. 2011). Some of the plant growth-promoting rhizobacteria that stimulate plant growth through nitrogen fixation are those belonging to the genera *Azoarcus*, *Burkholderia*, *Gluconacetobacter*, and *Herbaspirillum* (Vessey 2003).

10.2.3 Associative Nitrogen-Fixation

The association of plants with many free-living and/or associative plant growth-promoting rhizobacteria (PGPR) is the simplest nitrogen-fixing symbiosis (Fig. 10.1). Their mutual interactions in the rhizosphere are the determinants of plant health, productivity, and soil fertility (Souza et al. 2015). These diazotrophs respond to plant root exudation via chemotaxis and colonization in the plant rhizosphere, or biofilm formation on the surface of the root, without invading the host tissues (Mus et al. 2016). Many species of free-living rhizospheric nitrogen-fixing bacteria were found to stimulate plant growth and fitness, such as those belonging to the genera *Azospirillum*, *Azotobacter*, *Burkholderia*, *Bacillus*, and numerous species of Cyanobacteria (Vessey 2003).

10.3 Biological Nitrogen Fixation in Sustainable Agriculture

Agricultural production commonly depends on the large-scale use of chemical fertilizers (e.g., N fertilizers) to provide essential nutrients for plants (Souza et al. 2015). However, the heavy use of chemical N fertilizers is a global concern due to its economic and environmental costs. Long before, excessive use of N fertilizers poses great unanticipated environmental impacts (e.g., eutrophication, ecosystem damage, plant toxicity, excessive plant growth, and stratospheric ozone depletion) and adverse effect on human health (methemoglobinemia in infants, cancer, and respiratory illness) (Bohlool et al. 1992). In addition, excessive use of nitrogen fertilization from chemical fertilizers also leads to soil acidification (Goulding 2016) and potentially increases water contamination due to leakage of unabsorbed chemical fertilizers (Munroe and Isaac 2014). Thus, there is a need for a sustainable farming system that maintains soil fertility by using renewable, environmentally friendly, and cost-effective agricultural practices.

Alternative system options in sustainable agriculture are sought and exploited to achieve maximum cropping benefits in terms of fertilizer savings and better plant growth, which includes the application of microbial-based inoculation technology (Souza et al. 2015; Aggangan et al. 2019). Besides, BNF through diazotrophic microorganisms offers this alternative system (Soumare et al. 2020). A nitrogen-fixing system provides an economically attractive and ecologically sound means of reducing external inputs and improving internal resources. BNF improves nitrogen input that serves as an N fertilizer substitute and maintains soil N reserves to attain large crop production (Bohlool et al. 1992). Moreover, the changes in the composition and number of soil microbiota can change the rate and process of nitrogen transformation in the soil. Specifically, the number and nitrogen fixation activity of diazotrophs tend to increase when low mineral fertilizer is applied, providing a favorable soil condition for their development (Volkogon et al. 2021). Microbe-based inoculation and commercialization of these biofertilizers have shown to be effective and relevant in sustainable agriculture; however, some success-limiting factors against the universal utilization still exist as the efficiency of microbe-based biofertilizers depends on the targeted crop, edaphic, biotic, and climatic factors (Soumare et al. 2020). In agroforestry systems, intercropping of important non-legume crops like coffee and cacao with N_2-fixing plants is being employed in order to provide the crops with available N sources, shade, and protection from biological agents. In the next section, it will be discussed how the non-N_2-fixing plants like cacao and coffee acquire N sources from the environment.

10.4 Characteristics of Cacao (*Theobroma cacao* L.) and Coffee (*Coffea* spp.) Plants

Theobroma cacao L. or known as "cacao" is a diploid tree fruit species and its beans are widely known for being the raw material of cocoa and chocolate (Rusconi and Conti 2010; Argout et al. 2011). Cacao is a fast-growing tropical forest plant usually being cultivated in association with tall trees (e.g., nitrogen-fixing trees and fruit-bearing trees) that provide shade and N sources (Fig. 10.2a). Based on its nature and characteristics, cacao is grouped into three cultivars: Criollo, Forastero, and Trinitario (Rusconi and Conti 2010). The Criollo produces large fruits with a wrinkled, thin, or thick surface, which presents red or green color, and the seeds

Fig. 10.2 Nitrogen fixing trees intercropped with cacao and coffee crops. (**a**) *Theobroma cacao* L. intercropped with *Gliricidia sepium* and fruit-bearing plants at Dariano Cacao Farm in Silang, Cavite, Philippines. (**b**) *Coffea liberica* intercropped with *Albizia saman* (Jacq.) Merr. at National Coffee Research, Development and Extension Center (NCRDEC) in Cavite State University, Philippines

are large with a white or pale violet interior (de Souza et al. 2018). This variety has a nearly unique and homozygous genotype, which produces a more aromatic and finer flavor chocolate (Argout et al. 2011), but only 5–10% of chocolate is made (Rusconi and Conti 2010). On the other hand, the Forastero presents intensely pigmented seeds, with dark violet or blackish interior, green fruits when immature and yellow when ripe, ovoid shape, smooth surface, and wrinkled (de Souza et al. 2018). This variety has foreign genotypes being introduced to create hybrids that will overcome poor agronomic performance and disease susceptibility (Argout et al. 2011). Forastero trees produce cheaper cocoa beans, thus used for 80% of world chocolate production. Trinitario is a hybrid of Criollo and Forastero, which is used in about 10–15% of chocolate production (Rusconi and Conti 2010). The pods are less bumpy than Criollo and exhibit light to medium purple colors. Besides, as a result of several breeding programs, a cacao cultivar clone (CCN 51) was obtained and is recognized worldwide for its high yield and adaptability to different edaphoclimatic conditions (Jaimez et al. 2022). In addition, *T. cacao* also poses biomedical importance and is reported to have several biological activities, including anticancer, antioxidant, antimalarial, antidiabetic, and antihypertensive activities (Ishaq and Jafri 2017).

Coffea spp. locally known as "coffee," is among the most important agricultural commodities on the world market. These perennial shrubs develop into small trees maximum 3–3.5 m high, producing highly scented white flowers. It is commonly intercropped with trees (e.g., N_2-fixing and fruit-bearing trees) to provide shade, cover crops, and available nutrients like nitrogen (Fig. 10.2b). Coffee trees produce indehiscent drupes that contain two beans, the main source of aromatic coffee beverage (Simon-Gruita et al. 2019). Caffeine is the main constituent of coffee that is found in its beans, fruits, and leaves, which vary among the common commercially grown cultivars. This alkaloid stimulates the central nervous system and intestinal peristalsis, increases heart rate and blood pressure, and is bronchodilator and diuretic (Simon-Gruita et al. 2019). More than 124 species of coffee were identified, but two known species *Coffea arabica* (Arabica) and *Coffea canephora* (Robusta) have shown a great economic importance. Arabica coffee grows best in tropical highlands with elevations ranging from 700 m to 2200 m, thus better adapted to cooler temperatures (16–24 °C). It is self-pollinating and genetically complex carrying a tetraploid chromosome set, producing sweet and aromatic with a range of fine flavors and some desirable acidity. Arabica bean is oval-shaped, 7 mm to 12 mm long, and consists of a zig-zagged center cut pattern. On the other hand, Robusta coffee is better adapted to the lowlands (below 900 m altitude) and prefers a hotter climate (21–30 °C). This species is cross-pollinating (self-sterile) and genetically simpler that carries a diploid set of chromosomes, producing a strong and intense taste and a more pronounced bitter flavor. The Robusta bean is round, 5–8 mm long, and the center cut is almost straight (Bozzola et al. 2021). Other commercially grown varieties and cultivars are *Coffea liberica* var. *liberica* (Liberica) and *Coffea liberica* var. *dewevrei* (Excelsa), which showed differences in their quantitative traits as well as detected to have genetic differentiation (N'Diaye et al. 2005). Liberica coffee known as "Barako" is fairly abundant in Southeast Asia, especially in the Philippines and Malaysia. It grows as a large tree (up to 18 m high), which has larger leaves, cherries, and beans (N'Diaye et al. 2005; Gibson 2018). Liberica coffee has a strong flavor and sharp aroma (Department of Agriculture 2019). Meanwhile, Excelsa grows on large, hardy trees at medium altitudes and has a teardrop shape bean. It is more drought and pest resistant than other varieties. This variety produces a pronounced tart, fruity, and dark flavor coffee (Gibson 2018). Based on the 2017–2022 Philippine coffee industry roadmap, Robusta gains the highest percent share (69%) of total production of green coffee beans (GCB), followed by Arabica (24%), Excelsa (6%), and Liberica (1%) (Department of Agriculture 2019).

10.5 Biological Nitrogen Fixation in Cacao (*Theobroma cacao* L.) and Coffee (*Coffea* spp.)

10.5.1 Nitrogen Transfer from N_2-Fixing Plants to Cacao and Coffee Plants

Woody perennial crops like coffee (*Coffea* spp.) and cacao (*Theobroma cacao* L.) are often intercropped with N_2-fixing trees within multistrata agroforestry systems (Schroth et al. 2001; Nygren and Leblanc 2009). These N_2-fixing leguminous trees function as shade and cover crops to regulate the direct penetration of sunlight to cacao and coffee crops (Schroth et al. 2001; Munroe and Isaac 2014; Mus et al. 2016). In addition, the incorporation of a tree as an overstorey enhances nutrient and water cycling, organic matter accumulation, and pest regulation of crops (Munroe and Isaac 2014). In tropical agroforestry systems, N_2-fixing legume trees may also enhance soil N availability to the benefit of the non-legume plants, making them an alternative to N fertilizers (Kaba et al. 2019). The N transfer from N_2-fixing trees to non-N_2-fixing crops (e.g., cacao and coffee) could be attributed to the decomposition and mineralization of organic compounds (e.g., litter, prunings, roots, and nodules), root-to-root direct transfer via exudation, and common mycorrhizal networks (Munroe and Isaac 2014).

In mixed-stand agroforestry systems, with intercropped *Gliricidia sepium* and *Theobroma cacao* trees, the amount of N produced from gliricidia diminishes the need for N fertilizers for cocoa trees. Specifically, it was found that the leaves of gliricidia and cocoa trees growing in proximity had similar $\delta^{15}N$, whereas the foliar $\delta^{15}N$ value of gliricidia was lower than that of distant cocoa trees (Kaba et al. 2019). In addition, a study also demonstrated a lower $\delta^{15}N$ value in *T. cacao* below *Inga edulis* than *T. cacao* below mixed-species shade, indicating direct N transfer from *I. edulis* to *T.* cacao (Nygren and Leblanc 2009). *Erythrina poeppigiana* was also cultivated together with cacao plants to provide N supply and improve its growth yield (Somarriba and Beer 2011). On the other hand, Snoeck et al. (2000) demonstrated that in the field condition, roughly 30% of the nitrogen effectively fixed by a

Table 10.1 Summary of commonly intercropped N_2-fixing plants with cacao (*Theobroma cacao* L.) and coffee (*Coffea* spp.) crops for the provision of shade, cover crop, and N sources

Crop	Intercropped N_2-fixing plants	References
Cacao (*Theobroma cacao* L.)	*Inga edulis*	Nygren and Leblanc (2009)
	Erythrina poeppigiana	Somarriba and Beer (2011)
	Gliricidia sepium	Kaba et al. (2019)
Coffee (*Coffea* spp.)	*Flemingia macrophylla*	Snoeck et al. (2000)
	Desmodium intortum	Snoeck et al. (2000)
	Leucaena leucocephala	Snoeck et al. (2000)
	Leucaena diversifolia	Snoeck et al. (2000)
	Calliandra calothyrsus	Snoeck et al. (2000)
	Erythrina abyssinica	Snoeck et al. (2000)
	Cajanus cajan[a]	Mendonça et al. (2017)
	Crotalaria spectabilis[a]	Mendonça et al. (2017)
	Calopogonium mucunoides[a]	Mendonça et al. (2017)

[a]Green manure crops

legume was transferred to the associated coffee trees, this is based on the measured N from litter fall or soil. Mendonça et al. (2017) also observed higher rates of N transfer of green manure crops *Cajanus cajan* (55.8%), *Crotalaria spectabilis* (48.8%), and *Calopogonium mucunoides* (48.1%) to the coffee plants. These results suggest that the decomposed litters of leguminous trees increase nutritional status of the soil, resulting to optimum uptake and utilization by nearby crops (Zaharah and Bah 1999). In addition, direct N transfer from legumes to non-legume plants may also occur through root exudation and common mycelial networks (Jalonen et al. 2009). In the agroforestry system, Nygren and Leblanc (2015) observed that fine roots of the cacaos associated with *Inga* species contained ~35% of the N fixed from the atmosphere out of the total N measured. Overall, they demonstrated that approximately 20% of the fixed atmospheric N in the system was found in cacao, suggesting direct N transfer from *Inga* via recycling of its N-rich root exudates or a common mycelial network of mycorrhizal fungi. Lastly, in the review of Munroe and Isaac (2014), the common N_2-fixing trees growing in proximity with coffee and cacao crops were described based on their functions, N transfer estimation method, and percent fixed-N in receiver crop. The intercropping system is found to be an effective way to improve the growth and yield of economically important crops while minimizing or even omitting the use of chemical fertilizers. The commonly intercropped N_2-fixing plants with cacao and coffee crops are summarized in Table 10.1.

10.5.2 Nitrogen Fixation by Rhizospheric Microorganisms in Cacao and Coffee

There is an increased interest in exploiting the beneficial role of plant growth-promoting rhizobacteria as biofertilizers due to their ability to promote plant growth using their own metabolism to solubilize phosphates, produce hormones and siderophores, produce indoleacetic acid, and fix atmospheric nitrogen (Park et al. 2005, Glick 2012; Ji et al. 2014; dos Santos et al. 2020; Cortes et al. 2020; Zuluaga et al. 2020). Studies have shown that these rhizobacterial diazotrophs are shown to be effective in improving the growth of economically important crops, including the commonly consumed rice (Ji et al. 2014), maize (Breedt et al. 2017), wheat (Din et al. 2021), and tomato and lulo plants (Zuluaga et al. 2020). These bacteria are an excellent alternative for farmers to reduce chemical fertilization and pesticide input without compromising the environment (dos Santos et al. 2020). They are continuously exploited to replace the use of chemicals in agriculture, horticulture, silviculture, and environmental cleanup strategies (Glick 2012).

In perennial crops like cacao and coffee, few studies have focused on the direct effects of nitrogen-fixing microorganisms on their growth and development. In the study of Argüello-Navarro and Moreno-Rozo (2014), cocoa plants effectively responded to diazotrophic bacterial treatments by increasing their growth variables and foliar % N, suggesting that these rhizospheric microorganisms could be associated with the adaptability of cacao plant to adverse environmental conditions. Meanwhile, the growth of the coffee crop, especially Robusta, responded positively to the inoculation of endophytic and rhizobacterial isolates belonging to the genera *Bacillus* and *Pseudomonas,* suggesting to be a potential biofertilizer for the sustainable production of Robusta coffee (Asyiah et al. 2020; Nguyen et al. 2021). There are only a few published studies focusing on the use of diazotrophic rhizobacteria on the growth responses of cacao and coffee plants, thus it is recommended to continuously explore their potential as biofertilizer through direct application.

Several studies demonstrated the promising plant growth-promoting characteristics of certain rhizobacterial species obtained from the rhizosphere of cacao plants belonging to the genera *Bacillus*, *Burkholderia*, *Gluconacetobacter*, *Herbaspirillum*, *Rhizobium*, and *Ralstonia* (Argüello-Navarro and Moreno-Rozo 2014; Cortes et al. 2020). In addition, Simarmata et al. (2020) obtained plant growth-promoting endophytic bacteria from cacao plants with higher plant growth-promoting traits, which were identified as *Pantoea* sp. and *Brevibacillus brevis*. The presence of nitrogen-fixing PGPR community in the acidic cacao rhizosphere was recently demonstrated, suggesting their role in nitrogen cycling in cacao plants (Cortes et al. 2021). On the other hand, a metagenomic study in the coffee rhizosphere showed that the relative abundance of nitrogen-fixing bacteria ranged from three to six percent of the whole community, and *Bradyrhizobium* genus was found in all rhizospheric soil samples (Silva et al. 2020). The ability to fix atmospheric nitrogen was demonstrated for certain species of diazotrophic rhizobacteria associated with coffee plants belonging to the genera *Acetobacter*

Table 10.2 Rhizobacterial genera known to be associated with cacao and coffee crops that have shown plant growth-promoting traits

Plant	Associated diazotrophic bacterial genera with plant growth-promoting traits	References
Cacao (*Theobroma cacao* L.)	*Bacillus, Burkholderia, Rhizobium, Ralstonia*	Cortes et al. (2020)
	Gluconacetobacter, Herbaspirillum	Argüello-Navarro and Moreno-Rozo (2014)
	Pantoea sp., *Brevibacillus*	Simarmata et al. (2020)
Coffee (*Coffea* spp.)	*Acetobacter*	Jimenez-Salgado et al. (1997)
	Burkholderia	Estrada-De Los Santos et al. (2001)
	Bradyrhizobium	Silva et al. (2020)
	Rhizobium, Azotobacter, Azospirillum, Pseudomonas, Erwinia, Bacillus	Urgiles-Gómez et al. (2021)

(Jimenez-Salgado et al. 1997) and *Burkholderia* (Estrada-De Los Santos et al. 2001). Moreover, the genera *Rhizobium*, *Azotobacter*, *Azospirillum*, *Acetobacter*, *Pseudomonas*, *Erwinia*, *Bacillus*, and *Burkholderia* were consistently observed in the coffee rhizosphere, contributing significantly for biocontrol, biofertilization, biostimulation in coffee plants (Urgiles-Gómez et al. 2021). Thus, N_2-fixing bacterial isolates are potential biofertilizer candidates to improve nutrient acquisition of cacao and coffee crops and resistance to biotic and abiotic stresses. The associated plant growth-promoting rhizobacteria with coffee and cacao crops are shown in Table 10.2.

10.6 Challenges in Biological Nitrogen Fixation in Coffee and Cacao Areas

Due to climate change, some common N_2-fixing trees are experiencing losses in habitat suitability, particularly in cacao and coffee plantations. Across Mesoamerica, 25 of the 30 N_2-fixing trees assessed in coffee and 18 N_2-fixing trees in cocoa areas are expected to have >15% high losses in terms of habitat suitability. N_2-fixing trees like *Erythrina poeppigiana* and majority of *Inga* species losses 56% in suitable areas, whereas only two selected species, *Inga laurina* (guam) and *Senna atomaria* (vainillo) may expand their suitability in >26% across cocoa areas and 4% in future coffee areas. Thus, it is recommended to identify the best species of N_2-fixing trees when rethinking current agroforestry species composition in coffee and cocoa landscapes (de Sousa et al. 2019). Moreover, intensive crop production is accompanied by several environmental issues, including soil degradation, biodiversity loss, and pollution due to extensive use of agrochemical inputs and generated wastes (Duong et al. 2020). In response to climate change mitigation projects, Dawoe et al.

(2016) suggested the implementation of REDD+ (Reduced Emissions from Deforestation and Forest Degradation) interventions in cacao landscapes to diversify the income sources and create incentives for farmers by rolling out tree diversification strategies.

Microorganisms associated with coffee and cacao plants are being isolated and explored to highlight their plant growth-promoting characteristics. However, this review found limited available studies being performed on the direct effects of nitrogen-fixing bacteria on the growth of cacao and coffee plants. The application of promising rhizobacterial isolates, like diazotrophs, as biofertilizers yet to be strengthened to improve the growth and yield of cacao and coffee plants.

10.7 Conclusions

Perennial crops like *Theobroma cacao* and *Coffea* spp. are commonly cultivated by smallholder farmers, contributing to the cacao and coffee world market. Several agricultural strategies are being adopted by farmers to improve the available soil nitrogen while reducing the use of chemical fertilizers to stimulate the growth yield of cacao and coffee crops. Excessive chemical fertilization poses environmental and health threats due to leakage of unabsorbed substances, which can further lead to soil acidification and water contamination. Biological nitrogen fixation is one of the promising mechanisms that many agroforestry systems adopt due to their ability to provide N sources for plant nutrition. This minireview found two mechanisms where cacao and coffee plants may acquire N sources from the environment, either through intercropped N_2-fixing trees or direct association with rhizospheric diazotrophs, which fix atmospheric nitrogen. The former is extensively used in multistrata agroforestry systems and has been shown to effectively improve nitrogen acquisition of cacao and coffee plants. Despite the promising results obtained from the inoculation of biofertilizers on cacao and coffee, the long-term effects of these biofertilizers should be explored to attain sustainable agricultural farming, particularly for cacao and coffee plants.

Acknowledgement Not applicable.

Conflict of Interest Author(s) declares no conflict of interest.

References

Aggangan NS, Cortes AD, Reaño CE (2019) Growth response of cacao (*Theobroma cacao* L.) plant as affected by bamboo biochar and arbuscular mycorrhizal fungi in sterilized and unsterilized soil. Biocatal Agric Biotechnol 22:101347

Argout X, Salse J, Aury JM, Guiltinan MJ, Droc G, Gouzy J, Allegre M, Chaparro C, Legavre T, Maximova SN, Abrouk M (2011) The genome of *Theobroma cacao*. Nat Genet 43:101–108

Argüello-Navarro AZ, Moreno-Rozo LY (2014) Evaluation of potential biofertilizer diazotrophas bacteria isolated from cocoa crop (*Theobroma cacao* L.). Acta Agron 63(3):238–245

Asyiah IN, Mudakir I, Hoesain M, Pradana AP, Djunaidy A, Sari RF (2020) Consortium of endophytic bacteria and rhizobacteria effectively suppresses the population of *Pratylenchus coffeae* and promotes the growth of Robusta coffee. Biodiversitas 21(10):4702–4708

Backer R, Rokem JS, Ilangumaran G, Lamont J, Praslickova D, Ricci E, Subramanian S, Smith DL (2018) Plant growth-promoting rhizobacteria: context, mechanisms of action, and roadmap to commercialization of biostimulants for sustainable agriculture. Front Plant Sci 9:1473

Baliyan N, Dhiman S, Dheeman S, Kumar S, Arora NK, Maheshwari DK (2021) Optimization of gibberellic acid production in endophytic *Bacillus cereus* using response surface methodology and its use as plant growth regulator in chickpea. J Plant Growth Regul:1–11. https://doi.org/10.1007/s00344-021-10492-2

Bohlool BB, Ladha JK, Garrity DP, George T (1992) Biological nitrogen fixation for sustainable agriculture: a perspective. Plant Soil 141(1):1–1

Bozzola M, Charles S, Ferretti T, Gerakari E, Manson H, Rosser N, von der Goltz P (2021) The coffee guide. International Trade Centre, Geneva, Switzerland. https://digitalcollection.zhaw.ch/bitstream/11475/23941/3/2021_ITC_Coffee-Guide.pdf. Accessed April 11 2022

Breedt G, Labuschagne N, Coutinho TA (2017) Seed treatment with selected plant growth-promoting rhizobacteria increases maize yield in the field. Ann Appl Biol 171(2):229–236

Canarini A, Kaiser C, Merchant A, Richter A, Wanek W (2019) Root exudation of primary metabolites: mechanisms and their roles in plant responses to environmental stimuli. Front Plant Sci 10:157

Compant S, Duffy B, Nowak J, Clément C, Barka EA (2005) Use of plant growth-promoting bacteria for biocontrol of plant diseases: principles, mechanisms of action, and future prospects. Appl Environ Microbiol 9:4951–4959

Cooper JE, Scherer HW (2012) Nitrogen fixation. In: Marschner H (ed) Marschner's mineral nutrition of higher plants, 3rd edn. Academic Press, Cambridge, Massachusetts, pp 389–408

Cortes AD, Aggangan NS, Opulencia RB (2021) Taxonomic microbiome profiling and abundance patterns in the cacao (*Theobroma cacao* L.) rhizosphere treated with arbuscular mycorrhizal fungi and bamboo biochar. Philipp Agric Sci 104(1):19–33

Cortes AD, Opulencia RB, Aggangan NS (2020) Characterization of plant growth promoting diazotrophic bacteria isolated from cacao (*Theobroma cacao* l.) rhizosphere treated with bamboo biochar and arbuscular mycorrhizal fungi. Philipp J Sci 149(4):1063–1070

De Mandal S, Sonali Singh S, Hussain K, Hussain T (2021) Plant–microbe association for mutual benefits for plant growth and soil health. In: Yadav AN, Singh J, Singh C, Yadav N (eds) Current trends in microbial biotechnology for sustainable agriculture. Environmental and Microbial Biotechnology. Springer, Singapore. https://doi.org/10.1007/978-981-15-6949-4_5

Dawoe E, Asante W, Acheampong E, Bosu P (2016) Shade tree diversity and aboveground carbon stocks in *Theobroma cacao* agroforestry systems: implications for REDD+ implementation in a west African cacao landscape. Carbon Balance Manag 11(1):1–13

de Sousa K, van Zonneveld M, Holmgren M, Kindt R, Ordoñez JC (2019) The future of coffee and cocoa agroforestry in a warmer Mesoamerica. Sci Rep 9(1):1–9

de Souza PA, Moreira LF, Sarmento DH, da Costa FB (2018) Cacao—*Theobroma cacao*. In: Exotic fruits. Academic Press, Cambridge, Massachusetts, pp 69–76

Department of Agriculture (2019) 2017-2022 Philippine Coffee Industry Roadmap. https://www.da.gov.ph/wp-content/uploads/2019/06/Philippine-Coffee-Industry-Roadmap-2017-2022.pdf. Accessed March 18 2022

Din I, Khan H, Khan NA, Khil A (2021) Inoculation of nitrogen fixing bacteria in conjugation with integrated nitrogen sources induced changes in phenology, growth, nitrogen assimilation and productivity of wheat crop. J Saudi Soc Agric Sci 20(7):459–466

Dixon R, Kahn D (2004) Genetic regulation of biological nitrogen fixation. Nat Rev Microbiol 8: 621–631

dos Santos RM, Diaz PAE, Lobo LLB, Rigobelo EC (2020) Use of plant growth-promoting rhizobacteria in maize and sugarcane: characteristics and applications. Front Sustain Food Syst 4:136

Duan B, Li L, Chen G, Su-Zhou C, Li Y, Merkeryan H, Liu W, Liu X (2021) 1-Aminocyclopropane- 1-carboxylate deaminase-producing plant growth-promoting rhizobacteria improve drought stress tolerance in grapevine (*Vitis vinifera* L.). front. Plant Sci 12:706990

Duong B, Marraccini P, Maeght JL, Vaast P, Lebrun M, Duponnois R (2020) Coffee microbiota and its potential use in sustainable crop management: a review. Front Sustain Food Syst 4: 607935

Estrada-De Los Santos P, Bustillos-Cristales R, Caballero-Mellado J (2001) *Burkholderia*, a genus rich in plant-associated nitrogen fixers with wide environmental and geographic distribution. Appl Environ Microbiol 67(6):2790–2798

Ferreira T, Shuler J, Guimarães R, Farah A (2019) Introduction to coffee plant and genetics. In: Coffee: production, quality and chemistry. Royal Society of Chemistry, London, pp 1–25

Forman HJ, Fukuto JM, Miller T, Zhang H, Rinna A, Levy S (2008) The chemistry of cell signaling by reactive oxygen and nitrogen species and 4-hydroxynonenal. Arch Biochem Biophys 477(2): 183–195

Foyer CH, Noctor G, Hodges M (2011) Respiration and nitrogen assimilation: targeting mitochondria-associated metabolism as a means to enhance nitrogen use efficiency. J Exp Bot 62(4):1467–1482

Gibson M (2018) Chapter 18 – tea and coffee. In: Gibson M, Newsham P (eds) Food science and the culinary arts. Academic Press, Cambridge, Massachusetts, pp 353–372

Glick BR (2012) Plant growth-promoting bacteria: mechanisms and applications. Scientifica (Cairo) 2012:963401. 15 pages

Goulding KWT (2016) Soil acidification and the importance of liming agricultural soils with particular reference to the United Kingdom. Soil Use Manag 32(3):390–399

Gupta S, Pandey S (2019) ACC deaminase producing bacteria with multifarious plant growth promoting traits alleviates salinity stress in French bean (*Phaseolus vulgaris*) plants. Front Microbiol 10:1506

Igiehon NO, Babalola OO (2018) Rhizosphere microbiome modulators: contributions of nitrogen fixing bacteria towards sustainable agriculture. Int J Environ Res Public Health 15(4):574

International Cocoa Organization (2021) May 2021 Quarterly Bulletin of Cocoa Statistics. https://www.icco.org/may-2021-quarterly-bulletin-of-cocoa-statistics/. Accessed Feb 5 2022

International Coffee Organization (2021) Coffee Market Report. https://www.ico.org/documents/cy2021-22/cmr-1221-e.pdf. Accessed Feb 5 2022

Ishaq S, Jafri L (2017) Biomedical importance of cocoa (*Theobroma cacao*): significance and potential for the maintenance of human health. Mat Sci Pharma 1(1):1–5

Jaimez RE, Barragan L, Fernández-Niño M, Wessjohann LA, Cedeño-Garcia G, Cantos IS, Arteaga F (2022) *Theobroma cacao* L. cultivar CCN 51: a comprehensive review on origin, genetics, sensory properties, production dynamics, and physiological aspects. Peer J 10:e12676

Jalonen R, Nygren P, Sierra J (2009) Transfer of nitrogen from a tropical legume tree to an associated fodder grass via root exudation and common mycelial networks. Plant Cell Environ 32(10):1366–1376

Ji SH, Gururani MA, Chun SC (2014) Isolation and characterization of plant growth promoting endophytic diazotrophic bacteria from Korean rice cultivars. Microbiol Res 169(1):83–98

Jimenez-Salgado T, Fuentes-Ramirez LE, Tapia-Hernandez A, Mascarua-Esparza MA, Martinez-Romero E, Caballero-Mellado J (1997) *Coffea arabica* L.: a new host plant for *acetobacter diazotrophicus*, and isolation of other nitrogen-fixing *Acetobacteria*. Appl Environ Microbiol 63(9):3676–3683

Kaba JS, Zerbe S, Agnolucci M, Scandellari F, Abunyewa AA, Giovannetti M, Tagliavini M (2019) Atmospheric nitrogen fixation by Gliricidia trees (*Gliricidia sepium* (Jacq.) Kunth ex Walp.) intercropped with cocoa (*Theobroma cacao* L.). Plant Soil 435(1):323–336
Kaur M, Tak Y, Bhatia S, Asthir B, Lorenzo JM, Amarowicz R (2021) Crosstalk during the carbon–nitrogen cycle that interlinks the biosynthesis, mobilization and accumulation of seed storage reserves. Int J Mol Sci 22(21):12032
Kumar KK, Dara SK (2021) Fungal and bacterial endophytes as microbial control agents for plant-parasitic nematodes. Int J Environ Res Public Health 18:4269
Mahmud K, Makaju S, Ibrahim R, Missaoui A (2020) Current progress in nitrogen fixing plants and microbiome research. Plan Theory 9(1):97
Maheshwari DK, Dheeman S, Agarwal M (2015) Phytohormone-producing PGPR for sustainable agriculture. In: Maheshwari DK (ed) Bacterial metabolites in sustainable agroecosystem. Springer, Cham, pp 159–182
Masson-Boivin C, Giraud E, Perret X, Batut J (2009) Establishing nitrogen-fixing symbiosis with legumes: how many rhizobium recipes? Trends Microbiol 17(10):458–466
Mehta P, Walia A, Shirkot CK (2015) Functional diversity of phosphate solubilizing plant growth promoting rhizobacteria isolated from apple trees in the trans Himalayan region of Himachal Pradesh. India Biol Agric Hortic 31(4):265–288
Mendonça ED, Lima PC, Guimarães GP, Moura WD, Andrade FV (2017) Biological nitrogen fixation by legumes and N uptake by coffee plants. Rev Bras Ciênc Solo 41:e0160178
Meneses CH, Rouws LF, Simões-Araújo JL, Vidal MS, Baldani JI (2011) Exopolysaccharide production is required for biofilm formation and plant colonization by the nitrogen-fixing endophyte *Gluconacetobacter diazotrophicus*. Mol Plant-Microbe Interact 24:1448–1458
Munroe J, Isaac M (2014) N2-fixing trees and the transfer of fixed-N for sustainable agroforestry: a review. Agronomy for Sustainable Development, Springer Verlag/EDP Sciences/INRA 34(2): 417–427
Mus F, Crook MB, Garcia K, Garcia Costas A, Geddes BA, Kouri ED, Paramasivan P, Ryu MH, Oldroyd GE, Poole PS, Udvardi MK (2016) Symbiotic nitrogen fixation and the challenges to its extension to nonlegumes. Appl Environ Microbiol 82(13):3698–3710
N'diaye A, Poncet V, Louarn J, Hamon S, Noirot M (2005) Genetic differentiation between *Coffea liberica* var. *liberica* and *C. liberica* var. *dewevrei* and comparison with *C. canephora*. Plant Syst Evol 253(1):95–104
Nguyen DN, Wang SL, Nguyen AD, Doan MD, Tran DM, Nguyen TH, Ngo VA, Doan CT, Tran TN, Do VC, Nguyen VB et al (2021, 1887) Potential application of rhizobacteria isolated from the central highland of Vietnam as an effective biocontrol agent of robusta coffee nematodes and as a bio-fertilizer. Agron 11(9)
Nygren P, Leblanc HA (2009) Natural abundance of ^{15}N in two cacao plantations with legume and non-legume shade trees. Agrofor Syst 76(2):303–315
Nygren P, Leblanc HA (2015) Dinitrogen fixation by legume shade trees and direct transfer of fixed N to associated cacao in a tropical agroforestry system. Tree Physiol 35(2):134–147
Oldroyd GE, Murray JD, Poole PS, Downie JA (2011) The rules of engagement in the legume-rhizobial symbiosis. Annu Rev Genet 45:119–144
Park M, Kim C, Yang J, Lee H, Shin W, Kim S, Sa T (2005) Isolation and characterization of diazotrophic growth promoting bacteria from rhizosphere of agricultural crops of Korea. Microbiol Res 160(2):127–133
Prasad M, Srinivasan R, Chaudhary M, Mahawer SK, Jat LK (2020) Endophytic bacteria: role in sustainable agriculture. In: Microbial endophytes. Woodhead Publishing, Sawston, pp 37–60
Rusconi M, Conti A (2010) Theobroma cacao L., the food of the gods: a scientific approach beyond myths and claims. Pharmacol Res 61(1):5–13
Santi C, Bogusz D, Franche C (2013) Biological nitrogen fixation in non-legume plants. Ann Bot 111(5):743–767

Sayyed RZ, Chincholkar SB, Reddy MS, Gangurde NS, Patel PR (2013) Siderophore producing PGPR for crop nutrition and phytopathogen suppression. In: Bacteria in agrobiology: disease management. Springer, Berlin, Heidelberg, pp 449–471

Schroth G, Lehmann J, Rodrigues MR, Barros E, Macêdo JL (2001) Plant-soil interactions in multistrata agroforestry in the humid tropicsa. Agrofor Syst 53(2):85–102

Sellstedt A, Richau KH (2013) Aspects of nitrogen-fixing actinobacteria, in particular free-living and symbiotic Frankia. FEMS Microbiol Lett 342(2):179–186

Silva MD, Veloso TG, Entringer TL, Bullergahn VB, Pereira LL, Anastácio LM, Kasuya MC (2020) Diversity of nitrogen-fixing bacteria in coffee crops (*Coffea arabica* L.). Revista Ifes Ciência 6(3):12–21

Simarmata R, Widowati T, Dewi TK, Lekatompessy SJ, Antonius S (2020) Isolation, screening and identification of plant growth-promoting endophytic bacteria from *Theobroma cacao*. Biosaintifika: J Biol Biol Educ 12(2):155–162

Simon-Gruita A, Pojoga MD, Constantin N, Duta-Cornescu G (2019) Genetic engineering in coffee. In: Caffeinated and cocoa based beverages. Woodhead Publishing, Sawston, pp 447–488

Snoeck D, Zapata F, Domenach A-M (2000) Isotopic evidence of the transfer of nitrogen fixed by legumes to coffee trees. Biotechnol Agron Soc Environ 4:95–100

Somarriba E, Beer J (2011) Productivity of Theobroma cacao agroforestry systems with timber or legume service shade trees. Agrofor Syst 81(2):109–121

Soumare A, Diedhiou AG, Thuita M, Hafidi M, Ouhdouch Y, Gopalakrishnan S, Kouisni L (2020) Exploiting biological nitrogen fixation: a route towards a sustainable agriculture. Plan Theory 9(8):1011

Souza RD, Ambrosini A, Passaglia LM (2015) Plant growth-promoting bacteria as inoculants in agricultural soils. Genet Mol Biol 38:401–419

Urgiles-Gómez N, Avila-Salem ME, Loján P, Encalada M, Hurtado L, Araujo S, Collahuazo Y, Guachanamá J, Poma N, Granda K, Robles A (2021) Plant growth-promoting microorganisms in coffee production: from isolation to field application. Agronomy 11:1531

Vessey JK (2003) Plant growth promoting rhizobacteria as biofertilizers. Plant Soil 255:571–586

Volkogon VV, Dimova SB, Volkogon KI, Sidorenko VP, Volkogon MV (2021) Biological nitrogen fixation and denitrification in rhizosphere of potato plants in response to the fertilization and inoculation. Front Sustain Food Syst 5:606379

Zaharah AR, Bah AR (1999) Patterns of decomposition and nutrient release by fresh Gliricidia (*Gliricidia sepium*) leaves in an ultisol. Nutr Cycl Agroecosyst 55(3):269–277

Zhang JY, Cun Z, Chen JW (2020) Photosynthetic performance and photosynthesis-related gene expression coordinated in a shade-tolerant species *panax notoginseng* under nitrogen regimes. BMC Plant Biol 20(1):1–9

Zuluaga MYA, Lima Milani KM, Azeredo Goncalves LS, Martinez de Oliveira AL (2020) Diversity and plant growth-promoting functions of diazotrophic/N-scavenging bacteria isolated from the soils and rhizospheres of two species of solanum. PLoS One 15(1):e0227422

Chapter 11
Plant Growth-Promoting Bacteria and Nitrogen Fixing Bacteria: Sustainability of Non-legume Crops

Arshad Jalal, Marcelo Carvalho Minhoto Teixeira Filho, Edson Cabral da Silva, Carlos Eduardo da Silva Oliveira, Leandro Alves Freitas, and Vagner do Nascimento

Abstract Plant growth-promoting bacteria (PGPBs) and nitrogen-fixing bacteria (NFBs) are biologically increasing nitrogen (N) efficiency with significant impacts on nutrients transformation, soil organic matter mineralization, carbon dynamics for sustainable growth and productivity of non-legume crops. Plant growth-promoting bacteria may have direct or indirect impact on soil health, fertility and nutrients cycle, leading to better crop diversity, growth and productivity. Beneficial aspects of microbes were being ignored due to chemical fertilizers and leguminous symbiotic associations in recent decades. Moreover, there is research gap with use of PGPBs and NFBs to highlight their impacts as growth promoting and N-fixing bacteria on non-legume production. In this context, our chapter focused on inoculation/co-inoculation of species or strains of microbes as bio-fertilizer, bio-enhancer and bio-control agents for improving growth, metabolism and production of non-legumes in an economical and ecofriendly way. These beneficial microbes are natural growth stimulators for better production of non-legume crops like cereals, oil seed plants, vegetables, fruits, forages and other important crops. They provide outstanding opportunities and utilize a wide range of processes like solubilization, mineralization of nutrients and production of phytohormones while discouraging synthetic and chemical fertilizers as well as biotic resilience. This review tried to understand the reported mechanisms of PGPBs and NFBs that contributed to the

A. Jalal · M. C. M. T. Filho (✉) · C. E. da Silva Oliveira · L. A. Freitas
São Paulo State University (UNESP), Ilha Solteira, São Paulo, Brazil
e-mail: arshad.jalal@unesp.br; mcm.teixeira-filho@unesp.br; ces.oliveira@unesp.br; leandro.a.freitas@unesp.br

E. C. da Silva
Goiano Federal Institute (IF Goiano), Rio Verde, Goiás, Brazil
e-mail: edsoncabral@ifgoiano.edu.br

V. do Nascimento
São Paulo State University (UNESP), Dracena, São Paulo, Brazil
e-mail: vagner.nascimento@unesp.br

© The Author(s), under exclusive license to Springer Nature Singapore Pte Ltd. 2022

D. K. Maheshwari et al. (eds.), *Nitrogen Fixing Bacteria: Sustainable Growth of Non-legumes*, Microorganisms for Sustainability 36,
https://doi.org/10.1007/978-981-19-4906-7_11

above and below soil biome of non-legumes for better and sustainable crop plant-soil interactions and productivity.

Keywords Cereals · Biological nitrogen fixation · Biofertilizers · Ecofriendly climate · Plant processes

11.1 Introduction

The contribution of rhizospheric microorganisms to crop productivity, plant growth, and stresses without any devastating impact has been subjected to countless discussions in recent years (Salvo et al. 2018). There are several beneficial microbes in soil rhizosphere so far known as "plant growth-promoting bacteria" which may directly or indirectly be associated with plant health and growth (Zeffa et al. 2018). Plant growth-promoting bacteria (PGPBs) are usually free-living bacteria that can colonize in root rhizosphere with a beneficial impact on soil health, fertility and nutrients cycle which, therefore lead to better crop diversity and productivity. The non-legumes crops (for example maize, wheat, barely, sorghum, brassica, etc.) are associated with different species of PGPBs (*Pseudomonas* sp., *Burkholderia* sp., *Azospirillum* sp., *Rhizobium* sp., and *Bacillus* sp.) to combat biotic and abiotic stresses with sustainable crop production (Ramakrishna et al. 2019; Zaheer et al. 2019).

The knowledge of transcriptomics, metabolomics and proteomics of plants and PGPBs interaction optimized formulation of inoculants by improving nutrients uptake and crop productivity in profitable strategies. These microbial communities in the rhizosphere or within plant tissues adapted different mechanisms of action to promote plant growth processes. These PGPBs may promote plant growth by the synthesis of phytohormones such as indole-3-acetic acid (IAA), cytokinins, gibberellins and ethylene (Meza et al. 2015), plant growth regulators like abscisic acid (Cohen et al. 2008), nutrients (N, P, K, S and Zn) and beneficial elements (Si) availability (Teixeira Filho and Galindo 2019; Jalal et al. 2021; Galindo et al. 2021), and polyamines like spermidine and spermine (Cassán et al. 2009) while suppressing diseases and pathogen infestation (Corrêa et al. 2010). These microbes can help the plants in biological nitrogen fixation (BNF), regulating cell homeostasis and biomass production to suppress stressful environments. They improve the stability of leaf cell membranes and reduce leaf abscission rate under water stress conditions (Silva et al. 2019). Several PGPBs mediated defense-associated metabolites which increase nutrients acquisition and plant health therefore, replacing chemical fertilizers and pesticides while sustaining an ecofriendly environment (Rey and Dumas 2017).

The nutrition and growth of non-legume crops are known since long-time but their sustainability and balance is only possible with holistic approach of PGPBs inoculation. Several PGPBs are contributing to BNF in non-legumes crop species which can therefore, reduce use of mineral fertilizer (Santi et al. 2013). In addition, some other studies indicated that several PGPBs can improve nutrients (both macro and micronutrients) acquisition by enhancing their solubilization and availability. The concentration and uptake of other nutrients in non-legume crops can also be increased through root morphological modification and thus providing large root surface for the accumulation of nutrients like P, K, Cu, Zn, Fe and Mn (Behera et al. 2021). Therefore, inoculation of non-legume crops with PGPBs and nitrogen fixing bacteria (NFB) is a promising and effective alternative to promote their growth and productivity in cost-effective and environment-friendly conditions. In this context, the current chapter aims to highlight benefits of PGPBs and NFB to improve nutrition and productivity of non-legume crops through BNF, nutrients solubilization, phytohormones production in a sustainable environment.

11.2 Factors Affecting Plant Growth Promotion

Efficient crop establishment, management and production is the challenge of the day due to several stressful factors that may limit to a greater or lesser degree for plant growth and development. The edaphoclimatic factors of environmental stresses represent a large portion of alterations while some others are caused by living organisms like plants competition for survival, microorganisms (fungi, bacteria and viruses) and mankind through pollution, soil compaction, agrochemicals, fire, flood, etc. (Larcher 2006). The soil organisms respond differently to different stresses depending on the symptoms however, these responses may not be immediate (Meena et al. 2017).

Climate change is one of the main global events that emphasizes on environmental issues like irregular precipitation, frost occurrence (freezing temperatures), high temperature (global warming), and contamination of soils and water used in food production is becoming natural. Abiotic stresses (water, salinity, heavy metals and high and low temperatures) cause irreversible damage to cultivated plants that can lead to cell and tissue death, nutritional and hormonal imbalance, lower yield and even death of plants (He et al. 2018). The scarcity of water resources is a frequent concern in the world where quality of water has been destined for human consumption while forcing farmers to use low-quality water (saline or contaminated with heavy metals) to fulfil plant requirements (Oliveira et al. 2022). The farmers' community increased irrigated areas to circumvent drought stress while improving quantity and quality of agricultural crops and avoiding crop losses (Enebe and Babalola 2018).

Cultivation on saline soils (salinized by excessive use of fertilizers and saline water) has been one of the main environmental challenges that has limited agricultural production around the world (Cirillo et al. 2016). Excessive salinity causes ionic toxicity and water and nutrient deficiencies hence, inhibiting plant growth promotion (Acosta-Motos et al. 2017). The excessive salt concentration in soil and irrigation water solution adversely affects crop physiology, growth and productivity (Lima et al. 2016). In addition, the indiscriminated use of fertilizers, pesticides, compost, municipal waste, industrial wastes and metal mines contaminated countless cultivation areas (Yang et al. 2005) that suppress plant growth promotion. The toxicity of heavy metals in soil can be observed by several morphological, physiological and biochemical disorders. Such metals can be essential or non-essential elements for plant growth while their phytotoxicity depends on exposure period, concentration, plant species, and affected organ/tissues. This hyperaccumulation is possible with water and nutrient absorption in plant tissues from the soil and environment (Chong-qing et al. 2013). Several strategies have been studied to mitigate abiotic effects and improve crop growth, physiology and yield under adverse soil, climate and water conditions (Silva et al. 2019).

In case of biotic stresses, the induction of plant defence during pathogen attack is non-economic due to increasing photoassimilates demand and cause alterations in the primary metabolism. In addition, pathogen also manipulates carbohydrate metabolism to its own needs, removes nutrients and increases the demand of assimilates (Berger et al. 2007). The development of chlorotic and necrotic areas on leaf surface from infection lead to a consequent reduction in photosynthetically active area (Schultz et al. 2013). Another stress such as weeds competing with crop plants for vital resources such as water, CO_2, nutrients, radiation and space that can severely alter morphological and physiological characteristics of plants, which cause significant losses to crop yield and quality (Galon et al. 2013). Effective crop growth and production require early and accurate detection of different types of biotic stress. Thus, appropriate technologies have been used in precision agriculture to early check of weeds, diseases and pests in crops and adapt strategies to control them in an effective and sustainable manner (Behmann et al. 2015).

11.3 Plant Growth-Promoting Bacteria and Plant Processes

Plant growth-promoting bacteria (PGPBs) interact with plants to stimulate synthesis of different phytohormones and growth regulators, enzymatic activities, polyamide and increase absorption and translocation of nutrients for effective plant processes (Etesami and Maheshwari 2018).The microbial community regulates plant growth and development processes by producing phytohormones such as auxins (indole-3-acetic acid), cytokinins and gibberellins (Meza et al. 2015), growth regulators (abscisic acid, ethylene and ACC deaminase, etc.) to mitigate water stressful conditions (Cohen et al. 2008), nitric oxide (Fibach-Paldi et al. 2012) and polyamines like spermidine, spermine and cadaverine to avoid senescence (Cassán et al. 2009) and increase phosphate solubilization (Puente et al. 2004; Hungria et al. 2010), potassium solubilization (Etesami and Maheshwari 2018), sulphur oxidation (Dhiman et al. 2019), and zinc solubilization (Mumtaz et al. 2017). Greater secretion of secondary metabolites and siderophores also exhibit in rhizosphere (Neilands 1995), BNF (Pankievicz et al. 2015), and altered N use efficiency (Galindo et al. 2016; Hungria et al. 2016). The positive responses of plants to inoculation were observed multifarious plant microbe interactions (Salvo et al. 2018).

11.3.1 Auxins

PGPBs increase the synthesis of auxin in root rhizosphere that contributes to cell elongation and differentiation of stem, leaves and roots, also increasing root hairs initiation plant water status (Meza et al. 2015; Velasquez et al. 2016). Many of the PGPB genera are known for the synthesis of different auxins such as IAA and IBA. The bacteria-mediated IAA, particularly by rhizobia, develops root morphogenesis, architecture and surface area in non-legume crops (Dazzo and Yanni 2006) which could increase the root's ability to accumulate more nutrients and promote growth. Enhanced production of IAA corresponds to increased cell elongation and growth by stimulating synthesis of enzymes that promote cell wall softening, enabling cell distension (Majda and Robert 2018).

IAA is a colourless and soluble substance in organic solvents which stimulates cell elongation by increasing osmotic content and permeability of H_2O into cells while reducing cell wall pressure while encouraging flowering and fruiting, and delaying leaves abscission. Several PGPBs are synthesizing IAA which could interfere with plant growth processes (Singh et al. 2017; Singh et al. 2019). The PGPBs secrete IAA, which is important in nodulation, branching and elongation of root systems that trigger nutrient exchange and plant growth with better management of biotic diseases in an ecofriendly environment (Keswani et al. 2020). Inoculation of rice with *Gluconacetobacter diazotrophicus* PAL 5, *Azospirillum baldaniorum* Sp 245 and *Escherichia coli* DH10b produced IAA that had increased growth, diameter, volume and area of root with better nutrients acquisition and biomass accumulation (da Silva et al. 2022).

11.3.2 Abscisic Acid

Abscisic acid (ABA) is synthesized via terpenes or terpenoids and derived from mevalonic acid which is generated from three acetyl-CoA molecules or in methylerythritol phosphate (MEP). ABA increases tolerance to abiotic stresses like it alters the hormonal status of a plant when it exposes to water stress conditions. ABA could initially close stomata to prevent plants from dehydration by accompanying plant cytokinine. Several studies reported that ABA is one of the main factors that triggers a series of events as a result of stomata closure to increase tolerance to drought stress (Sussmilch and McAdam 2017). Abscisic acid is an early signalling hormone in roots during drought stress which is then transported into leaves. The ABA in wheat is antiproportional chlorophyll and photosynthesis efficiency and leads to pollen sterility (Dong et al. 2017; Gietler et al. 2020). Abscisic acid may interact with other hormones (methyl jasmonate and salicylic acid) to help plants to develop defence against biotic and abiotic stresses (Bharath et al. 2021). It is successfully reported to mitigate stress in cereal crops, improving resistance against salinity in rice through biosynthesis of 40 different proteins that could improve plant defence and metabolism (Liu et al. 2019). Treatment of wheat with ABA has improved leaves proline accumulation under osmotic stress (Pál et al. 2018).

Several PGPBs are being reported to enhance tolerance of cereal crops to stressful conditions by decreasing ABA accumulation and preserving photosynthetic efficiency (Barnawal et al. 2017; Shahzad et al. 2017). The PGPB such as strains of *Bacillus amyloliquefaciens* synthesized ABA and increased resistance of rice to drought stress (Shahzad et al. 2017). Several strains of *Azospirillum* and *Bacillus* were assigned to increase the biosynthesis of ABA in different crops under stomatal stress (Cohen et al. 2009; Ilyas and Bano 2010). The bacterial strains (*Rhodococcus* sp. P1Y and *Novosphingobium* sp. P6W) isolated from rice rhizosphere indicated that both strains use ABA as a source of carbon and also reduced ABA concentration in root and leaves of tomato crop which described that ABA-metabolizing bacteria interact with each other through ABA dependent mechanism (Belimov et al. 2014; Yuzikhin et al. 2021).

11.3.3 Cytokinins

Cytokinin plays a key role in the development of vascular system, as act on promotion of protoxylem differentiation and development of vascular cambium (Dettmer et al. 2009). The limited synthesis of cytokinin in root vascular system of plants could produce roots without phloem and metaxylem, and reduction of cell proliferation in procambial cells which may reduce vascular system (Argyros et al. 2008). In addition, cytokinins could also affect seed germination, shoot bud formation, breaking of apical dominance, inhibition of leaf senescence and regulation of cell division in leaves (Efroni et al. 2013). Auxin and cytokinin interact in a balanced way to promote plant growth (Hussain et al. 2021). The excessiveness of one of the hormones can directly inhibit the activity of the other. Thus, growth promotion is possible only by the adaptation and intrinsic hormonal regulation of plants (Kieber and Schaller 2014).

Cytokinins influence cell division and chloroplasts differentiation (Taiz et al. 2017). The biosynthesis induction of cytokinin through nitrate availability has been characterized in several non-legume crop species (Kamada-Nobusada et al. 2013). Several PGPBs strains are being reported for the production and biosynthesis of cytokinin in plants. Inoculation with *Bacillus subtilis* (AE016877) has increased cytokinin concentration in shoot of *Platycladus orientalis* (Liu et al. 2013) and therefore stimulates shoot growth and fruit formation. Some other bacterial strains like *Pseudomonas fluorescens* have increased total cytokinin concentration in the leave of *A. thaliana* and *Brassica napus* (Pallai et al. 2012; Großkinsky et al. 2016). These PGPBs biosynthesized cytokinin have the ability to alter homeostasis and adopting such mechanisms that stimulate growth and provide tolerance to plants against biotic and abiotic stresses. However, interaction of cytokinin with PGPBs is not adequately reported due to limited characterizations and mechanisms.

11.3.4 Gibberellins

Gibberellins (GAs) are mainly defined by their chemical structure rather than biological activities in plants. Gibberellins play an important role in mediating environmental effects and stimuli on plant development. Exogenous application of GAs increases stem growth of dwarf varieties to become similar to normal-growing varieties. Gibberellins stimulate stem growth and also increase fruit production in more bunch spaces in non-legume crops (Plackett and Wilson 2018). Gibberellins application in citrus plants delayed senescence and increased internodes elongation in sugarcane, leading to greater biomass (50 t ha^{-1}) and sugar (5 t ha^{-1}) production (Nguyen et al. 2019).

Plant growth-promoting bacteria alter and influence GA synthesis in plants much like other hormones. Gibberellin biosynthesized PGPBs such as *Bacillus cereus* MJ-1 (Joo and Chang 2005), *Promicromonospora* sp. SE188 (Kang et al. 2014), and *Leifsonia soli* SE134 and *Enterococcus faecium* LKE12 increased dry matter and length of root and shoot in rice. These PGPBs with the biosynthesis of GA can alter different biochemical functions of plants and act as an outstanding alternative for increasing tolerance against stresses (Kenneth et al. 2019).

11.3.5 Ethylene

Ethylene can be produced by almost all parts of higher plants, although the rate of production depends on tissue type and developmental stage. Ethylene levels increase during leaf abscission, flower senescence and fruit (tomato, citrus, cucumber, peanuts, etc.) ripening (Iqbal et al. 2017). The biosynthesis of ethylene is induced by mechanical damage and physiological stresses where plant tissues convert [14C]-Methionine into [14C]-Ethylene through an immediate precursor of 1-aminocyclopropane carboxylic acid (ACC) (Poyatos-Pertíñez et al. 2016). PGPB-mediated ACC are able to improve root initiation and morphology for greater nutrient acquisition in non-legume crops (Pieterse et al. 2009). Some of PGP bacterial strains like *R. leguminosarum* bv. viciae and *M. loti* enhanced lateral roots growth in *Arabidopsis thaliana* (Contesto et al. 2008). Several PGPBs can mitigate drought stress by lowering the ethylene production in root system of plants (Arshad et al. 2008; Saleem et al. 2018). These bacteria are lowering the enzymatic activity (1-aminocyclopropane-1-carboxylic acid (ACC) by the production of α-ketobutyrate and ammonia, which could constrain excessive synthesis of ethylene (an efficient mechanism) under several biotic stresses (del Carmen et al. 2020).

11.3.6 Polyamines

Polyamines can be found in vacuoles and chloroplasts that are mainly associated with cell walls in free or conjugated with phenolic acids. They can stabilize DNA and cell membranes by interacting with phosphorus residues and altering enzymes activities. The alteration in cell fluidity and membranes structure can be measured by polyamines (Masson et al. 2017). Polyamines stimulate synthesis of proteins, kinases and fructose-1,6-biphosphate which contribute to cell division as well as elongation of root and stem. In addition, polyamines are used as a substitute to auxin treatment which declare them as secondary messengers for this hormonal class. An increase in polyamine amounts can develop flowers while their decline can cause senescence (Nambeesan et al. 2010). The application of low concentration of polyamines in mono- or dicotyledonous plants prevents senescence-related processes (chlorophyll, proteins and RNA). In addition to these processes, polyamines are involved in the maturation of fruits and pollen grains formation, stem and root formation, and vascular differentiation (Alcázar et al. 2020). Polyamine is maintaining membrane integrity and gene expression, reducing production of ROS, and regulating accumulation of Na^+ and Cl^- ions for synthesis of osmolytes (Afzal et al. 2009; Roychoudhury et al. 2011). The over-expression of polyamine biosynthesis genes and other precursors like putrescine, spermidine, and spermine increased tolerance in rice, tobacco, and *Arabidopsis* against salt stress (Shevyakova et al. 2013). Two polyamines synthesized bacterial strains (*Enterobacter bugandensis* XY1 and *Serratia marcescens* X43) were observed for the effect of heavy metal accumulation and biomass production of spinach vegetables. It was noted that these polyamine strains decreased Cd and Pb accumulation due to metal ion chelation and cell absorption and provided resistance to their toxicity (Wang et al. 2022).

11.4 Linkage Among Non-legumes, Nitrogen and Soil Microbiota

Nitrogen is excessively used to improve non-legumes production and feed the world. However, its excessive application has hazardous environmental impacts. The alternative process of biological nitrogen fixation (BNF) is one of the most important economic, social and environmental processes, wherein nitrogenase enzyme complex catalyses and breaks down triple bond of atmospheric dinitrogen (N_2) and converting it into ammonia (NH_3) for plant uptake (Reis and dos Teixeira 2006; Franche et al. 2009; Santi et al. 2013). The recent studies (Schillaci et al. 2019, 2021; Raffi and Charyulu 2021 and others) have increased interest of researchers in association of non-legumes with different rhizospheric bacteria (diazotrophes, epiphytes, endophytes, etc.). However, this intention is still far less than legumes due to rhizobia and root nodules interaction.

Non-legume crops contribute almost 0–75 kg ha^{-1} $year^{-1}$ of N to environment through BNF (Reis et al. 2018). The most recent literature reported that BNF increased N concentration by 64% in sugarcane under field conditions (Martins et al. 2020). In addition, inoculation of wheat and corn with strain of *Pseudomonas protegens* Pf-5 X940 also increased N concentration under reduced N-fertilization in sorghum, switchgrass, wheat, maize, rice, etc. Van Deynze et al. (2018) reported that inoculation of corn with non-symbiotic diazotrophs bacteria fix up to 82% N in soil rhizosphere. The inoculation of non-symbiotic microbes in non-legume crops is grabbing attention of scientists and researchers in several countries, and with emphasis on corn and wheat crops (Salvo et al. 2018). Several other studies reported microbial benefits in non-legume crops such as rice (Long et al. 2018), sugarcane (Santos et al. 2017), sorghum (Wu et al. 2021) and pastures (Hungria et al. 2016).

11.4.1 Diazotrophic Microorganisms and Biological N-Fixation Capacity

The symbiotic relationship of legumes with some microorganisms in the soil has always been remained a point of interest to researchers due to its efficiency and economic values in agriculture. However, recent studies on soil microbial relationship with non-legume crops highlighted the possible contribution of the microbial community in growth and development of plants (Galindo et al. 2022; Jalal et al. 2022; Rosa et al. 2022). These studies indicated that some soil microorganisms can increase crop yields, reduce pathogen attacks, as well as reduce abiotic or biotic alterations without harming crop productivity (de Cassetari et al. 2016; Teixeira Filho and Galindo 2019; Mahmud et al. 2020). There are several genera and species that can carry out BNF, generally known as diazotrophic microorganisms.

The habitats and association with crops involved several N-fixing bacteria such as cyanobacteria (*Anabaena, Calothrix, Nostoc*), aerobic bacteria (*Azospirillum, Azotobacter, Beijerinckia, Derxia, Bacillus, Klebsiella*), and anaerobic bacteria (*Clostridium, Methanococcus, Chromatium, Rhodospirillum*) (Reis et al. 2018), mainly exhibited in all types of soil, salt or freshwater as well as in rhizosphere or lithosphere. There are several organisms that are capable of fixing N on association with grasses or non-legume species which are demonstrated in Table 11.1 (Reis and dos Teixeira 2006).

These microorganisms are also known as plant growth-promoting bacteria (PGPBs) with greater emphasis on genera *Azospirillum, Bacillus, Pseudomonas*, which may have endophytic or epiphytic characteristics and directly contributing to plant growth through BNF in soil (Pankievicz et al. 2015) or indirectly as a biological control agent of pests and diseases.

The intense interest in the possible association of free-living bacteria with non-legume such as sugarcane (*Saccharum* sp.), cover grasses and rice (*Oryza sativa*) exhibited the importance of bacterial inoculation. The meta-analysis of over last two decades under different environmental conditions and different non-legume crops exhibited that soil microorganisms (in especial diazotrophic bacteria) have the ability to increase crop productivity up to 30% (Fukami et al. 2018). Although, contribution of diazotrophes and non-legume crops to BNF is minimal in relation to N fixation as

Table 11.1 Host plants and microorganisms association of nitrogen-fixing with non-legumes crops

Gender	Species	Host plant	Reference
Azospirillum	*A. brasilense*	Grasses	Tarand et al. (1978)
	A. Lipoferum	Grasses	Tarand et al. (1978)
	A. Amazonian	Grasses	Magalhães et al. (1983)
	A. Halopraeferens	Kallar grass	Reinhold et al. (1987)
	A. Iraqi	Rice	Khammas et al. (1989)
	A. Doebereinerae	Miscanthus	Eckert et al. (2001)
Gluconacetobacter	*G. Diazotrophicus*	Sugarcane	Cavalcante and Dobereiner (1988)
		Pineapple, sweet potato and Eleusine Coracana	Yamada et al. (1997)
	G. johannae	Coffee rhizosphere	Fuentes-Ramirez et al. (2001)
Class 1: Proteobacteria—Alpha subdivision			
Kisses	*G. Nitrogen captans*	Coffee rhizosphere	Fuentes-Ramirez et al. (2001)
Derxia	*B. fluminensis*	Sugarcane	Krieg and Holt (1984)
	D. gummosa	Sugarcane	Krieg and Holt (1984)
Class 2: Proteobacteria—Beta subdivision			
Burkholderia	*B. Vietnamiensis*	Rice	Gillis et al. (1995)
	B. Kururiensis	Aquifer polluted with tri-chloroethylene (TCE)	Zhang et al. (2000)
	B. "brasilensis"	Rice, cassava, sweet potatoes, sugar cane and corn	Baldani et al. (1997)
	B. tropica	Sugarcane	Reis et al. (2004)
Herbaspirillum	*H. Seropediae*	Grasses	Baldani et al. (1986)
	H. Rubrissubalbicans	Sugarcane	Baldani et al. (1986)
	H. Frisingense	Pennisetum, miscanthus and spartina	Kirchhof et al. (2001)
Alkalines	*A. faecalis*	Rice	You et al. (1991)
	A. latus	Rice	Malik et al. (1997)
	A. paradoxus	Rice	Malik et al. (1997)
Azoarcus	*A. Indigenes*	Kallar grass	Reinhold et al. (1993)
	A. communis	Kallar grass	Reinhold et al. (1993)

(continued)

Table 11.1 (continued)

Gender	Species	Host plant	Reference
"Azovibrio"	*A. restrictus*	Kallar grass	Reinhold-Hurek and Hurek (2000)
"Azospira"	*A. oryzae*	Kallar grass	Reinhold-Hurek and Hurek (2000)
"Azonexus"	*A. Fungiphilus*	Kallar grass	Reinhold-Hurek and Hurek (2000)
Class 3: Proteobacteria—Gamma subdivision			
Azotobacter	*A. paspali*	Paspalum notatum	Krieg and Holt (1984)
Klebsiella	*K. pneumoniae*	Corn, sugar cane, sweet potatoes, wheat	Krieg and Holt (1984)
	K. Oxytoca	Rice, wheat	Kovtunovych et al. (1999)
	K. Planticola	Rice, wheat	Ladha et al. (1983)
	K. terrigena	Grasses	Haahtela et al. (1988)
Pantoea	*P. agglomerans*	Wheat	Ruppel et al. (1992)
Class 4: Phylum BX3—Firmicutes phy. Nov. class III "bacilli"			
Paenobacillus	*P. Nitrogenfixans*	Grasses	Seldin et al. (1984)

Adapted from (Reis and dos Teixeira 2006)

result of symbiotic bacteria and leguminous crops (Reis et al. 2018). Therefore, attention must be given to increase the contribution of BNF to non-legume crops to decrease consumption of N fertilizers with additional ecosystem services like mitigation of greenhouse gases (GHG), as well as reduction of N leaching to aquifers (Kaye and Quemada 2017).

The association of plants and microbiological community in root rhizosphere use available carbon and environmental energy to break down bacterial cells and release fixed N for the crops (James 2000; Samuel et al. 2013). The microbial community is influenced by several factors (crop genotype, soil texture, available nutrients, soil water retention capacity, and amount and types of exudates) which may affect availability of macro- and micronutrients, especially N and phosphorus (Mahmud et al. 2020). Nitrogen-fixing bacteria are competing with non-symbiotic bacteria for common available environmental resources. The association made by diazotrophs (*Beijerinckia* and *Azotobacter*) and free-living (*Klebsiella)* to fix N_2 may use carbohydrates and exudates (James 2000). Thus, PGPBs promote the availability of unavailable nutrients in soil while preventing pests and diseases attacks (Santoyo et al. 2016).

11.4.2 Interaction of Plant-Soil-Microbes and Environment

Agricultural systems are built on the interaction of plant-soil-atmosphere, and soil management systems. Research efforts have been devoted to characterize importance of microbiomes to soil health with extensive details on microbiological complexity and functions associated with soil, plants, animals and insects (Fierer 2017; Hartmann et al. 2019). However, the new interest is yet to understand the structure and function of microbiome as a whole, evaluate the heritability of host phenotypes and determine the aspects that govern stability and resilience to disturbance (Finkel et al. 2017; Toju et al. 2018; Sergaki et al. 2018).

Recent research showed interconnected nature of microbiomes to understand processes of their composition and function in agroecosystem (French et al. 2021). Integrated knowledge about the factors influencing microbiomes has direct implications on soil quality and long-term sustainability, including management practices and advances in technology.

11.4.2.1 Interaction Between Agricultural Practices and Microbiomes

Modern agriculture adapted different management and sustainable strategies including beneficial microbes that prevent transport and accumulation of pathogens in soil. Soil-associated microbial communities are studied as distinct sectors, without integration between disciplines. Recent studies demonstrated the interconnected nature of microbiological communities associated with plants, soil, insects and pollinators (Fig. 11.1) (Toju et al. 2018; Besset-Manzoni et al. 2018).

The traditional agricultural practices drive alterations across all domains of microbial communities/ microbiomes with beneficial impact and improvement on soil characteristics and quality, leading to healthy crops and productivity. There are several studies on association of microbial communities with plants and impact of management practices on microbiomes (Wolmarans and Swart 2014; Venter et al. 2016; Kim et al. 2020). However, most of the researchers focused on soil root interaction without studying the effects on all sectors of microbiomes. Little is known that how management practices regulate the dynamics of microbial communities associated with aboveground plant tissues (stems, leaves and flowers) in agroecosystem.

The interconnected nature of microbiomes with plants, insects and environment is presented in Fig. 11.1. The integrated management practices of these factors may have positive or negative influence on soil and crop quality (French et al. 2021). Microbes-mediated agricultural systems manipulate crop production by incorporation of traditional and emerging management practices to improve the ecosystem with greater crop growth, yield and resilience to stresses (Dubey et al. 2019). The main management practices and their impact on microbial community are summarized in Table 11.2.

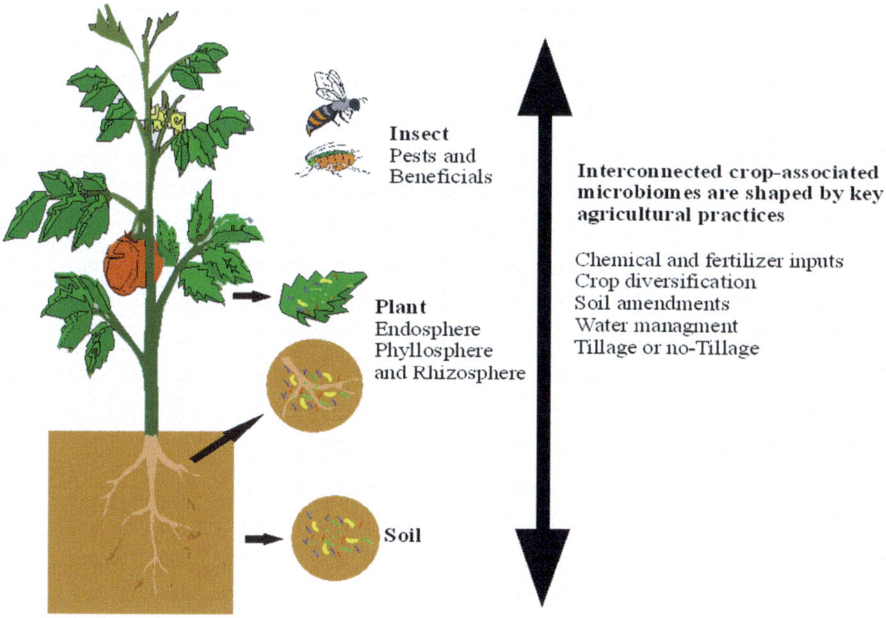

Fig. 11.1 Crop-associated microbiomes (soil, plant, insects) in association with different agricultural practices. The arrow highlights integration into microbiome in relation to agricultural management and practices and how this can affect microorganisms. (Source: This figure has adapted from French et al. (2021))

11.4.2.2 Environmental Factors

The soil microorganisms depend on the limits imposed by environmental conditions and their genetic factors. There are several environmental abiotic factors that can suppress survival and functions of microorganisms in the soil (Sandhya et al. 2010). The main abiotic factors that interfere with the interaction of diazotrophic bacteria and plants are: soil temperature, moisture, pH, energy sources and organic carbohydrates, available nutrients and toxic elements.

Soil temperature is considered one of the determining factors in survival, physiology and distribution of microbial diversity and activities in soil. The soil microorganisms are classified on the basis of temperature for their growth activities; psychrophiles, mesophiles and thermophiles with optimal temperature of 15, 37 and 60 °C while a tolerance range of −5 to 20, 15 to 45, 40 to 70 °C, respectively (Leite and Araújo 2007). It is also important to emphasize that temperature influences N-cycle, mineralization and nitrification processes. The low temperature has harsh effects on mineralization of organic matter and therefore, compromising diazotrophic activities by decreasing carbohydrates availability (Reis et al. 2018). Soil moisture is another indispensable factor that can also trigger several antagonistic physiological responses to plants and microbial communities, which can reduce

Table 11.2 Positives and negatives changes in management practices in microbiomes culture

Agricultural practice	Description	Effect on crop microbiomes	References
Chemical control	Application of chemicals to control pests (chewing and sap-sucking insects) and pathogens (harmful viruses, bacteria and fungi) alter microbial community dynamics through direct toxicity and by acting as an energy or nutrient resource. Using agrochemicals	↓ microbial diversity↑ microbial activity↑ or ↓ pathogen suppressiondisrupts relationships with beneficial microbeslinked to microbe-mediated insecticide resistancealtered microbial functioning	Hussain et al. (2009) Wolmarans and Swart (2014) Imfeld and Vuilleumier (2012) Tago et al. (2015) Zhang et al. (2016) Kakumanu et al. (2016) del Fernández et al. (2019)
Crop diversification	Diversification strategies that grow two or more crops in the same location are often designed to improve pest suppression and soil fertility, and are thus predicted to alter microbial processes linked to nutrient cycling and pathogen build-up. Practices include crop rotation, cover cropping and intercropping	↑ or ↓ microbial diversity ↑ soil fertility ↑ disease insect pest suppression ↑ beneficial microbe–plant interactions	Kim et al. (2020); Venter et al. (2016)
Fertilizer inputs	Chemical fertilizers (nitrogen, phosphorus and potassium) enhance crop yield but also alter soil properties (for example, pH) and plant physiology (for example, root exudate production) in ways that are predicted to impact the structure and functional diversity of crop-associated microbiomes	Microbial diversity; varies (for example, soil- versus root-associated) ↑ soil acidification negatively impacts microbial growth ↑ overall microbial biomass; depends on rate and amount ↑ abundance of plant-growth-promoting bacteria ↓ microbial benefits to plant growth with excessive nutrients	Bünemann et al. (2006) Zhang et al. (2016) Yeoh et al. (2016)
Organic soil amendments	Addition of organic matter contributes to soil fertility by enhancing water and nutrient availability to crops, counteracting soil erosion and modulating soil temperature and pH, which in turn are predicted to reshape microbial	↑ microbial diversity, abundance and metabolic activity ↑ suppression of soil-borne pathogens ↑ positive plant–soil feedback (for example, in maintaining soil structure)	Bünemann et al. (2006)

(continued)

Table 11.2 (continued)

Agricultural practice	Description	Effect on crop microbiomes	References
	community structure and function. Examples include residues, manure, biosolids, biochar and plant material residues		
Tillage	Tillage practices mechanically turn soil as a way to control weeds and pests. However, tillage ultimately causes physical disturbance that leads to changes in soil physiochemical properties and erosion, which in turn are predicted to affect soil microbial communities on many levels.	↓ microbial diversity and overall biomass	Wang et al. (2020a)
		↓ microbial functional diversity (for example, catabolic diversity)	
		Disrupts relationships with beneficial microorganisms	
Water management	Water can strongly influence soil microorganisms directly through soil structure and physiochemical properties such as pH. Management of soil moisture through irrigation practices is predicted to influence soil and rhizosphere microbiome communities	↓ associations with beneficial root microorganisms	Hartmann et al. (2017) Mavrodi et al. (2018)
		↓ presence of water-stress-tolerant microorganisms	
		↑ multidrug-resistant bacteria in soil (for example, wastewater irrigation)	
		↑ microbial activity	

Adapted from (French et al. 2021)

infection of endophytic bacteria, emergence of root hairs from host crop and also reduce N-fixation rate (de Cassetari et al. 2016).

The soil pH is also a determining factor in the distribution and activities of microbial communities. Soil pH is depending on a number of factors and processes such as source material and soil formation processes, frequency of fertilization, and mineralization of organic matter by soil microbiota (Remigi et al. 2016). Different microbes are adapted to different pH values as like fungi are generally adapted to pH values lower than 5.0 while bacteria and actinomycetes are adapted to pH values ranging from 6.0 to 8.0.

The soil pH had a great influence on the availability and toxicity of chemical elements such as Al^{3+}, Mn and Fe which can harm soil microbiota. In acidic soils, root development is inhibited by the presence of Al^{3+} which may affect nutritional absorption and also reducing root target area to be infected by bacteria (Leite and Araújo 2007; Remigi et al. 2016).

The available soil organic residues and applied organic fertilizers like cattle manure and chicken litter favour microbe's growth and activities while chemical fertilizers, pesticides and other pollutants can affect microorganisms and their biochemical processes. The chemoautotrophic microorganisms are using the energy of organic materials while the decomposition of organic substrates depends on the complexity of the carbon chain. Thus, materials with a complex (lignified) carbon chain have a greater resistance to decomposition whereas materials containing proteins and glucose could decompose faster (Redin et al. 2014). In addition, concentrations of N/lignin and N/polyphenols of decomposing organic material must be known. The decreasing ratio of N/lignin may lower the rate of organic material decomposition and also N availability (Doneda et al. 2012).

Generally, when it comes to non-legume species, organic material is difficult to decompose due to lower N and higher C content as well as higher N/lignin ratio than legume species. Therefore, to regulate nutritional availability, especially N for bacterial growth and diversity it is obvious to include leguminous species in intercropping or in crop rotation to increase mineralization of organic material and thus enhancing N availability for soil microbiota.

Soil are being the reservoir of several mineral nutrients (C, H, O, N, P and S) for microorganisms and is provided with organic matter. Mineral nutrients are important components of different physiological, metabolic and biochemical processes such as DNA and RNA structuring, cell wall stabilization, enzyme synthesis, cell division, mobility, symbiotic and associative interactions (Leite and Araújo 2007). In addition, the activities of soil microbiota are also dependent on the relationships of C/N, C/P and C/S that may influence dynamics of biogeochemical cycles of microorganisms. Therefore, soil management and practices are considered to be most conservative and important for a healthy microbiome.

11.5 Rhizospheric Nitrogen Fixation in Non-legume Plants

The cropping systems with non-legume crops are considered non-symbiotic where free-living bacteria have the capacity to fix N through energy provided by the environment. These free-living bacteria use carbon and energy obtained from the mineralization of soil organic matter and thus release N fixed for uptake of crops (Mahmud et al. 2020). These microorganisms do not have nodules and constitute an association with plants in two main ways:

(a) Rhizospheric association or between plants and free-living bacteria.
(b) Association between diazotrophic endophytic bacteria and plants.

(a) This group of microorganisms live in soil and can be associated with plant tissues whenever environmental conditions are favourable (Reis et al. 2018). They are considered pioneers in colonization and have the ability to fix N in nitrogen-poor niches under most whethered agroecosystems (de Cassetari et al. 2016). This group is composed of aerobic bacteria of the genus *Azospirillum, Azotobacter, Beijerinckia, and Derxia* (Reis et al. 2018). They are chemiorganotrophic microorganisms and have the capacity to use soil carbon as a source of energy for their survival and keeping their population active.

(b) The group of these microorganisms colonizes in the internal tissues of crops without any apparent external structure and pathogenicity to crops (Teixeira Filho and Galindo 2019). The microbial diversity of this group is classified into facultative or obligatory endophytic bacteria. They can be isolated in the rhizosphere or inside plants.

The precise localization of important microorganisms within non-legume plants can be identified by optical and electronic microscopy in addition to immunodeficiency and immunofluorescence labelling with the use of polyclonal or monoclonal (Fig. 11.2a and c) (Bashan and Levanony 1990; James 2000). The other methods such as green fluorescent protein (Egener et al. 1998), fluorescently labelled and RNA-specific oligonucleotide probes are more reliable analysis for identification and localization of microorganisms. For example, to identify N-fixing bacteria in field conditions, the microscopy labelling of associative interactions between bacteria and plants is considered very important in distinguishing and characterizing these individuals.

Some non-endophytic and diazotrophic associative strains of *Azospirillum* were identified on the root surface, root elongation zones or within root epidermis. However, endophytic diazotrophics such as *A. diazotrophicus, Azoarcus* spp., *Herbaspirillum* spp. and a strain of *A. brasiliense* (Sp245) colonized in root cortex or even endoderm can be translocated to aerial parts (James 2000). These diazotrophs colonized in the apoplast, xylem vessels (Fig. 11.2a and b), intercellular spaces (Fig. 11.2c) and lignified parenchyma (Fig. 11.2d) of non-legume crops.

The location of bacteria in microbiomes did not mention much about the association between plants and PGPBs. For example, a high concentration of bacteria in associative processes, or inside plants does not mean that the nitrogenase process is taking place. Also, the expression of nitrogenase genes or proteins cannot determine the availability of N to crops in association (James 2000). The determination of N in non-symbiotic associative relationships only occurs after the death and possible mineralization of bacteria (Mahmud et al. 2020). The amount of released N in this process is inefficient and possibly delayed compared to the process that occurs with symbiotic bacteria.

11.6 Inoculation with Bacteria in Non-legumes: Plant Nutrition, Yield and Fertilization Efficiency

The contribution of rhizospheric microorganisms to plant growth, crop productivity, controlling and/or reducing pathogens infestation and mitigating the effects of biotic and abiotic stresses without causing pathogenicity has been widely discussed. In addition to being an economic, ecological and sustainable alternative, these microorganisms are able to increase fertilizers use efficiency (García de Salamone 2011). They reduce N fertilizers through BNF by diazotrophic bacteria (Fukami et al. 2016; Galindo et al. 2020a). In addition, they also increased soil nutrients availability by solubilizing phosphate, converting insoluble P into soluble P by releasing organic acids, chelation and ion exchange (Saeid et al. 2018), and zinc solubilization through

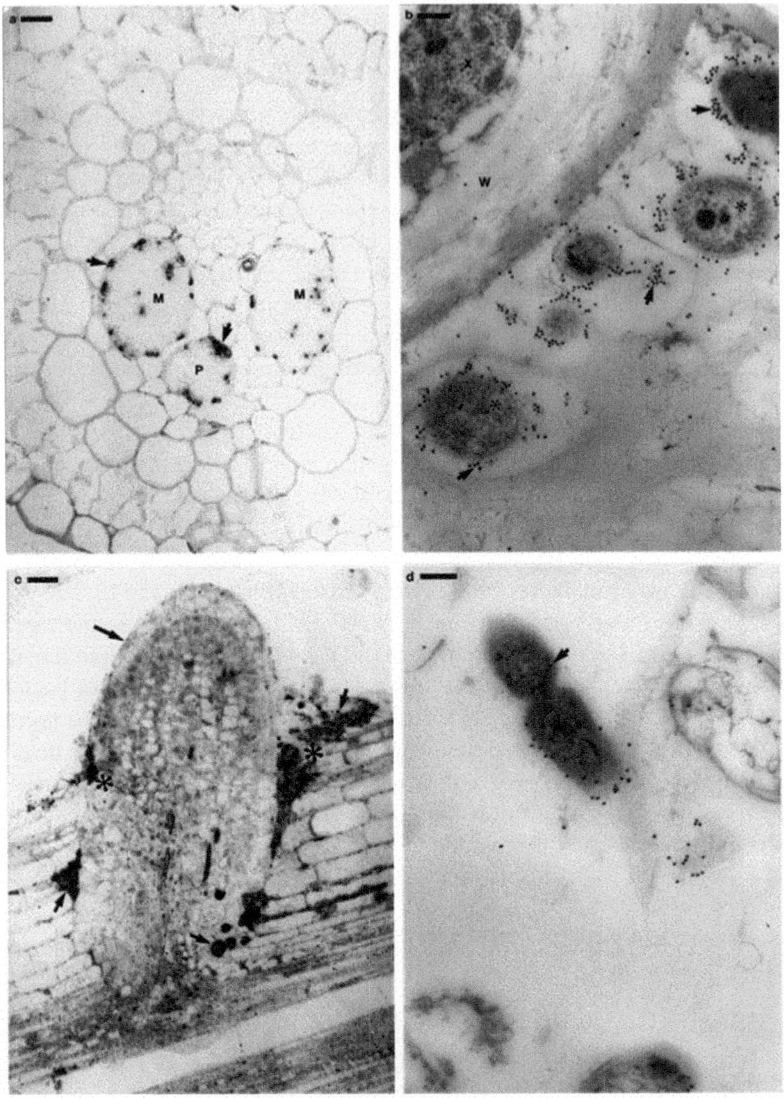

Fig. 11.2 (**a**) Optical micrograph of a sugarcane leaf cross sect. 7 days after inoculation with *A. diazotrophicus* strain. Bacteria can be seen as black bodies (arrows) within metaxylem (M) and protoxylem (P). (**b**) Transmission electronic micrograph of *A. diazotrophicus* (*) colonizing lignified parenchyma of sugarcane leaf. (**c**) Optical micrograph of a longitudinal section of rice root after 2 days inoculation with *Herbaspirillum seporopedicae* strain. The arrows indicate black bodies of bacterial colonies in addition to a large lateral arrow emerging from root. There is a concentration of bacteria in the intercellular spaces in which they are being colonized. These bacteria are entering through cracks at the junction between emerging lateral root (*). (**d**) Transmission electron micrograph of *Serratia* sp. (arrows) colonizing in the parenchyma of a rice stem. (Source: Adapted from James (2000))

production of chelators, secretion of organic acids, amino acids, vitamins and phytohormones, oxide-reducing systems and proton extrusion (Saravanan et al. 2003). This reflects better plant nutrition and quality in agricultural production around the world. The benefits of these beneficial microorganisms in non-legume cropping system are described below.

11.6.1 Cereals

Inoculation with plant growth-promoting bacteria (PGPBs) increased photosynthetic activities, crop yield and improved nutrient uptake in cereal crops. Inoculation with *Azospirillum brasilense* in maize increased the concentration of P, S, Cu, Mn and Zn, leaf chlorophyll index, grain yield and agronomic efficiency (Galindo et al. 2016; Jalal et al. 2020). Inoculation had also increased N use efficiency in maize (Galindo et al. 2019). Inoculation with *A. brasilense* via leaves and seeds of maize increased leaf Zn concentration. The inoculation of wheat with *Azospirillum brasilense* has increased leaf chlorophyll index, leaf N and Si concentration, shoot dry mass, number of ears per meter, number of grains per ear, grain yield and N use efficiency (Souza et al. 2019; Galindo et al. 2020b). Inoculation with *Burkholderia pyrrocinia* increased photosynthesis rate, stomatal conductance, intracellular CO_2 assimilation, leaf Si concentration and activities of catalase and ascorbate peroxidase enzymes in rice plants (Bueno et al. 2017).

The bacterial strains *Mycobacterium phlei* MbP18, *Pseudomonas alcaligenes* PsA15 and *Bacillus polymyxa* BcP26 boosted plant growth, absorption and accumulation of N, P and K in maize (Egamberdiyeva 2007). Inoculation of *A. brasilense* along with N application increased shoot N, K, Ca, B, and Si and grain N, S, Zn and Cu uptake in maize under tropical conditions (Galindo et al. 2021). The inoculation of *A. brasilense* increased Zn acquisition in maize and wheat plants as well as increased productivity and Zn use efficiency in both crops (Galindo et al. 2021). Maize and wheat plants inoculated with *A. brasilense* and *A. lipoferum* had increased grain yield, concentrations of N, P, K and Fe in leaves, while concentrations of P, K, Mg, S, Zn, Mn and Cu in grains (Hungria et al. 2010). Inoculation with *A. brasilense* provided wheat and barley plants with a greater yield component, grain yield and grain protein content (Ozturk et al. 2003). Inoculation with *Bacillus megaterium, Bacillus licheniformis, Paenibacillus polymyxa* and *Bacillus* OSU-142 in barley promoted growth and root biomass accumulation, greater BNF, concentration of N, Ca, Fe, Mn, Zn and Cu in aerial part of plants (Çakmakçi et al. 2007).

Inoculation with *Bacillus* sp. and *B. cereus* increased chlorophyll concentration, biometric attributes, grain yield and Zn solubilization in rice plants (Shakeel et al. 2015). Inoculation of maize with *A. brasilense* and *B. subtilis* increased P use efficiency while in combination with P_2O_5 doses have increased productivity by 39% under tropical conditions (Pereira et al. 2020). The cultivation of maize inoculated with *B. subtilis* and *B. aryabhattai* was observed with greater solubilization of Zn from insoluble sources with a lengthy root system, greater shoot and root dry mass, improved nutritional grains quality use efficiency of Zn and P (Mumtaz et al. 2017). Maize inoculation with bacterial strains of the genera *Azospirillum* and *Pseudomonas* increased absorption and accumulation of N, P, K, Zn, Cu, Fe and Mn in shoot, leaf area, plant and root length (Goteti et al. 2013). In addition, inoculation of sorghum with *Bacillus* strains favoured P concentration in shoots, roots and grains, increased shoot and root dry mass and grain yield (Mattos et al. 2020).

There are several other studies that report inoculation of different bacterial strains increased crop growth, productivity and nutrient accumulation in different cereal crops. For example, growth, productivity and P accumulation in maize was increased with *Pseudomonas tolaasii* (Viruel et al. 2014), *P. fluorescens* and *Enterobacter radicincitans* (Krey et al. 2013). In addition, rice growth, productivity, phytohormones production and phosphate solubilization were increased with inoculation of *Burkholderia* sp. strains (BRRh-2, BRRh-3 and BRRh-4), *Pseudochrobactrum* sp. strain (BRRh-1), and *Pseudomonas aeruginosa* strains (BRRh-5 and BRRh-6) (Khan et al. 2017). While *Burkholderia cepacia* strain (RRE25) increased uptake of N, K and P, production of indole acetic acid (AIA), leaf chlorophyll index and lateral roots in rice plants (Singh et al. 2013). Bacterial inoculation of wheat with different bacterial strains (*Klebsiella pneumoniae, Klebsiella varicola, Bacillus sp.* and *Agrobacterium tumefaciens*) increased availability of N, P and K, and leading to greater absorption and translocation of these nutrients to plant shoot with taller plants, greater shoot dry mass of wheat (Wang et al. 2020b). The isolated strains of bacteria (*Pseudomonas poae, Serratia marcescens, Bacillus pumilus, Pantoea agglomerans, Pseudomonas sp.* and *Microbacterium* sp.) increased the availability of N, P, K, Cu, Zn and Mn in soil that favoured these nutrients uptake of these nutrients in roots and leaf of sorghum plants (Sahib et al. 2020).

11.6.2 Oilseed Crops

Oilseed crops are the rich source of oil, raw material of oleo industries and also a source of renewable energy for power generation. These crops consisted of several crops (soybean, sesame, canola, safflower, sunflower, groundnut, castor, etc.) (Weiss 2000), while canola is ranked second among global oilseed crops. Several rhizobacteria/ PGPBs increased growth, productivity and quality of non-legume oilseed crops. Inoculation with *Herbaspirillum* sp. increased root growth, seedling vigor and dry mass, *Paenibacillus* sp. increased plant height and shoot dry mass production while *Bacillus* sp. increased nitrogenase enzyme activities and N uptake in canola plants (Islam et al. 2009). Rapeseed plants were benefited from the inoculation of *Bacillus pumilus* by increasing boron (B) concentration in leaves with greater growth and productivity under nutrients deficient soil (Masood et al. 2019). In addition, inoculation of rapeseed with strains of *Pseudomonas thivervalensis* (Y1-3-9), *Microbacterium oxydans* (JYC17) and *Burkholderia cepacia* (J62) in Cu-contaminated soil provided greater productivity, Cu accumulation and antioxidant contents such as ascorbic acid and glutathione (Ren et al. 2019).

The strain of *Bacillus* sp. (LTAD-52) had increased P solubilization, grain yield and dry weight of rapeseed plants by promoting root growth and nutrient uptake (Valetti et al. 2018).

Inoculation of canola with strains of *Rhizobium* sp. (8121) and A. *brasilense* (Ab-V5) root-shoot growth with greater dry mass (Gomes et al. 2018). Inoculation with four strains of *Pseudomonas* sp. (312, 642, 1313W and 9421) increased soil P contents through phosphate solubilization which had positive influence on root-shoot growth with greater root and shoot biomass accumulation of canola plants (Jamalzadeh et al. 2021). There was an increase in root-shoot growth, grain yield and nutritional quality of crambe (*Crambe abyssinica)* plants with inoculation of *Methylobacterium komagatae, Azomonas* sp. and *Rhizobium* sp. (Aquino et al. 2018).

Sunflower plants inoculated with *Bacillus* sp. (Ps-5) and *Alcaligenes faecalis* (Ss-2) had increased lactic acid production (related to phosphate solubilization), root proliferation and shoot growth, grain yield and oil concentration in grains (Shahid et al. 2015). The rhizobacteria of *Bacillus* sp. and *Enterobacter cloacae* promoted plant growth by increasing dry and fresh mass of shoots and roots, improved accumulation of K, N and P in sunflower shoots. In addition, concentrations of soluble carbohydrates, free amino acids and soluble proteins in leaves and roots of sunflower plants were increased (Santos et al. 2014). The inoculation of sunflower with *Acinetobacter* sp. strain (CC30) increased shoot-root dry weight, length, Cu content and photosynthetic pigment content in Cu-contaminated soil (Rojas-Tapias et al. 2012). *Paenibacillus polymyxa* through BNF increased leaf N content, shoot dry mass accumulation, pod dry mass and yield of canola plant (Padda et al. 2016).

11.6.3 Vegetables and Fruits

Vegetable crop species are relatively low nutrients use efficiency in comparison of arable crops and usually related to short growing periods and superficial rooting which are used for edible purposes (Tei et al. 2020). Inoculation of vegetable crops with different PGPBs stimulated several biochemical, physiological and metabolic processes to boost up their growth, productivity and nutritional values. Inoculation with *Gordonia rubripertincta* and *Pseudomonas stutzeri* provided greater growth of cucumber with greater dry mass of shoots and roots in hydroponic system (Corrêa et al. 2011). There was an increase in leaf chlorophyll index and leaf area of lettuce plants inoculated with *P. chlororaphis, B. subtilis* and *P. aphanidermatum* in a hydroponic system (Corrêa et al. 2010). Inoculation of *P. chlororaphis* increased root hair growth, root length, shoot fresh mass, number of leaves and plant length of lettuce in a hydroponic system (Lee et al. 2016). There are different PGPBs like *Bacillus* spp. and *Pseudomonas* sp., facilitated growth, yield attributes and yield of pepper and tomato plants in hydrophonic system (García et al. 2004; Kıdoğlu et al. 2009).

The inoculation of cucumber with different bacterial genera like *Acinetobacter baumannii, Cronobacter dublinensis, Enterobacter cloacae, Arthrobacter* sp. and *Cronobacter sakazakii* increased root length, root dry mass, plant height, shoot dry mass, and chlorophyll, proline, N and P contents (Kartik et al. 2021). Another study reported that inoculation of *Rhodobacter sphaeroides, Saccharomyces cerevisiae* and *Lactobacillus plantarum* had increased root-shoot fresh and dry mass with higher leaf chlorophyll index, concentration of Ca, K, Mg, P, and several amino acids like threonine, aspartic acid, serine, lysine, glutamic acid, glycine, tyrosine, isoleucine, alanine, cysteine, valine, methionine, leucine, phenylalanine, proline, histidine and arginine (Kang et al. 2015).

Cabbage plants inoculated with *Pseudomonas poae, Plantibacter flavus* and *Bacillus amyloliquefaciens* were observed with a greater number of leaves, leaf area, shoot dry mass, leaf chlorophyll index, flavonoids and yield along with higher concentrations of N, P, K, Ca, Mg, Fe, Cu and Zn in leaves (Helaly et al. 2020). The Chinese cabbage when inoculated with *Ensifer fredii* had been observed with higher concentrations of N, P, Ca, K, Mg, Zn, and Fe in leaves and greater leaves dry mass (Pongsilp and Nimnoi 2020). Inoculation with *Bacillus amyloliquefaciens* increased fresh and dry mass of roots and shoots, K concentration and leaf chlorophyll index in spinach plants. In addition, it also contributed to mitigate the toxic effects of lead (Pb) in soil by accumulating greater amounts in its roots (Zafar-ul-Hye et al. 2020). The inoculation of *Thiobacillus thiooxidans, Azotobacter chroococcum* and *Azospirillum lipoferum* in combination with N and S fertilization increased productivity, growth and accumulation of N, P, K, S, Zn and Mn in onion bulbs (Awad et al. 2011). The inoculation of *Pseudomonas proteins* with N doses provided greater root growth and yield of garlic with a reduction of 25% in N application as a top dressing (Wang et al. 2020c). Inoculation of potatoes with *Brevundimonas* sp. strain (TN37) promoted phosphate solubilization, stimulated absorption and accumulation of N and P in potato tubers with greater shoot and root growth (Naqqash et al. 2020). Leaf and seed inoculation of radish with *Kosakonia radicincitans* provided greater leaf fresh mass, tuber fresh mass, and tuber diameter with a 20% greater leaf and tuber yield (Berger et al. 2015).

The fruit diameter and productivity of cantaloupe melon had increased under an adequate supply of N along inoculation of *A. brasilense* (Vendruscolo et al. 2019). Inoculation of *Bacillus* spp. promoted shoot dry mass, stem diameter, leaf area, and concentration of N, P, K, Ca and Mg in melon plants and fruits (Vasileva et al. 2019). In addition, *B. subtilis* provided greater productivity, husk firmness, soluble solids content and fruit weight of melon (Abraham-Juárez et al. 2018). Bacterial inoculation of strawberries with *Enterobacter cloacae, A. brasilense* and *Burkholderia cepacia* reduced up to 50% N fertilizer application and increased shoot-root growth and dry mass with greater stem diameter (Andrade et al. 2019). Inoculation with *Azospirillum brasilense* and *Burkholderia vietnamiensis* increased leaf chlorophyll index, plant growth index and strawberry fruit yield (Lovaisa et al. 2015).

11.6.4 Forage Grasses

Inoculation with *R. tropici* CIAT 899 and *A. brasilense* Ab-V6 in association with N fertilizer increased accumulation of N, Ca, Fe, Mn and Zn in shoots, absorption of P and K by roots, and leaf chlorophyll index of *Panicum maximus* cv. BRS Zuri grasses (de Lima et al. 2020). Inoculation of *A. brasilense* and *B. subtilis* in combination with N fertilizer increased leaf dry mass and accumulation of K, N, P, Ca, S and Mg of Marandu grass (Sampaio et al. 2021). *A. brasilense, R. tropici and P. fluorescens* applied along with adequate N application increased shoot and root dry mass, tiller number, leaf chlorophyll index and N accumulation in shoot of Mavuno grass (Sá et al. 2019). Inoculation of *P. fluorescens* and *P. ananatis* increased leaf and root mass, crude protein concentration, and plant productivity of Ruziziensis grass under reduced use of N fertilizers in cover application (Duarte et al. 2020).

The inoculation of *Brachiaria brizantha* and *Brachiaria ruziziensis* grasses with *A. brasilense* reduced N application by 40 kg ha^{-1} through BNF, along with greater productivity and higher N concentration in the aerial parts of plants during 26 cuts in three different cities of Brazil (Hungria et al. 2016). Similarly, inoculation with *A. brasilense* mitigated water stress, increased productivity and reduced N fertilizers application in Marandu grass (Leite et al. 2019). *Brachiaria brizantha* cv. BRS Piatã inoculated alone or together with *Burkholderia pyrrocinia* and *P. fluorescens* were observed with larger leaf area, high leaf chlorophyll index and greater leaf fresh biomass along with high tolerance of darkness in forest intercropping (Lopes et al. 2018). Rocha and Costa (2018) reported that inoculation with *A. brasilense* increased shoot dry and fresh biomass, leaf chlorophyll index, number of tillers and leaf production in relation to stalks of *U. brizantha* cv. Paiaguás under 50% of mineral N is recommended for these grasses.

The forage plant-like sorghum growth, quality and productivity were boosted when inoculated with *P. fluorescens* strain 93 under adequate supply of P at sowing (Ehteshami et al. 2018). The use of *B. subtilis* and *P. fluorescens* in ryegrass (*Lolium perenne*) favoured greater fresh and dry mass accumulation in three harvests. This inoculation also increased diversity of natural beneficial bacteria in soil rhizosphere (Stamenov et al. 2012). The fresh and dry mass accumulation of Napier grass were increased with inoculation of *Micrococcus* sp. and *Arthrobacter* sp. strains which also increased plant tolerance to harmful effects of Cu and Cd excess in soil (Wiangkham and Prapagdee 2018). Different strains of *Herbaspirillum* sp. and *Azospirillum* sp. had promoted plant growth by increasing antioxidant activities, enzymatic activities and phytohormones production which therefore, lead to greater shoot dry and fresh mass of perennial ryegrass (Cortés-Patiño et al. 2021). The inoculation of pastures and forages for animal feed has become more frequent and attaining attention of the researchers to alleviate harmful effects of cultivation in marginal and contaminated areas, arid and semi-arid regions, as they favour growth in limiting water conditions and also reduce consumption of mineral fertilizers to provide sustainable animal production.

11.6.5 Other Crops

Mentha arvensis when inoculated with *Exiguobacterium oxidotolerans* and *Glomus fasciculatum* had increased root, leaf and shoot growth, greater fresh and dry leaf weight and oil yield (Bharti et al. 2016). Inoculation of *Pseudomonas fluorescens* WCS417 and *Bacillus amyloliquefaciens* GB03 improved the quality of essential oil, phenolic content and antioxidants along with greater growth and yield of *Mentha piperita* (Chiappero et al. 2019). The bacterial strains *Halomonas desertis* G11, *Pseudomonas rhizophila* S211 and *Oceanobacillus iheyensis* E9 provided higher productivity and citronellol compounds, and better-quality essential oil of *Pelargonium graveolens* (Riahi et al. 2020). The antioxidants, cannabinoids, phenolic compounds and plant growth of *Cannabis sativa* were increased under the effects of inoculation with native rhizobacteria (Pagnani et al. 2018). Inoculation of lemon balm (*Melissa officinalis* L.) with *P. fluorescens* and *P. putida* was observed with higher productivity, photosynthetic pigments, phenolic compounds, total soluble proteins and accumulation of fresh and dry mass of leaves (Hatami et al. 2021). In general, PGPBs increased root and aerial growth, productivity, nutrient absorption, hormone production, N, P and Zn solubilization while reducing the use of mineral fertilizers in plant cultivation to promote sustainable cultivation in the world.

11.7 Benefits of Co-Inoculation of Plant Growth Bacteria in Non-legumes

Due to multiple growth-promoting mechanisms and the potential of BNF in non-legume plants, several studies have sought to evaluate the use of microorganisms in co-inoculation with plant growth-promoting bacteria (PGPBs) to enhance or add up the individual beneficial effect of each bacterium. Therefore, reducing the application of synthetic fertilizers with greater development and productivity of non-legume crops.

The global inoculants market has been looking for new strains with new formulations and validation of application methods. In the last decade, the idea of combined application of different species of microorganisms contributed to different plant processes and is known as mixed inoculation or co-inoculation. Most of the studied co-inoculations include symbiotic rhizobia together with PGPBs. Currently, a variety of co-inoculants are present in the market for many crops (Santos et al. 2019). Treatments containing inoculation with *Azotobacter* sp. + *Azospirillum* sp. and *Bacillus* sp. + *Pseudomonas* sp. favoured the growth, the increase in the number of branches, the productivity of seeds, leaves and essential oil in the two cuts of the basil plants, contributing to the reduction in the use of mineral fertilizers (Tahami et al. 2017). Co-inoculation of *Andrographis paniculata* with *Azotobacter chroococcum, Bacillus megaterium, Pseudomonas monteilii* and *Glomus intraradices* improved soil chemical properties, acid phosphatase, plant growth, yield and desired herb quality (Khan et al. 2015).

11.7.1 Maize

Interestingly, the co-inoculation of maize showed prominent improvement in plant growth, biochemical properties and yield (Fukami et al. 2018) as compared to single inoculation. Picazevicz et al. (2017) observed that maize shoot dry mass was increased by 21.4 and 15.1% with co-inoculation of *A. brasilense* and *R. tropici* in the absence and presence of reduced N application, respectively. We conducted a field study on maize single and co-inoculation of *A. brasilense* with combinations of *A. brasilense* + *B. subtilis, A. brasilense* + *P. fluorescens, B. subtilis* + *P. fluorescens* and *A. brasilense* + *B. subtilis* + *P. fluorescens* in association with N doses. Our results indicated that greater maize grain yield was observed with co-inoculation under a 25% reduction in recommended N dose (Table 11.3) for the crop (Teixeira Filho and Galindo 2019).

Several other studies performed with co-inoculations of maize showed that co-inoculation of *R. tropici* and *A. brasilense* (Ab-V5 and Ab-V6) increased plant height, grain yield, and an interesting alternative to combat saline stress in maize (Fukami et al. 2017b). In another study, it was noted that co-inoculation of *A. brasilense* (strain Ab-V5) and *Herbaspirillum seropedicae* (strain SmR1) in maize increased shoot dry matter by 12% and yield by 7% in relation to the control (Dartora et al. 2013). Reis Júnior et al. Reis Júnior et al. (2008) described that combined inoculation of *A. amazonense* and *H. seropedicae* strains increased phytohormones production which therefore, increased root and shoot dry matter production by 40% as compared to control. A huge combination of plant growth-promoting bacteria (*A. brasilense, B. amyloliquefaciens, B. licheniformis, B. pumilus, B. subtilis* and *P. fluorescens*) was applied via maize seeds or jet directed at V3 stage. The results reported that *B. licheniformis* improved nutrient absorption

Table 11.3 Maize grain yield as a function of the combination of PGPBs and nitrogen topdressing

Treatment	Grain yield (kg ha^{-1})
0% N without inoculation	4995[b]
75% N without inoculation	5477[b]
100% N without inoculation	4872[b]
Azo and 75% N	6246[a]
Azo+bac and 75% N	7102[a]
Azo+pseud and 75% N	6449[a]
Bac + pseud and 75% N	6434[a]
Azo+bac + pseud and 75% N	6033[a]
Standard error	298.43
Overall mean	5951

Means followed by similar letters in column do not differ statistically from each other by Scott Knott test at 5% significance

Teixeira Filho and Galindo (2019), Extension and Education Research Station, São Paulo State University (UNESP), Ilha Solteira, Brazil while the co-inoculation of *B. licheniformis* and *P. fluorescens* increased grain yield can be reduced N fertilization by 25% when inoculated via seed or in a jet directed at V3 of maize (Gaspareto 2018).

11.7.2 Wheat

Several researches are being conducted with greater yield and nutritional values of wheat under co-inoculations. A study was conducted by Teixeira Filho and Galindo (2019) with inoculation of *A. brasilense* and *B. subtilis* alone and together showed promising results for concentrations of P and K in leaf tissue which was reflected in greater grain yield by 62% in relation to control (Table 11.4). In addition, co-application of *Azospirillum* and *mycorrhizae* had synergetic influence on efficiency N, P and K absorption in wheat (Ardakani et al. 2011).

11.7.3 Rice

The single and co-inoculation of rice with rhizobacteria under irrigated conditions raised growth yield attributes. Santos et al. (2019) verified that inoculation and co-inoculation of rhizobia and *A. brasilense* increased number of panicles and grain yield with a 60% reduction in N fertilizer doses. The co-inoculation of *P. putida* REN_5 and *P. fluorescens* REN_1 has the ability to increase rhizosphere colonization. They have a positive impact on root and stem height, root fresh mass and shoot dry mass and root branching to better uptake of N under field conditions (Etesami and Alikhani 2016). The co-inoculation of rice genotypes with *Trichoderma asperellum* and *Pseudomonas fluorescens was observed with better*

Table 11.4 Leaf N, P and K concentrations and wheat grain yield as a function of PGPBs

Inoculation	Leaf N concentration (g kg^{-1})	Leaf P concentration (g kg^{-1})	Leaf K concentration (g kg^{-1})	Grain yield (kg kg^{-1})
Control	41.87^{a}	3.08^{c}	23.33^{c}	3052^{c}
A. brasilense	41.06^{a}	3.30^{b}	23.75^{c}	4037^{b}
B. subtilis	41.25^{a}	3.41^{ab}	26.67^{b}	4397^{ab}
A. brasilense + *B. subtilis*	40.53^{a}	3.51^{a}	28.50^{a}	4947^{a}
LSD	2.18	0.21	1.78	657
Overall means	41.18	3.33	25.56	4221
CV (%)	5.39	6.51	7.10	19.00

Means followed by similar letters in the column do not differ by Tukey's test at 5% probability
Teixeira Filho and Galindo (2019), Extension and Education Research Station, São Paulo State University (UNESP), Ilha Solteira, Brazil

biochemical and physiological processes, and soil nutrient concentration that were reflected in the greater growth and development of rice plants (Singh et al. 2020).

11.7.4 Sugarcane

Several studies have been so far conducted on sugarcane interactions with N-fixing bacteria since 1950s (Döbereiner 1992). A consortium of diazotrophic bacterial strains contributed an approximately 30% of BNF to sugarcane planted in greenhouse (Oliveira et al. 2002); however, sugarcane productivity was increased under field conditions (Oliveira et al. 2006). The combination of diazotrophic bacteria (*Gluconacetobacter diazotrophics, Herbaspirillum seropedicae, Herbaspirillum rubrisubalbicans, Azospirillum amazonense* and *Burkholderia tropica*) in associated with N doses increased stalk yield, dry matter, N accumulation and N use efficiency. In addition, sugarcane was inoculated with *A. brasilense, B. subtilis* and *P. fluorescens* (single and combined) to increase stalk productivity (Fig. 11.3) up to 17% and reduce phosphate fertilization (Rosa et al. 2022). The researchers recommend further studies to improve the methodology with great potential to optimize this process on a commercial scale.

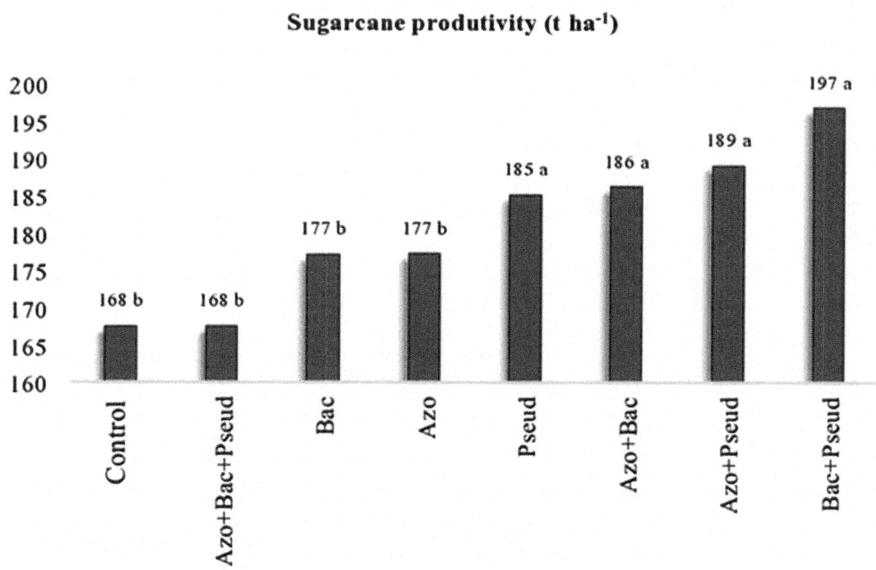

Fig. 11.3 Sugarcane productivity as a function of PGPBs inoculations or co-inoculations. [Abbreviations: Azo: *Azospirillum brasilense*, Bac: *Bacillus subtilis* and Pseud: *Pseudomonas fluorescens*; Means followed by similar letters in the column do not differ from each other by Scott-Knott test at 5% probability] (Rosa et al. 2022)

11.7.5 Pastures

The innovative and revolutionary multifunctional inoculation technology of pastures contributed to plant growth via different microbial processes. The elite strains of *Azospirillum* sp. and *Pseudomonas* sp. were selected and can be applied to seed at the time of pasture implantation, or by foliar spraying of established pastures. Inoculation of *Poaceas* plants with different strains of *Azospirillum* alone or combined contributed to plant growth hormones (Hungria et al. 2010; Fukami et al. 2017a) for better establishment and yield.

There are several other studies that indicated the beneficial impacts of PGPBs on pasture quality and biomass production. Co-inoculation of Mavuno grass with *R. tropici* and *A. brasilense* in association with N doses was observed with greater root-shoot dry mass (Sá et al. 2019). The co-inoculation of *P. fluorescens* with *R. tropici* + *A. brasilense* was observed for a relatively higher chlorophyll index in Zuri Guinea grass (de Lima et al. 2020).

11.8 Final Considerations

The PGPBs and NFBs improve plant nutrition, growth and production of non-legume crops like cereals, oil seed plants, vegetables, fruits, forages and other important crops through solubilization of plant promoting nutrients in mainly N and P and production of phytohormones from soil nutrients pool while discouraging synthetic fertilizers and chemicals. Due to the positive impacts of inoculation with these PGPBs for BNF and other nutrient acquisition, the growing use of these technologies has been observed in large agricultural crops of economic interest, such as corn, wheat, pastures, rice, sugarcane and sorghum, with emphasis on the genera *Azospirillum* sp., *Bacillus* sp. and *Pseudomonas* sp. The use of PGPBs tends to grow even more in the coming years due to their low cost (an average dose of inoculants is \$4.00 per hectare), ease in acquisition and application (via seed, planting furrow or foliar). In addition, their application is a non-polluting technology which is part of a sustainable context and has a potential of BNF in non-legumes in order to promote better nutrition, plant growth and agricultural yields. Several researches with inoculations or co-inoculations of PGPBs have shown that it is possible to obtain high yields while reducing N dose in coverage of main cultivated crops (cereals and grasses) by 25% and therefore, these microorganisms increase fertilization use efficiency. Furthermore, their occurrence and activities are not limited to host plants but have other environmental features such as bioindicators of environmental changes and are applicable in bioremediation processes, especially those related to the degradation of C-rich environmental pollutants. We believe that considering the interaction between environment x genotypes x strains (bacteria) is the key to success in developing new recommendations and applicability of inoculants in agriculture which yearns for low-cost sustainable technologies.

11.9 Future Prospective

The improvement of N fixation in non-legume crops is being considered as a dream in agronomic prospective for so many decades which needs to be addressed and realized on urgent basis. The scientists had been overlooked a natural solution due to the intense focus on legume crops as a result of their nodulation. However, the promising results of PGPBs in non-legumes, simultaneous application (co-inoculations) of several microorganisms in same cultivation cycle or similar cropping system as well as forms and times of application should be increased. The study of interaction between several NFB and PGPBs, applied together in the same production cycle aiming to aggregate positive characteristics of each bacterium in the development of plants. The interaction among PGPBs with microorganisms already existing in soil and their synergistic or antagonistic effects in different edaphoclimatic conditions needs to be focused and highlighted. Therefore, it is believed that this knowledge can lead to the development of new inoculants or techniques to protect bacteria against environmental effects such as solar radiation, high temperatures, water deficit and others for the practice of sustainable tropical agriculture. These PGPBs and NFBs need special focus to identify the most suitable consortium/ consortia, their mode of action and attribution to production plant hormones and non-hormones that may lead to increase root growth and uptake of N from soil, phosphate solubilization, siderophore production, and antagonism towards pathogens. The development of technologies like sequencing, gene editing and synthetic biology can also manipulate plants and microbes at a large scale. We believe that combination plant and microbial diversity along with genetic engineering should increase N fixation, plant nutrition and other metabolic and biochemical processes of non-legume crops in short and long cropping systems to feed the world in a more sustainable manner. Finally, it is hoped that the gains obtained with these technologies will sustain and encourage the exploration of countless other processes that occur in soil microbiology with a potential biotechnological application for increasing modern and sustainable agriculture.

Acknowledgement Not applicable.

Conflict of Interest Author(s) declare no conflict of interest.

References

Abraham-Juárez MR, Espitia-Vázquez I, Guzmán-Mendoza R, Olalde-Portugal V, Ruiz-Aguilar GML, García-Hernández JL, Herrera-Isidrón L, Núñez-Palenius HG (2018) Development, yield, and quality of melon fruit (*Cucumis melo* L.) inoculated with Mexican native strains of *Bacillus subtilis* (Ehrenberg). Agrociencia 52(1):91–102

Acosta-Motos JR, Ortuño MF, Bernal-Vicente A, Diaz-Vivancos P, Sanchez-Blanco MJ, Hernandez JA (2017) Plant responses to salt stress: adaptive mechanisms. Agronomy 7:1–18

Afzal I, Munir F, Ayub CM, Basra SMA, Hameed A, Nawaz A (2009) Changes in antioxidant enzymes, germination capacity and vigour of tomato seeds in response of priming with polyamines. Seed Sci Technol 37(3):765–770

Alcázar R, Strong AM, Tiburcio AF (2020) Polyamines in plant biotechnology, food nutrition and human health. Front Plant Science 11:120

Andrade FM, de Assis PT, Souza TP, Guimarães PHS, Martins AD, Schwan RF, Pasqual M, Dória J (2019) Beneficial effects of inoculation of growth-promoting bacteria in strawberry. Microbiol Res 223:120–128. https://doi.org/10.1016/j.micres.2019.04.005

Aquino GS, Ventura MU, Alexandrino RP, Michelon TA, de Araujo Pescador PG, Nicio TT, Watanabe VS, Diniz TG, de Oliveira ALM, Hata FT (2018) Plant-promoting rhizobacteria *Methylobacterium komagatae* increase crambe yields, root system and plant height. Ind Crop Prod 121:277–281. https://doi.org/10.1016/J.INDCROP.2018.05.020

Ardakani MR, Mazaheri D, Mafakheri S, Moghaddam A (2011) Absorption efficiency of N, P, K through triple inoculation of wheat (*Triticum aestivum* L.) by *Azospirillum brasilense, streptomyces* sp., *glomus intraradices* and manure application. Physiol Mol Biol Plants 17:181–192

Argyros RD, Mathews DE, Chiang YH, Palmer CM, Thibault DM, Etheridge N, Argyros DA, Mason MG, Kieber JJ, Schaller GE (2008) Type B response regulators of *Arabidopsis* play key roles in cytokinin signaling and plant development. Plant Cell 20:2102–2116

Arshad M, Shaharoona B, Mahmood T (2008) Inoculation with plant growth promoting rhizobacteria containing ACC-deaminase partially eliminates the effects of water stress on growth, yield and ripening of *Pisum sativum* L. Pedosphere 18(5):611–620

Awad NM, Abd El-Kader AA, Attia M, Alva AK (2011) Effects of nitrogen fertilization and soil inoculation of sulfur-oxidizing or nitrogen-fixing bacteria on onion plant growth and yield. Int J Agron 6:316856. https://doi.org/10.1155/2011/316856

Baldani JI, Baldani VLD, Seldin L, Döbereiner J (1986) Characterization of Herbaspirillum seropedicae gen. nov., a root-associated nitrogen-fixing bacterium. Int J Syst Bacteriol 36:86–93

Baldani VDL, Oliveira E, Balota E, Baldani JI, Kirchhof G, Döbereiner J (1997) Burkhoderia brasilensis sp. nov., uma espécie de bactéria diazotrófica endofítica. An Acad Brasi Ci 69:116

Barnawal D, Bharti N, Pandey SS, Pandey A, Chanotiya CS, Kalra A (2017) Plant growth-promoting rhizobacteria enhance wheat salt and drought stress tolerance by altering endogenous phytohormone levels and TaCTR1/TaDREB2 expression. Physiol Plant 161(4):502–514

Bashan Y, Levanony H (1990) Current status of *Azospirillum* inoculation technology: *Azospirillum* as a challenge for agriculture. Can J Microbiol 36:591–608. https://doi.org/10.1139/m90-105

Behera B, Das TK, Raj R, Ghosh S, Raza MB, Sen S (2021) Microbial consortia for sustaining productivity of non-legume crops: prospects and challenges. Agric Res 10:1–14. https://doi.org/10.1007/s40003-020-00482-3

Behmann J, Mahlein AK, Rumpf T, Römer C, Plümer L (2015) A review of advanced machine learning methods for the detection of biotic stress in precision crop protection. Precis Agric 16(3):239–260

Belimov AA, Dodd IC, Safronova VI, Dumova VA, Shaposhnikov AI, Ladatko AG, Davies WJ (2014) Abscisic acid metabolizing rhizobacteria decrease ABA concentrations in planta and alter plant growth. Plant Physiol Biochem 74:84–91

Berger B, Wiesner M, Brock AK, Schreiner M, Ruppel SK (2015) A beneficial bacteria that promotes radish growth under field conditions. Agron Sustain Dev 35:1521–1528. https://doi.org/10.1007/s13593-015-0324-z

Berger S, Sinha AK, Roitsch T (2007) Plant physiology meets phytopathology: plant primary metabolism and plant–pathogen interactions. J Exp Bot 58(16):4019–4026

Besset-Manzoni Y, Rieusset L, Joly P, Comte G, Prigent-Combaret C (2018) Exploiting rhizosphere microbial cooperation for developing sustainable agriculture strategies. Environ Sci Pollut Res 25:29953–29970. https://doi.org/10.1007/s11356-017-1152-2

Bharath P, Gahir S, Raghavendra AS (2021) Abscisic acid-induced stomatal closure: An important component of plant defense against abiotic and biotic stress. Front Plant Sci 12:324

Bharti N, Barnawal D, Shukla S, Tewari SK, Katiyar RS, Kalra A (2016) Integrated application of *Exiguobacterium oxidotolerans, Glomus fasciculatum*, and vermicompost improves growth, yield and quality of *Mentha arvensis* in salt-stressed soils. Ind Crop Prod 83:717–728. https://doi.org/10.1016/J.INDCROP.2015.12.021

Bueno ACSO, Castro GLS, Silva Junior DD, Pinheiro HA, Filippi MCC, Silva GB (2017) Response of photosynthesis and chlorophyll a fluorescence in leaf scald-infected rice under influence of rhizobacteria and silicon fertilizer. Plant Pathol 66:1487–1495

Bünemann EK, Schwenke GD, Van Zwieten L (2006) Impact of agricultural inputs on soil organisms—a review. Soil Res 44:379. https://doi.org/10.1071/SR05125

Çakmakçi R, Donmez MF, Erdogan U (2007) The effect of plant growth promoting rhizobacteria on barley seedling growth, nutrient uptake, some soil properties, and bacterial counts. Turk J Agric For 31:189–199

Cassán F, Maiale S, Masciarelli O, Vidal A, Luna V, Ruiz O (2009) Cadaverine production by *Azospirillum brasilense* and its possible role in plant growth promotion and osmotic stress mitigation. Eur J Soil Biol 45:12–19. https://doi.org/10.1016/j.ejsobi.2008.08.003

Cavalcante VA, Dobereiner J (1988) A new acid-tolerant nitrogen-fixing bacterium associated with sugarcane. Plant Soil 108(1):23–31

de Cassetari AS, SPM G, MCP S (2016) Fixação biológica de nitrogênio associativa e de vida livre. Microbiol do solo 221. https://doi.org/10.1160/69788586481567

Chiappero J, Cappellari LDR, Sosa Alderete LG, Palermo TB, Banchio E (2019) Plant growth rhizobacteria improves the antioxidant status in *Mentha piperita* grown under drought stress leading to an enhancement of plant growth and total phenolic content. Ind Crop Prod 139: 111553. https://doi.org/10.1016/J.INDCROP.2019.111553

Chong-qing W, Tao W, Ping M, Zi-chao L, Ling Y (2013) Quantitative trait loci for mercury tolerance in rice seedlings. Rice Sci 20:238–242. https://doi.org/10.1016/S1672-6308(13)60124-9

Cirillo C, Rouphael Y, Caputo R, Raimondi G, Sifola MI, De Pascale S (2016) Effects of high salinity and the exogenous of an osmolyte on growth, photosynthesis and mineral composition in two ornamental shrubs. J Hortic Sci Biotechnol 91:14–22

Cohen A, Bottini R, Piccoli P (2008) *Azospirillum brasilense* Sp 245 produces ABA in chemically-defined culture medium and increases ABA content in *Arabidopsis* plants. Plant Growth Regul 54:97–103. https://doi.org/10.1007/s10725-007-9232-9

Cohen AC, Travaglia CN, Bottini R, Piccoli PN (2009) Participation of abscisic acid and gibberellins produced by endophytic *Azospirillum* in the alleviation of drought effects in maize. Botany 87(5):455–462

Contesto C, Desbrosses G, Lefoulon C, Béna G, Borel F, Galland M, Gamet L, Varoquaux F, Touraine B (2008) Effects of rhizobacterial ACC deaminase activity on *Arabidopsis* indicate that ethylene mediates local root responses to plant growth-promoting rhizobacteria. Plant Sci 175:178–189

Corrêa EB, Bettiol W, Sutton JC (2010) Biological control of root rot (*Pythium aphanidermatum*) and growth promotion by *pseudomonas chlororaphis* 63-28 and *Bacillus subtilis* GB03 in hydroponic lettuce. Summa Phytopathol 36:275–281

Corrêa ÉB, Galvão JAH, Bettiol W (2011) Controle biológico da podridão radicular e promoção de crescimento em pepino hidropônico com microrganismos de manguezais. Pesq Agropec Bras 46:130–136

Cortés-Patiño S, Vargas C, Álvarez-Flórez F, Bonilla R, Estrada-Bonilla G (2021) Potential of *Herbaspirillum* and *Azospirillum* consortium to promote growth of perennial ryegrass under water deficit. Microorganisms 9(1):1–16. https://doi.org/10.3390/microorganisms9010091

da Silva MSR, Tavares OCH, de Oliveira ISR (2022) Stimulatory effects of defective and effective 3-indoleacetic acid-producing bacterial strains on rice in an advanced stage of its vegetative cycle. Braz J Microbiol 53:281–288. https://doi.org/10.1007/s42770-021-00651-8

Dartora J, Guimarães VF, Marini D, Sander G (2013) Adubação nitrogenada associada à inoculação com *Azospirillum brasilense* e *Herbaspirillum seropedicae* na cultura do milho. Rev Bras Eng Agríc Ambient 17:1023–1029

Dazzo FB, Yanni YG (2006) The natural *rhizobium*-cereal crop association as an example of plant-bacteria interaction. In: Uphoff N, Ball AS, Fernandes E, Herren H, Husson O, Laing M, Palm C, Pretty J, Sanchez P, Sanginga N, Thies J (eds) Biological approaches to sustainable soil systems. CRC Press, Boca Raton, pp 109–127

del Carmen O-MM, Glick BR, Santoyo G (2020) ACC deaminase in plant growth-promoting bacteria (PGPB): an efficient mechanism to counter salt stress in crops. Microbiol Res 235: 126439

Dettmer J, Elo A, Helariutta Y (2009) Hormone interactions during vascular development. Plant Mol Biol 69:347–360

Dhiman S, Dubey RC, Maheshwari DK, Kumar S (2019) Sulfur-oxidizing buffalo dung bacteria enhance growth and yield of *Foeniculum vulgare* mill. Can J Microbiol 65(5):377–386

Döbereiner J (1992) History and new perspectives of diazotrophs in association with non-leguminous plants. Symbiosis 13:1–13

Doneda A, Aita C, Giacomini SJ, Miola ECC, Giacomini DA, Schirmann J, Gonzatto R (2012) Fitomassa e decomposição de resíduos de plantas de cobertura puras e consorciadas. Rev Bras Cienc Solo 36:1714–1723. https://doi.org/10.1590/S0100-06832012000600005

Dong B, Zheng X, Liu H, Able JA, Yang H, Zhao H, Zhang M, Qiao Y, Wang Y, Liu M (2017) Effects of drought stress on pollen sterility, grain yield, abscisic acid and protective enzymes in two winter wheat cultivars. Front Plant Sci 8:1008

Duarte CFD, Cecato U, Hungria M, Fernandes HJ, Biserra TT, Mamédio D, Galbeiro S, Nogueira MA (2020) Inoculation of plant growth-promoting bacteria in *Urochloa Ruziziensis*. Res Soc Dev 9:e630985978. https://doi.org/10.33448/rsd-v9i8.5978

Dubey A, Malla MA, Khan F, Chowdhary K, Yadav S, Kumar A, Sharma S, Khare PK, Khan ML (2019) Soil microbiome: a key player for conservation of soil health under changing climate. Biodivers Conserv 28(8):2405–2429

Eckert B, Weber OB, Kirchhof G, Halbritter A, Stoffels M, Hartmann A (2001) Azospirillum doebereinerae sp. nov., a nitrogen-fixing bacterium associated with the C4-grass Miscanthus. Int J Syst Evol Microbiol 51(1):17–26

Efroni I, Han SK, Kim Hye J, Wu MF, Steiner E, Birnbaum KD, Hong JC, Eshed Y, Wagner D (2013) Regulation of leaf maturation by chromatin-mediated modulation of cytokinin responses. Dev Cell 24:438–445

Egamberdiyeva D (2007) The effect of plant growth promoting bacteria on growth and nutrient uptake of maize in two different soils. Appl Soil Ecol 36:184–189. https://doi.org/10.1016/j.apsoil.2007.02.055

Egener T, Hurek T, Reinhold-Hurek B (1998) Use of green fluorescent protein to detect expression of nif genes of *Azoarcus* sp. BH72, a grass-associated diazotroph on rice roots. Mol Plant-Microbe Interact 11:71–75. https://doi.org/10.1094/MPMI.1998.11.1.71

Ehteshami SMR, Khavazi K, Asgharzadeh A (2018) Forage sorghum quantity and quality as affected by biological phosphorous fertilization. Grass Forage Sci 73(4):926–937. https://doi.org/10.1111/gfs.12388

Enebe MC, Babalola OO (2018) The influence of plant growth-promoting rhizobacteria in plant tolerance to abiotic stress: a survival strategy. Appl Microbiol Biotechnol 102:7821–7835. https://doi.org/10.1007/s00253-018-9214-z

Etesami H, Alikhani HA (2016) Co-inoculation with endophytic and rhizosphere bacteria allows reduced application rates of N-fertilizer for rice plant. Rhizosphere 2:5–12. https://doi.org/10.1016/j.rhisph.2016.09.003

Etesami H, Maheshwari DK (2018) Use of plant growth promoting rhizobacteria (PGPRs) with multiple plant growth promoting traits in stress agriculture: action mechanisms and future prospects. Ecotoxicol Environ Safe 156:225–246 del Fernández M, Meeus I, Billiet A, Van Nieuwerburgh F, Deforce D, Vandamme P, Viñuela E, Smagghe G (2019) Influence of microbiota in the susceptibility of parasitic wasps to abamectin insecticide: deep sequencing, esterase and toxicity tests. Pest Manag Sci 75:79–86. https://doi.org/10.1002/ps.5195

Fibach-Paldi S, Burdman S, Okon Y (2012) Key physiological properties contributing to rhizosphere adaptation and plant growth promotion abilities of *Azospirillum brasilense*. FEMS Microbiol Lett 326:99–108. https://doi.org/10.1111/j.1574-6968.2011.02407.x

Fierer N (2017) Embracing the unknown: disentangling the complexities of the soil microbiome. Nat Rev Microbiol 15:579–590. https://doi.org/10.1038/nrmicro.2017.87

Finkel OM, Castrillo G, Herrera Paredes S, Salas González I, Dangl JL (2017) Understanding and exploiting plant beneficial microbes. Curr Opin Plant Biol 38:155–163. https://doi.org/10.1016/j.pbi.2017.04.018

Franche C, Lindström K, Elmerich C (2009) Nitrogen-fixing bacteria associated with leguminous and non-leguminous plants. Plant Soil 321:35–59. https://doi.org/10.1007/s11104-008-9833-8

French E, Kaplan I, Iyer-Pascuzzi A, Nakatsu CH, Enders L (2021) Emerging strategies for precision microbiome management in diverse agroecosystems. Nat Plants 7:256–267. https://doi.org/10.1038/s41477-020-00830-9

Fuentes-Ramírez LE, Bustillos-Cristales R, Tapia-Hernández A, Jiménez-Salgado T, Wang ET, Martínez-Romero E, Caballero-Mellado J (2001) Novel nitrogen-fixing acetic acid bacteria, Gluconacetobacter johannae sp. nov. and Gluconacetobacter azotocaptans sp. nov., associated with coffee plants. Int J Syst Evol Microbiol 51(4):1305–1314

Fukami J, de la Osa C, Ollero FJ, Megías M, Hungria M (2017b) Co-inoculation of maize with *Azospirillum brasilense* and *rhizobium tropici* as a strategy to mitigate salinity stress. Funct Plant Biol 45:328–339

Fukami J, Nogueira MA, Araujo RS, Hungaria M (2016) Accessing inoculation methods of maize and wheat with *Azospirillum brasilense*. AMB Express 6:3–16

Fukami J, Ollero FJ, De La Osa C, Valderrama-Fernandez R, Nogueira MA, Megías M, Hungria M (2018) Antioxidant activity and induction of mechanisms of resistance to stresses related to the inoculation with *Azospirillum brasilense*. Arch Microbiol 200:1191–1203. https://doi.org/10.1007/s00203-018-1535-x

Fukami J, Ollero FJ, Megías M, Hungria M (2017a) Phytohormones and induction of plant-stress tolerance and defense genes by seed and foliar inoculation with *Azospirillum brasilense* cells and metabolites promote maize growth. AMB Express 7:153–166. https://doi.org/10.1186/s13568-017-0453-7

Galindo FS, Buzetti S, Rodrigues WL, Boleta EHM, Silva VM, Tavanti RFR, Fernandes GC, Biagini ALC, Rosa PAL, Teixeira Filho MCM (2020b) Inoculation of *Azospirillum brasilense* associated with silicon as a liming source to improve nitrogen fertilization in wheat crops. Sci Rep 6160

Galindo FS, Teixeira Filho MCM, Buzetti S, Santini JMK, Alves CJ, Nogueira LM, Ludkiewicz MGZ, Andreotti M, Bellotte JLM (2016) Corn yield and foliar diagnosis affected by nitrogen fertilization and inoculation with *Azospirillum brasilense*. Rev Bras Ciênc Solo 40:1–18

Galindo FS, Teixeira Filho MC, Buzetti S, Pagliari PH, Santini JM, Alves CJ, Megda MM, Nogueira TA, Andreotti M, Arf O (2019) Maize yield response to nitrogen rates and sources associated with Azospirillum brasilense. Agron J 111(4):1985–1997

Galindo FS, Teixeira Filho MCM, Da Silva EC, Buzetti S, Fernandes GC, Rodrigues WL (2020a) Technical and economic viability of cowpea co-inoculated with *Azospirillum brasilense* and *Bradyrhizobium* spp. and nitrogen doses. Rev Bras Eng Agric Ambient 24:305–312. https://doi.org/10.1590/1807-1929/agriambi.v24n5p305-312

Galindo FS, Bellotte JL, Santini JM, Buzetti S, Rosa PA, Jalal A, Teixeira Filho MCM (2021) Zinc use efficiency of maize-wheat cropping after inoculation with Azospirillum brasilense. Nutr Cycl Agroecosyst 120:205–221

Galindo FS, Rodrigues WL, Fernandes GC, Boleta EHM, Jalal A, Rosa PAL, Buzetti S, Lavres J, Teixeira Filho MCM (2022) Enhancing agronomic efficiency and maize grain yield with

Azospirillum brasilense inoculation under Brazilian savannah conditions. Euro J Agron 134: 126471

Galon L, Concepto G, Ferreira EA, Aspiazu I, Silva AFD, Giacobbo CL, Andres A (2013) In: Dubinsky Z (ed) Influence of biotic and abiotic stress factors on physiological traits of sugarcane varieties. InTech, Photosynthesis. Rijeka, pp 185–208. https://doi.org/10.5772/55255

García De Salamone IE (2011) Microorganisms of suelo and sustainability of agroecosystems. Rev Argent Microbiol 43:1–3

García JAL, Probanza A, Ramos B, Palomino MR, Manero FJG (2004) Effect of inoculation of *Bacillus licheniformis* on tomato and pepper. Agronomy 24:169

Gaspareto RN (2018) Formas de inoculação com bactérias promotoras de crescimento na nutrição e desempenho agronômico de milho no cerrado. Dissertação (mestrado) – Universidade Estadual Paulista, Faculdade de Engenharia de Ilha Solteira, p. 82 (Portuguese)

Gietler M, Fidler J, Labudda M, Nykiel M (2020) Abscisic acid—enemy or savior in the response of cereals to abiotic and biotic stresses? Int J Mol Sci 21(13):4607. https://doi.org/10.3390/ijms21134607

Gillis M, Van Van T, Bardin R, Goor M, Hebbar P, Willems A, Segers P, Kersters K, Heulin T, Fernandez MP (1995) Polyphasic taxonomy in the genus Burkholderia leading to an emended description of the genus and proposition of Burkholderia vietnamiensis sp. nov. for N2-fixing isolates from rice in Vietnam. Int J Syst Evol Microbiol 45(2):274–289

Gomes DG, Radi AJ, Aquino GS (2018) Growth-promoting bacteria change the development of aerial part and root system of canola. Semina: Cienc Agrar 39(6):2375–2384. https://doi.org/10.5433/1679-0359.2018v39n6p2375

Goteti PK, Emmanuel LDA, Desai S, Shaik MHA (2013) Prospective zinc solubilizing bacteria for enhanced nutrient uptake and growth promotion in maize (*Zea mays* L.). Int J Microbiol:1–8. https://doi.org/10.1155/2013/869697

Großkinsky DK, Tafner R, Moreno MV, Stenglein SA, García de Salamone IE, Nelson LM, Novák O, Strnad M, van der Graaff E, Roitsch T (2016) Cytokinin production by *Pseudomonas fluorescens* G20-18 determines biocontrol activity against *pseudomonas syringae* in *Arabidopsis*. Sci Rep 6(1):1–11. https://doi.org/10.1038/srep23310

Haahtela K, Laakso T, Nurmiaho-Lassila EL, Korhonen TK (1988) Effects of inoculation of Poa pratensis and Triticum aestivum with root-associated, N2-fixing Klebsiella, Enterobacter and Azospirillum. Plant Soil 106(2):239–248

Hartmann A, Fischer D, Kinzel L, Chowdhury SP, Hofmann A, Baldani JI, Rothballer M (2019) Assessment of the structural and functional diversities of plant microbiota: achievements and challenges – a review. J Adv Res 19:3–13. https://doi.org/10.1016/j.jare.2019.04.007

Hartmann M, Brunner I, Hagedorn F, Bardgett RD, Stierli B, Herzog C, Chen X, Zingg A, Graf-Pannatier E, Rigling A, Frey B (2017) A decade of irrigation transforms the soil microbiome of a semi-arid pine forest. Mol Ecol 26:1190–1206. https://doi.org/10.1111/mec.13995

Hatami M, Khanizadeh P, Bovand F, Aghaee A (2021) Silicon nanoparticle-mediated seed priming and *pseudomonas* spp. inoculation augmentation growth, physiology and antioxidant metabolic status in *Melissa officinalis* L. plants. Ind Crops Prod 162:113238. https://doi.org/10.1016/J.INDCROP.2021.113238

He M, He CQ, Ding NZ (2018) Abiotic stresses: general defenses of land plants and chances for engineering multistress tolerance. Front Plant Sci 871:1–18. https://doi.org/10.3389/fpls.2018.01771

Helaly AA, Hassan SM, Craker LE, Mady E (2020) Effects of growth-promoting bacteria on growth, yield and nutritional value of collard plants. Annal Agric Sci 65:77–82. https://doi.org/10.1016/j.aoas.2020.01.001

Hungria M, Campo RJ, Souza EM, Pedrosa FO (2010) Inoculation with selected strains of *Azospirillum brasilense* and *A. lipoferum* improves yields of maize and wheat in Brazil. Plant Soil 331:413–425. https://doi.org/10.1007/s11104-009-0262-0

Hungria M, Nogueira MA, Araujo RS (2016) Inoculation of *Brachiaria* with plant growth-promoting bacterium *Azospirillum brasilense*: An environment-friendly component in the

reclamation of degraded pastures in the tropics. Agric Ecosyst Environ 221:125–131. https://doi.org/10.1016/J.AGEE.2016.01.024
Hussain S, Siddique T, Saleem M, Arshad M, Khalid A (2009) Impact of pesticides on soil microbial diversity, enzymes, and biochemical reactions. Adv Agron 102:159–200. https://doi.org/10.1016/S0065-2113(09)01005-0
Hussain S, Nanda S, Zhang J, Rehmani MIA, Suleman M, Li G, Hou H (2021) Auxin and cytokinin interplay during leaf morphogenesis and phyllotaxy. Plan Theory 10(8):1732. https://doi.org/10.3390/plants10081732
Ilyas N, Bano A (2010) *Azospirillum* strains isolated from roots and rhizosphere soil of wheat (*Triticum aestivum* L.) grown under different soil moisture conditions. Biol Fertil Soil 46(4): 393–406
Imfeld G, Vuilleumier S (2012) Measuring the effects of pesticides on bacterial communities in soil: a critical review. Eur J Soil Biol 49:22–30. https://doi.org/10.1016/j.ejsobi.2011.11.010
Iqbal N, Khan NA, Ferrante A, Trivellini A, Francini A, Khan MIR (2017) Ethylene role in plant growth, development and senescence: interaction with other phytohormones. Front Plant Sci 8: 475. https://doi.org/10.3389/fpls.2017.00475
Islam MR, Madhaiyan M, Deka Boruah HP, Yim W, Lee G, Saravanan VS, Fu Q, Hu H, Sa T (2009) Characterization of plant growth-promoting traits of free-living diazotrophic bacteria and their inoculation effects on growth and nitrogen uptake of crop plants. J Microbiol Biotechnol 19(10):1213–1222. https://doi.org/10.4014/jmb.0903.3028
Jalal A, Azeem K, Teixeira Filho MCM, Khan A (2020) Enhancing soil properties and maize yield through organic and inorganic nitrogen and diazotrophic bacteria. Sustainable crop production. IntechOpen, London, pp 165–178
Jalal A, Galindo FS, Boleta EHM, Oliveira CEDS, Reis ARD, Nogueira TAR, Moretti Neto MJ, Mortinho ES, Fernandes GC, Teixeira Filho MCM (2021) Common bean yield and zinc use efficiency in association with diazotrophic bacteria co-inoculations. Agronomy 11(5):959. https://doi.org/10.3390/agronomy11050959
Jalal A, Oliveira C, Freitas L, Galindo F, Lima B, Boleta E, da Silva E, Nascimento V, Nogueira TAR, Buzetti S, Teixeira Filho MCM (2022) Agronomic biofortification and productivity of wheat with soil zinc and diazotrophic bacteria under tropical savannah. Crop Pasture Sci 73(7–8):817–830. https://doi.org/10.1071/CP21457
Jamalzadeh A, Darvishnia M, Khodakaramian G, Bazgir E, Zafari D (2021) Genetic diversity and plant growth. Egyptian J Biol Pest Contr 31:1–9. https://doi.org/10.1186/s41938-021-00442-1
James EK (2000) Nitrogen fixation in endophytic and associative symbiosis. Field Crops Res 65: 197–209. https://doi.org/10.1016/S0378-4290(99)00087-8
Joo HS, Chang CS (2005) Production of protease from a new alkalophilic Bacillus sp. I-312 grown on soybean meal: optimization and some properties. Process Biochem 40(3–4):1263–1270
Kakumanu ML, Reeves AM, Anderson TD, Rodrigues RR, Williams MA (2016) Honey bee gut microbiome is altered by in-hive pesticide exposures. Front Microbiol 7:1255. https://doi.org/10.3389/fmicb.2016.01255
Kamada-Nobusada T, Makita N, Kojima M, Sakakibara H (2013) Nitrogen-dependent regulation of de novo cytokinin biosynthesis in rice: the role of glutamine metabolism as an additional signal. Plant Cell Physiol 54(11):1881–1893. https://doi.org/10.1093/pcp/pct127
Kang SM, Khan AL, Waqas M, You YH, Kim JH, Kim JG, Hamayun M, Lee IJ (2014) Plant growth-promoting rhizobacteria reduce adverse effects of salinity and osmotic stress by regulating phytohormones and antioxidants in Cucumis sativus. J Plant Interact 9(1):673–682
Kang SM, Waqas M, Shahzad R, You YH, Asaf S, Khan A, Lee KE, Joo GJ, Kim SJ, Lee IJ (2015) Mechanism of plant growth promotion elicited by *bacillus* sp. LKE15 in oriental melon. Acta Agric Scand B Soil Plant Sci 65:36–44. https://doi.org/10.1080/09064710.2014.960889
Kartik VP, Jinal HN, Amaresan N (2021) Inoculation of cucumber (*Cucumis sativus* L.) seedlings with salt-tolerant plant growth promoting bacteria improves nutrient uptake, plant attributes and physiological profiles. J Plant Growth Regul 40:1728–1740. https://doi.org/10.1007/s00344-020-10226-w

Kaye JP, Quemada M (2017) Using cover crops to mitigate and adapt to climate change. A review. Agron Sustain Dev 37:4. https://doi.org/10.1007/s13593-016-0410-x

Kenneth OC, Nwadibe EC, Kalu AU, Unah UV (2019) Plant growth promoting rhizobacteria (PGPR): a novel agent for sustainable food production. Am J Agric Biol Sci 14:35–54. https://doi.org/10.3844/ajabssp.2019.35.54

Keswani C, Singh SP, Cueto L, Garcia-Estrada C, Mezaache-Aichour S, Glare TR, Borriss R, Singh SP, Blázquez MA, Sansinenea E (2020) Auxins of microbial origin and their use in agriculture. Appl Microbiol Biotechnol 104:8549–8565. https://doi.org/10.1007/s00253-020-10890-8

Khammas KM, Ageron E, Grimont PA, Kaiser P (1989) Azospirillum irakense sp. nov., a nitrogen-fixing bacterium associated with rice roots and rhizosphere soil. Res Microbiol 140(9):679–693

Khan K, Pankaj U, Verma SK, Gupta AK, Singh RP, Verma RK (2015) Bio-inoculants and vermicompost influence on yield, quality of *Andrographis paniculata*, and soil properties. Ind Crop Prod 70:404–409. https://doi.org/10.1016/J.INDCROP.2015.03.066

Khan MMA, Haque E, Paul NC, Khaleque MA, Al-Garni SMS, Rahman M, Islam MT (2017) Enhancement of growth and grain yield of rice in nutrient deficient soils by rice probiotic bacteria. Rice Sci 24(5):264–273. https://doi.org/10.1016/j.rsci.2017.02.002

Kıdoğlu F, Gül A, Tüzel Y, Özaktan H (2009) Yield enhancement of hydroponically grown tomatoes by rhizobacteria. Horticulture 807:475–480

Kieber JJ, Schaller GE (2014) Cytokinins. Arabidopsis Book 11:e0168. https://doi.org/10.1199/tab.0168

Kim N, Zabaloy MC, Guan K, Villamil MB (2020) Do cover crops benefit soil microbiome? A meta-analysis of current research. Soil Biol Biochem 142:107701. https://doi.org/10.1016/j.soilbio.2019.107701

Kirchhof G, Eckert B, Stoffels M, Baldani JI, Reis VM, Hartmann A (2001) Herbaspirillum frisingense sp. nov., a new nitrogen-fixing bacterial species that occurs in C4-fibre plants. Int J Syst Evol Microbiol 51(1):157–168

Kovtunovych G, Lar O, Kamalova S, Kordyum V, Kleiner D, Kozyrovska N (1999) Correlation between pectate lyase activity and ability of diazotrophic Klebsiella oxytoca VN 13 to penetrate into plant tissues. Plant Soil 215(1):1–6

Krey T, Vassilev N, Baum C, Eichler-Löbermann B (2013) Effects of long-term phosphorus application and plant-growth promoting rhizobacteria on maize phosphorus nutrition under field conditions. Eur J Soil Biol 55:1124–1130

Krieg NR, Holt JG (1984) Bergey's manual of systematic bacteriology, 1st edn. Williams and Wiilkins, Baltimore, MD

Larcher W (2006) Plant ecophysiology. RiMa, São Carlos, SP, p 531

Ladha JK, Barraquio WL, Watanabe I (1983) Isolation and identification of nitrogen-fixing Enterobacter cloacae and Klebsiella planticola associated with rice plants. Can J Microbiol 29 (10):1301–1308

Lee S, An R, Grewal P, Yu Z, Borherova Z, Lee J (2016) High-performing window farm hydroponic system: transcriptomes of fresh produce and microbial communities in response to beneficial bacterial treatment. Mol Plant-Microbe Interact 29:965–976

Leite LFC, Araújo ASF (2007) Ecologia Microbiana do Solo. Embrapa Meio-Norte 24. (Portuguese)

Leite RDC, Dos Santos JGD, Silva EL, Alves CRCR, Hungaria M, Leite RDC, Dos Santos AC (2019) Productivity increase, reduction of nitrogen fertiliser use and drought-stress mitigation by inoculation of Marandu grass (*Urochloa brizantha*) with *Azospirillum brasilense*. Crop Pest Sci 70(1):61–67. https://doi.org/10.1071/CP18105

de Lima GC, Hungary M, Nogueira MA, Teixeira Filho MCM, Moreira A, Heinrichs R, Soares Filho CV (2020) Yield, yield components and nutrients uptake in Zuri Guinea grass inoculated with plant growth-promoting bacteria. Int J Innov Educ Res 8(4):103–124. https://doi.org/10.31686/ijier.vol8.iss4.2268

Lima GS, Santos JB, Soares LAA, Gheyi HR, Noble RG, Pereira RF (2016) Irrigation with saline water and application of foliar proline in "All Big" pepper crop. Comun Sci 7:513–522

Liu X-L, Zhang H, Jin Y-Y, Wang M-M, Yang H-Y, Ma H-Y, Jiang C-J, Liang ZW (2019) Abscisic acid primes rice seedlings for enhanced tolerance to alkaline stress by upregulating antioxidant defense and stress tolerance-related genes. Plant Soil 438:39–55. https://doi.org/10.1007/s11104-019-03992-4

Liu F, Xing S, Ma H, Du Z, Ma B (2013) Cytokinin-producing, plant growth-promoting rhizobacteria that confer resistance to drought stress in *Platycladus orientalis* container seedlings. Appl Microbiol Biotechnol 97(20):9155–9164. https://doi.org/10.1007/s00253-013-5193-2

Long XE, Yao H, Huang Y, Wei W, Zhu YG (2018) Phosphate levels influence the utilisation of rice rhizodeposition carbon and the phosphate-solubilising microbial community in a paddy soil. Soil Biol Biochem 118:103–114. https://doi.org/10.1016/j.soilbio.2017.12.014

Lopes MJS, Dias-Filho MB, Castro THR, Silva GB (2018) Light and plant growth-promoting rhizobacteria effects on *Brachiaria brizantha* growth and phenotypic plasticity to shade. Grass Forage Sci 73(2):493–499. https://doi.org/10.1111/gfs.12336

Lovaisa NC, Guerrero Molina MF, Delaporte Quintana PAG, Salazar SM (2015) Response of strawberry plants inoculated with *Azospirillum* and *Burkholderia* at field conditions. Rev Agron Noroeste Argent 35:33–36

Magalhães FM, Baldani JI, Souto SM, Kuykendall JR, Dobereiner J (1983) New acid-tolerant Azospirillum species. Anais-Academia Brasileira de Ciências

Mahmud K, Makaju S, Ibrahim R, Missaoui A (2020) Current progress in nitrogen fixing plants and microbiome research. Plan Theory 9:1–17. https://doi.org/10.3390/plants9010097

Majda M, Robert S (2018) The role of auxin in cell wall expansion. Int J Mol Sci 19(4):951. https://doi.org/10.3390/ijms19040951

Malik KA, Bilal R, Mehnaz S, Rasul G, Mirza MS, Ali S (1997) Association of nitrogen-fixing, plant-growth-promoting rhizobacteria (PGPR) with kallar grass and rice. In: Opportunities for biological nitrogen fixation in rice and other non-legumes. Springer, Dordrecht, pp 37–44

Martins DS, Reis VM, Schultz N, Alves BJR, Urquiaga S, Pereira W, Sousa JS, Boddey RM (2020) Both the contribution of soil nitrogen and of biological N_2 fixation to sugarcane can increase with the inoculation of diazotrophic bacteria. Plant Soil 454:155–169. https://doi.org/10.1007/s11104-020-04621-1

Masood S, Zhao XQ, Shen RF (2019) *Bacillus pumilus* increases boron uptake and inhibits rapeseed growth under boron supply irrespective of phosphorus fertilization. AoB Plants 11(4):36. https://doi.org/10.1093/aobpla/plz036

Masson PH, Takahashi T, Angelini R (2017) Editorial: molecular mechanisms underlying polyamine functions in plants. Front Plant Sci 8:14

Mattos BB, Marriel IE, Sousa SM, Lana UGP, Schaffert RE, Gomes EA, Oliveira CA (2020) Sorghum genotypes response to inoculation with phosphate solubilizing bacteria. Braz J Corn Sorghum 19:e1177

Mavrodi DV, Mavrodi OV, Elbourne LDH, Tetu S, Bonsall RF, Parejko J, Yang M, Paulsen IT, Weller DM, Thomashow LS (2018) Long-term irrigation affects the dynamics and activity of the wheat rhizosphere microbiome. Front Plant Sci 9:345. https://doi.org/10.3389/fpls.2018.00345

Meena KK, Sorty AM, Bitla UM, Choudhary K, Gupta P, Pareek A, Singh DP, Prabha R, Sahu PK, Gupta VK, Singh HB (2017) Abiotic stress responses and microbe-mediated mitigation in plants: the omics strategies. Front Plant Sci 8:172

Meza B, de Bashan LE, Bashan Y (2015) Involvement of indole-3-acetic acid produced by *Azospirillum brasilense* in accumulating intracellular ammonium in *Chlorella vulgaris*. Res Microbiol 166(2):72–83. https://doi.org/10.1111/pce.12874

Mumtaz ZM, Ahmada M, Jamil M, Hussain T (2017) Zinc solubilizing *bacillus* spp. potential candidates for biofortification in maize. Microbiol Res 202:51–60. https://doi.org/10.1016/j.micres.2017.06.001

Nambeesan S, Datsenka T, Ferruzzi MG, Malladi A, Mattoo AK, Handa AK (2010) Overexpression of yeast spermidine synthase impacts ripening, senescence and decay symptoms in tomato. Plant J 63:836–847

Naqqash T, Imran A, Hameed S, Shahid M, Majeed A, Iqbal J, Hanif MK, Ejaz S, Malik KA (2020) First report of diazotrophic *Brevundimonas* spp. as growth enhancer and root colonizer of potatoes. Sci Rep 10(1):1–14. https://doi.org/10.1038/s41598-020-69782-6

Neilands JB (1995) Siderophores - structure and function of microbial iron transport compounds. J Biol Chem 270:26723–26726. https://doi.org/10.1074/jbc.270.45.26723

Nguyen CT, Dang LH, Nguyen DT, Tran KP, Giang BL, Tran NQ (2019) Effect of GA3 and Gly plant growth regulators on productivity and sugar content of sugarcane. Agriculture 9(7):1–13. https://doi.org/10.3390/agriculture9070136

Oliveira ALM, Canuto EDL, Urquiaga S, Reis VM, Baldani JI (2006) Yield of micropropagated sugarcane varieties in different soil types following inoculation with diazotrophic bacteria. Plant Soil 284:23–32

Oliveira ALM, Urquiaga S, Döbereiner J, Baldani JI (2002) The effect of inoculating endophytic N_2-fixing bacteria on micropropagated sugarcane plants. Plant Soil 242:205–215

Oliveira CES, Zoz T, Seron CC, Boleta EHM, Lima BH, Souza LRR, Pedrinho DR, Matias R, Lopes CS, Oliveira Neto SS, Teixeira Filho MCM (2022) Can saline irrigation improve the quality of tomato fruits? Agron J 114:1–15. https://doi.org/10.1002/agj2.21003

Ozturk A, Caglar O, Sahin F (2003) Yield response of wheat and barley to inoculation of plant growth promoting rhizobacteria at various levels of nitrogen fertilization. J Plant Nutri Soil Sci 166(2):262–266. https://doi.org/10.1002/jpln.200390038

Padda KP, Puri A, Chanway CP (2016) Plant growth promotion and nitrogen fixation in canola (*Brassica napus*) by an endophytic strain of *Paenibacillus polymyxa* and its GFP-tagged derivative in a long-term study. Botany 94(12):1209–1217. https://doi.org/10.1139/cjb-2016-0075

Pagnani G, Pellegrini M, Galieni A, D'Egidio S, Matteucci F, Ricci A, Stagnari F, Sergi M, Lo Sterzo C, Pisante M, Del Gallo M (2018) Plant growth-promoting rhizobacteria (PGPR) in *Cannabis sativa* 'Finola' cultivation: An alternative fertilization strategy to improve plant growth and quality characteristics. Ind Crop Prod 123:75–83. https://doi.org/10.1016/J.INDCROP.2018.06.033

Pál M, Tajti J, Szalai G, Peeva V, Vegh B, Janda T (2018) Interaction of polyamines, abscisic acid and proline under osmotic stress in the leaves of wheat plants. Sci Rep 8(1):12839. https://doi.org/10.1038/s41598-018-31297-6

Pallai R, Hynes RK, Verma B, Nelson LM (2012) Phytohormone production and colonization of canola (*Brassica napus* L.) roots by *Pseudomonas fluorescens* 6-8 under gnotobiotic conditions. Can J Microbiol 58(2):170–178

Pankievicz VCS, Do Amaral FP, Santos KFDN, Agtuca B, Xu Y, Schueller MJ, Arisi ACM, Steffens MBR, De Souza EM, Pedrosa FO, Stacey G, Ferrieri RA (2015) Robust biological nitrogen fixation in a model grass-bacterial association. Plant J 81:907–919. https://doi.org/10.1111/tpj.12777

Pereira SIA, Abreu D, Moreira H, Vega A, Castro PML (2020) Plant growth-promoting rhizobacteria (PGPR) improve the growth and nutrient use efficiency in maize (*Zea mays* L.) under water deficit conditions. Heliyon 6(10):e05106

Picazevicz AA, Kusdra JF, Moreno ADL (2017) Maize growth in response to *Azospirillum brasilense, rhizobium tropici*, molybdenum and nitrogen. Rev Bras Eng Agríc Ambien 21:623–627. https://doi.org/10.1590/1807-1929/agriambi.v21n9p623-627

Pieterse CM, Leon-Reyes A, Van der Ent S, Van Wees SC (2009) Networking by small-molecule hormones in plant immunity. Nat Chem Biol 5(5):308–316

Plackett AR, Wilson ZA (2018) Gibberellins and plant reproduction. Annl plant rev online, pp 323–358

Pongsilp N, Nimnoi P (2020) Inoculation of *Ensifer fredii* strain LP2/20 immobilized in agar results in growth promotion and alteration of bacterial community structure of Chinese kale planted soil. Sci Rep 10:15857. https://doi.org/10.1038/s41598-020-72986-5

Poyatos-Pertíñez S, Quinet M, Ortiz-Atienza A, Yuste-Lisbon FJ, Pons C, Giménez E (2016) A factor linking floral organ identity and growth revealed by characterization of the tomato mutant unfinished flower development. Front Plant Sci 7:1648. https://doi.org/10.3389/fpls.2016.01648

Puente M, Li C, Bashan Y (2004) Microbial populations and activities in the rhizoplane of rock-weathering desert plants II: growth promotion of cactus seedlings. Plant Biol 6:643–650. https://doi.org/10.20546/ijcmas.2019.801.013

Raffi MM, Charyulu PBBN (2021) *Azospirillum*-biofertilizer for sustainable cereal crop production: current status. In: Vishwanath B (ed) Recent developments in applied microbiology and biochemistry. Academic Press, Elsevier B.V., London UK, pp 193–209

Ramakrishna W, Yadav R, Li K (2019) Plant growth promoting bacteria in agriculture: two sides of a coin. Appl Soil Ecol 138:10–18. https://doi.org/10.1016/j.apsoil.2019.02.019

Redin M, Recous S, Aita C, Dietrich G, Skolaude AC, Ludke WH, Schmatz R, Giacomini SJ (2014) How the chemical composition and heterogeneity of crop residue mixtures decomposing at the soil surface affects C and N mineralization. Soil Biol Biochem 78:65–75. https://doi.org/10.1016/j.soilbio.2014.07.014

Reinhold B, Hurek T, Fendrik I, Pot B, Gillis M, Kersters K, Thielemans S, De Ley J (1987) Azospirillum halopraeferens sp. nov., a nitrogen-fixing organism associated with roots of Kallar grass (Leptochloa fusca (L.) Kunth). Int J Syst Evol Microbiol 37(1):43–51

Reinhold Hurek B, Hurek T, Gillis M, Hoste B, Vancanneyt M, Kersters K, Ley J (1993) Azoarcus gen. nov., nitrogen-fixing Proteobacteria associated with roots of kallar grass (Leptochloa fusca (L.) Kunth), and description of two species, Azoarcus indigens sp. nov. and Azoarcus communis sp. nov. Int J Syst Bacteriol 43:574–584

Reinhold-Hurek B, Hurek T (2000) Reassessment of the taxonomic structure of the diazotrophic genus Azoarcus sensu lato and description of three new genera and new species, Azovibrio restrictus gen. nov., sp. nov., Azospira oryzae gen. nov., sp. nov. and Azonexus fungiphilus gen. nov., sp. nov. Int J Syst Evol Microbiol 50(2):649–659

Reis VM, Estrada-De Los Santos P, Tenorio-Salgado S, Vogel J, Stoffels M, Guyon S, Mavingui P, Baldani VL, Schmid M, Baldani JI, Balandreau J (2004) Burkholderia tropica sp. nov., a novel nitrogen-fixing, plant-associated bacterium. Int J Syst Evol Microbiol 54(6):2155–2162

Reis VM, Teixeira KRS (2005) Fixação biológica do nitrogênio – Estado da arte. In: Aquino AM, Assis RL (eds) Processos biológicos no sistema solo-planta ferramentas para uma agricultura sustentável. Embrapa Informação Tecnológica, Brasília, pp 151–180. (Portuguese)

Reis Júnior, F. B.; de Machado, C. T.; Machado, A. T.; de Mendes, I.; Mehta, A. (2008) Isolamento, caracterização e seleção de estirpes de *Azospirillum amazonense* e *Herbaspirillum seropedicae* associadas a diferentes variedades de milho cultivadas no Cerrado. Embrapa Cerrados-Boletim de Pesquisa e Desenvolvimento (INFOTECA-E). (Portuguese) Planaltina: Empresa Brasileira de Pesquisa Agropecuária 36

Reis VM, Jesus E, Schwab S, Oliveira A, Olivares F, Baldani V, JI Baldani (2018) Fixação biológica de nitrogênio simbiótica e associativa. In: Fernandes MS, de Souza SR, Santos LA. Nutrição Mineral de Plantas. 2nded. Viçosa: Sociedade Brasileira de Ciência do Solo 1: 279–308. (Portuguese)

Reis VM, dos Teixeira KR (2006) Fixação Biológica de Nitrogênio – Estado da Arte. Miolo_Biotapmd, 151–180. (Portuguese)

Remigi P, Zhu J, Young JPW, Masson-Boivin C (2016) Symbiosis within Symbiosis: evolving nitrogen-fixing legume symbionts. Trends Microbiol 24:63–75. https://doi.org/10.1016/j.tim.2015.10.007

Ren XM, Guo SJ, Tian W, Chen Y, Han H, Chen E, Chen ZJ (2019) Effects of plant growth-promoting bacteria (PGPB) inoculation on the growth, antioxidant activity, cu uptake, and bacterial community structure of rape (*Brassica napus* L.) grown in cu-contaminated agricultural soil. Front Microbiol 10:1455. https://doi.org/10.3389/fmicb.2019.01455

Rey T, Dumas B (2017) Plenty is no plague: *streptomyces* symbiosis with crops. Trends Plant Sci 22(1):30–37. https://doi.org/10.1016/j.tplants.2016.10.008

Riahi L, Cherif H, Miladi S, Neifar M, Bejaoui B, Chouchane H, Masmoudi AS, Cherif A (2020) Use of plant growth growing bacteria as an efficient biotechnological tool to enhance the biomass and secondary metabolites production of the industrial crop *Pelargonium graveolens* L'Hér. Under semi-controlled conditions. Ind Crop Prod 154:112721. https://doi.org/10.1016/J.INDCROP.2020.112721

Rocha AFS, Costa RRGF (2018) Performance of *Urochloa brizantha* cv. Paiaguás inoculated with *Azospirillum brasilense* and different nitrogen doses. Global Sci Technol 11(3):177–186

Rojas-Tapias DF, Bonilla RR, Dussán J (2012) Effect of inoculation with plant growth-promoting bacteria on the growth and uptake of copper by sunflowers. Water Air Soil Pollut 223:643–654. https://doi.org/10.1007/s11270-011-0889-3

Roychoudhury A, Basu S, Sengupta DN (2011) Amelioration of salinity stress by exogenously applied spermidine or spermine in three varieties of indica rice differing in their level of salt tolerance. J Plant Physiol 168(4):317–328

Rosa PA, Galindo FS, Oliveira CE, Jalal A, Mortinho ES, Fernandes GC, Marega EM, Buzetti S, Teixeira Filho MCM (2022) Inoculation with plant growth-promoting bacteria to reduce phosphate fertilization requirement and enhance technological quality and yield of sugarcane. Microorganisms 10(1):192.[D1]

Ruppel S, Hecht-Buchholz C, Remus R, Ortmann U, Schmelzer R (1992) Settlement of the diazotrophic, phytoeffective bacterial strain Pantoea agglomerans on and within winter wheat: an investigation using ELISA and transmission electron microscopy. Plant Soil 145(2):261–273

Sá GCR, Carvalho CLM, Moreira A, Hungria M, Nogueira MA, Heinrichs R, Soares Filho CV (2019) Biomass yield, nitrogen accumulation and nutritive value of Mavuno grass inoculated with plant growth-promoting bacteria. Commun Soil Sci Plant Anal 50(15):1931–1942. https://doi.org/10.1080/00103624.2019.1648498

Saeid A, Prochownik E, Dobrowolska-Iwanek J (2018) Phosphorus solubilization by *bacillus* species. Molecules 23(11):2897

Sahib MR, Pervaiz ZH, Williams MA, Saleem M, DeBolt S (2020) Rhizobacterial species richness improves sorghum growth and soil nutrient synergism in a nutrient-poor greenhouse soil. Sci Rep 10:15454. https://doi.org/10.1038/s41598-020-72516-3

Saleem AR, Brunetti C, Khalid A, Della Rocca G, Raio A, Emiliani G, De Carlo A, Mahmood T, Centritto M (2018) Drought response of *Mucuna pruriens* (L.) DC. inoculated with ACC deaminase and IAA producing rhizobacteria. PLoS One 13(2):e0191218. https://doi.org/10.1371/journal.pone.0191218

Salvo LP, Ferrando L, Fernandéz-Scavino A, Salamone IEG (2018) Microorganisms reveal what plants do not: wheat growth and rhizosphere microbial communities after *Azospirillum brasilense* inoculation and nitrogen fertilization under field conditions. Plant Soil 424:405–417

Sampaio FAR, Teixeira Filho MCM, da Silva Oliveira CE, Jalal A, Boleta EHM, de Lima BH, Rosa PAL, Galindo FS, de Souza JS (2021) Nitrogen supply associated with rhizobacteria in the first productive cycle of Marandu grass. J Crop Sci Biotechnol 24:429–439. https://doi.org/10.1007/s12892-021-00091-8

Samuel S, Scott PT, Gresshoff PM (2013) Nodulation in the legume biofuel feedstock tree *Pongamia pinnata*. Agric Res 2:207–214. https://doi.org/10.1007/s40003-013-0074-6

Sandhya V, Ali SZ, Grover M, Reddy G, Venkateswarlu B (2010) Effect of plant growth promoting *pseudomonas* spp. on compatible solutes, antioxidant status and plant growth of maize under drought stress. Plant Growth Regul 62:21–30. https://doi.org/10.1007/s10725-010-9479-4

Santi C, Bogusz D, Franche C (2013) Biological nitrogen fixation in non-legume plants. Annal Bot 111(5):743–767. https://doi.org/10.1093/aob/mct048

Santos FL, da Silva FB, de Sá EL, Vian AL, Muniz AW, dos Santos RN (2019) Inoculation and co-inoculation of growth promoting rhizobacteria in irrigated rice plants. Rev Bras Ciênc Agrár 14(3):1–5

Santos JF, Sacramento BL, do Mota KNAB, Souza JT, de Azevedo Neto AD (2014) Sunflower growth as a function of seed inoculation with endophytic bacteria. Trop Agric Res 44(2): 142–150. https://doi.org/10.1590/s1983-40632014000200008

Santos SG, Ribeiro FS, Fonseca CS, Pereira W, Santos LA, Reis VM (2017) Development and nitrate reductase activity of sugarcane inoculated with five diazotrophic strains. Arch Microbiol 199(6):863–873. https://doi.org/10.1007/s00203-017-1357-2

Santoyo G, Moreno-Hagelsieb G, del Carmen O-MM, Glick BR (2016) Plant growth-promoting bacterial endophytes. Microbiol Res 183:92–99. https://doi.org/10.1016/j.micres.2015.11.008

Saravanan VS, Subramoniam SR, Raj SA (2003) Assessing in vitro solubilization of different zinc solubilizing bacterial (ZBS) strains. Braz J Microbiol 34:121–125

Schillaci M, Arsova B, Walker R, Smith P, Nagel KA, Roessner U, Watt M (2021) Time-resolution of the shoot and root growth of the model cereal *Brachypodium* in response to inoculation with *Azospirillum* bacteria at low phosphorus and temperature. Plant Growth Regul 93(1):149–162. https://doi.org/10.1007/s10725-020-00675-4

Schillaci M, Gupta S, Walker R, Roessner U (2019) The role of plant growth-promoting bacteria in the growth of cereals under abiotic stresses. In: Ohyama T (ed) Root biology: growth, physiology, and functions. IntechOpen, London, pp 45–66. https://doi.org/10.5772/intechopen.87083

Schultz JC, Appel HM, Ferrieri AP, Arnold TM (2013) Flexible resource allocation during plant defense responses. Front Plant Sci 4:324. https://doi.org/10.3389/fpls.2013.00324

Seldin L, Van Elsas JD, Penido EG (1984) Bacillus azotofixans sp. nov., a nitrogen-fixing species from Brazilian soils and grass roots. Int J Syst Evol Microbiol 34(4):451–456

Sergaki C, Lagunas B, Lidbury I, Gifford ML, Schäfer P (2018) Challenges and approaches in microbiome research: from fundamental to applied. Front Plant Sci 9:1205. https://doi.org/10.3389/fpls.2018.01205

Shahid M, Hameed S, Tariq M, Zafar M, Ali A, Ahmad N (2015) Characterization of mineral phosphate-solubilizing bacteria for enhanced sunflower growth and yield-attributing traits. Ann Microbiol 65:1525–1536. https://doi.org/10.1007/s13213-014-0991-z

Shahzad R, Khan AL, Bilal S, Waqas M, Kang SM, Lee IJ (2017) Inoculation of abscisic acid-producing endophytic bacteria enhances salinity stress tolerance in *Oryza sativa*. Environ Exp Bot 136:68–77

Shakeel M, Rais A, Hassan MN, Hafeez FY (2015) Root associated *bacillus* sp. improves growth, yield and zinc translocation for basmati Rice (*Oryza sativa*) varieties. Front Microbiol 6:1286. https://doi.org/10.3389/fmicb.2015.01286

Shevyakova NI, Musatenko LI, Stetsenko LA, Vedenicheva NP, Voitenko LP, Sytnik KM, Kuznetsov VV (2013) Effects of abscisic acid on the contents of polyamines and proline in common bean plants under salt stress. Russian J Plant Physiol 60(2):200–211

Silva ER, Zoz J, Oliveira CDS, Zuffo AM, Steiner F, Zoz T, Vendruscolo EP (2019) Can coinoculation of *Bradyrhizobium* and *Azospirillum* alleviate adverse effects of drought stress on soybean (*Glycine max* L. Merrill.)? Arch Microbiol 201:325–335. https://doi.org/10.1007/s00203-018-01617-5

Singh DP, Singh V, Shukla R, Sahu P, Prabha R, Gupta A, Sarma BK, Gupta VK (2020) Stage-dependent concomitant microbial fortification improves soil nutrient status, plant growth, antioxidative defense system and gene expression in rice. Microbiol Res 239:126538. https://doi.org/10.1016/j.micres.2020.126538

Singh HB, Keswani C, Reddy MS, Sansinenea E, García-Estrada C (2019) Secondary metabolites of plant growth promoting rhizomicroorganisms: discovery and applications. Springer-Nature, Singapore, p 392

Singh HB, Sarma BK, Keswani C (2017) Advances in PGPR research. CABI, UK, p 408

Singh RK, Malik N, Singh S (2013) Improved nutrient use efficiency increases plant growth of rice with the use of IAA-overproducing strains of endophytic *Burkholderia cepacia* strain RRE25. Microb Ecol 66:375–384. https://doi.org/10.1007/s00248-013-0231-2

Souza EM, Galindo FS, Teixeira Filho MCM, Silva PRT, Santos AC, Fernandes GC (2019) Does nitrogen application associated with *Azospirillum brasilense* inoculation influence with nutrition and yield? Rev Bras Eng Agríc 23:53–59. https://doi.org/10.1590/1807-1929/agriambi.v23n1p53-59

Stamenov D, Jarak M, Đurić S, Milošev D, Hajnal-Jafari T (2012) Plant growth promoting rhizobacteria in the production of English ryegrass. Plant Soil Environ 58:477–480

Sussmilch FC, McAdam SA (2017) Surviving a dry future: abscisic acid (ABA)-mediated plant mechanisms for conserving water under low humidity. Plan Theory 6(4):54. https://doi.org/10.3390/plants6040054

Tago K, Kikuchi Y, Nakaoka S, Katsuyama C, Hayatsu M (2015) Insecticide applications to soil contribute to the development of *Burkholderia* mediating insecticide resistance in stinkbugs. Mol Ecol 24:3766–3778. https://doi.org/10.1111/mec.13265

Tahami MK, Jahan M, Khalilzadeh H, Mehdizadeh M (2017) Plant growth growing rhizobacteria in an ecological cropping system: a study on basil (*Ocimum basilicum* L.) essential oil production. Ind Crop Prod 107:97–104. https://doi.org/10.1016/J.INDCROP.2017.05.020

Taiz L, Zeiger E, Moller IM, Murphy A (2017) Plant physiology and development, 6th edn. Sinauer Associates, Sunderland, p 761

Tarand JJ, Krieg NR, Döbereiner J (1978) A taxonomie study of the Spirillum lipoferum group with description of a new genus, Azospirillum gen. nov. and two species, Azospirillum lipoferum comb. nov. and Azospirillum brasilense sp. nov. Can J Microbiol 24:967–980

Tei F, De Neve S, de Haan J, Kristensen HL (2020) Nitrogen management of vegetable crops. Agric Water Manag 240:106316. https://doi.org/10.1016/j.agwat.2020.106316

Teixeira Filho MCM, Galindo FS (2019) Inoculation of bacteria with a focus on biological nitrogen fixation and plant growth promotion. In: Severiano EC, Moraes MF, Paula AM (eds) Topics in soil science, 1st edn. Brazilian Society of Soil Science, p 728

Toju H, Peay KG, Yamamichi M, Narisawa K, Hiruma K, Naito K, Fukuda S, Ushio M, Nakaoka S, Onoda Y, Yoshida K, Schlaeppi K, Bai Y, Sugiura R, Ichihashi Y, Minamisawa K, Kiers ET (2018) Core microbiomes for sustainable agroecosystems. Nat Plants 4:247–257. https://doi.org/10.1038/s41477-018-0139-4

Valetti L, Iriarte L, Fabra A (2018) Growth promotion of rapeseed (*Brassica napus*) associated with the inoculation of phosphate solubilizing bacteria. Appl Soil Ecol 132:1–10. https://doi.org/10.1016/J.APSOIL.2018.08.017

Van Deynze A, Zamora P, Delaux PM, Heitmann C, Jayaraman D, Rajasekar S, Graham D, Maeda J, Gibson D, Schwartz KD, Berry AM (2018) Nitrogen fixation in a landrace of maize is supported by a mucilage-associated diazotrophic microbiota. PLoS Biol 16(8):e2006352. https://doi.org/10.1371/journal.pbio.2006352

Vasileva K, Ivanova ZH, Tringovska I (2019) Evaluation of bacterial isolates for stimulating the growth of young melon (*Cucumis melo* L.) plants. Mech Agric Conserv Resour 6:217–219

Velasquez SM, Barbez E, Kleine-Vehn J, Estevez J (2016) Auxin and cellular elongation. Plant Physiol 170. https://doi.org/10.1104/pp.15.01863

Vendruscolo EP, Oliveira PR, Campos LFC, Seleguini A, Lima SF (2019) Planting method, nitrogen fertilization and inoculation for diazotrophic bacteria for cantaloupe melon plants. Rev Colomb Cienc Hortic 13(1):e9720. https://doi.org/10.17584/rcch.2019v13i1.9720

Venter ZS, Jacobs K, Hawkins H-J (2016) The impact of crop rotation on soil microbial diversity: a meta-analysis. Pedobiologia 59:215–223. https://doi.org/10.1016/j.pedobi.2016.04.001

Viruel E, Erazzú LE, Martínez Calsina L, Ferrero MA, Lucca ME, Siñeriz F (2014) Inoculation of maize with phosphate solubilizing bacteria: effect on plant growth and yield. J Soil Sci Plant Nutri 14(4):819–831. https://doi.org/10.4067/S0718-95162014005000065

Wang J, Li R, Zhang H, Wei G, Li Z (2020a) Beneficial bacteria activate nutrients and promote wheat growth under conditions of reduced fertilizer application. BMC Microbiol 20:38. https://doi.org/10.1186/s12866-020-1708-z

Wang M, Bian Z, Shi J, Wu Y, Yu X, Yang Y, Ni H, Chen H, Bian X, Li T, Zhang Y (2020b) Effect of the nitrogen-fixing bacterium Pseudomonas protegens CHA0-ΔretS-nif on garlic growth under different field conditions. Ind Crop Prod 145:111982. https://doi.org/10.1016/J.INDCROP.2019.111982

Wang Z, Li Y, Li T, Zhao D, Liao Y (2020c) Conservation tillage decreases selection pressure on community assembly in the rhizosphere of arbuscular mycorrhizal fungi. Sci Total Environ 710:136326. https://doi.org/10.1016/j.scitotenv.2019.136326

Wang X, Cai D, Ji M, Chen Z, Yao L, Han H (2022) Isolation of heavy metal-immobilizing and plant growth-promoting bacteria and their potential in reducing cd and pb uptake in water spinach. Sci Total Environ 819:153242. https://doi.org/10.1016/j.scitotenv.2022.153242

Weiss EA (2000) Oil seed crops. In: World Agriculture Science, 2nd edn. Black Well Science Ltd., Paris, Berlin, p 365

Wiangkham N, Prapagdee B (2018) Potential of Napier grass with cadmium-resistant bacterial inoculation on cadmium phytoremediation and its possibility to use as biomass fuel. Chemosphere 201:511–518. https://doi.org/10.1016/J.CHEMOSPHERE.2018.03.039

Wolmarans K, Swart WJ (2014) Influence of glyphosate, other herbicides and genetically modified herbicide-resistant crops on soil microbiota: a review. S Afr J Plant Soil 31:177–186. https://doi.org/10.1080/02571862.2014.960485

Wu AL, Jiao XY, Wang JS, Dong EW, Guo J, Wang LG, Sun AQ, Hu HW (2021) Sorghum rhizosphere effects reduced soil bacterial diversity by recruiting specific bacterial species under low nitrogen stress. Sci Total Environ 770:144742. https://doi.org/10.1016/j.scitotenv.2020.144742

Yamada Y, Hoshino K, Ishikawa T (1997) The phylogeny of acetic acid bacteria based on the partial sequences of 16S ribosomal RNA: the elevation of the subgenus Gluconoacetobacter to the generic level. Biosci Biotechnol Biochem 61:1244–1251

Yang X, Feng Y, He Z, Stoffell PJ (2005) Molecular mechanisms of heavy metal hyperaccumulation and phytoremediation. J Trace Elem Med Biol 18:339–353. https://doi.org/10.1016/j.jtemb.2005.02.007

Yeoh YK, Paungfoo-Lonhienne C, Dennis PG, Robinson N, Ragan MA, Schmidt S, Hugenholtz P (2016) The core root microbiome of sugarcanes cultivated under varying nitrogen fertilizer application. Environ Microbiol 18:1338–1351. https://doi.org/10.1111/1462-2920.12925

You CB, Song W, Wang HX, Li JP, Lin M, Hai WL (1991) Association of Alcaligenes faecalis with wetland rice. Plant Soil 137(1):81–85

Yuzikhin OS, Gogoleva NE, Shaposhnikov AI, Konnova TA, Osipova EV, Syrova DS, Ermakova EA, Shevchenko VP, Nagaev IY, Shevchenko KV, Myasoedov NF, Safronova VI, Shavarda AL, Nizhnikov AA, Belimov AA, Gogolev YV (2021) Rhizosphere bacterium *Rhodococcus* sp. P1Y metabolizes abscisic acid to form dehydrovomifoliol. Biomolecules 11(3):345. https://doi.org/10.3390/biom11030345

Zafar-ul-Hye M, Tahzeeb-ul-Hassan M, Abid M, Fahad S, Brtnicky M, Dokulilova T, Datta R, Danish S (2020) Potential role of compound mixed biochar with rhizobacteria in mitigating lead toxicity in spinach. Sci Rep 10:12159. https://doi.org/10.1038/s41598-020-69183-9

Zaheer MS, Raza MAS, Saleem MF, Khan IH, Ahmad S, Iqbal R, Manevski K (2019) Investigating the effect of *Azospirillum brasilense* and *rhizobium pisi* on agronomic traits of wheat (*Triticum aestivum* L.). Arch Agron Soil Sci 65:1554–1564. https://doi.org/10.1080/03650340.2019.1566954

Zeffa DM, Fatin LH, Santos OJAP, Oliveira ALM, Canteri MG, Scapim CA, Gonçalves LSA (2018) The influence of topdressing nitrogen on *Azospirillum* spp. inoculation in maize crops through meta-analysis. Bragantia 77:493–500. https://doi.org/10.1590/1678-4499.2017273

Zhang H, Hanada S, Shigematsu T, Shibuya K, Kamagata Y, Kanagawa T, Kurane R (2000) Burkholderia kururiensis sp. nov., a trichloroethylene (TCE)-degrading bacterium isolated from an aquifer polluted with TCE. Int J Syst Evol Microbiol 50(2):743–749

Zhang J, Pan Y, Zheng C, Gao X, Wei X, Xi J, Peng T, Shang Q (2016) Rapid evolution of symbiotic bacteria populations in spirotetramat-resistant *Aphis gossypii* glover revealed by pyrosequencing. Comp Biochem Physiol Part D Genomics Proteomics 20:151–158. https://doi.org/10.1016/j.cbd.2016.10.001

Chapter 12
Harnessing Cereal–Rhizobial Interactions for Plant Growth Promotion and Sustainable Crop Production

Swati Tyagi, Kedharnath Reddy, Koj Haniya, Karivaradharajan Swarnalakshmi, Murugesan Senthilkumar, Upendra Kumar, and Kannepalli Annapurna

Abstract Rhizobia are known to establish symbiotic association with legume crops, and develop root nodules, a specific niche for N_2 fixation. The interaction between rhizobia and cereal crops does not elicit to nodulation or nitrogen fixation but found to exhibit the plant growth-promoting characteristics and positively influence growth and yield by direct and indirect means. They can directly promote plant growth in cereal crops by producing plant hormones such as auxin, gibberellin, abscisic acid, and cytokinin, as well as lowering plant ethylene levels by producing the enzyme 1-aminocyclopropane-1-carboxylic acid (ACC) deaminase and providing bioavailable phosphorus and iron for plant uptake. They can also indirectly promote the plant growth of cereal crops by inhibiting the growth of pathogens by removing the iron in the rhizosphere with siderophore production, by releasing the antibiotics, and/or by producing cell wall degrading enzymes. Rhizobia forms endophytic association with cereal crops without forming any structure such as nodules or causing any symptoms of the disease. They enter through crack entry and colonize the intercellular space and xylem tissues. Inoculation of rhizobia imparts more tolerant toward biotic and abiotic stress and helps sustainable cereal crop production.

Keywords *Rhizobium* · Cereals · Endophyte · Biocontrol · Plant growth promotion

---

S. Tyagi · K. Reddy · K. Haniya · K. Swarnalakshmi · K. Annapurna (✉)
Division of Microbiology, ICAR-Indian Agricultural Research Institute, New Delhi, India

M. Senthilkumar
Division of Basic Sciences, ICAR-Indian Institute of Pulse Research, Kanpur, Uttar Pradesh, India

U. Kumar
Crop Production Division, ICAR-National Rice Research Institute, Cuttack, Odisha, India

© The Author(s), under exclusive license to Springer Nature Singapore Pte Ltd. 2022

D. K. Maheshwari et al. (eds.), *Nitrogen Fixing Bacteria: Sustainable Growth of Non-legumes*, Microorganisms for Sustainability 36,
https://doi.org/10.1007/978-981-19-4906-7_12

12.1 Introduction

Field crops are becoming more intensive to fulfill human food demands and save renewable energy supplies. According to the UN Food and Agriculture Organization (FAO), global agricultural product demand would increase to 60% by 2030. To meet an ever-growing need for food, the world has relied on increased agricultural yields for almost half a century. Application of balanced fertilizer on a timely basis will improve sustainable food production. Nitrogen fertilizer is one of the most essential factors in producing high yields from cereal crops. As a result, farmers are using large amounts of fertilizers, which are expensive and detrimental to the environment, especially when used in an indiscriminate manner. The use of biofertilizers containing predominantly N-fixing microorganism(s) may minimize the need for synthetic nitrogen fertilizers by increasing plant N absorption. These inoculants increase the nitrogen availability of cereal crops via biological nitrogen fixation (BNF) (Ladha and Reddy 1995; Rogers and Oldroyd 2014). Most common free-living N-fixing (diazotrophs) microorganisms that form natural associations with cereal crops include *Azotobacter* (Gerlach and Vogel 1902), *Azospirillum* (Boddey et al. 1986), *Beijerinckia* spp. (Govedarica 1990), *Herbaspirillum* (Baldani et al. 2000), *Gluconacetobacter diazotrophicus* (Boddey et al. 1991), *Azoarcus* (Reinhold-Hurek and Hurek 1997), and N contribution by diazotrophs in cereal budget is only 10% (Ladha et al. 2016).

The levels of N_2 fixed by associative diazotrophs in cereals are modest and inefficient while compared to biologically fixed nitrogen contributed by legume–rhizobia interaction under favorable conditions (Lupwayi et al. 2006; Saikia and Jain 2007; Swarnalakshmi et al. 2020). Rhizobia form a nitrogen-fixing symbiosis with legume plants belonging to diverse groups of α- and β-proteobacteria and fix nitrogen in a host-specific manner. *Bradyrhizobium* inoculation significantly increases fixed nitrogen in sweet corn and cotton (McInroy and Kloepper 1995) as bradyrhizobial strains are able to fix nitrogen in free-living conditions. Plant growth-promoting *Rhizobium leguminosarum* bv. *phaseoli*, *Bradyrhizobium japonicum*, *Rhizobium leguminosarum* bv. *trifolii*, *Rhizobium leguminosarum* bv. *viciae*, and *Sinorhizobium* are known to colonize cereal crops (Antoun et al. 1998). Parasponia, a nonlegume woody member of the dicotyledonous elm family, is confined to rhizobial infection and develops an efficient nodular nitrogen-fixing symbiosis (Webster et al. 1995). *Azorhizobium caulinodans* ORS571, a diazotrophic bacteria that forms nodulation in *Sesbania*, can colonize rapeseed (O'Callaghan et al. 2000), which emphasize that the plant–rhizobial interaction is highly promiscuous.

Rhizobia also act as a PGPR (plant growth-promoting rhizobacteria), enhance nonleguminous plant development by producing phytohormones such as auxin, gibberellic acid, and cytokinin, which improve root architecture and stimulate water and mineral intake from the soil (Gopalakrishnan et al. 2015; Vargas et al. 2017; Jaiswal et al. 2021). As a PGPR, rhizobium also increases the solubility and availability of nutrients by phosphate solubilization, organic acid production, and siderophore production (Bardin et al. 1996; Marra et al. 2012; Hu et al. 2018).

Rhizobia also inhibit many soilborne pathogens through various biocontrol mechanisms. Biocontrol potential of several rhizobial genera including *Sinorhizobium meliloti*, *Bradyrhizobium japonicum*, *Rhizobium leguminosarum* bv. *phaseoli*, *Rhizobium leguminosarum* bv. *trifolii* R39, *Rhizobium* sp. NBR19513 against *Macrophomina phaseolina*, *Phytophthora megasperma*, *Fusarium oxysporum*, *Sclerotium rolfsii*, *Rhizoctonia bataticola*, *Pythium* sp., *Fusarium* sp., *Helminthosporium sativum*, *Gaeumannomyces graminis* have been reported (Nautiyal 1997; Deshwal et al. 2003). Production of siderophores, HCN, toxins (antibiotics), hydrolytic enzymes (chitinase degrade the cell wall of the pathogenic fungi) suppresses the broad spectrum of pathogens (Ahmad et al. 2008; Igiehon et al. 2019). The current focus is on figuring out how rhizobia might help to improve the growth and productivity of cereals, as well as the processes involved in nonlegume–rhizobia interactions. This chapter will cover the rhizobium–nonlegume interaction, their plant growth-promoting mechanisms, and applications of this group of organisms in cereals, and the synergism involved in this plant–microbe interaction.

12.2 Rhizobium and Non-legume Interaction

Rhizobia are rod-shaped soil bacteria that are gram-negative, chemolithotrophic in nature (Werner 1992). Beijerinck (1988) isolated the first bacterium nodulating the legume, named *Bacillus radicicola* and then renamed as *Rhizobium leguminosarum*. Rhizobia belong to the Rhizobiaceae family, which are a perfect example of mutualism. *Allorhizobium*, *Azorhizobium*, *Mesorhizobium*, *Bradyrhizobium*, *Rhizobium*, and *Sinorhizobium* are common genera of rhizobia (Young 1996). They are aerobic motile and nonspore producing, having the exclusive ability to infect legumes and form root nodules to fix the atmospheric nitrogen via symbiotic relationship (Schloter et al. 1997). The interaction between plant and rhizobia may either closely associated or loosely associated. They would be rhizosphere (present in the rhizosphere), phyllosphere (resides on leaves, stem, fruits), and endophytic (exist in the internal tissue of plant). In nonleguminous plants, rhizobia colonize as endophytes and live inside the root tissue for the rest of their lives without causing any disease symptoms (Chabot et al. 1996; Hussain et al. 2009; Matiru et al. 2000; Hilali et al. 2001; Yanni et al. 2001; Peng et al. 2002; Lupwayi et al. 2004).

Rhizobia have been encountered as an endophyte in diverse range of cereal crops. They form endophytic association with rice (Roger and Watanabe 1986; Ladha et al. 1989; Singh et al. 2006), wheat (Hilali et al. 2001; Lupwayi et al. 2004), maize (Schloter et al. 1997; Gutierrez-Zamora and Martınez-Romero 2001; Rosenblueth and Martinez-Romero 2004; Mehboob et al. 2008), corn (Cassan et al. 2009), barley (Lupwayi et al. 2004; Peix et al. 2001) and promote plant growth and productivity. The practice of cereal–legume rotation might have promoted endophytic colonization of indigenous rhizobia in nonlegume crops. *Rhizobium leguminosarum* bv. *trifolii* found to form endophytic association with rice roots under field conditions when the crop has been grown in rotation with the clover for many generations

(Yanni et al. 2001). A positive correlation between rhizobial population on wheat roots with N accumulation and yield was observed when the crop was grown after pea (Lupwayi et al. 2004). Similarly, *Rhizobium etli* bv. *phaseoli* found to colonize maize roots and promote plant growth under maize–bean rotation (Gutierrez-Zamora and Martınez-Romero 2001). Photosynthetic *Bradyrhizobium* was found to be a true endophyte of rice when the crop was grown in the same wetland site of aquatic legume (Chaintreuil et al. 2000). Inoculation of *Azorhizobium caulinodans* accumulated increased levels of phytohormones and improved root biomass in rice (Chi et al. 2005). As *Azorhizobium* and photosynthetic *Bradyrhizobium* invade their legume host by nod independent pathway, they dwell as endophyte in cereal crops. A lot of experimental evidence is available on rhizobial-associated yield enhancement in cereal crops (Table 12.1).

12.3 Root Colonization and Nodule Formation of *Rhizobium* in Nonlegume Plants

The presence of rhizobia at the time of germination and root development performs similarly with nonlegumes and legumes (Pena-Cabriales and Alexander 1983). Rhizobia not only colonize the rhizoplane of cereals but also colonize the intercellular space of the root cap (Wiehe et al. 1994). However, a penetration of rhizobia in the cortex cells, within the xylem and in root meristem of wheat, was reported by Sabry et al. (1997). Rhizobia and nonlegume interaction vary from variety to variety due to their variation in the root-exudate composition and different soil ecology. The process of colonization starts from the rhizosphere to the apoplastic region of root, then further colonizes the intercellular space before colonizing the vessels where the main colonization takes place (James 2000). *Azorhizobium* spp. can directly enter into the intercellular region of the cortex by the cuts and wounds present in the lateral roots of rice (Cocking et al. 1994; Jain and Gupta 2003). Transmission electron microscopy with high magnification and resolution explains the colonization of azorhizobia with the rice roots in the intercellular spaces of the root cortex (Reddy et al. 1997). Rice-adapted rhizobial strains tagged with the green fluorescent protein (GFP) confirmed that their mode of entry is via the lateral root emergence site, and interior of the root epidermal cells, followed by the movement up to stem and leaf sheath (Chi et al. 2005). *Rhizobium leguminosarum* bv. *phaseoli* strains tagged with lux genes also showed in situ colonization of bacteria in maize roots (Chabot et al. 1996). Intercellular colonization *Azorhizobium caulinodans* ORS571 at later root cracks (LRC) and xylem tissues of rice was detected using *lacZ* reporter gene system (Gopalaswamy et al. 2000).

The interaction between the rhizobia and nonleguminous plants results in the production of metabolites that help in the improvement of seed germination, root elongation, root architecture, shoot growth, photosynthetic activity, leaf area, grain yield, nutrient uptake, and tolerance to abiotic stress (Hilali et al. 2001; Hafeez et al.

Table 12.1 Influence of rhizobial inoculation on plant growth and yield of cereal crops

Host plants	Rhizobial genera	Percentage increase in growth/yield	References
Rice	*Bradyrhizobium*	23–59% root/shoot weight	Chaintreuil et al. (2000)
		15–22% grain yield	Bhattacharjee et al. (2012)
			Greetatorn et al. (2019)
Rice	*Rhizobium leguminosarum* bv. *trifolii*	18% biomass	Yanni et al. (1997, 2001)
		43% yield	Biswas et al. (2000),
			Afify et al. (2019)
Rice	*Rhizobium leguminosarum*	Shoot dry matter by 24%	Hussain et al. (2009)
			Jha et al. (2020)
Rice	*Azorhizobium caulinodans* ORS 571, *Sinorhizobium meliloti* 1021, and *Mesorhizobium huakui* 93	Improved growth and yield	Chi et al. (2005)
Rice	*Bradyrhizobium*, *Rhizobium*	Improved early growth and seedling vigor	Tan et al. (2014)
Rice	*Sinorhizobium meliloti* 1021	Accelerating cell division and expansion in seedlings	Wu et al. (2018)
Wheat	*Rhizobium leguminosarum* bv. *trifolii*	34% yield	Hilali et al. (2001)
	Rhizobium sp.		Adnan et al. (2014)
			Ullah et al. (2017)
Maize	*Rhizobium leguminosarum* bv. *trifolii*	34.3–43.4% dry weight	Riggs et al. (2001)
	Sinorhizobium sp.		
Maize	*Rhizobium radiobacter*	15–25.73% dry weight	Singh et al. (2020)
	Rhizobium leguminosarum bv. *phaseoli*	30% Yield	Chabot et al. (1996)
Barley	*Mesorhizobium mediterraneum*	Plant biomass by 56%, P uptake by 100%	Peix et al. (2001)
Sorghum	*Bradyrhizobium japonicum*	8–55% shoot dry weight	Matiru et al. (2005)
			Wasai-Hara et al. (2020)

2004; Siddiqui 2007; Reimann et al. 2008). The interaction of rice and rhizobia is differed with variety due to the secretion of the specific type of root exudates. The rhizobial infection disrupts cytoplasmic membrane, which induces the production of phenolic compounds in rice roots. These phenolic compounds such as gallic, tannic, ferulic, and cinnamic acids accumulate in leaves and actively participate in the stress response in vivo. The flavanone naringenin improves the intercellular colonization in rice roots by *Azorhizobium caulinodans*. The application of naringenin at a low concentration (10^{-5} M) increased the lateral roots in rice and rhizobial colonization in xylem (Gopalaswamy et al. 2000). Inoculation of wheat plants with *Azorhizobium caulinodans* ORS571 induced 2,4-dichlorophenoxyacetic acid and formed paranodules (Liu et al. 2017). Tagging this strain with GFP label showed the rhizobial infection in the paranodules.

Al-Mallah et al. (1989) successfully performed the induction of nodule in rice by treating 2-day-old rice seedling roots with cell wall degrading enzyme followed by rhizobium inoculation in the presence of polyethylene glycol. The interaction of genetically engineered Rhizobium transconjugants with maize and rice seedlings resulted in root hair curling but not nodulation (Plazinski et al. 1985). When rice seedling roots are treated with a cellulase–pectolyase enzyme mixture and infected with either *Rhizobium* or *Bradyrhizobium* in the presence of polyethylene glycol, nodular formations appeared on the roots. Rhizobia with nodulated legumes are the foremost example of endosymbiosis and it is the most significant nitrogen fixation mechanism in agriculture (Dent and Cocking 2017). The nodule-like structure formed by rhizobia could not help in the nitrogen fixation in cereal crops (Al-Mallah et al. 1989; Bender et al. 1990; De Bruijn et al. 1995; Jing et al. 1992; Li et al. 1991; Rolfe and Bender 1990; Trinick and Hadobas 1995; Naidu et al. 2004).

12.4 Nitrogen Fixation in Nonlegumes

Nitrogen is the most vital nutrient that is required for metabolic function as well as for optimal growth and yield. The phenomenon of biological nitrogen fixation (BNF) in cereals and leguminous plants is well known. N contribution of free-living diazotrophs in cereal budget is low due to poor bacterial nitrogen release. Endophytic diazotrophs have an advantage over free-living N fixers, in which the former can enjoy direct provision to carbon substrates within the host (Dobereiner 1992; Boddey et al. 1995a, b). The low pO_2 factor also facilitates better expression of nitrogenase and N fixation (Baldani et al. 1997). In sugarcane, endophytic diazotrophic bacteria contribute 30–80 kg N ha^{-1} year^{-1} (Boddey et al. 1995a, b). Endophytic *Azoarcus* sp. isolated from Kallar grass, profusely colonize and express *nif* genes in rice roots (Engelhard et al. 2000) and the endophytic diazotrophs contribute 16–24% of the total nitrogen in cereal crops (Ladha et al. 2016; Keymer and Kent 2014). Several rhizobial inoculation tests with nonlegumes failed to reveal a significant contribution of biological nitrogen fixation on plant growth and development. Although it was observed that shoot-N and grain-N in rhizobium inoculated rice plants, majority of the increased combined nitrogen is derived from soil mineral nitrogen rather than biological nitrogen fixation (Biswas et al. 2000; Yanni et al. 2001). Sabry et al. (1997) reported increased dry weight and nitrogen contents in wheat inoculated with *Azorhizobium caulinodans*. In a hydroponic experiment, Naidu et al. (2004) discovered nitrogenase activity in rice plants inoculated with *Azorhizobium caulinodans*. Various attempts to extend *Rhizobium*'s host range beyond legumes to nonlegumes through plant genetic modification have had little or no success in inducing symbiosis between cereals and diazotrophs (Saikia and Jain 2007). According to Velazquez et al. (2005), the coexistence of symbiosis and pathogenicity-determining genes was found in *Rhizobium rhizogenes* strains, allowing them to create nodules or tumors depending on plant species. Rice roots were treated with a cell wall degrading enzyme combination including 1% cellulase YC, 0.1% pectolyase Y23, and 8% mannitol, and then inoculated with rhizobia in the presence of polyethylene glycol to produce nodules. Naringenin, a flavonoid, influenced the colonization of *Azorhizobium caulinodans* in rice roots systems (Shamala et al. 2018). The criteria for successful nitrogen fixation are most likely to be met only in endophytic systems (Quispel 1991). Another straightforward strategy to improve BNF in cereal crops is based on the finding that naturally occurring nonrhizobial nitrogen-fixing bacteria that fix nitrogen under N-deficient conditions are reported in sugarcane. Such nonrhizobial endophytic diazotrophs can infect the root systems of cereals and form intracellular symbiosis with host plants and fix nitrogen (Dent and Cocking 2017). However, in contrast to root-nodulating rhizobia, nonrhizobial diazotrophs assimilate ammonia for their growth instead of providing N-rich compounds to the host (James 2000). The use of mutant strains such as ammonium excreting *Azospirillum* (Schnabel and Sattely 2021) and *Azotobacter* (Bageshwar et al. 2017) improved N availability and growth in cereal crops. The genetically engineered recombinant N-fixing *Pseudomonas protegens* Pf-5, able to fix nitrogen constitutively, increased soil ammonium in maize under N-deficient conditions (Setten et al. 2013).

12.5 Plant Growth Enhancement: Mode of Action

Endophytic rhizobacteria act as biofertilizers and bioenhancer for a variety of nonlegumes and enhance plant growth and yield in cereal crops plants (Peix et al. 2001; Lupwayi et al. 2004; Yanni and Dazzo 2010; Wu et al. 2018). *Rhizobium leguminosarum* isolated from red clover nodules increases the growth and production of nonlegumes such as maize, wheat, barley, and radish by 10%, 8% 16%, and 21%, respectively (Hoflich 2000). *Rhizobium* significantly increases the plant growth of cereals by improving seed germination, radicular growth and aerial portion, radical elongation, plumule length, cumulative leaf and root areas, and grain yield (Yanni and Dazzo 2010; Hemissi et al. 2011). It was found that rice plants inoculated with rhizobia had increased N uptake (Yanni et al. 1997), raising

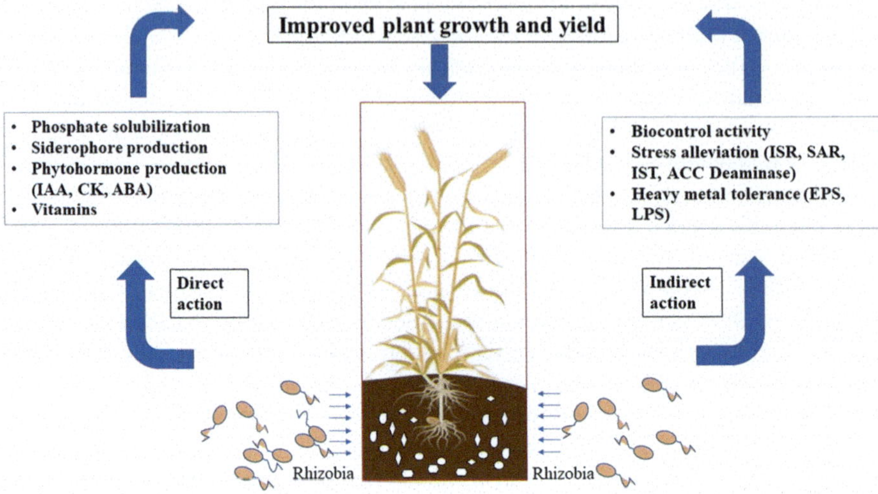

Fig. 12.1 Rhizobial association improves plant growth and yield in cereal crops

the question of whether this rhizobial benefit is due to their associative N_2 fixing activity and/or their ability to change the phytohormone balance, which impacts growth physiology.

Certain mechanisms are postulated toward *Rhizobium*, which may be involved in growth-promoting activities, i.e., mobilization and efficient uptake of nutrients, enhancement of stress tolerance, solubilization of insoluble phosphate, induction of systematic disease resistance, production of phytohormones, vitamins, and siderophores (Biswas et al. 2000; Mayak et al. 2004; Alikhani et al. 2006; Dakora et al. 2015; Dobbelaere et al. 2003). The beneficial contribution of rhizobia–cereal crop growth and yield has been shown through multiple mechanisms that influence growth physiology and yield of cereals (Fig. 12.1). The basic mechanisms behind the rhizobia–nonlegume interaction which improved the plant growth and yield production were through phytohormone production and nutrient (P and Fe) solubilization, as well as biocontrol potential (Franks et al. 2006). Antimicrobial metabolites, cell wall disintegrating enzymes, siderophores, and nutritional competition have all been involved in the biocontrol process.

Cellular and molecular bases for beneficial cereal–rhizobia interaction were revealed by several researchers, and plant growth responses were induced by the rhizobia in cereals via bacterial synthesis of plant growth-stimulating substances (Wu et al. 2018). Phenotypic analyses revealed that rice seedlings inoculated with live cells of *Sinorhizobium meliloti* strain 1021 had improved plant growth by mediating long-distance signaling at early stages of plant growth. Significant cytological differences including enlargement of parenchyma cells and reduction in shape complexity were observed in rice–rhizobial association. Transcriptomic analysis of shoots showed that upregulation of 46 differentially expressed genes (DEGs) involved in phytohormone production, photosynthetic efficiency, carbohydrate metabolism, cell division, and wall expansion. These cellular changes are in consistent with the observed phenotypic changes in rice cell morphology and shoot growth (Wu et al. 2018). These findings suggest the involvement of molecular crosstalk during rhizobial colonization in rice. However, the rhizobial–cereal interaction may support or inhibit the crop growth or remain as commensal, without causing any effect on the nonleguminous plants. The strains of plant growth-promoting rhizobacteria improve the plant growth and yield in nonlegumes by direct and indirect modes of action that depicted below.

12.5.1 Direct Growth Promotion in Plants

Phyto-effective metabolites produced by the rhizobacteria cause the direct growth promotion in nonlegumes in the absence of pathogens (Lugtenberg and Kamilova 2009). Better root colonization, synthesis of phytohormones, siderophores, organic acids, enhanced nutrient absorption, and induced systemic resistance might be linked to plant-beneficial traits of rhizobium species. Rhizobia can directly promote nonleguminous plant growth by producing plant hormones such as auxins, gibberellins, abscisic acid, and cytokinins, as well as lowering plant ethylene levels by producing the enzyme 1-aminocyclopropane-1-carboxylic acid (ACC) deaminase, providing bioavailable phosphorus for plant uptake, conserving iron for plants via siderophores production, and other useful compounds such as lipo-chitooligosaccharides (LCOs) and riboflavin.

12.5.1.1 Plant Growth Hormone Production

Rhizobial phytohormones triggered changes in root architecture and physiology, resulting in higher nutrient and water intake from the soil (Mantelin and Touraine 2004). Phytohormones are synthesized endogenously by plants and have beneficial effects on its growth and development. Five major groups of phytohormones are auxins, gibberellins, cytokinins, ethylene, and abscisic acid (ABA). Indole-3-acetic acid, cytokinins, gibberellins, and abscisic acid are produced by rhizobia in response to seed or root inoculation with nonleguminous plants. Early research revealed that the auxin indoleacetic acid (IAA) was secreted by rhizobial endophyte in gnotobiotic rice culture. Several species of *Rhizobium* are capable of synthesizing IAA, which affect plant functions such as cell elongation and cell division, apical dominance, root initiation, vascular tissue differentiation, ethylene biosynthesis mediation of tropic responses, and altering the expression of specific genes to influence root development (Warren and Warren 1993). IAA-producing rhizobial strains belong to *Bradyrhizobium japonicum*, *Azorhizobium caulinodans*, *Bradyrhizobium elkanii*, *Rhizobium japonicum*, *Mesorhizobium loti*, *Rhizobium leguminosarum*, *Rhizobium lupine*, *Rhizobium meliloti*, *Rhizobium phaseoli*, *Rhizobium trifolii*, and *Sinorhizobium* spp. (Ullah et al. 2019; Weyens et al. 2009). IAA improves shoot and root growth and seedling vitality and promotes nutrient absorption by increasing root surface area. As a result, the roots release chemicals in the form of root exudates into the rhizosphere and promote microbial population growth and its interaction with plant roots for disease suppression and plant stimulation (Glick 2012). Inoculation of *Rhizobium* and L-Tryptophan-a precursor for IAA synthesis resulted in the presence of IAA at rhizosphere and support for the development of stronger root system in maize (Qureshi et al. 2013). The cereal–rhizobia interaction may have some deleterious effects on plants due to the overproduction of phytohormones and some toxic metabolites. The overproduction of IAA and related compounds suppresses plant growth (Antoun et al. 1998).

Different *Rhizobium* strains associated with rice also produce cytokinins (CK) and gibberellic acid (GA) (Phillips and Torrey 1972; Molla et al. 2001; Jarzyniak et al. 2021). *Bradyrhizobium japonicum* 61A68 (Sturtevant and Taller 1989) has been found to release cytokinin into the medium in pure culture. Bacterial CK appeared to shorten root length, but increase total root mass, indicating a swelling of root axes (Vogel et al. 1998). Cytokinins were quickly loaded into the xylem vessels directly and accumulated mostly in shoots of inoculated plants and not in roots. Cytokinin can interact with plant tissues and stimulate cell division and induce polyploid mitoses (Caba et al. 2000), and rhizobial-secreted CK are essential for nodulation (Heckmann et al. 2011; Kisiala et al. 2013). However, for Nod-independent symbiosis in *Aeschynomene* plants, CK-synthesized *Bradyrhizobium* sp. ORS285 is not essential (Podlesakova et al. 2013).

GA is secreted by a broad range of bacterial strains of *Rhizobium* and *Sinorhizobium meliloti*, and its relevance in plant cell elongation and seed germination was revealed (Boiero et al. 2007). Abscisic acid (ABA) is produced by *Rhizobium* sp. and *Bradyrhizobium japonicum* when they colonize plant root systems (Boiero et al. 2007). Abscisic acid travels via the xylem and phloem, stimulating root development, inhibiting shoot growth, inducing proteinase inhibitors, and thereby activating the defense system (Mauch-Main and Mauch 2005).

In the absence of auxins and cytokinins, rhizobia release lipo-chitooligosaccharides (LCOs) that can repair or continue cell division and embryogenesis in nonlegumes. Application of LCOs to the rhizosphere at low concentrations (10^{-7} to 10^{-9} M) can improve seed germination, early seedling development, root mass, and root length in nonlegumes, whereas application of LCOs to the leaves at micromolar concentrations (10^{-6}, 10^{-8}, or 10^{-10} μM) can boost photosynthate production and grain yield in nonlegumes (Mehboob et al. 2012).

Besides phytohormones, riboflavin produced by *Sinorhizobium meliloti* and *Rhizobium leguminosarum* bv. *viciae* is converted into lumichrome either enzymatically or photochemically (Yang et al. 2002; Dakora et al. 2002). Lumichrome, a by-product of riboflavin degradation, is a signaling molecule found in the culture filtrate of *Sinorhizobium meliloti* that can stimulate the growth of nonlegumes by increasing root respiration (Dakora et al. 2015).

12.5.1.2 Phosphate Solubilization

Depending on the soil and pH, a considerable amount of organic and inorganic form of soil phosphorus, as well as applied phosphorus immobilized in soil, becomes inaccessible to plants. Waterlogged conditions of rice cultivation resulted in poor soil-phosphorus (<8 ppm) availability to the crop. In situ solubilization of complex forms of phosphate by rhizobacteria enhances the phosphate availability to cereal crops. The organic form of bound phosphorus is mineralized by rhizobial strains through releasing phosphatases and inorganic bound form is solubilized by releasing organic acids, such as 2-ketogluconic acid, glutamic acid, sulphuric acid, nitric acid, and carbonic acids (Alori et al. 2017). Organic acids bind to the cation of the phosphate complexes and make soluble form of phosphate through hydroxyl and carboxyl groups. Phosphate mineralization requires microbial enzymes such as acid phosphatases, phosphohydrolases, phytase, phosphonoacetate hydrolase, D-a-glycerophosphate, and C–P lyase. *Rhizobium leguminosarum*, *Rhizobium meliloti*, *Mesorhizobium mediterraneum*, *Bradyrhizobium japonicum*, and *Bradyrhizobium* sp. are among the rhizobial species capable to solubilize or mineralize phosphate in the rhizosphere of nonlegumes. Inoculation of phosphate-solubilizing rhizobia in cereal crops may aid on improving P-acquisition and use efficiency in agriculture system where naturally or synthetically generated P resources are used.

12.5.1.3 Siderophore Production

In an iron-deficient environment, rhizobial retention of iron by producing siderophore is considered as a source of available iron for plants. Plants require Fe for chlorophyll production; however, Fe is usually found in the environment in the form of ferric hydroxide, which is highly insoluble. Siderophores are low-molecular-weight compounds that are utilized to mobilize iron and are used to retain Fe^{3+} due to their high Fe^{3+} affinity constants (Plessner et al. 1993). Siderophore production may solve a dual purpose for PGP: increasing plant Fe absorption and inhibiting rhizosphere pathogens that are unable to use the Fe–siderophore complex. Siderophore production by *Rhizobium* is strain specific (Smith and Neilands 1984) and it provides competitive advantage to *Bradyrhizobium* under iron-limited environment (Fuhrmann and Wollum 1989). In addition, siderophore production also reduces the availability of rhizosphere iron and inhibits the pathogen growth. Rhizobia and plants can cross-utilize siderophore produced by other species (Plessner et al. 1993). Experiments by Yanni et al. (2001) found no siderophore production on chrome azurol sulfonate (CAS) agar, leaving the role of siderophore in *Rhizobium*–rice interaction undetermined. However, siderophore-producing rhizobial isolates were discovered in the root system or the vicinity of the rice rhizosphere. A variety of *Rhizobium* strains were found to produce siderophores that bind with insoluble Fe^{3+} and convert them into plant available form (Rajkumar et al. 2010).

12.5.1.4 ACC Deaminase

Production of high-level 1-aminocyclopropane-1-carboxylic acid (ACC) deaminase enzyme is considered to decrease plants' ethylene level that promotes the growth mechanism in plants (Glick 2012). Alpha-ketobutyrate and NH_3 are produced from 1-aminocyclopropane-1-carboxylic acid, which is the source of nitrogen and carbon and plays a key role in plant development. Rhizobial strains such as *Rhizobium leguminosarum* bv. *viciae*, *Rhizobium hedysari*, *Rhizobium japonicum*, *Bradyrhizobium japonicum*, *Bradyrhizobium elkanii*, *Rhizobium gallicum*, *Mesorhizobium loti*, and *Sinorhizobium meliloti* are known to produce ACC deaminase (Duan et al. 2009; Hafeez et al. 2008) and increase nodulation and nitrogen fixation (Ma et al. 2003; Murset et al. 2012). Inoculation of ACC deaminase-producing *Rhizobium leguminosarum* bv. *trifolii* SN10 colonizes rice roots and imparts stress tolerance (Bhattacharjee et al. 2012).

12.5.2 Indirect Growth Promotion in Plants

Rhizobia indirectly enhance the growth of nonleguminous plants by a different mechanism that reduces or avoids the harmful effect of pathogenic organisms and abiotic stress factors. Rhizobia can secrete secondary metabolites such as antibiotics, volatile substances, HCN, and fungal cell wall degrading enzymes that help to inhibit the growth of plant pathogens. Rhizobium successfully competes for nutrient resources with pathogens, as well as induces plant's defensive mechanisms through induced systemic resistance (ISR). Siderophore production is another mechanism to inhibit the growth of a pathogen by making limited iron sources unavailable for the pathogen. Colonization of roots with rhizobial species induces physiological immunity in nonleguminous plants such as rice, sunflower, okra, and potato against viral, bacterial, and fungal infections (Ehteshamul-Haque and Ghaffar 1993).

12.5.2.1 Biocontrol Activity

Several rhizobial species have been reported to lead disease resistance while promoting plant biomass and yield proliferation. The biocontrol properties of rhizobia can be associated with lytic enzymes and the production of an antimicrobial metabolites, especially when it is associated with diseases affecting plant roots. In addition to the action of antifungal cells, the suppression of plant diseases may be associated with the promotion of rhizobial plant growth and/or symbiotic activity. In addition, rhizobia have been found to cause systemic resistance to plant vaccines. Inoculation of *Sinorhizobium meliloti*, *Rhizobium leguminosarum* bv. *viciae*, and *Bradyrhizobium japonicum* reduced infection of *Macrophomina phaseolina*, *Rhizoctonia solani*, and *Fusarium* spp. in both leguminous and nonleguminous plants

(Ehteshamul-Haque and Ghaffar 1993). *Rhizobium leguminosarum* bv. *phaseoli* RRE6 and *Rhizobium leguminosarum* bv. *trifolii* ANU843 have successfully prevented the sheath blight disease caused by *Rhizoctonia solani* in rice (Chandra et al. 2007; Mishra et al. 2006). Some rhizobial strains of *Rhizobium leguminosarum* bv. *phaseoli*, *Rhizobium leguminosarum* bv. *trifolii*, *Rhizobium leguminosarum* bv. *viciae*, and *Mesorhizobium loti* can inhibit the growth of pathogenic microorganisms by producing HCN. The overproduction of HCN in plant-microbe interaction was reported by O'Sullivan and O'Gara (1992) and Alström and Burns (1989), showed inhibitor effect on the root pathogen as well as plant growth and development. *Sinorhizobium meliloti 1021* produces a bacteriocin-like substance that inhibits rice growth (Perrine-Walker et al. 2009).

12.6 PGPR Effects on Abiotic and Biotic Stress

Plant growth-promoting rhizobacteria make the plants more tolerant of biotic and abiotic stress. Drought, soil salinity, acidity in the soil, and chemical stress are some of the challenges for the cultivation of cereals. Kulkarni and Nautiyal (2000) described the capacity of *Rhizobium* and *Bradyrhizobium* to attenuate abiotic stress in nonlegume crops. Heavy metals have a deleterious influence on plant and microbial growth in the environment. Certain microbes have developed unique mechanisms for using heavy metals and reducing heavy metal pollution in the environment. The harmful heavy metals in the soil are reduced to harmless forms by these bacteria. Heavy metal tolerance in *Rhizobium* spp. has been aided by determinants such as extracellular polymeric substances (EPS) and lipopolysaccharides (LPS) (Liu et al. 2001). Heavy metal-tolerant *Rhizobium* species include *Rhizobium etli*, *Rhizobium meliloti*, *Rhizobium leguminosarum* bv. *trifolii*, *Rhizobium leguminosarum* bv. *viciae*, *Bradyrhizobium japonicum*, and *Bradyrhizobium* sp. (Kinkle et al. 1994).

The pH of the soil is a significant component in determining the microbiota in rhizosphere. Low pH denotes a high proton concentration in the soil, which affects plant colonization by microbes and reduces the crop yield. Most beneficial microorganisms are sensitive to soil acidity. Acidity reduces the concentration of calcium and phosphate in soil, which inhibits plant development and reduces the grain yield. Some rhizobial strains belong to *Azorhizobium*, *Rhizobium*, and *Bradyrhizobium* can withstand acidic soil and aid plants in combating acid stress. *Rhizobium leguminosarum* bv. *trifolii* accumulates potassium (K) and phosphorous (P) that impart acid tolerance (Watkin et al. 2003). Glutathione produced by *Rhizobium tropici* helps to survive under acidic pH (Muglia et al. 2007).

Drought and salinity stress in plants can affect morphological, physiological, and molecular responses because of severe and rapid global climate change. Although not all rhizobial isolates are effective in increasing growth and yield in dry or semiarid environments, some have demonstrated outstanding positive characteristics of these beneficial microorganisms that incorporated in rice agriculture's long-term sustainability. *Rhizobium phaseoli*, *Rhizobium leguminosarum*, and *Mesorhizobium ciceri* produce LPS (lipopolysaccharides), EPS (extracellular polysaccharides), catalase, and heat shock proteins that aid rhizobia in surviving in arid climates by increasing the available nutrients and water retention capacity at rhizosphere that indirectly allow the plants to cope with stress.

12.7 Conclusion

Cereals suffer from a mismatch of their available nutrients supplied through chemical fertilizer due to substantial loss of the applied fertilizer. Around 50–70% of applied fertilizer always vanishes from the plant–soil system. The high input of commercially available fertilizers has led to the degradation of air, soil, and water quality with the exhaustion of natural resources such as nutrients and water. Nitrogen fixation and plant growth promotion by plant beneficial bacteria are important criteria for an effective biofertilizer. Rhizobial interaction in cereals such as rice, wheat, corn, barley grains, and other grains as endophytic association without forming any structure such as nodules or causing any symptoms of the disease showed that rhizobia can be used as a biofertilizer in nonlegumes crops. Rhizobial inoculation in cereal crops improved plant nutrients such as P, K, Ca, Mg, and Fe accumulation, apart from imparting biotic and abiotic stress tolerance. Cereal–legumes crop rotation should be employed to capitalize beneficial plant–microbe association if higher levels of production and sustainability are to be realized. Future research in this area will be able to develop a rhizobial technology for large and natural grain production systems.

Acknowledgment Not applicable.

Conflict of Interest Author(s) declares no conflict of interest.

References

Adnan M, Shah Z, Khan A, Khan GA, Ali A, Khan NA, Saleem N, Nawaz S, Akbar S, Samreen S, Zaib K (2014) Integrated effects of rhizobial inoculum and inorganic fertilizers on wheat yield and yield components. Am J Plant Sci 5(13):47504

Afify AH, Hauka FIA, El-Sawah AM, Yanni YG, El-Saadany AY (2019) Inoculation with single, dual or consortia of *Rhizobium leguminosarum* bv. *trifolii*, *Pseudomonas stutzeri* and *Anabaena* sp. and their effect on yield components of rice plant. J Agric Chem Biotechnol 10(9):189–193

Ahmad F, Ahmad I, Khan MS (2008) Screening of free-living rhizospheric bacteria for their multiple plant growth promoting activities. Microbiol Res 163(2):173–181

Alikhani HA, Saleh-Rastin N, Antoun H (2006) Phosphate solubilization activity of rhizobia native to Iranian soils. In: Velázquez E, Rodríguez BC (eds) First international meeting on microbial phosphate solubilization. Springer, Dordrecht, pp 35–41

Al-Mallah MK, Davey MR, Cocking EC (1989) Formation of nodular structures on rice seedlings by rhizobia. J Exp Bot 40(4):473–478

Alori ET, Glick BR, Babalola OO (2017) Microbial phosphorus solubilization and its potential for use in sustainable agriculture. Front Microbiol 8:971

Alström S, Burns RG (1989) Cyanide production by rhizobacteria as a possible mechanism of plant growth inhibition. Biol Fertil Soils 7(3):232–238

Antoun H, Beauchamp CJ, Goussard N, Chabot R, Lalande R (1998) Potential of *Rhizobium* and *Bradyrhizobium* species as plant growth promoting rhizobacteria on non-legumes: effect on radishes (*Raphanus sativus* L.). Plant Soil 204:57–67

Bageshwar UK, Srivastava M, Pardha-Saradhi P, Paul S, Gothandapani S, Jaat RS, Shankar P, Yadav R, Biswas DR, Kumar PA, Padaria JC, Mandal PK, Annapurna K, Dasa HK (2017) An environment friendly engineered *Azotobacter* can replace substantial amount of urea fertilizer and yet sustain same wheat yield. Appl Environ Microbiol 83(15):17. https://doi.org/10.1128/AEM.00590-17

Baldani JI, Caruso L, Baldani VLD, Goi SR, Dobereiner J (1997) Recent advances in BNF with non-legume plants. Soil Biol Biochem 29:911–922

Baldani VLD, Baldani JI, Dobereiner J (2000) Inoculation of rice plants with the endophytic diazotrophs *Herbaspirillum seropedicae* and *Burkholderia* spp. Biol Fertil Soils 30(5):485–491

Bardin S, Dan S, Osteras M, Finan TM (1996) A phosphate transport system is required for symbiotic nitrogen fixation by *Rhizobium meliloti*. J Bacteriol 178(15):4540–4547. https://doi.org/10.1128/JB.178.15.4540-4547.1996

Beijerinck MW (1988) Die Bacterian der Papillionaceenknollchen. Bot Zeit 46:726–735

Bender GL, Preston L, Barnard D, Rolfe BG (1990) Formation of nodule-like structures on the roots of the non-legumes rice and wheat. In: Nitrogen fixation: achievements and objectives. Springer, New York, NY, p 825

Bhattacharjee RB, Jourand P, Chaintreuil C, Dreyfus B, Singh A, Mukhopadhyay SN (2012) Indole acetic acid and ACC deaminase-producing *Rhizobium leguminosarum* bv. *trifolii* SN10 promote rice growth, and in the process undergo colonization and chemotaxis. Biol Fertil Soils 48(2): 173–182. https://doi.org/10.1007/s00374-011-0614-9

Biswas JC, Ladha JK, Dazzo FB (2000) Rhizobia inoculation improves nutrient uptake and growth of lowland rice. Soil Sci Soc Am J 64(5):1644–1650

Boddey RM, Baldani VLD, Baldani JI, Dobereiner J (1986) Effect of inoculation of *Azospirillum* spp. on nitrogen accumulation by field-grown wheat. Plant Soil 95:109–121

Boddey RM, Urquiaga S, Reis VM, Dobereiner J (1991) Biological nitrogen fixation associated with sugarcane. Plant Soil 37:111–117

Boddey RM, De Oliveira OC, Urquiaga S, Reis VM, De Olivares FL, Baldani VLD, Dobereiner J (1995a) Biological nitrogen fixation associated with sugarcane and rice: contributions and prospects for improvement. In: Ladha JK, Peoples MB (eds) Management of biological nitrogen fixation for the development of more productive and sustainable agricultural systems. Springer, Dordrecht, pp 195–209

Boddey RM, De Oliveira OC, Urquiaga S, Reis VM, Olivares FL, Baldani VLD, Dobereiner J (1995b) Biological nitrogen fixation associated with sugar cane and rice: contributions and prospects for improvement. Plant Soil 174:195–209

Boiero L, Perrig D, Masciarelli O, Penna C, Cassán F, Luna V (2007) Phytohormone production by three strains of *Bradyrhizobium japonicum* and possible physiological and technological implications. Appl Microbiol Biotechnol 74(4):874–880

Caba JM, Centeno ML, Fernandez B, Gresshoff PM, Ligero F (2000) Inoculation and nitrate alter phytohormone levels in soybean roots: differences between a super nodulating mutant and the wild type. Planta 211(1):98–104

Cassan F, Perrig D, Sgroy V, Masciarelli O, Penna C, Luna V (2009) *Azospirillum brasilense* Az39 and *Bradyrhizobium japonicum* E109, inoculated singly or in combination, promote seed germination and early seedling growth in corn (*Zea mays* L.) and soybean (*Glycine max* L.). Eur J Soil Biol 45(1):28–35

Chabot R, Antoun H, Cescas MP (1996) Growth promotion of maize and lettuce by phosphate-solubilizing *Rhizobium leguminosarum* biovar. *phaseoli*. Plant Soil 184(2):311–321

Chaintreuil C, Giraud E, Prin Y, Lorquin J, Ba A, Gillis M, De Lajudie P, Dreyfus B (2000) Photosynthetic bradyrhizobia are natural endophytes of the African wild rice *Oryza breviligulata*. Appl Environ Microbiol 66(12):5437–5447. https://doi.org/10.1128/AEM.66.12.5437-5447.2000

Chandra S, Choure K, Dubey RC, Maheshwari DK (2007) Rhizosphere competent *Mesorhizobium loti* MP6 induces root hair curling, inhibits *Sclerotinia sclerotiorum* and enhances growth of Indian mustard (*Brassica campestris*). Braz J Microbiol 38(1):124–130

Chi F, Shen SH, Cheng HP, Jing YX, Yanni YG, Dazzo FB (2005) Ascending migration of endophytic rhizobia from roots to leaves inside rice plants and assessment of benefits to rice growth physiology. Appl Environ Microbiol 71(11):7271–7278

Cocking EC, Webster G, Batchelor CA, Davey MR (1994) Nodulation of non-legume crops. A new look. Agro Food Indus Hi Tech:21–24

Dakora FD, Matiru V, King M, Phillips DA (2002) Plant growth promotion in legumes and cereals by lumichrome, a rhizobial signal metabolite. Nitrogen fixation: global perspectives. CABI publishing, Wallingford, pp 321–322

Dakora FD, Matiru V, Kanu AS (2015) Rhizosphere ecology of lumichrome and riboflavin, two bacterial signal molecules eliciting developmental changes in plants. Front Plant Sci 6(700): 1–11. https://doi.org/10.3389/fpls.2015.00700

De Bruijn FJ, Jing Y, Dazzo FB (1995) Potential and pitfalls of trying to extend symbiotic interactions of nitrogen-fixing organisms to presently non-nodulated plants, such as rice. In: Ladha JK, Peoples MB (eds) Management of biological nitrogen fixation for the development of more productive and sustainable agricultural systems. Springer, Dordrecht, pp 225–240

Dent D, Cocking E (2017) Establishing symbiotic nitrogen fixation in cereals and other non-legume crops: the Greener Nitrogen Revolution. Agric Food Secur 6(1):1–9

Deshwal VK, Pandey P, Kang SC, Maheshwar DK (2003) Rhizobia as a biological control agent against soil borne plant pathogenic fungi. Indian J Exp Biol 41(10):1160–1164

Dobbelaere S, Vanderleyden J, Okon Y (2003) Plant growth-promoting effects of diazotrophs in the rhizosphere. Crit Rev Plant Sci 22(2):107–149

Dobereiner J (1992) History and new perspectives of diazotrophs in association with non-leguminous plants. Symbiosis 13:1–13

Duan J, Muller KM, Charles TC, Vesely S, Glick BR (2009) 1-aminocyclopropane-1-carboxylate (ACC) deaminase genes in rhizobia from southern Saskatchewan. Microb Ecol 57(3):423–436

Ehteshamul-Haque S, Ghaffar A (1993) Use of rhizobia in the control of root rot diseases of sunflower, okra, soybean and mungbean. J Phytopathol 138(2):157–163

Engelhard M, Hurek T, Reinhold-Hurek B (2000) Preferential occurrence of diazotrophic endophytes, *Azoarcus* spp., in wild rice species and land races of *Oryza sativa* in comparison with modern races. Environ Microbiol 2(2):131–141

Franks A, Ryan RP, Abbas A, Mark GL, O'Gara F (2006) Molecular tools for studying plant growth promoting rhizobacteria (PGPR). Molecular techniques for soil and rhizosphere microorganisms. CABI Publishing, Wallingford, pp 116–131

Fuhrmann J, Wollum AG (1989) Nodulation competition among *Bradyrhizobium japonicum* strains as influenced by rhizosphere bacteria and iron availability. Biol Fertil Soils 7(2): 108–112. https://doi.org/10.1007/BF00292567

Gerlach M, Vogel I (1902) Stickstoffsammelden bakterien. Zeit Bakter II 8:669

Glick BR (2012) Plant growth-promoting bacteria: mechanisms and applications. Scientifica 2012: 963401. https://doi.org/10.6064/2012/963401

Gopalakrishnan S, Sathya A, Vijayabharthi R, Varshney RK, Gowda CLL, Krishnamurthy L (2015) Plant growth promoting rhizobia: challenges and opportunities. Biotechnology 5(4): 355–377

Gopalaswamy G, Kannaiyan S, O'Callaghan KJ, Davey MR, Cocking EC (2000) The Xylem of rice (*Oryza sativa*) is colonized by *Azorhizobium caulinodans*. Proc R Soc Lond 267:103–107

Govedarica M (1990) Specific relationship between *Beijerinckia derx* strains and some maize hybrids. Zem Biljka 39(2):125–132

Greetatorn T, Hashimoto S, Sarapat S, Tittabutr P, Boonkerd N, Uchiumi T, Teaumroong N (2019) Empowering rice seedling growth by endophytic *Bradyrhizobium sp.* SUTN 9-2. Lett Appl Microbiol 68(3):258–266

Gutierrez-Zamora MT, Martınez-Romero E (2001) Natural endophytic association between *Rhizobium etli* and maize (*Zea mays* L.). J Biotechnol 91(2-3):117–126

Hafeez FY, Safdar ME, Chaudhry AU, Malik KA (2004) Rhizobial inoculation improves seedling emergence, nutrient uptake and growth of cotton. Aus J Exp Agric 44(6):617–622

Hafeez FY, Hassan Z, Naeem F, Bashir A, Kiran A, Khan SA, Malik KA (2008) *Rhizobium leguminosarum* bv *viciae* strain LC-31: analysis of novel bacteriocin and ACC deaminase gene (s). In: Dakora FD, Chimphango SBM, Valentine AJ, Elmerich C, Newton WE (eds) Biological nitrogen fixation: towards poverty alleviation through sustainable agriculture. Springer, Dordrecht, pp 247–248

Heckmann AB, Sandal N, Bek AS, Madsen LH, Jurkiewicz A, Nielsen MW, Tirichine L, Stougaard J (2011) Cytokinin induction of root nodule primordia in *Lotus japonicus* is regulated by a mechanism operating in the root cortex. Mol Plant-Microbe Interact 24(11):1385–1395

Hemissi I, Gargouri S, Sifi B (2011) Attempt of wheat protection against *Fusarium culmorum* using *Rhizobium* isolates. Tunis J Plant Prot 6(7586):32

Hilali A, Prevost D, Broughton WJ, Antoun H (2001) Effects of inoculation with *Rhizobium leguminosarum* biovar *trifolii* on wheat cultivated in clover crop rotation agricultural soil in Morocco. Can J Microbiol 47(6):590–593

Hoflich G (2000) Colonization and growth promotion of non-legumes by Rhizobium bacteria. In: Bell CR, Brylinsky M, Johnson-Green P (eds) Microbial Biosystems: New Frontiers Proceedings of the 8th International Symposium on Microbial Ecology, Atlantic Canada Society for Microbial Ecology, Halifax, Canada, pp 827–830

Hu Y, Jiao J, Liu LX, Sun YW, Chen WF, Sui XH, Chen WX, Tian CF (2018) Evidence for phosphate starvation of rhizobia without terminal differentiation in legume nodules. Mol Plant-Microbe Interact 31(10):1060–1068. https://doi.org/10.1094/MPMI-02-18-0031-R

Hussain MB, Mehboob I, Zahir ZA, Naveed M, Asghar HN (2009) Potential of *Rhizobium* spp. for improving growth and yield of rice (*Oryza sativa* L.). Soil Environ 28(1):49–55

Igiehon NO, Babalola OO, Aremu BR (2019) Genomic insights into plant growth promoting rhizobia capable of enhancing soybean germination under drought stress. BMC Microbiol 19(1):159. https://doi.org/10.1186/s12866-019-1536-1

Jain V, Gupta K (2003) The flavonoid naringenin enhances intercellular colonization of rice roots by *Azorhizobium caulinodans*. Biol Fertil Soils 38(2):119–123

Jaiswal SK, Mohammed M, Ibny FY, Dakora FD (2021) Rhizobia as a source of plant growth-promoting molecules: potential applications and possible operational mechanisms. Front Sustain Food Syst 4:311

James EK (2000) Nitrogen fixation in endophytic and associative symbiosis. Field Crop Res 65(2–3):197–209

Jarzyniak K, Banasiak J, Jamruszka T, Jamruszka T, Pawela A, Donato MD, Novak O, Geisler MM, Michal Jasinski M (2021) Early stages of legume–rhizobia symbiosis are controlled by ABCG-mediated transport of active cytokinins. Nat Plant 7(4):428–436. https://doi.org/10.1038/s41477-021-00873-6

Jha PN, Gomaa AB, Yanni YG, El-Saadany AEY, Stedtfeld TM, Stedtfeld RD, Gantner S, Chai B, Cole J, Hashsham SA, Dazzo FB (2020) Alterations in the endophyte-enriched root-associated microbiome of rice receiving growth-promoting treatments of urea fertilizer and *Rhizobium* biofertilizer. Microb Ecol 79(2):367–382

Jing Y, Li G, Shan X (1992) Development of nodule-like structure on rice roots. In: Khush GS, Bennett J (eds) Nodulation and nitrogen fixation in rice potential and prospect. IRRI, Manila, pp 123–126

Keymer DP, Kent AD (2014) Contribution of nitrogen fixation to first year *Miscanthus* × *giganteus*. GCB Bioenergy 6:577–586. https://doi.org/10.1111/gcbb.12095

Kinkle BK, Sadowsky MJ, Johnstone K, Koskinen WC (1994) Tellurium and selenium resistance in rhizobia and its potential use for direct isolation of *Rhizobium meliloti* from soil. Appl Environ Microbiol 60(5):1674–1677

Kisiala A, Laffont C, Emery RJN, Frugier F (2013) Bioactive cytokinins are selectively secreted by *Sinorhizobium meliloti* nodulating and non-nodulating strains. Mol Plant-Microbe Interact 26(10):1225–1231

Kulkarni S, Nautiyal CS (2000) Effects of salt and pH stress on temperature-tolerant *Rhizobium* sp. NBRI330 nodulating *Prosopis juliflora*. Curr Microbiol 40(4):221–226

Ladha JK, Reddy PM (1995) Extension of nitrogen fixation to rice—necessity and possibilities. Geo J 35(3):363–372

Ladha JK, Garcia M, Miyan S, Padre AT, Watanabe I (1989) Survival of *Azorhizobium caulinodans* in the soil and rhizosphere of wetland rice under *Sesbania rostrata*-rice rotation. Appl Environ Microbiol 55(2):454–460

Ladha J, Tirol-Padre A, Reddy CK, Cassman KG, Verma S, Powlson DS, Van Kessel D, Ritcher DB, Chakraborty D, Pathak H (2016) Global nitrogen budgets in cereals: a 50-year assessment for maize, rice and wheat production systems. Sci Rep 6:19355. https://doi.org/10.1038/srep19355

Li G, Jing Y, Shan X, Wang H, Guan C (1991) Identification of rice nodules that contain *Rhizobium* bacteria. Chinese J Bot 3:8–17

Liu Y, Lam MC, Fang HH (2001) Adsorption of heavy metals by EPS of activated sludge. Water Sci Technol 43(6):59–66

Liu H, Wang X, Qi H, Wang Q, Chen Y, Li Q, Zhang Y, Qui L, Fontana JE, Zhang B, Wang W, Xie Y (2017) The infection and impact of *Azorhizobium caulinodans* ORS571 on wheat (*Triticum aestivum* L.). PLoS One 12(11):e0187947. https://doi.org/10.1371/journal.pone.0187947

Lugtenberg B, Kamilova F (2009) Plant-growth-promoting rhizobacteria. Annu Rev Microbiol 63: 541–556

Lupwayi NZ, Clayton GW, Hanson KG, Rice WA, Biederbeck VO (2004) Endophytic rhizobia in barley, wheat and canola roots. Can J Plant Sci 84(1):37–45

Lupwayi NZ, Clayton GW, Rice WA (2006) Rhizobial inoculants for legume crops. J Crop Improv 15(2):289–321

Ma W, Sebestianova SB, Sebestian J, Burd GI, Guinel FC, Glick BR (2003) Prevalence of 1-aminocyclopropane-1-carboxylate deaminase in *Rhizobium* spp. Antonie Van Leeunwenhoek 83:285–291. https://doi.org/10.1023/A:1023360919140

Mantelin S, Touraine B (2004) Plant growth-promoting bacteria and nitrate availability: impacts on root development and nitrate uptake. J Exp Bot 55(394):27–34

Marra LM, Soares CRFS, DeOliveira SM, Ferreira PAA, Soares BL, De Fraguas CR, DeLima JM, DeSouza-Moreira FM (2012) Biological nitrogen fixation and phosphate solubilization by bacteria isolated from tropical soils. Plant Soil 357:289–307. https://doi.org/10.1007/s11104-012-1157-z

Matiru V, Jaffer MA, Dakora FD (2000) Rhizobial colonization of roots of African landraces of sorghum and millet and the effects of sorghum growth and P nutrition. In: Proceedings of the 4th Congress of the African Association for Biological Nitrogen Fixation: Imperatives for BNF Research and Application in Africa for the 21st Century. African Association for Biological Nitrogen Fixation, Nairobi, Kenya, pp 99–100

Matiru VN, Jaffer MA, Dakora FD (2005) Rhizobial infection of African landraces of sorghum (*Sorghum bicolor* L.) and finger millet (*Eleucine coracana* L.) promotes plant growth and alters tissue nutrient concentration under axenic conditions. Symbiosis 40:7–15

Mauch-Main B, Mauch F (2005) The role of abscisic acid in plant–pathogen interactions. Curr Opin Plant Biol 8(4):409–414. https://doi.org/10.1016/j.pbi.2005.05.015

Mayak S, Tirosh T, Glick BR (2004) Plant growth-promoting bacteria confer resistance in tomato plants to salt stress. Plant Physiol Biochem 42(6):565–572

McInroy JA, Kloepper JW (1995) Survey of indigenous bacterial endophytes from cotton and sweet corn. Plant Soil 173(2):337–342

Mehboob I, Zahir ZA, Mahboob A, Shahzad SM, Jawad A, Arshad M (2008) Preliminary screening of *Rhizobium* isolates for improving growth of maize seedlings under axenic conditions. Soil Environ 27:64–71

Mehboob I, Naveed M, Zahir ZA, Ashraf M (2012) Potential of rhizobia for sustainable production of non-legumes. In: Ashraf M, Ozturk M, Ahmad MSA, Aksoy A (eds) Crop production for agricultural improvement. Springer, Dordrecht, pp 659–704

Mishra RP, Singh RK, Jaiswal HK, Kumar V, Maurya S (2006) *Rhizobium*-mediated induction of phenolics and plant growth promotion in rice (*Oryza sativa* L.). Curr Microbiol 52(5):383–389

Molla AH, Shamsuddin ZH, Halimi MS, Morziah M, Puteh AB (2001) Potential for enhancement of root growth and nodulation of soybean co-inoculated with *Azospirillum* and *Bradyrhizobium* in laboratory systems. Soil Biol Biochem 33(4–5):457–463

Muglia CI, Grasso DH, Aguilar OM (2007) *Rhizobium tropici* response to acidity involves activation of glutathione synthesis. Microbiology 153(4):1286–1296

Murset V, Hennecke H, Pessi G (2012) Disparate role of rhizobial ACC deaminase in root-nodule symbioses. Symbiosis 57(1):43–50

Naidu VSGR, Panwar JDS, Annapurna K (2004) Effect of synthetic auxins and *Azorhizobium caulinodans* on growth and yield of rice. Indian J Microbiol 44:211–213

Nautiyal CS (1997) Rhizosphere competence of *Pseudomonas* sp. NBRI9926 and *Rhizobium* sp. NBRI9513 involved in the suppression of chickpea (*Cicer arietinum* L.) pathogenic fungi. FEMS Microbiol Ecol 23(2):145–158. https://doi.org/10.1111/j.1574-6941.1997.tb00398.x

O'Callaghan KJ, Stone PJ, Hu X, Griffiths DW, Davey MR, Cocking EC (2000) Effects of glucosinolates and flavonoids on colonization of the roots of *Brassica napus* by *Azorhizobium caulinodans* ORS571. Appl Environ Microbiol 66(5):2185–2191. https://doi.org/10.1128/AEM.66.5.2185-2191.2000

O'Sullivan DJ, O'Gara F (1992) Traits of fluorescent *Pseudomonas* spp. involved in suppression of plant root pathogens. Microbiol Rev 56(4):662–676

Peix A, Rivas-Boyero AA, Mateos PF, Rodriguez-Barrueco C, Martınez-Molina E, Velazquez E (2001) Growth promotion of chickpea and barley by a phosphate solubilizing strain of *Mesorhizobium mediterraneum* under growth chamber conditions. Soil Biol Biochem 33(1): 103–110

Pena-Cabriales JJ, Alexander M (1983) Growth of *Rhizobium* in soil amended with organic matter. Soil Sci Soc Am J 47(2):241–245

Peng S, Biswas JC, Ladha JK, Gyaneshwar P, Chen Y (2002) Influence of rhizobial inoculation on photosynthesis and grain yield of rice. Agron J 94(4):925–929

Perrine-Walker FM, Hynes MF, Rolfe BG, Hocart CH (2009) Strain competition and agar affect the interaction of rhizobia with rice. Can J Microbiol 55(10):1217–1223

Phillips DA, Torrey JG (1972) Studies on cytokinin production by *Rhizobium*. Plant Physiol 49(1): 11–15

Plazinski JA, Innes RW, Rolfe BG (1985) Expression of *Rhizobium trifolii* early nodulation genes on maize and rice plants. J Bacteriol 163(2):812–815

Plessner O, Klapatch T, Guerinot ML (1993) Siderophore utilization by *Bradyrhizobium japonicum*. Appl Environ Microbiol 59(5):1688–1690

Podlesakova K, Fardoux J, Patrel D, Bonaldi K, Novak O, Strnad M, Giraud E, Spichal L, Nouwen N (2013) Rhizobial synthesized cytokinins contribute to but are not essential for the symbiotic interaction between photosynthetic *Bradyrhizobia* and *Aeschynomene* legumes. Mol Plant-Microbe Interact 26(10):1232–1238. https://doi.org/10.1094/MPMI-03-13-0076-R

Quispel A (1991) A critical evaluation of the prospects for nitrogen fixation with non-legumes. Plant Soil 137:1–11

Qureshi MA, Shahzad H, Imran Z, Mushtaq M, Akhtar N, Ali MA, Mujeeb F (2013) Potential of *Rhizobium* species to enhance growth and fodder yield of maize in the presence and absence of l-tryptophan. J Anim Plant Sci 23(5):1448–1454

Rajkumar M, Ae N, Prasad MNV, Freitas H (2010) Potential of siderophore-producing bacteria for improving heavy metal phytoextraction. Trends Biotechnol 28(3):142–149

Reddy PM, Ladha JK, So RB, Hernandez RJ, Ramos MC, Angeles OR, Dazzo FB, De Bruijn FJ (1997) Rhizobial communication with rice roots: induction of phenotypic changes, mode of invasion and extent of colonization. Plant Soil 194(1):81–98

Reimann S, Hauschild R, Hildebrandt U, Sikora RA (2008) Interrelationships between *Rhizobium etli* G12 and *Glomus intraradices* and multitrophic effects in the biological control of the root-knot nematode *Meloidogyne incognita* on tomato. J Plant Dis Prot 115(3):108–113

Reinhold-Hurek B, Hurek T (1997) *Azoarcus* spp. and their interactions with grass roots. In: Ladha JK, De Bruijn FJ, Malik KA (eds) Opportunities for biological nitrogen fixation in rice and other non-legumes. Springer, Dordrecht, pp 57–64

Riggs PJ, Chelius MK, Iniguez AL, Kaeppler SM, Triplett EW (2001) Enhanced maize productivity by inoculation with diazotrophic bacteria. Funct Plant Biol 28(9):829–836

Roger PA, Watanabe I (1986) Technologies for utilizing biological nitrogen fixation in wetland rice: potentialities, current usage, and limiting factors. In: De Datta SK, Patrick WH Jr (eds) Nitrogen economy of flooded rice soils. Springer, Dordrecht, pp 39–77

Rogers C, Oldroyd GE (2014) Synthetic biology approaches to engineering the nitrogen symbiosis in cereals. J Exp Bot 65(8):1939–1946

Rolfe BG, Bender GL (1990) Evolving a *Rhizobium* for non-legume nodulation. In: Gresshoff PM, Roth LE, Stacey G, Newton WE (eds) Nitrogen fixation. Springer, Boston, MA, pp 779–780

Rosenblueth M, Martinez-Romero E (2004) *Rhizobium etli* maize populations and their competitiveness for root colonization. Arch Microbiol 181(5):337–344

Sabry SR, Saleh SA, Batchelor CA, Jones J, Jotham J, Webster G, Kothari SL, Davey MR, Cocking EC (1997) Endophytic establishment of *Azorhizobium caulinodans* in wheat. Proc R Soc Lond Ser B Biol Sci 264(1380):341–346

Saikia SP, Jain V (2007) Biological nitrogen fixation with non-legumes: an achievable target or a dogma? Curr Sci 92(3):317–322

Schloter M, Wiehe W, Assmus B, Steindl H, Becke H, Hoflich G, Hartmann A (1997) Root colonization of different plants by plant-growth-promoting *Rhizobium leguminosarum* bv. *trifolii* R39 studied with monospecific polyclonal antisera. Appl Environ Microbiol 63(5): 2038–2046

Schnabel T, Sattely E (2021) Improved stability of engineered ammonia production in the plant-symbiont *Azospirillum brasilense*. ACS Synth Biol 10(11):2982–2996

Setten L, Soto G, Mozzicafreddo M, Fox AR, Lisi C, Cuccioloni M, Angeletti M, Pagano E, Paleo AD, Ayub ND (2013) Engineering *Pseudomonas protegens* Pf-5 for nitrogen fixation and its application to improve plant growth under nitrogen-deficient conditions. PLoS One 8(5): e63666. https://doi.org/10.1371/journal.pone.0063666

Shamala T, Thilini AP, Anuradha S, Kulasooriya SA, Seneviratne G (2018) The effect of flavonoid naringenin coupled with the developed biofilm *Azorhizobium Caulinodans-Aspergillus* spp. on increase in rice yields in conventionally and organically grown rice. Int J Plant Stud 1(1):1–6

Siddiqui ZA (2007) Biocontrol of *Alternaria triticina* by plant growth promoting rhizobacteria on wheat. Arch Phytopathol Plant Prot 40(4):301–308

Singh RK, Mishra RP, Jaiswal HK, Kumar V, Pandey SP, Rao SB, Annapurna K (2006) Isolation and identification of natural endophytic rhizobia from rice (*Oryza sativa* L.) through rDNA PCR-RFLP and sequence analysis. Curr Microbiol 52(5):345–349

Singh NP, Patel AK, Banjare U, Pandey AK (2020) *Rhizobium radiobacter*: a unique maize endophyte with high level of stress tolerance and multiple plant growth promoting properties. Plant Arch 20(1):2483–2488

Smith MJ, Neilands JB (1984) Rhizobactin a siderophore from *Rhizobium meliloti*. J Plant Nutr 7: 449–458

Sturtevant DB, Taller BJ (1989) Cytokinin production by *Bradyrhizobium japonicum*. Plant Physiol 89(4):1247–1252

Swarnalakshmi K, Yadav V, Tyagi D, Dhar DW, Kannepalli A, Kumar S (2020) Significance of plant growth promoting rhizobacteria in grain legumes: growth promotion and crop production. Plants 9(11):1596

Tan KZ, Radziah O, Halimi MS, Khairuddin AR, Habib SH, Shamsuddin ZH (2014) Isolation and characterization of rhizobia and plant growth-promoting rhizobacteria and their effects on growth of rice seedlings. Am J Agric Biol Sci 9(3):342–360

Trinick MJ, Hadobas PA (1995) Formation of nodular structures on the non-legumes *Brassica napus*, *B. campestris*, *B. juncea* and *Arabidopsis thaliana* with *Bradyrhizobium* and *Rhizobium* isolated from *Parasponia* spp. or legumes grown in tropical soils. Plant Soil 172(2):207–219

Ullah S, Khan MY, Asghar HN, Akhtar MJ, Zahir ZA (2017) Differential response of single and co-inoculation of *Rhizobium leguminosarum* and *Mesorhizobium ciceri* for inducing water deficit stress tolerance in wheat. Ann Microbiol 67(11):739–749

Ullah S, Ashraf M, Asghar HN, Iqbal Z, Ali R (2019) A review: plant growth promoting rhizobacteria-mediated amelioration of drought in crop plants. Soil Environ 38(1):1–20

Vargas LK, Volpiano CG, Lisboa BB, Giongo A, Beneduzi A, Passaglia LMP (2017) Potential of rhizobia as plant growth-promoting rhizobacteria. In: Zaidi A, Khan M, Musarrat J (eds) Microbes for legume improvement. Springer, Cham, pp 153–174. https://doi.org/10.1007/978-3-319-59174-2_7

Velazquez E, Peix A, Zurdo-Piniro JL, Palomo JL, Mateos PF, Rivas R, Munoz-Adelantado E, Toro N, Garcia-Benavides P, Martinez-Molina E (2005) The coexistence of symbiosis and pathogenicity-determining genes in *Rhizobium rhizogenes* strains enables them to induce nodules and tumours or hairy roots in plants. Mol Plant-Microbe Interact 18(12):1325–1332

Vogel JP, Woeste KE, Theologis A, Kieber JJ (1998) Recessive and dominant mutations in the ethylene biosynthetic gene ACS5 of *Arabidopsis* confer cytokinin insensitivity and ethylene overproduction, respectively. Plant Biol 95:4766–4771

Warren WJ, Warren WP (1993) Mechanisms of auxin regulation of structural and physiological polarity in plants, tissues, cells and embryos. Funct Plant Biol 20(5):555–571

Wasai-Hara S, Hara S, Morikawa T, Sugawara M, Takami H, Yoneda J, Tokunaga T, Minamisawa K (2020) Diversity of *Bradyrhizobium* in non-leguminous sorghum plants: *B. ottawaense* isolates unique in genes for N_2O reductase and lack of the type VI secretion system. Microbes Environ 35(1):ME19102

Watkin EL, O'Hara GW, Glenn AR (2003) Physiological responses to acid stress of an acid-soil tolerant and an acid-soil sensitive strain of *Rhizobium leguminosarum* biovar *trifolii*. Soil Biol Biochem 35(4):621–624

Webster G, Davey MR, Cocking EC (1995) *Parasponia* with rhizobia: a neglected non-legume nitrogen-fixing symbiosis. Agbiotech News Information 7:119–124

Werner D (1992) Symbiosis of plants and microbes (No. SB731 W49). Chapman & Hall, London

Weyens N, van der Lelie D, Taghavi S, Newman L, Vangronsveld J (2009) Exploiting plant–microbe partnerships to improve biomass production and remediation. Trends Biotechnol 27(10):591–598

Wiehe W, Hecht-Buchholz CH, Hoflich G (1994) Electron microscopic investigations on root colonization of *Lupinus albus* and *Pisum sativum* with two associative plant growth promoting rhizobacteria, *Pseudomonas fluorescens* and *Rhizobium leguminosarum* bv. *trifolii*. Symbiosis 17:15–35

Wu Q, Peng X, Yang M, Zhang W, Dazzo FB, Uphoff N, Jing Y, Shen S (2018) Rhizobia promote the growth of rice shoots by targeting cell signalling, division and expansion. Plant Mol Biol 97(6):507–523

Yang G, Bhuvaneswari TV, Joseph CM, King MD, Phillips DA (2002) Roles for riboflavin in the *Sinorhizobium*-alfalfa association. Mol Plant-Microbe Interact 15(5):456–462

Yanni YG, Dazzo FB (2010) Enhancement of rice production using endophytic strains of *Rhizobium leguminosarum* bv. *trifolii* in extensive field inoculation trials within the Egypt Nile delta. Plant Soil 336(1):129–142

Yanni YG, Rizk RY, Corich V, Squartini A, Ninke K, Philip-Hollingsworth S, Orgambide G, Bruijn FD, Stoltzfus J, Buckley D, Schmidt TM (1997) Natural endophytic association between *Rhizobium leguminosarum* bv. *trifolii* and rice roots and assessment of its potential to promote rice growth. In: Ladha JK, De Bruijn FJ, Malik KA (eds) Opportunities for biological nitrogen fixation in rice and other non-legumes. Springer, Dordrecht, pp 99–114

Yanni YG, Rizk RY, Abd El-Fattah FK, Squartini A, Corich V, Giacomini A, de Bruijn F, Rademaker J, Maya-Flores J, Ostrom P, Vega-Hernandez M (2001) The beneficial plant growth-promoting association of *Rhizobium leguminosarum* bv. *trifolii* with rice roots. Funct Plant Biol 28(9):845–870

Young JPW (1996) Phylogeny and taxonomy of rhizobia. Plant Soil 186(1):45–52

Part III
Application to Sustainable Agriculture

Chapter 13
Ecology of Nitrogen-Fixing Bacteria for Sustainable Development of Non-legume Crops

Shrivardhan Dheeman and Dinesh Kumar Maheshwari

Abstract Plants provide a substantial ecological niche for bacteria. The symbiotic association between legume and rhizobia contributed World's largest share of biologically fixed nitrogen. This review explains the rational of using nitrogen-fixing bacteria in sustainable agriculture particularly from the genetic engineering in non-legumes for root nodule development to rhizobia and non-legume interaction covering mode of entry. Associative and entophytic role of nitrogen fixation bacteria in various cereal and non-cereal crops is well established and their functional molecules are covered in the chapter. This review attempts to discuss present challenges, future visions and missions to achieve improvement in soil fertility and crop production.

Keywords Non-legume · N-fixation · Sustainable agriculture · Plant–microbe interaction · Rhizobacteria

13.1 Introduction

Second Green Revolution from Green Revolution can be channelled via adopting micro-irrigation system, organic farming, precision farming, green agriculture, eco-agriculture, white agriculture, straw revolution and in all use of plant growth-promoting rhizobacteria. The value of crop rotation in improving the crop field was well known to Greeks, as they have practiced cultivation of corn followed by the cultivation of legume, for enhanced crop productivity of corn. It gives an idea of beneficial role of legume cultivation, increases soil fertility and nutrient balance. Though, in early days, it was not concluded, but understood later by the pioneer

S. Dheeman (✉)
Department of Microbiology, School of Allied Health Sciences, MVN University, Palwal, Haryana, India

D. K. Maheshwari
Department of Botany and Microbiology, Gurukula Kangri (Deemed to be University), Haridwar, Uttarakhand, India

© The Author(s), under exclusive license to Springer Nature Singapore Pte Ltd. 2022

D. K. Maheshwari et al. (eds.), *Nitrogen Fixing Bacteria: Sustainable Growth of Non-legumes*, Microorganisms for Sustainability 36,
https://doi.org/10.1007/978-981-19-4906-7_13

work of Gilbert (1891), in which they postulated "legume has inherent ability to add nitrogen in soil". Further, Hellriegel and Wilfarth in 1988 observed nitrogen gain in pea plant (*Pisum sativum*) due to the presence of microorganisms in its rhizosphere, also forms root nodule (Möllerová 1990). The discovery by Martinus W. Beijerinck became famous and successful, elucidating the role of root nodule bacteria in nitrogen fixation, and named it as *Bacillus radicicola* (Beijerinck 1901).

The plant root system significantly contributes to the establishment of the microbiome in the rhizosphere populated with diverse array of microorganisms; therefore, harnessing benefits from microbial processes and properties, a crucial determinant to support functional agriculture. The microbe–plant interaction in the rhizosphere is dynamic and characterized as symbiotic relationship and free-living relationship. The excellent examples of plant–microbe interaction are beneficial association on above-ground parts of the plant, i.e. development of stem nodule by *Azorhizobium* in *Aeschynomene americana*, *Sesbania aculeate*, *Sesbania rostrata* and *Neptunia* sp.

Among three basic classes of nitrogen-fixing bacteria, free-living N-fixers, associative N-fixers and symbiotic N-fixers, the first two are most often applied to non-legume crops (Fig. 13.1). The last group can be found in the rhizosphere of legume crops, establishing one of the most studied mutual interaction and forms root nodule. In the last few years, significant efforts have been made to extend nitrogen fixation to crops particularly in cereals and non-legume crops (Beyan et al. 2018). For this

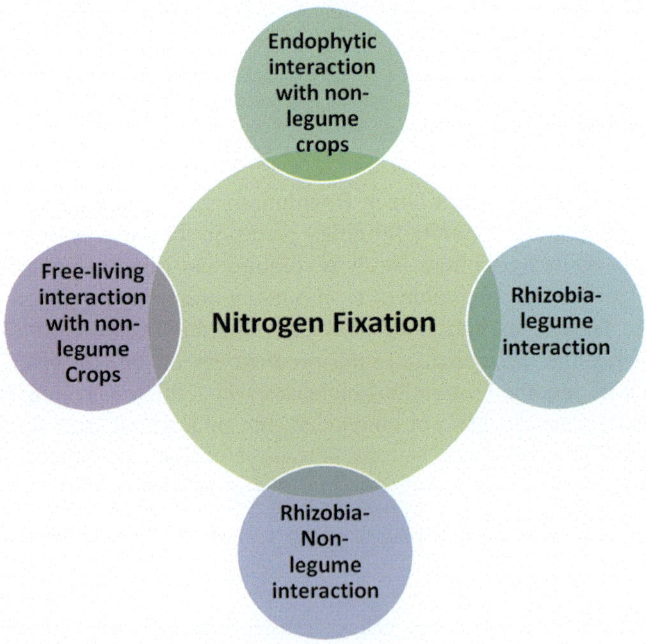

Fig. 13.1 Various type of bacterial associations including rhizobia, free-living bacteria, legumes and non-legumes relates with the biological nitrogen fixation

exclusive concern, bioformulations of certain free-living and associative N-fixer have been developed and applied as biofertilizers to the soil ecology and crop ecosystem.

Free-living bacteria such as *Bacillus* sp. are believed to enhance the plant growth through synthesis of plant growth regulators such as auxins (indole-3-acetic acid), nutrient mobilization, etc. Effective colonization of plant roots by plant growth-promoting rhizobacteria (PGPR) plays an important role in growth promotion irrespective of the mechanism of action, i.e. production of metabolites, antibiotics against pathogens, etc. It is now common knowledge that bacteria in natural environments survive by forming biofilms (Davey and O'Toole 2000). Many *Bacillus* strains are considered as natural factories of cyclic lipopeptides, including iturins, fengycins and surfactins, and their involvement in control of plant microbial diseases has been proved (Agarwal et al. 2017). The management of nitrogen in soil, particularly during the cultivation of non-legume crops, is only possible with nitrogen-fixing bacteria (NFB) compatible with non-legume crops.

In terrestrial ecosystem, non-symbiotic (free-living and associative) N-fixers contribute about 30% share of biologically fixed nitrogen, but have less agronomic importance than symbiotic nitrogen-fixing bacteria (Smercina et al. 2019). In the category of non-symbiotic bacteria, *Azospirillum* develops intimate relationship with certain non-legume crops and is called as associative symbiont. A few examples of such association are *Azotobacter paspali* with roots of tropical grass, *Beijerinckia* with sugarcane roots and *Azospirillum* with corn, wheat and sorghum roots. As associative symbiont, it does not form root nodule but causes root hair deformation, invasion in root's cortical and vascular tissues as well as enhancement of lateral root hairs. Other example of nitrogen-fixing bacteria form nodule in non-legume plants is *Frankia*. *Frankia* is a symbiont of actinorhizal plants, similar to the *Rhizobium* found in the root nodules of legumes.

The understanding of ecological factors may compensate with this exclusive task and can control biological nitrogen fixation systems in the field, which is quite essential for determining successful adoption of newer technologies of sustainable farming and so as to evaluate the fertilization efficiency of biofertilizers under different agro-climatic conditions. This review highlights the ecological and agro-economic importance of nitrogen-fixing bacteria for non-legume crops.

13.2 Genetic Engineering in Non-legumes for Root Nodule Development

Transferring traits of biological nitrogen fixation to non-legumes, especially in cereals, still remain elusive (Griesmann et al. 2018; Bueno Batista and Dixon 2019). The recent trend of genetic analysis has allowed a tremendous progress towards N-fixation in non-legumes. A decade ago, Rogers and Oldroyd (2014) attempted to engineer cereal plant by transferring gene responsible for root nodule but as nitrogenase enzyme requires anaerobic environment within the cell, it is difficult to deal with the oxygen toxicity issues and nodule development. In the process of gene transfer to *Escherichia coli* and *Saccharomyces cerevisiae*, plastids of tobacco provide a new ray of hope in this field for the near future to implement these approaches for the betterment of the agricultural systems (Mabrouk et al. 2018). The introduction of nitrogenase enzyme into cereal plants, so that plants can synthesize nitrogen for their needs without bacterial association, has proven difficult due to the complexity of biosynthetic pathway and oxygen sensitivity. Further, two mitochondria and root plastids of eukaryotic cell are considered to offer a low oxygen environment and expression of nitrogenase enzyme, hence can overcome the obstacles of oxygen sensitivity (Ivleva et al. 2016; Wardhani et al. 2019). Ivleva et al. (2016) suggest similarity in these organs to prokaryotes. Approaches in this field in recent years are adopted to improve N-fixation pathway in diazotrophic, endophytic, associative, symbiotic microorganisms, which are in relation with plants by using different strategies.

13.3 Rhizobia and Non-legume Interactions

Nitrogen-fixing endophytic bacteria belong to a very small group of the total number of endophytes. These bacteria are found in seeds and roots of different paddy cultivars (Mano and Morisaki 2008). A few of them are isolated and identified from cereal crops (Mia and Shamsuddin 2010). Certain non-leguminous dicots taxons possess root nodules formed by rhizobia in soil. These plants belong to 18 genera with about 175 species distributed in nine different families and do not bear pods. These are found in different habitats and have wide variations in their morphological forms such as prostrate herb (e.g. *Dryas* spp.), shrub (e.g. *Colletia* spp.) and some woody species (e.g. *Casuarina* spp.). Most of these are symbiotic with actinorhiza (e.g. *Frankia* spp.) but genus *Parasponia* of family Ulmaceae is associated with *Rhizobium* spp. It forms effective root nodules and exhibits nitrogen fixation (Trinick 1973). Root nodules in *Parasponia* recorded in three species, namely *P. rugosa*, *P. parviflora* and *P. andersonic*. Rhizobial nodulation has also been observed in xerophytic plants of family *Zygophyllaceae* (Sabet 1946), mainly *Zygophyllum coccineum*, *Z. album*, *Z. decumbens* and *Z. simplex*. Nodules have been seen in *Fagonia Arabica*, *Tribulus alatus* and *Tribulus terrestris* (Mostafa and Mahmoud 1951). The isolated strains of rhizobia showed cross-infectivity in *Trifolium alexandrinum* and *Arachis hypogaea*. Later, Becking (1982) also confirmed nodular structures in *Z. coccineum*.

A number of agro-biologically significant species namely *Rubus ellipticus* of horticultural importance, others such as *Ceanothus* spp., *Alnus* spp., etc. play a major role in re-afforestation and involve their functional role in plant succession ecosystem. Nodulation and N-fixation also occur in *Trema aspera*, now known as *Trema cannabina*. Later, it has been accommodated with *Parasponia* sp. (Akkermans et al. 1978). The difference in both plants lies in the presence of its intricate perianth and intra-petiolar connate stipules in the terminal bud. Long ago, scientists reported a non-legumes tree called Angqrung (Java) or Kuranj (Sudan) that bore nodules and have the ability to trap nitrogen.

The root nodules formed are of two types: (1) determinate and (2) indeterminate in nature. The later type involved active meristem having infections zone and nitrogen-fixing senescent zone. While studying, the internal root-nodular tissues showed similarity to that of nodular tissues of legume plants. More precisely, root nodules in non-legumes bear a central vascular bundle with bacteria-like structure of different shapes, which form a horse-shoe-shaped zone around it. The nodules of *Parasponia* possess an apical meristemate zone which provides for the continuous elongation of the nodules (van Velzen et al. 2018). The infection thread enters through root hair, persists and penetrates the host cells behind. The apical meristem consists of intercellular spaces between the cells and such cells become hypertrophied (cell enlargement), resulting in nodule formation. However, the *Rhizobium* cells in *Parasponia* species comprised mainly inside the cells and rarely released from the infection thread (Op den Camp et al. 2012), a distinct feature of this genus. In such cases, more than two-thirds of the cells showed infection. Probably, infection threads are the N-fixation sites which play similar role to that of bacteroids (in legumes) enclosed in a membrane envelope. Further, transmission electron microscope (TEM) studies showed the variations in the thickness of the thread walls, and these are observed without a rigid cell wall only enveloped by the cell membrane. The difference between the root nodules of *Parasponia* to that of *Alnus* lies in the structure, which showed attachment to the root quite thin but rhizobia exhibited of normal rod-like structure and do not show any distortion as seen in legumes (Soyano et al. 2021). The bacteroid enveloped with double-layered cell membrane, which is generally estimated varying number of bacteria. The poly-β-hydroxybutyrate is present as reserve food material similar to that of leguminous nodular cells.

13.3.1 Mode of Entry

In the roots of non-legumes, rhizobia have different modes of entry as observed by using various classical and modern techniques such as the use of green fluorescent marker (GFP). Perrine-Walker et al. (2007) reported *Rhizobium* strain ANU843, E4 and R8 on the root surface of rice. These strains adhere to the root surface for quite some time. Some endophytic bacteria proved aggressive colonizer and remain in root-rhizosphere (Verma et al. 2004). The curled root hair in *Brassica campestris* (Chandra et al. 2007) and GFP-labelled *Rhizobium trifolii* inoculated plants also showed curly root hair in rice (Perrine-Walker et al. 2007). In non-legumes, mode of entry of rhizobia generally occurs from root-tip, lateral root cracks of the emergence of roots, damaged tissues of epidermal cells and stomata (James et al. 2002; Sevilla et al. 2001; Perrine-Walker et al. 2007). *Herbaspirillum seropedicae* at entry point induces to emerge lateral root in *Arabidopsis thaliana* (James et al. 2002). From histological perspective, the inter-cellular and cortex region of the root showed initial colonization of endophytes, which is further spread in the intercellular space of the cortical region to the xylem (vessels) as reported by Gyaneshwar et al. (2002), Roncato-Maccari et al. (2003) and Yang and Lin (2005). The involvement of various lingo-cellulolytic enzymes namely endo-β-glucanases, exo-β-glucanase, β-glucosidase cellulose complex and pectolytic enzymes such as polygalacturonases and pectinolyase produced by *R. leguminosarum* bv. *trifolii*, *H. seropedicae*, *H. rubrisubalbicans*, *Pantoea agglomerans* assisted the entry (invasion) and dissemination of the bacteria in their host plants (Yanni et al. 2001). Compant et al. (2005) reported endo-β-glucanase and polygalacturonase cell wall-degrading enzymes by *Burkholderia* sp. infect *Vitis vinifera*.

For the effective process of rhizobia–plant interaction, it is imperative that the bacteria must attain a definite number for their establishment that helps them to show their benefits to the host plants. It is likely that some biofilm formation and quorum sensing (QS) phenomenon might also exist for successful relationship. Plant defence phenomenon is utmost requirement for the effective plant–microbe interaction of mutual benefits. When rhizobia enter and establish inside the root, plants produce certain phenolics, phyto-toxins, etc., which are defensive in nature. The increase in phenolics such as gallic acid, tannic acid, ferulic acid, cinnamic acid, *o*-dihydroxyphenols, etc. imparts resistance to pathogenic stress-related character in plants against entry of rhizobia and rhizobacteria-mediated induced systemic resistance (ISR). In sugarcane, ethylene signalling occurred due to the inoculation of a beneficial N-fixing bacteria *Acetobacter diazotrophicus* to the host cell, i.e. sugarcane, due to involvement of glycoproteins. Similarly, the significance of lipopolysaccharides secreted by rhizobia in maize rhizosphere contributes to aggressive bacterial colonization in *Zea mays*.

13.4 Associative and Endophytic Nitrogen Fixation in Wheat, Rice, Maize and Other Crops

Few taxa of the family Poaceae do not form symbiotic nodules due to rhizobia inoculation but they trap free atmospheric nitrogen and convert into NH_3. In such cases, nitrogen fixed (%) is quite small in comparison to that of legumes nodule forms by rhizobia, although increase in plant growth and yield have been observed significantly (Verma et al. 2010; Bhattacharyya and Jha 2012; Santi et al. 2013). Plant growth and health-promoting bacteria (Maheshwari 2010) have the ability to enhance growth and development of several plants due to their aggressive root colonization and biofilm formation with different plants including grasses. Their association with roots usually called their "associative" nitrogen-fixing bacteria (Elmerich 2007). However, few bacteria designated as "endophyte" can live inside different plant parts, having no visible deleterious effect on the plants-fixing nitrogen. The biology and biotechnology (Maheshwari 2017; Maheshwari et al. 2017) and their role in mineral nutrient management in different host plants have been incorporated in different volumes published earlier (Dhiman et al. 2021). However, more precisely, diazotrophic rhizobacteria have been reported in several bacterial genera of alpha- and beta-proteobacteria including *Acetobacter*, *Azoarcus*, *Azospirillum*, *Azotobacter*, *Burkholderia*, *Enterobacter*, *Herbaspirillum*, *Gluconobacter* and *Pseudomonas* (Cocking 2003; Richardson et al. 2009).

There are very few bacterial genera namely *Azoarcus* spp., *H. seropedicae* and *Gluconobacter* associated with various cells and tissues of maize, rice and wheat but do not live freely in soil (Reinhold-Hurek and Hurek 1998). Root exudates contain different primary and secondary metabolites. The primary metabolites containing amino acids, sugar, vitamins, etc. act as nutrient sources to the microbial community present in their vicinity, which depend upon the nature of root exudates. Differences in its comparison are obvious due to hosts. Some studies between legumes and rhizobia showed that the flavonoids seem to be important plant signals for the host interaction with bacteria as also evidenced in wheat by *Azospirillum brasilense*. The formation of microcolony or biofilm structures occurs on the root surface (Reinhold et al. 1994). Diazotrophic bacteria have an ability to outcompete the other indigenous bacteria and provide homeostasis (Pandey et al. 2005). This results in chemotaxis due to the presence of motility, twitching, etc. based on the mechanism, which includes extrusion, surface attachment of the pilus tip and retraction of pilus to facilitate the bacterial cell to the point of attachment (Böhm et al. 2007). In case of rice, genus *Azoarcus*, type IV pili, involved in adherence to the plant surface and the *pilA*, *pilB* and *pilI* genes required for the root surface colonization (Krause et al. 2006).

Participation of *A. brasilense* exo-polysaccharides (EPS) and lipopolysaccharides (LPS) in the maize root colonization occurred as evidenced by *Tn5* mutant (Balsanelli et al. 2010). In addition to the LPS, a major outer membrane protein from *A. brasilense* strain showed binding to roots of wheat, maize and sorghum seedlings in vitro (Burdman et al. 2001). Most of the endophytic bacteria passively enter the root system using disrupted endodermal cell layers resulting from the emergence of developing lateral roots. Various workers have demonstrated to measure nitrogenase activity by acetylene reduction assay, adopting different techniques ^{15}N dilution studies, immunogold labelling with antibodies against the iron protein of nitrogenase, expression of transcriptional fusion between *nifH* and reporter genes and RT-PCR on transcripts encoding the nitrogenase complex. Sevilla et al. (2001) used this technique in several endophytes tightly associated with grasses including *Gluconacetobacter diazotrophicus*, while Hurek et al. (2002) applied in *Azoarcus* sp. and *Herbaspirillum* spp. in rice (James et al. 2002) and *Klebsiella* in wheat and maize (Chelius and Triplett 2000; Iniguez et al. 2004). Van Puyvelde et al. (2011) studied transcriptome analysis in *A. brasilense* interfering with indole-3-acetic acid (IAA) biosynthesis led to transcriptional changes. Thus, IAA is a necessary signalling molecule that is responsible for plant–bacteria communication process.

13.5 Functional Molecules of N-Fixing Bacteria and PGPR

Symbiotic, associative and free-living N-fixing bacteria participate in multifarious activities including nutrient acquisition and assimilation, improvement in soil texture and modulating functional molecules (intracellular and extracellular). These molecules include hormones, secondary metabolites, antibiotics, organic acids, vitamins, etc. besides intermediates of various metabolic pathways. The N-fixing bacteria confer advantage in minimizing or diminishing the application of N-fertilizers and improve crop tolerance to abiotic and biotic stresses. Several N-fixing bacteria have plant growth promotion of non-legume hosts. The bacteria synthesize and release hormones, auxins, gibberellins, cytokines, ethylene, lumichrome, riboflavin, lipo-chitooligosaccharides, rhizobitoxins, jasmonic acids, brassinosteroids, enzyme 1-aminocyclopropane-1-carboxylic acid (ACC) deaminase, etc., which can directly or indirectly support non-legumes hosts.

Phyto-hormones play major role in plant growth promotion sustainable in nature. As stated earlier, several N-fixing bacteria secrete IAA that imparts variable effects on host plants (Spaepen and Vanderleyden 2011). The rhizospheric free-living N-fixing and endophytic bacteria associated with non-legumes also produce gibberellic acids (GA). Baliyan et al. (2021) recently observed the endophytic GA-secreting *Bacillus cereus* in growth promotion of *Cicer arietinum* (chickpea). Phytohormones deficiency in general and gibberellic acids in particular had a major impact on the level of several major dicarboxylate supplied to rhizobia by the plant and also led to a significant deficit in the amino acids involved in glutamine–aspartate transamination, consistent with the limited bacteroid development and low fix rate of gibberellin-deficient mutant nodules. On the other hand, in contrast, no major effects of brassinosteroid deficiency or ethylene insensitivity on the key metabolites in these pathways were found. Therefore, although the three enzymes influenced interaction and nodulation, only gibberellic acid proved important for the establishment of a functional nodule metabolome (McGuiness et al. 2019).

In general, endophytic bacteria have been considered to hold more growth potential for sustaining phytohormone production demand in plants to that of free-living N-fixing bacteria (Dheeman et al. 2017) but the endophytic bacteria act well in comparison to free-living soil bacteria, e.g. *A. brasilense* has been suitably utilized as bioinoculants in improving N management for wheat production (Galindo et al. 2022). In fact, free-living bacteria of the genus *Azospirillum* host more than 100 angiospermic plants and improve their growth and development (Pedrosa et al. 2021) mainly by the action of secreting beneficial functional molecules in increasing root surface that leads to more nutrient acquisition from soil and water (Caires et al. 2021) as well as improve N use efficiency integrated with chemical-based nitrogenous fertilizer. *A. brasilense* excrete nitric oxide that regulates phytohormones so as to maintain/regulate hormonal balance in the plants, thus assists in facilitating growth (Cassán et al. 2020; Barbosa et al. 2021). Biological nitrogen fixation was the first biochemical phenomenon wherein *A. brasilense* induced growth of both legumes and non-legumes (Day and Döbereiner 1976; Okon et al.

1983; Pedrosa et al. 2020). On the other hand, Dakora et al. (2015) studied the ecology of rhizobial molecules lumichrome and riboflavin in symbiotic legumes. Their functional role in non-legumes to rhizobial hosts, e.g. *Parasponia* spp., is yet to be established. Involvement of ethylene and its regulation in plants have been observed by several workers (Glick et al. 1998; Glick 2003; Pandey et al. 2005). Dubois et al. (2018) described the pivotal role of enzyme ACC deaminase-producing rhizobia PGPR in favour of the enhancement of plant growth and development. Earlier, Ma et al. (2003) observed improve in nodulation and plant growth by involvement of rhizobitoxine via the bio-synthesis of ethylene production. Further, the influence of ethylene, gibberellin and brassinosteroids on energy and nitrogen fixation metabolites in nodular tissues was reported (McGuiness et al. 2019).

13.6 Importance of PGPR/Nitrogen-Fixing Bacteria as Biofertilizer

Biofertilizers are the biological preparations of living microorganisms and considered to be the basic input of nutrients to promote the plant growth and organic farming. PGPR enhance the soil productivity as they transform various nutrients and major geochemicals from inutile to highly available forms without harming the natural environment (Kloepper et al. 1980). The efficient PGPR within the rhizosphere are activated through seed or soil interaction and make the nutrients available to the plants (Choudhary et al. 2007). It is also important to understand the concept of consortia due to the fact that it exerts beneficial effects in plants and plays a major role in the management of plant diseases that affect the single species. Once the microbial consortia are introduced into the soil, it benefits the plant by enhancing the defence mechanism against multiple phytopathogens. Hence, it can also be used as a biofertilizer and act as bio-control agents due to their extraordinary ability of increasing the yield of the crops (Yi et al. 2013; Bourion et al. 2018).

The importance of PGPR or biofertilizers is increasing day by day as it proved a boon for agriculture production and that's why the whole world is seeking to adopt organic farming to meet the demand of global population and to avoid the negative effects of chemical fertilizers, which damage the soil quality and fertility (Kumar and Goh 2002). India has gone through a vital change over the last few decades by emphasizing on sustainable agriculture system. In recent years, the worldwide researchers paid attention to this idea of replacing the agrochemicals (fertilizers and pesticides) with PGPR due to its extraordinary applications in biotechnological industries such as pharmaceuticals and mining and also considered as environmentally and economically beneficial (Udvardi and Poole 2013). The use of PGPR as biofertilizer depends upon crop selection, for instance, legume crops are biofertilized by nitrogen-fixing symbiotic bacteria, such as *Rhizobium*, *Bradyrhizobium*, *Mesorhizobium*, *Ensifer (Sinorhizobium)*, etc., while for nitrogen fixation in other non-legume crops, such as rice and wheat, *Pseudomonas*, *Azotobacter*, etc. are utilized (Gourion et al. 2015). *Pseudomonas* of the family Pseudomonadaceae distributed in the environment also inhabits in soil. It is a Gram-negative, rod-shaped, non-spore forming, aerobic, catalase–oxidase positive bacteria. Some of the *Pseudomonas* spp. is considered as one of the most promising groups of bacteria amongst PGPR as it is known to have various beneficial plant growth-promoting traits (Etesami 2022). Recently, seed inoculants of some specific strain of *Pseudomonas fluorescens* have been used on crop plants to increase crop yield and soil fertility (Swarnalakshmi et al. 2020).

PGPR as microbial inoculants are used to apply on several crops differently, in the form of liquid and solid formulations. In addition, the mode of inoculation may vary from crop to crop, depending upon their interaction with the host plants (Sindhu et al. 1999). For example, a rhizobia bioinoculant must be inoculated, often in liquid formulation, so as to colonize in the rhizosphere rapidly and develop root nodule in the initial of rhizosphere genesis. However, several reports suggest that solid bioformulation in certain carrier materials has a positive influence on vegetative and reproductive parameters of growth and increases crop productivity (Egamberdieva et al. 2013). Besides, the interaction of rhizobacteria with plants is ever considered as a central and focal phenomenon to decide the beneficial impact of bioinoculants on the plant growth promotion.

13.7 Constraints, Challenges and Future

The major constraints have arisen due to failure of performance in the farmer's field. It indicates the results obtained at laboratory scale are not adopted in field soil. The nitrogen content of the soil, its texture and chemical composition also remain challenging. Furthermore, host plant age, variety and other characteristics affect the performance of nitrogen-fixing bacteria. The recommended dose of N fertilizer is detrimental to the bacteria growth and survival; therefore, optimization of sub-lethal dose is utmost necessary to evaluate before inoculating the nitrogen-fixing bacteria in soil or coating on the seed. In some non-legumes, e.g. in sugarcane, the high N-fertilized soil (NH_3) reduces root colonization of *H. seropedicae*. Similarly, Ca^{2+} $(PO_4)^{3-}$ also inhibit colonization efficiency of *Azospirillum* in wheat. Hence, chemical adaptive nitrogen-fixing bacteria (NFB) variants counteract the derogatory effect of N fertilizer and allow them to colonize aggressively. In addition, edaphic, climatic and other environmental factors contribute to the variable effects in diazotrophic bacteria. Hence, proper research is required to sustain the benefits and application of NFB in non-legumes.

13.8 Conclusion

Associative including free-living and endophytic bacteria are exclusively beneficial for non-legume crops with respect to nitrogen fixation. However, these exhibit other beneficial roles, considering their roles as plant growth-promoting rhizobacteria (PGPR). On the other hand, rhizobia–non-legume interactions are still in the nutshell and require putting more effort to emerge these bacteria as bioinoculants or biofertilizers. However, having a successful bioinoculant depends upon various aspects during plant–microbe interaction, hence selecting demand-based biofertilizers for future agriculture is an inevitable approach that can lead us to ever-sustainable agriculture.

Acknowledgement Not applicable.

Conflict of Interest Author(s) declares no conflict of interest.

References

Agarwal M, Dheeman S, Dubey RC, Kumar P, Maheshwari DK, Bajpai VK (2017) Differential antagonistic responses of *Bacillus pumilus* MSUA3 against *Rhizoctonia solani* and *Fusarium oxysporum* causing fungal diseases in *Fagopyrum esculentum* Moench. Microbiol Res 205:40–47

Akkermans AD, Abdulkadir S, Trinick MJ (1978) N_2-fixing root nodules in *Ulmaceae*: *Parasponia* or (and) *Trema* spp. Plant Soil 49(3):711–715

Baliyan N, Dhiman S, Dheeman S, Kumar S, Arora NK, Maheshwari DK (2021) Optimization of gibberellic acid production in endophytic *Bacillus cereus* using response surface methodology and its use as plant growth regulator in chickpea. J Plant Growth Regul 1:11. https://doi.org/10.1007/s00344-021-10492-2

Balsanelli E, Serrato RV, De Baura VA, Sassaki G, Yates MG, Rigo LU et al (2010) *Herbaspirillum seropedicae* rfbB and rfbC genes are required for maize colonization. Environ Microbiol 12(8):2233–2244

Barbosa JZ, Hungria M, da Silva Sena JV, Poggere G, dos Reis AR, Corrêa RS (2021) Meta-analysis reveals benefits of co-inoculation of soybean with *Azospirillum brasilense* and *Bradyrhizobium* spp. in Brazil. Appl Soil Ecol 163:103913

Becking JH (1982) Nitrogen fixation in nodulated plants other than legumes. Adv Agric Microbiol 1:89–110

Beijerinck MW (1901) Über oligonitrophile mikroben, centralblatt für bakteriologie parasitenkunde, infektionskrankheiten und hygiene. Abteilung II 7:561–582

Beyan SM, Wolde-Meskel E, Dakora FD (2018) An assessment of plant growth and N_2 fixation in soybean genotypes grown in uninoculated soils collected from different locations in Ethiopia. Symbiosis 75(3):189–203

Bhattacharyya PN, Jha DK (2012) Plant growth-promoting rhizobacteria (PGPR): emergence in agriculture. World J Microbiol Biotechnol 28(4):1327–1350

Böhm M, Hurek T, Reinhold-Hurek B (2007) Twitching motility is essential for endophytic rice colonization by the N2-fixing endophyte *Azoarcus* sp. strain BH72. Mol Plant-Microbe Interact 20(5):526–533

Bourion V, Heulin-Gotty K, Aubert V, Tisseyre P, Chabert-Martinello M, Pervent M et al (2018) Co-inoculation of a pea core-collection with diverse rhizobial strains shows competitiveness for nodulation and efficiency of nitrogen fixation are distinct traits in the interaction. Front Plant Sci 8:2249

Bueno Batista M, Dixon R (2019) Manipulating nitrogen regulation in diazotrophic bacteria for agronomic benefit. Biochem Soc Trans 47(2):603–614

Burdman S, Dulguerova G, Okon Y, Jurkevitch E (2001) Purification of the major outer membrane protein of *Azospirillum brasilense*, its affinity to plant roots, and its involvement in cell aggregation. Mol Plant-Microbe Interact 14(4):555–561

Caires EF, Bini AR, Barão LFC, Haliski A, Duart VM, Ricardo KDS (2021) Seed inoculation with *Azospirillum brasilense* and nitrogen fertilization for no-till cereal production. Agron J 113(1): 560–576

Cassán F, Coniglio A, López G, Molina R, Nievas S, de Carlan CLN, Donadio F, Torres D, Rosas S, Pedrosa FO, de Souza E (2020) Everything you must know about *Azospirillum* and its impact on agriculture and beyond. Biol Fertil Soils 56(4):461–479

Chandra S, Choure K, Dubey RC, Maheshwari DK (2007) Rhizosphere competent *Mesorhizobium loti* MP6 induces root hair curling, inhibits *Sclerotinia sclerotiorum* and enhances growth of Indian mustard (*Brassica campestris*). Braz J Microbiol 38(1):124–130

Chelius MK, Triplett EW (2000) Immunolocalization of dinitrogenase reductase produced by *Klebsiella pneumoniae* in association with *Zea mays* L. Appl Environ Microbiol 66(2):783–787

Choudhary DK, Prakash A, Johri BN (2007) Induced systemic resistance (ISR) in plants: mechanism of action. Indian J Microbiol 47(4):289–297

Cocking EC (2003) Endophytic colonization of plant roots by nitrogen-fixing bacteria. Plant Soil 252(1):169–175

Compant S, Duffy B, Nowak J, Clément C, Barka EA (2005) Use of plant growth-promoting bacteria for biocontrol of plant diseases: principles, mechanisms of action, and future prospects. Appl Environ Microbiol 71(9):4951–4959

Dakora FD, Matiru V, Kanu AS (2015) Rhizosphere ecology of lumichrome and riboflavin, two bacterial signal molecules eliciting developmental changes in plants. Front Plant Sci 6:700

Davey ME, O'Toole GA (2000) Microbial biofilms: from ecology to molecular genetics. Microbiol Mol Biol Rev 64(4):847–867

Day JM, Döbereiner J (1976) Physiological aspects of N_2-fixation by *Azospirillum* from Digitaria roots. Soil Biol Biochem 8(1):45–50

Dheeman S, Maheshwari DK, Baliyan N (2017) Bacterial endophytes for ecological intensification of agriculture. In: Maheshwari DK (ed) Endophytes: biology and biotechnology. Springer, Cham, pp 193–231

Dhiman S, Kumar S, Baliyan N, Dheeman S, Maheshwari DK (2021) Cattle dung manure microbiota as a substitute for mineral nutrients and growth management practices in plants. In: Maheshwari DK, Dheeman S (eds) Endophytes: mineral nutrient management, vol 3. Springer, Cham, pp 77–103

Dubois M, Broeck LV, Inzé D (2018) The pivotal role of ethylene in plant growth. Trends Plant Sci 23:311–323

Egamberdieva D, Jabborova D, Wirth S (2013) Alleviation of salt stress in legumes by coinoculation with *Pseudomonas* and *Rhizobium*. In: Arora NK (ed) Plant microbe symbiosis: fundamentals and advances. Springer, New Delhi, pp 291–303

Elmerich C (2007) Historical perspective: from bacterization to endophytes. In: Associative and endophytic nitrogen-fixing bacteria and cyanobacterial associations. Springer, Dordrecht, pp 1–20

Etesami H (2022) Root nodules of legumes: a suitable ecological niche for isolating nonrhizobial bacteria with biotechnological potential in agriculture. Curr Res Biotech 4:78–86

Galindo FS, Rodrigues WL, Fernandes GC, Boleta EHM, Jalal A, Rosa PAL et al (2022) Enhancing agronomic efficiency and maize grain yield with *Azospirillum brasilense* inoculation under Brazilian savannah conditions. Eur J Agron 134:126471

Gilbert JH (1891) Results of experiments at Rothamsted on the question of the fixation of free nitrogen. Nature 45:32–33

Glick BR (2003) Phytoremediation: synergistic use of plants and bacteria to clean up the environment. Biotechnol Adv 21(5):383–393

Glick BR, Penrose DM, Li J (1998) A model for the lowering of plant ethylene concentrations by plant growth-promoting bacteria. J Theor Biol 190(1):63–68

Gourion B, Berrabah F, Ratet P, Stacey G (2015) *Rhizobium*–legume symbioses: the crucial role of plant immunity. Trends Plant Sci 20(3):186–194

Griesmann M, Chang Y, Liu X, Song Y, Haberer G, Crook MB et al (2018) Phylogenomics reveals multiple losses of nitrogen-fixing root nodule symbiosis. Science 361(6398):eaat1743

Gyaneshwar P, Naresh Kumar G, Parekh LJ, Poole PS (2002) Role of soil microorganisms in improving P nutrition of plants. Plant Soil 245(1):83–93

Hurek T, Handley LL, Reinhold-Hurek B, Piché Y (2002) *Azoarcus* grass endophytes contribute fixed nitrogen to the plant in an unculturable state. Mol Plant-Microbe Interact 15(3):233–242

Iniguez AL, Dong Y, Triplett EW (2004) Nitrogen fixation in wheat provided by *Klebsiella pneumoniae* 342. Mol Plant-Microbe Interact 17(10):1078–1085

Ivleva NB, Groat J, Staub JM, Stephens M (2016) Expression of active subunit of nitrogenase via integration into plant organelle genome. PLoS One 11(8):e0160951

James EK, Gyaneshwar P, Mathan N, Barraquio QL, Reddy PM, Iannetta PPM, Olivares FL, Ladha JK (2002) Infection and colonization of rice seedlings by the plant growth-promoting bacterium *Herbaspirillum seropedicae* Z67. Mol Plant-Microbe Interact 15(9):894–906

Kloepper JW, Leong J, Teintze M, Schroth MN (1980) Enhanced plant growth by siderophores produced by plant growth-promoting rhizobacteria. Nature 286(5776):885–886

Krause A, Ramakumar A, Bartels D, Battistoni F, Bekel T, Boch J et al (2006) Complete genome of the mutualistic, N_2-fixing grass endophyte *Azoarcus* sp. strain BH72. Nat Biotechnol 24(11): 1384–1390

Kumar K, Goh KM (2002) Management practices of antecedent leguminous and nonleguminous crop residues in relation to winter wheat yields, nitrogen uptake, soil nitrogen mineralization and simple nitrogen balance. Eur J Agron 16(4):295–308

Ma W, Sebestianova SB, Sebestian J, Burd GI, Guinel FC, Glick BR (2003) Prevalence of 1-aminocyclopropane-1-carboxylate deaminase in *Rhizobium* spp. Antonie Van Leeuwenhoek 83(3):285–291

Mabrouk Y, Hemissi I, Salem IB, Mejri S, Saidi M, Belhadj O (2018) Potential of rhizobia in improving nitrogen fixation and yields of legumes. Symbiosis 107:73495

Maheshwari DK (2010) Plant growth and health promoting bacteria, vol 18. Springer Science & Business Media

Maheshwari DK (2017) Conclusion. In: Maheshwari DK (ed) Endophytes: biology and biotechnology. Sustainable development and biodiversity, vol 15. Springer, Cham

Maheshwari DK, Dheeman S, Annapurna K (2017) Endophytes as contender of plant productivity and protection: an introduction. In: Maheshwari DK, Annapurna K (eds) Endophytes: crop productivity and protection. Sustainable development and biodiversity, vol 16. Springer, Cham

Mano H, Morisaki H (2008) Endophytic bacteria in the rice plant. Microbes Environ 23(2):109–117

McGuiness PN, Reid JB, Foo E (2019) The role of gibberellins and brassinosteroids in nodulation and arbuscular mycorrhizal associations. Front Plant Sci 10:269

Mia MB, Shamsuddin ZH (2010) *Rhizobium* as a crop enhancer and biofertilizer for increased cereal production. Afr J Biotechnol 9(37):6001–6009

Möllerová J (1990) Nitrogen fixation: hundred years after. Folia Geobot Phytotax 25:256. https://doi.org/10.1007/BF02913022

Mostafa MA, Mahmoud Z (1951) Bacterial isolates from root nodules of *Zygophyllaceae*. Nature 167:446–447

Okon Y, Heytler PG, Hardy RWF (1983) N_2 fixation by *Azospirillum brasilense* and its incorporation into host *Setaria italica*. Appl Environ Microbiol 46(3):694–697

Op den Camp RH, Polone E, Fedorova E, Roelofsen W, Squartini A, Op den Camp HJ, Bisseling T, Geurts R (2012) Non-legume *Parasponia andersonii* deploys a broad rhizobium host range strategy resulting in largely variable symbiotic effectiveness. Mol Plant-Microbe Interact 25(7): 954–963

Pandey P, Kang SC, Maheshwari DK (2005) Isolation of endophytic plant growth promoting *Burkholderia* sp. MSSP from root nodules of *Mimosa pudica*. Curr Sci 89:177–180

Pedrosa FO, Oliveira ALM, Guimarães VF, Etto RM, Souza EM, Furmam FG et al (2020) The ammonium excreting *Azospirillum brasilense* strain HM053: a new alternative inoculant for maize. Plant Soil 451(1):45–56

Pedrosa KM, de Almeida HA, Ramos MB, de Faria Lopes S (2021) Plants with similar characteristics drive their use by local populations in the semi-arid region of Brazil. Environ Dev Sustain 23(11):16834–16847

Perrine-Walker FM, Gartner E, Hocart CH, Becker A, Rolfe BG (2007) Rhizobium-initiated rice growth inhibition caused by nitric oxide accumulation. Mol Plant-Microbe Interact 20(3): 283–292

Reinhold BB, Chan SY, Reuber TL, Marra A, Walker GC, Reinhold VN (1994) Detailed structural characterization of succinoglycan, the major exopolysaccharide of *Rhizobium meliloti* Rm1021. J Bacteriol 176:1997–2002

Reinhold-Hurek B, Hurek T (1998) Life in grasses: Diazotrophic endophytes. Trends Microbiol 6(4):139–144

Richardson AE, Barea JM, McNeill AM, Prigent-Combaret C (2009) Acquisition of phosphorus and nitrogen in the rhizosphere and plant growth promotion by microorganisms. Plant Soil 321(1):305–339

Rogers C, Oldroyd GE (2014) Synthetic biology approaches to engineering the nitrogen symbiosis in cereals. J Exp Bot 65(8):1939–1946

Roncato-Maccari LD, Ramos HJO, Pedrosa FO, Alquini Y, Chubatsu LS, Yates MG, Rig L, Steffens MB, Souza EM (2003) Root colonization, systemic spreading and contribution of *Herbaspirillum seropedicae* to growth of rice seedling. Symbiosis 35:261–270

Sabet YS (1946) Bacterial root nodules in the *Zygophyllaceae*. Nature 157(3994):656–657

Santi C, Bogusz D, Franche C (2013) Biological nitrogen fixation in non-legume plants. Ann Bot 111(5):743–767

Sevilla M, Burris RH, Gunapala N, Kennedy C (2001) Comparison of benefit to sugarcane plant growth and 15N2 incorporation following inoculation of sterile plants with *Acetobacter diazotrophicus* wild-type and *nif* mutant strains. Mol Plant-Microbe Interact 14(3):358–366

Sindhu SS, Gupta SK, Dadarwal KR (1999) Antagonistic effect of *Pseudomonas* spp. on pathogenic fungi and enhancement of growth of green gram (*Vigna radiata*). Biol Fertil Soils 29(1): 62–68

Smercina DN, Evans SE, Friesen ML, Tiemann LK (2019) To fix or not to fix: controls on free-living nitrogen fixation in the rhizosphere. Appl Environ Microbiol 85(6):e02546–e02518

Soyano T, Liu M, Kawaguchi M, Hayashi M (2021) Leguminous nodule symbiosis involves recruitment of factors contributing to lateral root development. Curr Opin Plant Biol 59:102000

Spaepen S, Vanderleyden J (2011) Auxin and plant-microbe interactions. Cold Spring Harb Perspect Biol 3(4):a001438

Swarnalakshmi K, Yadav V, Tyagi D, Dhar DW, Kannepalli A, Kumar S (2020) Significance of plant growth promoting rhizobacteria in grain legumes: growth promotion and crop production. Plants 9(11):1596

Trinick MJ (1973) Symbiosis between *Rhizobium* and the non-legume, Trema aspera. Nature 244(5416):459–460

Udvardi M, Poole PS (2013) Transport and metabolism in legume-rhizobia symbioses. Annu Rev Plant Biol 64:781–805

Van Puyvelde S, Cloots L, Engelen K, Das F, Marchal K, Vanderleyden J, Spaepen S (2011) Transcriptome analysis of the rhizosphere bacterium *Azospirillum brasilense* reveals an extensive auxin response. Microb Ecol 61(4):723–728 van Velzen R, Holmer R, Bu F, Rutten L, van Zeijl A, Liu W et al (2018) Comparative genomics of the nonlegume Parasponia reveals insights into evolution of nitrogen-fixing rhizobium symbioses. Proc Natl Acad Sci 115(20):E4700–E4709

Verma SC, Singh A, Chowdhury SP, Tripathi AK (2004) Endophytic colonization ability of two deep-water rice endophytes, *Pantoea* sp. and *Ochrobactrum* sp. using green fluorescent protein reporter. Biotechnol Lett 26(5):425–429

Verma JP, Yadav J, Tiwari KN, Lavakush S, Singh V (2010) Impact of plant growth promoting rhizobacteria on crop production. Int J Agric Res 5(11):954–983

Wardhani TA, Roswanjaya YP, Dupin S, Li H, Linders S, Hartog M et al (2019) Transforming, genome editing and phenotyping the nitrogen-fixing tropical Cannabaceae tree *Parasponia andersonii*. J Vis Exp 150:e59971

Yang ZR, Lin Q (2005) Comparative morphology of the leaf epidermis in Schisandra (Schisandraceae). Bot J Linn Soc 148(1):39–56

Yanni YG, Rizk RY, El-Fattah FKA, Squartini A, Corich V, Giacomini A, de Bruijn F, Rademaker J, Maya-Flores J, Ostrom P, Vega-Hernandez M, Hollingsworth RI, Martinez-Molina E, Mateos P, Velazquez E, Wopereis J, Triplett E, Umali-Garcia M, Anarna JA, Rolfe BG, Ladha JK, Hill J, Mujoo R, Ng PK, Dazzo FB (2001) The beneficial plant growth-promoting association of *Rhizobium leguminosarum* bv. *trifolii* with rice roots. Aust J Plant Physiol 28:845–870

Yi HS, Yang JW, Ryu CM (2013) ISR meets SAR outside: additive action of the endophyte *Bacillus pumilus* INR7 and the chemical inducer, benzothiadiazole, on induced resistance against bacterial spot in field-grown pepper. Front Plant Sci 4:122

Chapter 14
Role of Bacterial Secondary Metabolites in Modulating Nitrogen-Fixation in Non-legume Plants

Asadullah and Asghari Bano

Abstract Nitrogen is one of the essential elements which plays a key role in plant growth and development. Beneficial microorganisms are eco-friendly and sustainable source of agroindustry. Bacteria either free-living or as endophytes in plant roots can fix atmospheric nitrogen to nitrogenous compounds in the soil and make them readily available to plant roots. Extensive research was carried out to unravel the mechanisms involved in rhizobia–legume interaction. However, little attention was paid to nitrogen-fixation in non-legume plants including important cereals like maize, wheat, rice, sugar cane and other actinorhizal plants. Both the associative symbiosis in non-legumes and those in legumes are modulated by the presence of secondary metabolites, exuded by both the partners (microbes and the host plants). This chapter will highlight on signalling molecules involved in establishing symbiotic association between non-legume plants and their partner with particular emphasis on growth-promoting rhizobacteria. Their role in improving nitrogen use efficiency in plants and the effects of abiotic factors will also be discussed. The identification of key secondary metabolites involve in cross talk for better N-fixation is necessary to engineer N-fixer non-legume plants.

Keywords Non-legume · N-fixation · Rhizobacteria · Symbiosis · PGPR

14.1 Introduction

Besides the inflated rate of fertilizers, their losses and subsequent contamination of ground water and eutrophication of rivers led scientists to search for an alternative effective eco-friendly approach, i.e. biological nitrogen-fixation.

Asadullah
The Peace Group of Schools and Colleges, Charsadda, KP, Pakistan

A. Bano (✉)
Department of Biosciences, University of Wah, Wah Cantt, Pakistan
e-mail: asghari.bano@uow.edu.pk

© The Author(s), under exclusive license to Springer Nature Singapore Pte Ltd. 2022
D. K. Maheshwari et al. (eds.), *Nitrogen Fixing Bacteria: Sustainable Growth of Non-legumes*, Microorganisms for Sustainability 36,
https://doi.org/10.1007/978-981-19-4906-7_14

14.2 Diversity of Nitrogen-Fixing Bacteria

The members of bacteria, archaea, cyanobacteria and mycorrhizal fungi are involved in nitrogen-fixation and occur in a wide range of habitat based on their growth requirements. They are found in multifaceted association: free-living in soil, symbiotic association with legumes, in associative symbiosis with actinorhizal plants and *Frankia*, and cyanobacterial association with mosses, ferns, angiosperms and gymnosperms. The symbiotic association found between cyanobacteria and cycads, *Frankia* and actinorhizal plants, rhizobia and legume plants is governed by the release of flavonoids and activation of nod factor. The bacteria exchange fixed nitrogen with host cell in response to carbon source from host. This constitutes a major group of nitrogen-fixing microbes. The associative nitrogen fixers (*Azospirillum*, *Herbaspirillum*, *Klebsiella*, etc.) have been reported to found on the histoplane or as endophytes in roots of maize, wheat and sugarcane. Their exudates contain carbon substrate, which serves as source of energy for nitrogen-fixation. They derive energy from oxidation of organic molecules for nitrogen-fixation. Free-living nitrogen fixers can fix about one-tenth of the total atmospheric nitrogen fixed by the symbiotic association (Dahal 2016).

Actinorhizal plants have the ability to develop endosymbiosis with *Frankia*, a nitrogen-fixing soil actinomycetes (Martin et al. 2017). This endosymbiosis is established by root nodule in which *Frankia* provides fixed N to plant roots in response to reduced carbon. In contrast to rhizobia-nodule legume association, *Frankia* can protect oxygen-sensitive nitrogenase enzyme complex. In some cases, e.g. cereals, these bacteria can colonize the plant roots system intracellularly and fix nitrogen inside the cells utilizing photo assimilates (such as glucose) as a source of energy.

The blue-green algae (cyanobacteria) establish symbiotic in a wide range of plant species, viz. angiosperm *Gunnera*, the water fern *Azolla*, the bryophytes *Blasia* and *Anthoceros* and the gymnosperm *Cycas*. In case of *Azolla*, it is an integral part of the host throughout the life and inherit to next generation (Adams et al. 2013; Chang et al. 2019; Warshan 2017). Mycorrhizal fungi are the most abundant fungal symbionts of plants that have colonized plant roots since 400 million years ago (Selosse et al. 2015). They are associated with roots of 70–90% of land plants, including trees, grasses and many crops. They are classified into two broad categories, referred to as ectomycorrhiza and endomycorrhiza. Endomycorrhiza are further divided into orchid, ericoid and arbuscular mycorrhizae (AM) fungi. Examples of AM fungi include *Funneliformis*, *Claroideoglomus*, *Rhizoglomus*, *Gigaspora*, *Acaulospora*, *Glomus*, *Diversispora*, *Septoglomus* and *Scutellospora* (Lara-Capistran et al. 2021). The AM fungi make association with *Coffea arabica*, *Hordeum vulgare*, *Triticum aestivum*, *Sorghum bicolor* and *Zea mays* (Chang et al. 2019; Lara-Capistran et al. 2021).

14.3 Diversity of Non-legume Plants

In non-legume plants, the symbiosis of root nodulation is limited to single clade Fabids, which include four orders: Fabales, Fagales, Rosales and Cucurbitales. Together these orders are commonly termed as nitrogen-fixing clade (NFC) (Li et al. 2015). The diversification of nitrogen-fixing trait into four orders finds its origin about 110 million years ago. According to the single gain massive loss hypothesis, two major switches in microsymbionts had occurred, resulting in the evolution of rhizobia from most primitive *Frankia* ancestor, as *Frankia* have intrinsic characteristic to protect nitrogenase from oxidation and can fix N in free form. In contrast to rhizobia, which depend on the mechanism provided by plants. N-fixing nodulation traits are distributed amongst ten lineage of NFC orders, despite the occurrence of many non-nodulating species. The greatest diversity of root nodule symbiosis is found in order Fabales, which contain legume family (Leguminosae). Actinorhizal plants represent a diverse group of about 220 species belonging to eight plant families distributed in three orders, Fagales, Rosales and Cucurbitales. They establish a N-fixing nodule symbiosis with diazotrophic Actinobacteria of the genus *Frankia* (Li et al. 2015). The remaining two lineage, legumes (Fabales) and *Parasponia* (Rosales), establish a nodule symbiosis with rhizobia (Li et al. 2015). *Parasponia* is the only non-legume lineage to have evolved a rhizobia symbiosis. *Parasponia andersonii* is able to modulate nitrogen-fixing rhizobia symbionts under low nitrogen conditions (Dupin et al. 2020). A study was conducted on *Sierra mixe* and *Maize landrace* to explore the mechanism of nitrogen-fixing microbial association with cereals in nitrogen-depleted field. The plant was characterized by the extensive development of aerial root system that secretes a large amount of mucilage. Analysis of mucilage showed that it was enriched with nitrogen-fixing microbiota. Three main functionalities were identified in maize mucilage responsible for productive diazotrophic association and these were fructose, arabinose and galactose (Van Deynze et al. 2018).

14.4 Signal Cross-Talk for Synergetic Symbiosis

The establishment of symbiosis for N-fixation is a complex event and requires coordinated regulation of corresponding genes expression and release of signalling molecules into the rhizosphere. The exchange of such signalling molecules varies according to the microbe–host interaction. Plants, the recruiter of efficient symbionts, secrete a complex mixture of secondary metabolites such as amino acids and amide, benzoxazinoids, coumarins, enzymes, growth factors, sugars, organic acids, phenolic acids, flavonoids and strigolactones to reshape and drive resilient N-fixer partner for symbiosis (Jacoby et al. 2020; Zhalnina et al. 2018).

Flavonoids are the major class of secondary metabolites (phlobaphene, flavones, flavanones, flavanols, aurones, isoflavonoids, anthocyanins and condensed tannins), which have prime role in biological nitrogen-fixation (Dong and Song 2020). Flavonoids activate nod box genes to encode enzymes responsible for the synthesis of lipo-chitooligosaccharides (LCOs), well known nod factors perceived by legumes and *Parasponia* that trigger a symbiotic signalling cascade (Granqvist et al. 2015). The control mechanism of N-fixation by rhizobia is not exclusive to the legume clade only. Sequenced genome of *Frankia* spp. showed the presence of homologues *nod*ABC genes that encode proteins required for the synthesis of core Nod factor structure (Nguyen et al. 2019). Root hair deformation factor, a signalling molecule secreted by *Frankia* to initiate deformation (curling) of root hair for infection, is chemically distinct from chitin-based rhizobia Nod factor (Cissoko et al. 2018). For *Frankia–Alnus* symbiosis, amino acids (citrulline, arginine, aspartate, glutamate, beta-aminobutyric acid and alanine) are involved in promoting nodule organogenesis and stimulating nitrogen-fixation (Hay et al. 2020). The microbial community (*Frankia* and non-*Frankia* plant growth-promoting bacteria) of the nodule in non-legumes appears to be shaped by different bioclimates, with being less abundant under more arid environments (Ghodhbane-Gtari et al. 2021). There are four taxonomic clusters of *Frankia*. Clusters I, II and III engage in root nodule symbiosis. Cluster II species activate symbiosis by producing LCOs, whereas clusters I and III do not produce LCOs as they lack the canonical *nod* genes. Initial characterization showed that molecules of clusters I and II alternative signal to LCOs are hydrophilic, thermoresistant and resistant to chitinase digestion, indicating structural differences from LCOs (Cissoko et al. 2018).

A class of terpenoids, strigolactones, has been recently identified as suitable candidate for AM fungi association with plant roots (Mishra et al. 2017). This novel phytohormone stimulates fungal metabolism and hyphae branching under nitrogen-deficient environment. AM fungi use Myc factors (a combination of chitin oligomers and LCOs) for symbiosis, which are structurally very similar to nod factor.

14.5 Mechanism of Symbiosis

All nitrogen-fixing symbiosis require the activation of common symbiotic signalling pathway (CSSP). Therefore, both Myc and Nod factors are perceived by lysine-motif-domain containing receptor-like kinase (LysM-RLKs) present on the plasma membrane of plant cells (Bozsoki et al. 2017). Upon binding, this also interacts with the leucine-rich repeat-type symbiosis receptor kinase, thereby activating the CSSP. In case of rhizobium symbiosis, the CSSP is activated by root exudates containing flavonoids in response to LCOs. Studies in the non-legume rhizobium nodulator *Parasponia andersonii* and the two actinorhizal plant species, *Casuarina glauca* and *Alnus glutinosa*, revealed that instead of LCOs, the CSSP pathway is regulated by calcium oscillations (Huisman and Geurts 2020). Nuclear Ca^{2+} oscillations are a hallmark of symbiotic signalling induced by rhizobia, *Frankia* and AM fungi. Ca^{2+} spiking is decoded by calcium-calmodulin-dependent kinase (CCaMK), leading to activation of MtNFP/LjNFR5/PanNFP2 (Van velzen et al. 2018). It results in the transcriptional activation of NIN (nodule organogenesis, which is also activated by cytokinin), ERN (intracellular infection) and RAM1 (AM symbiosis programme) (Huisman and Geurts 2020).

The general pattern of plant–cyanobacterial symbiosis involves two phases. The early phase comprised chemical signalling between partners, and in the second phase the physical association with host is achieved. The host plant releases chemical signal known as hormogonia-inducing factor (HIF) into rhizosphere soil, which induces the conversion of vegetative cyanobacterial filament into transient motile stage termed hormogonia (Warshan 2017). The motile hormogonia then migrate into the host's internal cells. Various chemo-attractants are believed to be released by host plant to trigger this phenomenon. One such compound is 1-palmitoyl-2-linoleoyl-sn-glycerol, isolated and characterized from coralloid roots of *Cycas revoluta* (Hashidoko et al. 2019). This sugar base molecule has proven to be attractive for hormogonia to permit their entry into host root cells. Once a cyanobiont successfully enters its host, the plant partner releases hormogonia repressing factor (HRF), which hinders HIF release. Besides HIF and HRF, phenolics are also known to participate in cyanobacteria–host symbiosis. The next stage is the formation of heterocyst, the site for N-fixation.

14.6 Abiotic and Biotic Factor Effects on the N-Fixation in Non-legumes

Abiotic factors like drought, salinity, temperature and light exert significant impact on N-uptake by plants as well on nitrogen solubilization in soil. An investigation was carried out to find out the effect of drought on nitrogen availability to fine roots and N-uptake by accessing the saplings of *Acer pseudoplatanus*, *Fagus sylvatica*, *Quercus petraea*, *Abies alba*, *Picea abies* and *Pinus sylvestris* (Joseph et al. 2021). Study evidences the role of drought in reduction of nitrogen uptake in plants, thereby affects the net biomass and overall N content in plants. Drought increased mineral N content (+31%) but reduced N-mineralization rate (−5.7%) and nitrification rate (−13.8%), and thus left total N unchanged (Deng et al. 2021). In contrast, high water availability also impacts nitrogen transfer as nitrates leach down below root zone and become less available to plants (Thilakarathna 2016). Global warming has negative impact on N-uptake by plants and their translocation from roots to shoots. Jayawardena et al. (2017) performed an experiment on *Lycopersicum esculentum*, which was grown at two levels of CO_2 (400 or 700 ppm) and two temperature regimes (30 or 37 °C) with NO_3^- or NH_4^+ as the N source. Elevated CO_2 plus warming decreased N assimilation by roots, and their translocation to shoots. In another experiment, it was suggested that global warming reduced fertilizer N use efficiency and increased N loss to environment in the paddy rice (Yang et al. 2019). Similarly, salinity affects different stages of N-metabolism including N-uptake, ammonia reduction and ammonium assimilation, leading to severe decline in crop productivity (Ashraf et al. 2018). Maize plants were treated with NaCl and grown in two separate N forms molecules, the first one contains NO_3^- while the other contains NH_4^+ as N source (Hessini et al. 2019). Results showed that nitrate was partly replaced by Cl^- ion during root uptake; however, NH_4^+ favoured maize growth. The spatial arrangement of root system has an association with efficient N transfer. The close contact of root system with N root exudates decreases the distance for mass flow, but as the distance increases the N translocation also decreases, because majority of N (95%) is located in top soil layer (Laberge et al. 2011; Rasmussen et al. 2013). The composition of plant community also directs the flow of N from soil to plant biomass. A mix crop stands of legume and non-legume plants enhance the N-uptake from soil (Li et al. 2015). Phenology of plants also affects N transfer. In mung bean, it was reported that N transfer slightly increased from pod setting (7.6%) to maturity (9.7%) (Zang et al. 2015).

14.7 Effects of N-Fixing Bacteria on Amount of Nitrogen-Fixed in Non-legumes

The effective role of bacteria in sustaining N-metabolism, as well as their uptake and effect on plant growth, is summarized in Table 14.1. A comprehensive study of BNF was carried out in *Saccharum* spp., inoculated with *Bacillus megaterium* (CY5) and *Bacillus mycoides* (CA1). The expression of the *nif*H gene in both sugarcane varieties (GT11 and GXB9) inoculated with CY5 and CA1 was confirmed (Singh et al. 2020). Nitrogen-fixation was also confirmed using N-balance and $^{15}N_2$ isotope dilution in different plant parts of sugarcane. This was the first report of *Bacillus mycoides* as a nitrogen-fixing rhizobacterium in sugarcane (Singh et al. 2020). In another study, it was reported that bacteria contribute up to 65% of N content of two sugarcane cultivars under field conditions (Martins et al. 2020). *Azospirillum brasilense* (HM053), an ammonium excreting strain, strongly promoted maize growth, increasing the production from 460.5 to 1769.3 kg ha^{-1} (Pedrosa et al. 2019). An experiment was conducted to evaluate whether N-metabolism affects the bacterial contribution to plant growths. In one study, *Setaria viridis* plants were inoculated with *A. brasilense* FP2 and its *ntr*C mutant (Kukolj et al. 2020); in another study, *S. viridis* plants were challenged with *Herbaspirillum seropedicae* SmR1 and its mutant *nif*A (Agtuca et al. 2020). Both the studies concluded that plants inoculated with wild and mutant strains could increase plant biomass, demonstrating that nitrogen-fixation is not essential for plant growth promotion, at least under laboratory conditions. A study was conducted to investigate the uptake of N and their impact on two actinorhizal plants: *Alnus incana* infected with *Frankia* strain Ar13 and *Hippophae rhamnoides* infected with *Frankia* strain T_1 (Gentili and Huss-Danell 2019). Result showed that inoculated plants exhibit increases in N-fixing rate and higher plant dry matter at all harvesting times. *Pennisetum*

Table 14.1 Effect of beneficial bacteria and N-metabolism and yield of plants

Microorganisms	Host plants	Effects on N-uptake and yield	References
Azotobacter chroococcum, *Klebsiella variicola*	Wheat (var. Xiaoyan)	Enhance plant height, dry weight and N content by 97.7%	Wang et al. (2020)
Stenotrophomonas maltophilia, *Enterobacter* sp., *Bacillus* sp., *Ochrobactrum haematophilum*	Rice	Higher germination percentage, seed vigour index and total dry biomass	Mir et al. (2021)
Paenibacillus beijingensis BJ-18	Wheat, maize and cucumber	The total N were increased by 49.1–92.3% under low N and by 13–15.5% under high N	Li et al. (2019)
Pantoea dispersa AA7, *Enterobacter asburiae* BY4	Sugarcane	Contributed 21–35% of plant N by biological N-fixation	Singh et al. (2021)
Rhodopseudomonas palustris, *Rhodobacter sphaeroides*, *Saccharomyces cerevisiae*, *Streptomyces griseoviridis*	*Brassica* 'Zahoor' cultivar	Application with N and P at the rate of 180 kg ha^{-1} and 130 kg ha^{-1}, increased yield	Hazratullah et al. (2021)
Frankia species	*Parasponia* species	Root nodules in the non-legume plant *Parasponia* sp.	Van Velzen et al. (2018)
Azospirillum brasilense DSM 1690, *A. brasilense* DSM 2298, *Pseudomonas* sp. DSM 25356	*Lactuca sativa*	Increased N leaf concentration particularly when combined with N at 30 or 60 kg ha^{-1}	Consentino et al. (2022)
Pseudomonas sp. Y1	Switchgrass	Containing ACC deaminase and increased the total root length, root surface, shoot length and fresh and dry weight	Chen et al. (2022)
Pseudomonas protegens Pf-5	*Arabidopsis thaliana*	Improved N-fixation	Jing et al. (2020)

purpureum is a perennial elephant grass. It receives 70% N through biological nitrogen-fixation (De Morais et al. 2017). Similarly, feather moss receives 50% of total N input from its partner cyanobiont (Rousk et al. 2013).

14.8 Conclusion

Enhancement in biological nitrogen-fixation in non-legume (cereals and cash crops) is imperative to boost crop production and quality of cereal grains and is an alternative effective eco-friendly approach to minimize fertilizer usages. A mix crop stands of legume and non-legume plants enhance the N-uptake from soil; phenological stage of plants also plays dominant role in this context. Plant growth-promoting rhizobacteria (PGPR) used as bioinoculants and exhibiting associative symbioses may be more effective in terms of N-metabolism as they also produce phytohormones, which modulate the process of N-fixation and plant growth under abiotic stresses. To unravel signalling mechanism, identification of more secondary metabolites and their cross talk appears an effective way to modulate biological N-fixation and metabolism in non-legumes.

14.9 Future Perspectives

Most crops belong to non-legumes, so understanding the molecular mechanism mediated by secondary metabolites of plant and bacterial origin is of prime importance for agronomic implications that enable the use of N-fertilizers to be reduced. Indeed, the short-term goal is to exploit the diversity of N-fixing bacteria, but the long-term goal is to transfer the endosymbiotic association of major non-legume crops. Thus, the identification of key secondary metabolites involved in cross talk for better N-fixation is necessary to engineer N-fixer non-legume plants.

Acknowledgement Not applicable.

Conflict of Interest Author(s) declares no conflict of interest.

References

Adams DG, Bergman B, Nierzwicki-Bauer SA, Duggan PS, Rai AN, Schubler A (2013) Cyanobacterial-plant symbioses. In: Rosenberg E, DeLong EF, Lory S, Stackebrandt E, Thompson F (eds) The prokaryotes – prokaryotic biology and symbiotic associations. Springer, Berlin, pp 359–400. https://doi.org/10.1007/978-3-642-30194-0_17

Agtuca BJ, Stopka SA, Tuleski TR, Do Amaral FP, Evans S, Liu Y, Xu D, Monteiro RA, Koppenaal DW, Pasa-Toli CL, Anderton CR, Vertes A, Stacey G (2020) In-situ metabolomic analysis of *Setaria viridis* roots colonized by beneficial endophytic bacteria. Mol Plant-Microbe Interact 33:272–283

Ashraf M, Shahzad SM, Imtiaz M, Rizwan MS (2018) Salinity effects on nitrogen metabolism in plants–focusing on the activities of nitrogen metabolizing enzymes: a review. J Plant Nutr 41(8): 1065–1081

Bozsoki Z, Cheng J, Feng F, Gysel K, Vinther M, Andersen KR, Stougaard J (2017) Receptor-mediated chitin perception in legume roots is functionally separable from Nod factor perception. Proc Natl Acad Sci 114(38):E8118–E8127

Chang ACG, Chen T, Li N, Duan J (2019) Perspectives on endosymbiosis in coralloid roots: association of cycads and cyanobacteria. Front Microbiol 10:1888. https://doi.org/10.3389/fmicb.2019.01888

Chen Z, Zhou W, Sui X, Xu N, Zhao T, Guo Z, Wang Q (2022) Plant growth-promoting rhizobacteria with ACC deaminase activity enhance maternal lateral root and seedling growth in switchgrass. Front Plant Sci 12:800783. https://doi.org/10.3389/fpls

Cissoko M, Hocher V, Gherbi H, Gully D, Carré-Mlouka A, Sane S, Pignoly S, Champion A, Ngom M, Pujic P, Fournier P, Gtari M, Swanson E, Pesce C, Tisa LS, Sy MO, Svistoonoff S (2018) Actinorhizal signaling molecules: *Frankia* root hair deforming factor shares properties with NIN inducing factor. Front Plant Sci 9:1494. https://doi.org/10.3389/fpls.2018.01494

Consentino BB, Aprile S, Rouphael Y, Ntatsi G, De Pasquale C, Iapichino G, Sabatino L (2022) Application of PGPB combined with variable N doses affects growth, yield-related traits, N-fertilizer efficiency and nutritional status of lettuce grown under controlled condition. Agronomy 12(2):236

Dahal B (2016) Diversity of free-living nitrogen fixing bacteria in the Badlands of South Dakota. Master's Thesis. South Dakota State University. https://openprairie.sdstate.edu/etd/688

De Morais RF, Quesada DM, Reis VM, Urquiaga S, Alves BJR, Boddey RM (2017) Contribution of biological nitrogen fixation to elephant grass (*Pennisetum purpureum*). Plant Soil 356:23–34

Deng L, Peng C, Kim DG, Li J, Liu Y, Hai X, Kuzyakov Y (2021) Drought effects on soil carbon and nitrogen dynamics in global natural ecosystems. Earth-Sci Rev 214:103501

Dong W, Song Y (2020) The significance of flavonoids in the process of biological nitrogen-fixation. Int J Mol Sci 21(16):5926

Dupin SE, Geurts R, Kiers ET (2020) The non-legume *Parasponia andersonii* mediates the fitness of nitrogen-fixing rhizobial symbionts under high nitrogen conditions. Front Plant Sci 10:1779

Gentili FG, Huss-Danell K (2019) The $\delta^{15}N$ value of N_2 fixing actinorhizal plants and legumes grown with N_2 as the only nitrogen source. Symbiosis 79:213–219

Ghodhbane-Gtari F, D'Angelo T, Gueddou A, Ghazouani S, Gtari M, Tisa LS (2021) Alone yet not alone: *Frankia* lives under the same roof with other bacteria in actinorhizal nodules. Front Microbiol 12:749760. https://doi.org/10.3389/fmicb.2021.749760

Granqvist E, Sun J, den Camp RO, Pujic P, Hill L, Normand P et al (2015) Bacterial-induced calcium oscillations are common to nitrogen-fixing associations of nodulating legumes and nonlegumes. New Phytol 207:551e558. https://doi.org/10.1111/nph.13464

Hashidoko Y, Nishizuka H, Tanaka M, Murata K, Murai Y, Hashimoto M (2019) Isolation and characterization of 1-palmitoyl-2-linoleoyl-sn-glycerol as a hormogonium-inducing factor (HIF) from the coralloid roots of *Cycas revoluta* (Cycadaceae). Sci Rep 9(1):1–12

Hay AE, Herrera-Belaroussi A, Rey M, Fournier P, Normand P, Boubakri H (2020) Feedback regulation of N-fixation in *Frankia*-Alnus symbiosis through amino acids profiling in field and greenhouse nodules. Mol Plant-Microbe Interact 33(3):499–508

Hazratullah, Muhammad A, Alam M, Ahmad I, Jalal A (2021) Role of beneficial microbes with nitrogen and phosphorous levels on canola productivity. Braz J Biol 82:e227703. https://doi.org/10.1590/1519-6984.227703

Hessini K, Issaoui K, Ferchichi S, Saif T, Abdelly C, Siddique KH, Cruz C (2019) Interactive effects of salinity and nitrogen forms on plant growth, photosynthesis and osmotic adjustment in maize. Plant Physiol Biochem 139:171–178

Huisman R, Geurts R (2020) A roadmap toward engineered nitrogen-fixing nodule symbiosis. Plant Commun 1(1):100019

Jacoby RP, Chen L, Schwier M, Koprivova A, Kopriva S (2020) Recent advances in the role of plant metabolites in shaping the root microbiome. F1000Research 9:F1000 Faculty Rev-151. https://doi.org/10.12688/f1000research.21796.1

Jayawardena DM, Heckathorn SA, Bista DR, Mishra S, Boldt JK, Krause CR (2017) Elevated CO_2 plus chronic warming reduce nitrogen uptake and levels or activities of nitrogen-uptake and assimilatory proteins in tomato roots. Physiol Plant 159(3):354–365

Jing X, Cui Q, Li X, Yin J, Ravichandran V, Pan D, Zhang Y (2020) Engineering *Pseudomonas protegens* Pf-5 to improve its antifungal activity and nitrogen fixation. Microb Biotechnol 13(1): 118–133

Joseph J, Luster J, Bottero A, Buser N, Baechli L, Sever K, Gessler A (2021) Effects of drought on nitrogen uptake and carbon dynamics in trees. Tree Physiol 41(6):927–943

Kukolj C, Pedrosa FO, de Souza GA, Sumner LW, Lei Z, Sumner B, Do Amaral FP, Juexin W, Trupti J, Huergo LF, Monteiro RA, Valdameri G, Stacey G, de Souza EM (2020) Proteomic and metabolomic analysis of *Azospirillum brasilense ntrC* mutant under high and low nitrogen conditions. J Proteome Res 19:92–105

Laberge G, Haussmann BIG, Ambus P, Høgh-Jensen H (2011) Cowpea N rhizodeposition and its below-ground transfer to a co-existing and to a subsequent millet crop on a sandy soil of the Sudano-Sahelian eco-zone. Plant Soil 340:369–382. https://doi.org/10.1007/s11104-010-0609-6

Lara-Capistran L, Zulueta-Rodriguez R, Murillo-Amador B, Preciado-Rangel P, Verdecia-Acosta DM, Hernandez-Montiel LG (2021) Biodiversity of AM fungi in coffee cultivated on eroded soil. Agronomy 11(3):567

Li HL, Wang W, Mortimer PE, Li RQ, Li DZ, Hyde KD, Xu JC, Soltis DE, Chen ZD (2015) Large-scale phylogenetic analyses reveal multiple gains of actinorhizal nitrogen-fixing symbioses in angiosperms associated with climate change. Sci Rep 5:14023

Li Y, Li Y, Zhang H, Wang M, Chen S (2019) Diazotrophic *Paenibacillus beijingensis* BJ-18 provides nitrogen for plant and promotes plant growth, nitrogen uptake and metabolism. Front Microbiol 10:1119

Martin FM, Uroz S, Barker DG (2017) Ancestral alliances: plant mutualistic symbioses with fungi and bacteria. Science 356(6340):4501

Martins DS, Reis VM, Schultz N, Alves BJR, Urquiaga S, Pereira W, Sousa JS, Boddey RM (2020) Both the contribution of soil nitrogen and of biological N2 fixation to sugarcane can increase with the inoculation of diazotrophic bacteria. Plant Soil 454:155–169

Mir MI, Hameeda B, Quadriya H, Kumar BK, Ilyas N, Zuan ATK, Sayyed RZ (2021) Multifarious indigenous diazotrophic rhizobacteria of rice (*Oryza sativa* L.) rhizosphere and their effect on plant growth promotion. Front Nutr 8:781764

Mishra S, Upadhyay S, Shukla RK (2017) The role of strigolactones and their potential cross-talk under hostile ecological conditions in plants. Front Physiol 7:691

Nguyen TV, Wibberg D, Vigil-Stenman T, Berckx F, Battenberg K, Demchenko KN, Blom J, Fernandez MP, Yamanaka T, Berry AM, Kalinowski J, Brachmann A, Pawlowski K (2019) *Frankia*-enriched metagenomes from the earliest diverging symbiotic *Frankia* cluster: they come in teams. Genome Biol Evol 11:2273–2291

Pedrosa FO, Oliveira ALM, Guimarães VF, Etto RM, Souza EM, Furmam FG, Gonçalves DRP, Santos OJAP, Gonçalves LSA, Battistus AG, Galvao CW (2019) The ammonium excreting *Azospirillum brasilense* strain HM053: a new alternative inoculant for maize. Plant Soil 451:45–56

Rasmussen J, Gylfadóttir T, Loges R, Eriksen J, Helgadóttir Á (2013) Spatial and temporal variation in N transfer in grass-white clover mixtures at three Northern European field sites. Soil Biol Biochem 57:654–662. https://doi.org/10.1016/j.soilbio.2012.07.004

Rousk K, Jones DL, DeLuca TH (2013) Moss-cyanobacteria associations as biogenic sources of nitrogen in boreal forest ecosystems. Front Microbiol 14:150

Selosse MA, Strullu-Derrien C, Martin FM, Kamoun S, Kenrick P (2015) Plants, fungi and oomycetes: a 400-million years affair that shapes the biosphere. New Phytol 206:501–506. https://doi.org/10.1111/nph.13371

Singh RK, Singh P, Li HB, Song QQ, Guo DJ, Solanki MK, Li YR (2020) Diversity of nitrogen-fixing rhizobacteria associated with sugarcane: a comprehensive study of plant-microbe interactions for growth enhancement in *Saccharum* spp. BMC Plant Biol 20:1–21

Singh P, Singh RK, Li HB, Guo DJ, Sharma A, Lakshmanan P, Li YR (2021) Diazotrophic bacteria *Pantoea dispersa* and *Enterobacter asburiae* promote sugarcane growth by inducing nitrogen uptake and defense-related gene expression. Front Microbiol 11:600417. https://doi.org/10.3389/fmicb.2020.600417

Thilakarathna RMMS (2016) Genotypic variability among diverse red clover cultivars for nitrogen fixation and transfer. Ph.D. Thesis. Dalhousie Killam library

Van Deynze A, Zamora P, Delaux PM, Heitmann C, Jayaraman D, Rajasekar S, Bennett AB (2018) Nitrogen fixation in a landrace of maize is supported by a mucilage-associated diazotrophic microbiota. PLoS Biol 16(8):e2006352

Van Velzen R, Holmer R, Bu F, Rutten L, van Zeijl A, Liu W, Geurts R (2018) Comparative genomics of the nonlegume *Parasponia* reveals insights into evolution of nitrogen-fixing rhizobium symbioses. Proc Natl Acad Sci 115(20):E4700–E4709

Wang J, Li R, Zhang H, Wei G, Li Z (2020) Beneficial bacteria activate nutrients and promote wheat growth under conditions of reduced fertilizer application. BMC Microbiol 20(1):1–12

Warshan D (2017) Cyanobacteria in symbiosis with Boreal Forest Feathermosses: from genome evolution and gene regulation to impact on the ecosystem. Ph.D. Dissertation, Stockholm University, Stockholm

Yang T, Zeng Y, Sun Y, Zhang J, Tan X, Zeng Y, Pan X (2019) Experimental warming reduces fertilizer nitrogen use efficiency in a double rice cropping system. Plant Soil Environ 65(10): 483–489

Zang H, Yang X, Feng X, Qian X, Hu Y, Ren C, Zeng Z (2015) Rhizodeposition of nitrogen and carbon by mungbean (*Vigna radiata* L.) and its contribution to intercropped oats (*Avena nuda* L.). PLoS One 10:e0121132. https://doi.org/10.1371/journal.pone.0121132

Zhalnina K, Louie KB, Hao Z, Mansoori N, da Rocha UN, Shi S, Brodie EL (2018) Dynamic root exudate chemistry and microbial substrate preferences drive patterns in rhizosphere microbial community assembly. Nat Microbiol 3(4):470–480

Chapter 15
Progress of N_2 Fixation by Rice–*Rhizobium* Association

Ali Tan Kee Zuan, Amir Hamzah Ahmad Ghazali, and M. A. Baset Mia

Abstract Large amounts of nitrogen fertilizer are required for successful rice production, which is very costly and hazardous to the environment. The biofertilizer application is an alternative source of chemical fertilizer that can create an environment-friendly sustainable rice production system. They are microbial inoculum often used for boosting crop productivity. Recently, the application of rhizobia biofertilizer has gained prominence for a sustainable rice production system. A large body of researches has been performed to develop a suitable rhizobia biofertilizer to increase rice production. However, the success rate is not satisfactory, and this chapter discusses the progress and challenges of developing suitable biofertilizers for rice cultivation.

Keywords Rice · *Rhizobium* · N_2 fixation · Nodulation · Root growth

15.1 Introduction

Rice is the staple food of most people around the world, and its demand is increasing gradually. The world will require 60% more rice than current production to meet the teeming hungry population (Ladha and Reddy 2003). Additionally, the demand for rice will be increased by 70% over the next 30 years, mainly due to rapid population

A. T. K. Zuan
Department of Land Management, Faculty of Agriculture, Universiti Putra Malaysia, Serdang, Selangor, Malaysia
e-mail: tkz@upm.edu.my

A. H. A. Ghazali
School of Biological Sciences, Universiti Sains Malaysia, Gelugor, Pulau Pinang, Malaysia
e-mail: amirhg@usm.my

M. A. B. Mia (✉)
Department of Crop Botany, Bangabandhu Sheikh Mujibur Rahman Agricultural University, Gazipur, Bangladesh
e-mail: miabaset@bsmrau.edu.bd

© The Author(s), under exclusive license to Springer Nature Singapore Pte Ltd. 2022
D. K. Maheshwari et al. (eds.), *Nitrogen Fixing Bacteria: Sustainable Growth of Non-legumes*, Microorganisms for Sustainability 36,
https://doi.org/10.1007/978-981-19-4906-7_15

growth. Consequently, the world has to fight against hunger and produce more food to feed people. Two approaches should be taken to enhance rice production at a significant level: increasing the land area and improving the management practices where the former is almost impossible for developing countries as the land area decreases gradually due to rapid industrialization and urbanization. Therefore, we have to move forward to the second approach, i.e., increase the management practices including higher application of fertilizers. The application of adequate fertilizers is essential to boost production and the development of high-yielding rice varieties. However, excessive use of chemical fertilizer especially nitrogen is expensive and may cause hazards to the environment. Hence, it is imperative to explore alternative cost-effective and environment-friendly solutions. Developing a continuous supplier vis-à-vis an automated system of nitrogen incorporation to the rice plant is required for an economical and environmentally friendly rice production system. Additionally, rice variety possessing N_2-fixing capacity will be naturally beneficial and appropriate for broader acceptance. Plant scientists have been questing for a long time to achieve this automated nitrogen incorporation system, which could be obtained through biofertilizers. Nowadays the application of biofertilizers to crop plants is gaining prominence, an internationally accepted and clean practice of crop production systems. A large body of research has been performed and remarkable progress has been achieved in developing biofertilizers and bioenhancer to boost the world's sustainable crop production. The biofertilizers are microbial inoculum that can add essential nutrients, especially the N, P, K, to the plants and the soil. Among the biofertilizer system, rhizobial inoculum has been utilized to boost legume production for a long time. *Rhizobium*–legume symbiosis is the well-established association for beneficial interaction between host and microbes accepted throughout the globe. *Rhizobium*, *Bradyrhizobium*, and *Azorhizobium* are efficient nitrogen (N_2) fixers for various leguminous crops. These three genera are collectively called Rhizobia and the family is Rhizobiaceae (Elkan and Bunn 1992; Young 1992). This symbiosis system is powerful for a successful beneficial interaction. Remarkable advancements have been obtained in inoculum development and their commercial application have been well documented. However, soil microbiologists have long been interested in using this rhizobial inoculum for rice production. Recently, the use of *Rhizobium* to nonlegumes has been initiated for cereal crop production, and much attention is being given to creating a successful arena in developing rhizobial biofertilizers for rice production systems. Although the development of biofertilizers by associative and free-living bacteria in rice has progressed to a certain extent, the beneficial effects, especially in field application, are very inconsistent. Interestingly, the rice rhizosphere harbor diazotrophic microbes both endo and ectophytically created serious interest to the microbiologist for utilizing as N_2 fixers and developing as biofertilizers. However, after an enormous effort, no consistent results and remarkable successes have been achieved in developing rhizobial biofertilizer for rice production. There are various challenges for achieving successful results, and this chapter discusses the present status, progress, and future perspective in the development of rhizobial biofertilizer for rice production (Table 15.1).

Table 15.1 List of *Rhizobium* and *Bradyrhizobium* spp., which are beneficial to rice under controlled and field conditions

Name of species	Name of strain	Beneficial effects	References
Rhizobium taibaishanense	CCNWSX 0483T	Growth enhancement	Zhao et al. (2017)
Rhizobium vitis	NCPPB	Stimulate plant growth	Eastwell et al. (2006)
Rhizobium oryziradicis sp. nov.		Increase plant growth	Zhao et al. (2017)
Rhizobium pseudoryzae sp. nov.	J3-A127	Stimulate plant growth	Zhang et al. (2011)
Rhizobium radiobacter		Stimulate rice plant growth	Tan et al. (2014)
Rhizobium leguminosarum bv. *Trifolii*		Enhanced plant growth	Yanni et al. (1997)
Bradyrhizobium liaoningense		Stimulate rice plant growth	Tan et al. (2001)
Rhizobium spp.	UPMR30, UPMR31	Increased plant growth of rice	Ali-Tan et al. (2017)
Rhizobium undicola	Ouran110	Increased plant growth	Singh et al. (2018)
Rhizobium spp.	Rf7, Rpr1, Rpr2, and Rpr11	Increased shoot growth	Hernández Forte and Nápoles García (2019)
Azorhizobium caulinodans	ORS571	Stimulate rice growth	Chi et al. (2005)
Sinorhizobium meliloti	1021	Enhanced plant growth	Chi et al. (2005)
Mesorhizobium huakui	93	Enhanced plant growth	Chi et al. (2005)
Rhizobium leguminosarum	PEPV16, LS1	Increased plant growth of lettuce and carrot, increased growth and yield of rice	Husssain et al. (2009)
Rhizobium leguminosarum bv. *viciae*	BICC635	Solubilize rock phosphate	Halder et al. (1990)
Rhizobium sp.	Rn1	Increased plant growth	Mirza et al. (2007)
Rhizobium phaseoli	A2	Increased growth and yield	Husssain et al. (2009)
Azorhizobium caulinodans	ORS571	Fixation of N_2	Liu et al. (2017)

15.2 Nitrogen in Rice

Nitrogen (N) is a primary macronutrient element most frequently used, and it often becomes a limiting factor for crop growth and development, especially rice. Although the N requirement is not equal and uniform throughout the whole life cycle of the rice plant, starting from vegetative to reproductive phases. It is a monocarpic plant, and the final product is the grain rice, and the grain:straw ratio of modern rice is almost 1:1. A large amount of N is required for growth, development, and grain production, and generally, rice plants utilize 14–20 kg N ha^{-1} to produce 1 ton of grain (Choudhury et al. 2013). Around 10 million tons of N-fertilizers are being used for rice production, and it is costly and causes an environmental hazard when used excessively. Presently, N-fertilizer especially urea is manufactured artificially through the Haber-Bosch process (Appl 1982), and is significantly an energy-consuming process and also creates a hazard to the environment as given below:

$$N_2 + 3H_2 \rightarrow 2NH_3, \quad \Delta H_0 - 91.8\,\mathrm{kJ/mol}$$

This process was discovered in the early twentieth century and has made commercialization since then (Goyal et al. 2021). To boost the rice production to an ambitious level, i.e., 13 ton ha^{-1}, N fertilizer application is increased to 400 kg ha^{-1} from its 220 kg ha^{-1}. To overcome the problem of N supply to the rice plants, two approaches may be taken viz. increased supply or automated incorporation through the biological nitrogen fixation (BNF) process. The latter is more acceptable as it is cost-effective and environmentally friendly (Reddy et al. 1997). The BNF contribution is very high, ranging from 40 to 70 Tg N $year^{-1}$, and this huge amount is almost 50% of the global production of N fertilizer (Galloway et al. 2008; Herridge et al. 2008).

The rhizosphere of rice plants harbors a million-microbe having both beneficial and harmful among the beneficial microbes, many of them do colonize both ecto- and endophytically. The endophytic microbes especially the bacteria create a conducive environment for beneficial interactions. However, associative and free-living diazotrophic microbes are less efficient as success rate is limited compared to an established symbiosis system like legume–*Rhizobium* symbiosis.

15.3 Nitrogen Fixation in Rice

The ever-increasing demand for boosting rice production by resource-poor marginal farmers and apprehension for environmental hazards due to excessive use of inorganic N-fertilizer has increased the interest in alternative approaches like autonomous incorporation of N to plant biology (Mia 2015). Biological nitrogen fixation (BNF) is a potential alternative system of accumulation of N to the plant system for their utilization, and this is a significant contribution of N incorporation to the biosphere amounting to nearly 30–50% of the total N in the crop field (Ormeño-Orrillo et al. 2013; Mia et al. 2013; Mia and Shamsuddin 2013). This process is mediated by the activity of an enzyme known as nitrogenase, and it is an energy-expensive process. Unfortunately, the higher plants do not contain this enzyme; however, bacteria and cyanobacteria possess the enzyme. The N_2-fixing bacteria are known as diazotroph, generally Gram-negative α-proteobacteria, Firmicutes, and cyanobacteria. They are mainly found in the rhizosphere via ectophytically, sometimes found in the stem or a leaf as endophytically. Among the diazotroph, *Rhizobium* and *Bradyrhizobium* are smart enough to fix N_2 in association with plants of the Fabaceae family, i.e., Leguminosae. Frank (1889) first described the genus *Rhizobium*, and the genus has great economic importance for boosting crop productivity, especially pulse crops. However, species of this genus *Rhizobium* have been found in various non-legumes namely rice, wheat, maize, sugarcane, cotton, carrot, and rape (Antoun et al. 1998; Chaintreuil et al. 2000; McInroy and Kloepper 1995; Gutiérrez-Zamora and Martínez-Romero 2001; Peng et al. 2008; Sharma et al. 2005; Yanni et al. 1997; Zhang et al. 2011). Egyptian scientist Yanni et al. (1997) succeeded in isolating *Rhizobium leguminosarum* bv. *trifolii* from clover in the Nile valley of rice-growing field and found a significant positive effect on growth promotion in rice after inoculation.

15.4 Beneficial Effects of *Rhizobium* on Rice

The *Rhizobium leguminosarum* strains LSI-23, LSI-26, LSI-29, LSI-30 and *Mesorhizobium ciceri* strains CRI-28, CRI-31, CRI-32, CRI-38 have been found to colonize on the roots of rice cv. Super Basmati. Inoculation process increased the number of the tiller (46%); grain yield (43%), total dry matter (18%), straw dry weight (45%), and 1000-grain (25%) improved maximally by the strain LSI-29 over uninoculated control (Husssain et al. 2009). Combined inoculation of rhizobial strains and *brasilense* increased the yield attributes of wetland rice and *Rhizobium* strains that could successfully colonize the root surface and lateral roots (Hahn et al. 2016a, b). Similarly, inoculation of rhizobial strains could form colonies endophytically and enhance plant growth through the synthesis of phytohormone like indole-3-acetic acid (IAA), solubilized complex phosphate into available phosphate, and are also able to act as biocontrol agents as reported by several scientists (Yanni et al. 2001; Singh et al. 2006; Yanni and Dazzo 2010; Chen et al. 2005; Bhattacharjee et al. 2012; Dutta et al. 2007). Inoculation of *Rhizobium* enhanced the vigor seedling in rice as reported by Mia and Shamsuddin (2009) (Fig. 15.1).

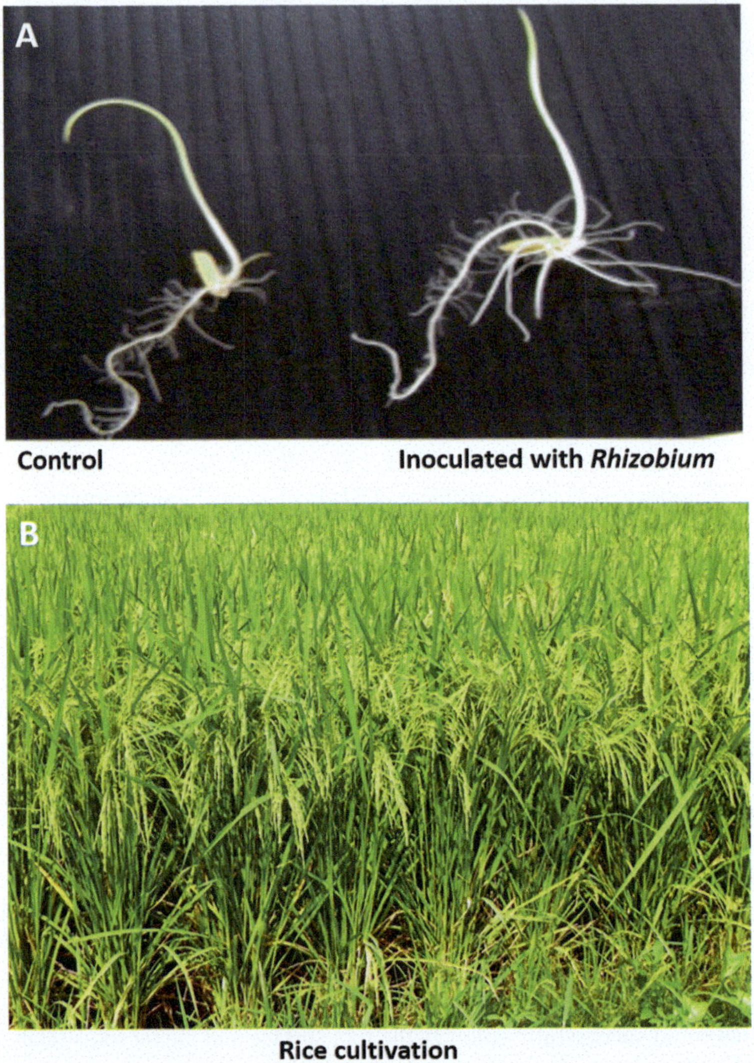

Fig. 15.1 Application of *Rhizobium* in rice: (**a**) control and *Rhizobium*-inoculated rice seedling, (**b**) rice field

15.5 Mechanism of Beneficial Effects

Rhizobial inoculation of rice promotes growth and yield through diversified ways of mechanisms (Hahn et al. 2016a, b; Mia and Shamsuddin 2013, 2010; Mia et al. 2013). The rhizobia exert beneficial effects upon inoculation of rice in various dimensions like fixation and incorporation of N_2 into rice plant, root stimulation through the production of growth hormone especially the auxin, enhanced nutrient

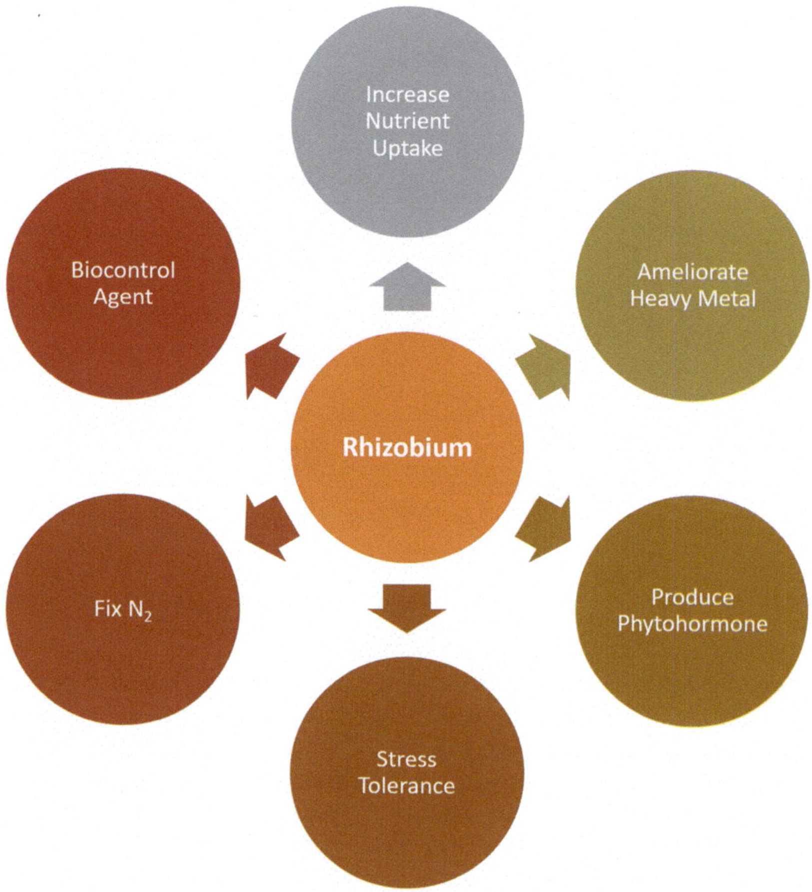

Fig. 15.2 Overview representation of the mechanisms of beneficial effects of *Rhizobium* inoculation in rice

accumulation of various nutrients like Fe, P, K, and finally act as a biocontrol agent (Fig. 15.2).

15.5.1 Colonization of Rhizobia on Rice Roots

Rhizobial strains could form colonies successfully on the roots of rice where the bacteria enter into the internal portion of roots through the junction of lateral roots, apoplastic area, and crack of the roots (Fig. 15.3). The colonization occurs mainly in the plant tissues of the root, stem, and leaf but not the real nodule but forming nodules-like structure or swollen structure on the root surface, which is known as paranodule/hypertrophies (Dakora 1995; Tchan and Kennedy 1989; Al-Mallah et al.

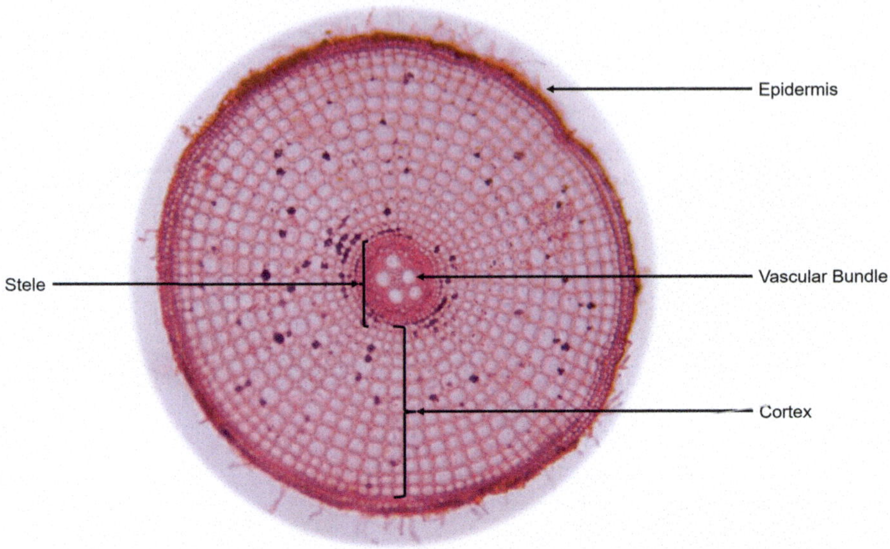

Fig. 15.3 Transverse section of rice roots showing tissue system of rice roots

1989; Rolfe and Bender 1990; De Bruijn et al. 1995). However, surface colonization could be carried out through the normal invasion process. Perrine-Walker et al. (2007) confirmed the inoculation process by GFP-labeled technique marked with bacteria. Aggregated type of colonization on the root surface of Malaysian rice roots cv. Mayang Segumpal has been reported by Naher et al. (2009). Among the colonization pattern, the endophytic behavior confers a significant ecological advantage over ectophytic (Santos et al. 2018). The pattern and mode of colonization as indicated lacZ reporter gene of azorhizobia indicated mainly on the site of lateral root initiation and lateral roots cracks. Colonization could be stimulated by applying flavanone naringenin in the apoplastic area and the site of lateral root formation (Fig. 15.4). Interestingly, to note that common *Sym* genes consisting of *nod*, *nif*, and *fix* genes are not required for internal colonization of rice roots by rhizobia (Chen and Zhu 2013).

15.5.2 N_2 Fixation of N_2 by Rhizobium

The *Rhizobium* could form colonies successfully and fix a significant amount of N_2 in associated with rice roots. Naher et al. (2009) and Ali-Tan et al. (2017) found enhanced tissue N concentration of Malaysian wetland rice cv. Mayang Segumpal by the inoculation of *Rhizobium* sp. strain SB16. Several researchers reported the fixation of N_2 by *Azorhizobium caulinodans* (ORS571) in association with rice roots by applying cellulase and pectolase enzymes (Liu et al. 2017).

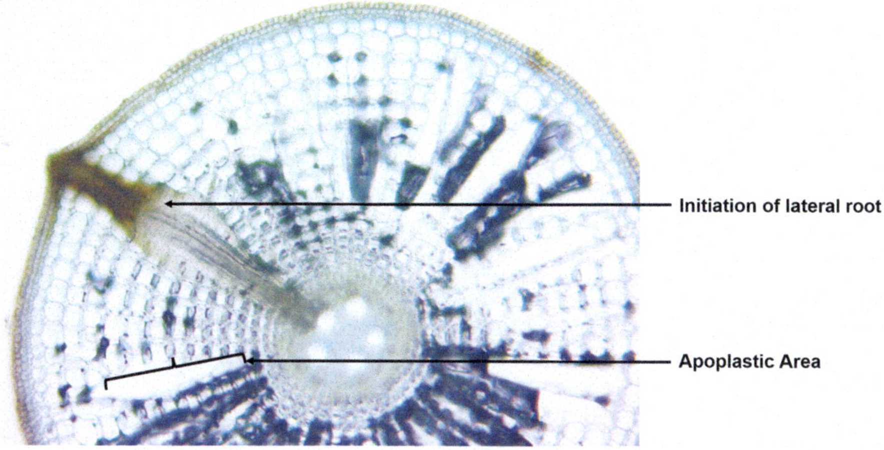

Fig. 15.4 Transverse section of rice roots showing lateral root initiation and apoplastic area

15.5.3 Bioremediation of Toxic Elements

Rhizobium has the potential to bioremediate several toxic compounds viz. polycyclic aromatic hydrocarbon, which are responsible for ecological and environmental pollution (Parshetti et al. 2010; Tan et al. 2001; Yanni et al. 1997; Zhang et al. 2012). This bacterium is a powerful tool for the bioremediation of heavy metals in soil (Hao et al. 2012). They also possess the ecological and biochemical capacity to degrade organic chemicals and to decrease the risk of heavy metals (Teng et al. 2015). Numerous findings have been obtained on the remediation of heavy metals by applying *Rhizobium* in different crop plants (Fagorzi et al. 2018).

15.5.4 Enhanced Uptake of Nutrients

Rhizobium leguminosarum bv. *trifolii* E11, *Rhizobium* sp. IRBG74, and *Bradyrhizobium* sp. IRBG271 strains inoculation on rice increased the uptake of N, P, and K by 10–28% as reported by Biswas et al. (2000a). Mia and Shamsuddin (2010) found that a higher accumulation of P, K, Ca, and Mg occurred when inoculated by *Rhizobium* spp. in Malaysian rice despite N_2 fixation.

15.5.5 Enhanced Uptake of Fe

Inoculation of endophytic rhizobial strains isolated from a rice field in Cuba could produce siderophore and biofilm when inoculated to Cuban rice consequently increasing the uptake of Fe (Hernández et al. 2021). Similarly, Biswas et al. (2000b) found 15–64% more Fe uptake by rice variety Pankaj under Filipino condition.

The N, P, and K uptake were increased by 10–28% due to rhizobial inoculation, and ^{15}N isotope-based studies indicated that the increased N uptake was not due to biological N_2 fixation (BNF). Inoculation also increased Fe uptake in rice by 15–64%, and IAA accumulated in the external root environment of rice plants when grown gnotobiotically with rhizobia. Certain strains of rhizobia can promote rice growth and yield, most likely through mechanisms that involve changes in growth physiology or root morphology rather than BNF.

Recently, *Rhizobium oryzihabitans* sp. nov. isolated from rice roots showed biofertilizer activity such as production of siderophore 1-aminocyclopropane-1-carboxylate (ACC) deaminase and IAA after inoculation to rice (Zhao et al. 2020). Inoculation of *Rhizobium* to rice could increase the availability of P through solubilization of unavailable and insoluble complex-P (Halder et al. 1990; Johri et al. 2003; Sashidhar and Podile 2010; Rodriguez and Fraga 1999).

15.5.6 Enhanced Total Dry Matter

Inoculation of endophytic rhizobial strains, namely Rpr2 and Rdp16, increased plant height and total dry matter of Cuban rice under glasshouse conditions (Hernández et al. 2021). Shahdi (2021) found increased grain number and rice yield by the inoculation of *Rhizobium* in mixed cultivation with clover. Nahar et al. (2021) isolated an astonishing number of *Rhizobium* spp. like *Rhizobium tropici*, *R. leguminosarum*, *R. freirei*, and *R. oryzae* from the rice rhizosphere, and inoculation of those selected rhizobial strains enhanced plant growth and yields, which could be brought significant benefit for the farmer and are recognized as an environment-friendly sustainable agricultural system. Inoculation of strains of *Rhizobium* sp., *Bradyrhizobium* sp., and *Achromobacter* sp. on rice resulted in plant growth-promoting traits like phytohormone production, namely IAA, and showed the P-solubilizing capacity (Satyanandam et al. 2010). On the other hand, inoculation of *Bradyrhizobium japonicum* E109 and *B. elkanii* SEMIA 587 increased the growth and yield of rice under field and laboratory conditions (Padukkage et al. 2021).

15.5.7 Enhanced Tolerance of Abiotic Stress

Despite plant growth-promoting rhizobacteria (PGPR) activity, inoculation of *alamii* significantly increased rice plants' growth and enhanced the host plants' capacity to overcome abiotic stress like drought by increasing the root growth, which helps to absorb water from deeper soil profiles. The rhizobia spp. can adjust to saline conditions through the intracellular gathering of low molecular weight organic solutes known as osmolytes. *Rhizobium meliloti*, *Rhizobium fredii*, and *Ensifer fredii* can adjust to these osmotic shocks where K^+ controls Mg^{2+} flux during osmotic shock (Botsford and Lewis 1990; Jian et al. 1993; Le Rudulier and Bernard 1986; Smith et al. 1994a, b; Zahran and Sprent 1986). This accumulation of osmolytes counteract the dehydration effect of low molecular structure or functions.

15.5.8 Nodulation Process

A large body of research has been conducted on the formation of nodule/nodule-like structure or paranodule formation on rice roots by *Rhizobium/Bradyrhizobium* spp. However, very little success has been achieved. The possible reasons are that the root exudates of rice roots cannot activate the nodulation gene. The species of genus *Rhizobium* form nodules in the roots of legume crops through the process of endosymbiosis. However, it is interesting to note that they have also found a positive association with the roots of non-legumes, namely rice, wheat, and maize; successful nodule formation is performed by the *Parasponia* (Cocking et al. 1990). The Rhizobia isolated from *Parasponia* and other tropical legumes can colonize rice roots and perform the beneficial activities endophytically. The path of entry of bacteria to the roots through the opening of lateral roots or crack of roots and a higher level of colonization have been found through the crack of roots, mainly found in the intercellular space. Despite nodule formation, the paranodule is different from the legume nodule formed by the inoculation of *Rhizobium* (Tchan and Kennedy 1989). The induction of paranodule is enhanced by applying IAA 2-4-D and tryptophan. These have been able to infect and colonize the rice roots despite nodule formation, and common *Sym* genes are not required for efficient colonization (Chen and Zhu 2013). Regardless of being unable to induce nodulation, rhizobia can infect and colonize the roots of non-legumes such as rice. One interesting question is whether the establishment of such associations requires the common symbiosis (*Sym*) genes that are essential for infection of plant cells by mycorrhizal fungi and rhizobia in legumes. Here, we demonstrated that the common *Sym* genes are not required for endophytic colonization of rice roots by nitrogen-fixing rhizobia. The cell structures of pseudo-nodules on barley roots were very fine, with organelles and infection threads from which the rhizobial bacteria were released to the cytoplasm. The bacteria enclosed in peribacteroid membranes become the bacteroids, and these morphological structures are similar to those of the legume nodules with seemingly symbiotic characteristics, but the bacteroids were vesiculated. In the pseudonodules on rice roots formed by *Rhizobium sesbania* sp. cultured in the medium containing rice extract, the bacteria were distributed in intercellular spaces and cells themselves. The structures of the infected cells were coarse and bacteria were not enclosed by the peribacteroid membrane, which is completely different from those of the legume nodules. However, no nitrogenase activity was detected, but plant growth was better than the control.

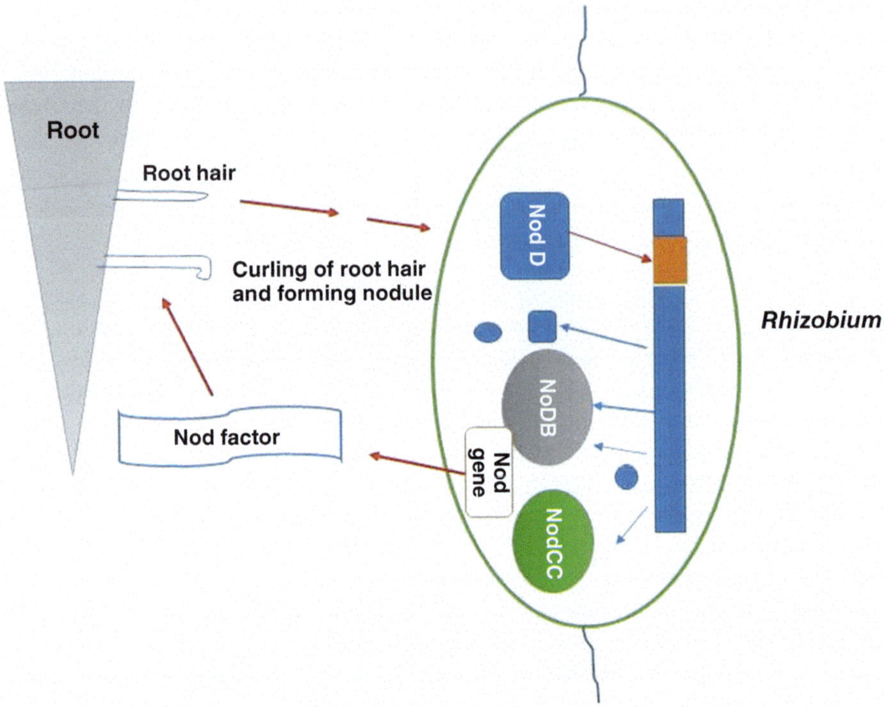

Fig. 15.5 Activation of nod gene for nodule formation

It is reported that certain strains of *Bradyrhizobium* could form nodule-like structures on the roots of *Arabidopsis* and *Brassica* spp. under controlled conditions, and the size of the nodule is relatively larger. Similarly, *Rhizobium* strains NGR234 and NGR76 isolated from *Lablab purpureus* and *Phaseolus vulgaris* respectively formed similar nodule-like structures on *Brassica* spp. Still, the form and shape of nodules vary on the strains inoculated. However, the formation of a nodule in rice roots by the inoculation of *Bradyrhizobium* or *Rhizobium* was not successful. The initiation of any nodular structure could not be observed through light or electron microscopy. The nodule formation by *Rhizobium* in plant roots is controlled by nodulin genes, which are either constitutive or inducible or even repressible, and N-1 is the largest and N-30 is the smallest one. The nod genes are categorized as *nodA*, *nodB*, *nodC*, and *nodD* (Sofi and Wani 2007). The nodulation process starts with the exchange of signals between the host plant and rhizobia (Fig. 15.5). The plant secretes a chemical substance known as flavonoid, a potential inducer of nod genes, into the rhizosphere, which activates the nod gene *nodD*. The nod factor binds to a specific plant kinase, initiating a signaling pathway that leads to root hair curling and trapping of the rhizobia (Khush and Bennett 1992). In the case of rice, the infection process of *Rhizobium/Bradyrhizobium* is not dependent on nod genes and even does not involve the formation of infection threads. The endophytic colonization is mostly confined within the apoplastic area either by schizogenously or lysigenously. The fiber of sclerenchymatous tissue of root hypodermis is the main barrier to forming deeper colonization in the roots (Reddy et al. 1997).

15.5.9 Genetic Regulation of N_2 Fixation by Rhizobia

The role of the *Nif* gene is vital for N incorporation into the rice plant as the commercial N-fertilizer is expensive. The genes regulation for symbiotic N_2 fixation by rhizobia are broadly recognized as nod gene, *fix* gene, and *nif* gene where the former is responsible for nodulation and the latter two are responsible for N_2 fixation (Dasgupta et al. 2021). Studies on genomic with the high-output capability of data can play a substantial role in the judicious mechanism of communication occupation in N_2-fixing ability (Goyal et al. 2021).

15.6 Conclusions and Future Perspective

Varieties of research have been performed throughout the globe for achieving an effective rhizobial inoculum for rice cultivation. The rhizobial application enhances root and shoot growth consequently, and the higher yield accumulation of N, P, Fe is also found by several researchers. Amelioration of various abiotic stresses has also been observed. However, the overall progress is not satisfactory. This approach for developing a sustainable rhizobial inoculum is a great challenge, which could be overcome through various ways, including the transfer to the manifestation of N_2 fixation (*nif, fix*) genes in the rice itself (Dixon et al. 1997). Alternatively, substantial steps should be taken to isolate naturally occurring endophytic N_2-fixing bacteria and modify them genetically for increasing the efficiency of N_2-fixing capacity (Barraquio et al. 1997; Colnaghi et al. 1997; Kennedy et al. 1997; Kirchhof et al. 1997; Stoltzfus et al. 1997). Another approach may be taken to modify rice and rhizobia for a functional symbiosis system that produces either nodule or paranodule. Appropriate knowledge is required to get a strategic design for extending this symbiosis to rice and other cereals (De Bruijn et al. 1995). To create a functional symbiosis system, genetically manipulated rice and/or N_2-fixing rhizobia should be developed, the interaction between rice and rhizobia at the cellular and molecular levels should be established, and evaluation should be performed in rice gene similar to nodulin gene of legume. A functional symbiosis could be achieved by gathering much information and genetic engineering tools to build genetically modified rice. This might be attained by cellular and molecular features of interaction between rice and rhizobia evaluation of rice gene similar to nodulin genes of legumes and revelation of their roles, and valuation of the expression of legume nodulin genes in a rice background and response to rhizobial inoculation or Nod factors. Therefore, a long way to move through a systematic approach and conduct molecular experimentation for developing effective rhizobial inocula for a successful rice production system.

Acknowledgment Not applicable.

Conflict of Interest Author(s) declares no conflict of interest.

References

Ali-Tan KZ, Radziah O, Halimi MS, Rahim KBA, Abdullah MZ, Shamsuddin ZH (2017) Growth and yield responses of rice cv. MR219 to rhizobial and plant growth-promoting rhizobacterial inoculations under different fertilizer-N rates. Bangladesh J Bot 46(1):481–488

Al-Mallah MK, Davey MR et al (1989) Formation of nodular structures on rice seedlings by rhizobia. J Exp Bot 40:473–478. https://doi.org/10.1007/BF00011323

Antoun H, Beauchamp CJ, Goussard N et al (1998) Potential of *Rhizobium* and *Bradyrhizobium* species as plant growth-promoting rhizobacteria on non-legumes: effect on radishes (*Raphanus sativus* L.). Plant Soil 204:57–67. https://doi.org/10.1023/A:1004326910584

Appl M (1982) The Haber–Bosch process and the development of chemical engineering. In: A century of chemical engineering. Plenum Press, New York, NY, pp 29–54

Barraquio WL, Revilla L, Ladha JK (1997) Isolation of endophytic diazotrophic bacteria from wetland rice. Plant Soil 194:15–24

Bhattacharjee RB, Jourand P, Chaintreuil C, Dreyfus B, Singh A, Mukhopadhyay SN (2012) Indole acetic acid and ACC deaminase-producing *Rhizobium leguminosarum* bv. *trifolii* SN10 promote rice growth, and in the process undergo colonization and chemotaxis. Biol Fertil Soils 48:173–182. https://doi.org/10.1007/s00374-011-0614-9

Biswas JC, Ladha JK et al (2000a) Rhizobial inoculation influences seedling vigor and yield of rice. Agron J 92:880–886. https://doi.org/10.2134/agronj2000.925880x

Biswas JC, Ladha JK et al (2000b) Rhizobial inoculation improves nutrient uptake and growth of lowland rice. Soil Sci Soc Am J 64:1644–1650. https://doi.org/10.2136/sssaj2000.6451644x

Botsford JL, Lewis TA (1990) Osmoregulation in *Rhizobium meliloti*: production of glutamic acid in response to osmotic stress. Appl Environ Microbiol 56:488–494

Chaintreuil C, Giraud E, Prin Y, Lorquin J, Ba A, Gillis M et al (2000) Photosynthetic bradyrhizobia are natural endophytes of the African wild rice *Oryza breviligulata*. Appl Environ Microbiol 66:5437–5447

Chen C, Zhu H (2013) Are common symbiosis genes required for endophytic rice-rhizobial interactions? Plant Signal Behav 8(9):e25453. https://doi.org/10.4161/psb.25453

Chen X, Feng J, Hou B, Li F, Li Q, Hong G (2005) Modulating DNA bending affects NodD-mediated transcriptional control in *Rhizobium leguminosarum*. Nucleic Acids Res 33:2540–2548. https://doi.org/10.1093/nar/gki537

Chi F, Shen S-H, Cheng H-P, Jing Y-X, Yanni YG, Dazzo FB (2005) Ascending migration of endophytic rhizobia, from roots to leaves, inside rice plants and assessment of benefits to rice growth physiology. Appl Environ Microbiol 71:7271–7278

Choudhury ATMA, Saleque MA, Zaman SK, Bhuiyan NI, Shah AL, Rahman MS (2013) Nitrogen fertilizer management strategies for rice production in Bangladesh. Biol Sci PJSIR 56(3): 167–174

Cocking EC, Al-Mallah MK, Benson E, Davey MR (1990) Nodulation of non-legumes by rhizobia. In: Gresshoff PM, Roth LE, Stacey G, Newton WE (eds) Nitrogen fixation: achievements and objectives. Chapman and Hall, New York, NY, pp 813–823

Colnaghi R, Green A, He L, Rudnick P, Kennedy C (1997) Strategies for increased ammonium production in free-living or plant associated nitrogen fixing bacteria. Plant Soil 194:145–154

Dakora FD (1995) Plant flavonoids: biological molecules for useful exploitation. Aust J Plant Physiol 22:7–99

Dasgupta D, Panda AK, Mishra R, Mahanty A, de Mandal S, Bisht SS (2021) *Nif* genes: tools for sustainable agriculture. Rec Adv Microb Biotechnol 2021:413–434

De Bruijn FJ, Jing Y, Dazzo FB (1995) Potential and pitfalls of trying to extend symbiotic interactions of nitrogen-fixing organisms to presently non-nodulated plants, such as rice. Plant Soil 172:207–219

Dixon R, Cheng Q, Shen G-F, Day A, Dowson-Day M (1997) *Nif* gene transfer and expression in chloroplasts: prospects and problems. Plant Soil 194:193–203. https://doi.org/10.1007/s11104-008-9668-3

Dutta S, Mishra AK, Kumar BSD (2007) Induction of systemic resistance against fusarial wilt in pigeon pea through interaction of plant growth promoting rhizobacteria and rhizobia. Soil Biol Biochem 40:452–461. https://doi.org/10.1016/j.soilbio.2007.09.009

Eastwell KC, Sholberg PL, Sayler RJ (2006) Characterizing potential bacterial biocontrol agents for suppression of *Rhizobium vitis*, causal agent of crown gall disease in grapevines. Crop Prot 25(11):1191–1200

Elkan GH, Bunn CR (1992) The rhizobia. In: Balows A et al (eds) The prokaryotes, vol III, 2nd edn. Springer, New York, NY

Fagorzi C, Checcucci A, diCenzo G, Debiec-Andrzejewska K, Dziewit L, Pini F et al (2018) Harnessing rhizobia to improve heavy-metal phytoremediation by legumes. Genes 9:542. https://doi.org/10.3390/genes9110542

Frank B (1889) Ueber die Pilzsymbiose der Leguminosen. Ber Dtsch Bot Ges 7:332–346

Galloway JN, Townsend AR, Erisman JW, Bekunda M, Cai Z, Freney JR, Martinelli LA, Seitzinger SP, Sutton MA (2008) Transformation of the nitrogen cycle: recent trends, questions, and potential solutions. Science 320(5878):889–892. https://doi.org/10.1126/science.1136674

Goyal RK, Schmidt MA, Hynes MF (2021) Molecular biology in the improvement of biological nitrogen fixation by *Rhizobia* and extending the scope to cereals. Microorganisms 9(1):125. https://doi.org/10.3390/microorganisms9010125

Gutiérrez-Zamora ML, Martínez-Romero E (2001) Natural endophytic association between *Rhizobium etli* and maize (*Zea mays* L.). J Biotechnol 91:117–126

Hahn L, Sá EL, Osório BD, Machado RG, Damasceno RG, Giongo A (2016a) Rhizobial inoculation, alone or coinoculated with *Azospirillum brasilense*, promotes growth of wetland rice. Rev Brasileira Ciên Solo 40:e0160006. https://doi.org/10.1590/18069657rbcs20160006

Hahn MW, Jezberová J, Koll U, Saueressig-Beck T, Schmidt J (2016b) Complete ecological isolation and cryptic diversity in *Polynucleobacter* bacteria not resolved by 16S rRNA gene sequences. ISME J 10:1642–1655. https://doi.org/10.1038/ismej.2015.237

Halder AK, Mishra AK, Bhattacharyya P (1990) Solubilization of rock phosphate by *Rhizobium* and *Bradyrhizobium*. J Gen Appl Microbiol 36(2):81–92

Hao X, Lin Y, Johnstone L, Baltrus D, Miller S, Wei G, Rensing C (2012) Draft genome sequence of plant growth-promoting rhizobium *Mesorhizobium amorphae*, isolated from zinc-lead mine tailings. J Bacteriol 194:736–737

Hernández Forte I, Nápoles García MC (2019) Rhizobia promote rice (*Oryza sativa* L.) growth: first evidence in Cuba. In: Zúñiga-Dávila D et al (eds) Microbial probiotics for agricultural systems. Sustainability in plant and crop protection. Springer, Cham. https://doi.org/10.1007/978-3-030-17597-9_10

Hernández I, Taulé C, Pérez-Pérez R, Battistoni F, Fabiano E, Rivero D, Nápoles MC (2021) Endophytic rhizobia promote the growth of Cuban rice cultivar. Symbiosis 85(2):175–190. https://doi.org/10.1007/s13199-021-00803-2

Herridge DF, Peoples MB, Boddey RM (2008) Global inputs of biological nitrogen fixation in agricultural systems. Plant Soil 311(1):1–18

Husssain MB, Mehboob I, Zahir ZA, Naveed M, Asghar HN (2009) Potential of *Rhizobium* spp. for improving growth and yield of rice (*Oryza sativa* L.). Soil Environ 28(1):49–55
Jian W, Susheng Y, Jilun L (1993) Studies on the salt tolerance of *Rhizobium meliloti*. Acta Microbiol Sin 33:260–267
Johri BN, Sharma A, Virdi JS (2003) Rhizobacterial diversity in India and its influence on soil. Adv Biochem Eng Biotechnol 4:49–89
Kennedy IR, Pereg-Gerk L, Wood C, Deaker R, Gilchrist K, Katupitiya S (1997) Biological nitrogen fixation in non-leguminous field crops: facilitating the evolution of an effective association between *Azospirillum* and wheat. Plant Soil 194:65–79
Khush GS, Bennett J (1992) Nodulation and nitrogen fixation in rice: potential and prospects. Documentation. International Rice Research Institute, Los Banos
Kirchhof G, Reis VM, Baldani JI, Eckert B, Döbereiner J, Hartmann A (1997) Occurrence, physiological and molecular analysis of endophytic diazotrophic bacteria in gramineous energy plants. Plant Soil 194:45–55
Ladha JK, Reddy PM (2003) Nitrogen fixation in rice systems: state of knowledge and future prospects. Plant Soil 252:151–167
Le Rudulier D, Bernard T (1986) Salt tolerance in *Rhizobium*: a possible role for betaines. FEMS Microbiol Rev 39:67–72
Liu H, Wang X, Qi H, Wang Q, Chen Y, Li Q, Zhang Y, Qiu L, Fontana JE, Zhang B, Wang W, Xie Y (2017) The infection and impact of *Azorhizobium caulinodans* ORS571 on wheat (*Triticum aestivum* L.). PLoS One 12(11):e0187947. https://doi.org/10.1371/journal.pone.0187947
McInroy JA, Kloepper JW (1995) Survey of indigenous bacterial endophytes from cotton and sweet corn. Plant Soil 173:337–342. https://doi.org/10.1007/BF00011472
Mia MAB (2015) Nutrition of crop. In: Plants. Nova Science Publisher, New York, NY
Mia MAB, Shamsuddin ZH (2009) Enhanced emergence and vigour seedling production of rice through plant growth promoting bacterial inoculation. Res J Seed Sci 2:96–104
Mia MAB, Shamsuddin ZH (2010) Nitrogen fixation and transportation by rhizobacteria: a scenario of rice and banana. Int J Bot 6:235–242
Mia MAB, Shamsuddin ZH (2013) Biofertilizer for banana production. Lambert Academic Publisher, Germany
Mia MAB, Hossain M, Shamsuddin ZH, Islam MT (2013) Plant-associated bacteria in nitrogen nutrition in crops, with special reference to rice and banana. In: Maheshwari DK (ed) Bacteria in agrobiology: crop productivity. Springer, Berlin
Mirza BS, Mirza MS, Bano A, Malik KA (2007) Coinoculation of chickpea with *Rhizobium* isolates from roots and nodules and phytohormone-producing *Enterobacter* strains. Aust J Exp Agric 47:1008–1015
Nahar S, Hasan MF, Sikdar B et al (2021) Effect of rhizosphere microbiome on different crop growing fields in various rice cultivars and its molecular approaches for sustainable agroecosystem. J Crop Sci Biotechnol 24(5):521–531. https://doi.org/10.1007/s12892-021-00099-0
Naher UA, Radziah O, Shamsuddin ZH, Halimi MS, Razi IM (2009) Isolation of diazotrophs from different soils of Tanjong Karang rice growing area in Malaysia. Int J Agric Biol 11:547–552
Ormeño-Orrillo E, Servín-Garcidueñas LE, Imperial J, Rey L, Ruiz-Argueso T, Martinez RE (2013) Phylogenetic evidence of the transfer of *nod*Z and *nol*L genes from *Bradyrhizobium* to other rhizobia. Mol Phylogenet Evol 67(3):626–630
Padukkage D, Geekiyanage S, Reparaz JM et al (2021) *Bradyrhizobium japonicum*, *B. elkanii* and *B. diazoefficiens* interact with Rice (*Oryza sativa*), promote growth and increase yield. Curr Microbiol 78:417–428. https://doi.org/10.1007/s00284-020-02249-z
Parshetti GK, Telke AA, Kalyani DC, Govindwar SP (2010) Decolorization and detoxification of sulfonated azo dye methyl orange by *Kocuria rosea* MTCC 1532. J Hazard Mater 176:1–3
Peng M, Wang D, Jiang Y (2008) An institution-based view of international business strategy: a focus on emerging economies. J Int Bus Stud 39:920–936. https://doi.org/10.1057/palgrave.jibs.8400377

Perrine-Walker FM, Prayitno J, Rolfe BG, Weinman JJ, Hocart CH (2007) Infection process and the interaction of rice roots with rhizobia. J Exp Bot 58(12):3343–3350. https://doi.org/10.1093/jxb/erm181

Reddy PM, Ladha JK, So RB, Hernandez RJ, Ramos MC, Angeles OR, Dazzo FB, de Bruijn FJ (1997) Rhizobial communication with rice roots: induction of phenotypic changes, mode of invasion and extent of colonization. Plant Soil 194:81–98. https://doi.org/10.1023/A:1004243915997

Rodriguez H, Fraga R (1999) Phosphate solubilizing bacteria and their role in plant growth promotion. Biotechnol Adv 17(4):319–339

Rolfe BG, Bender GL (1990) Evolving a *Rhizobium* for non-legume nodulation. In: Gresshoff PM et al (eds) Nitrogen fixation. Springer, Boston, MA. https://doi.org/10.1007/978-1-4684-6432-0_65

Santos ML, Berlitz DL, Wiest SL, Schünemann R, Knaak N, Fiuza LM (2018) Benefits associated with the interaction of endophytic bacteria and plants. Braz Arch Biol Technol 61:e18160431. https://doi.org/10.1590/1678-4324-2018160431

Sashidhar B, Podile AR (2010) Mineral phosphate solubilization by rhizosphere bacteria and scope for manipulation of the direct oxidation pathway involving glucose dehydrogenase. J Appl Microbiol 109:1–12

Satyanandam T, Babu K, Suneeta D, Bhaskararao CH, Rosaiah G, Vijayalakshmi M (2010) Isolation and screening of indigenous rhizobia from black gram cultivated in fallow rice soils for plant growth promoting traits. Malaysian J Soil Sci 25:125–142

Shahdi KA (2021) Effect of *Rhizobium trifolii*, *Pseudomonas fluorescens* and *Azotobacter chroococcum* on growth and yield of crimson clover and rice in a rice-clover rotation. J Crop Prod Process 10(4):17–31

Sharma P, Asztalos Z, Ayyub C, de Bruyne M, Dornan AJ, Gomez-Hernandez A, Keane J, Killeen J, Kramer S, Madhavan M, Roe H et al (2005) Isogenic autosomes to be applied in optimal screening for novel mutants with viable phenotypes in *Drosophila melanogaster*. J Neurogenet 19(2):57–85

Singh RK, Mishra RPN, Jaiswal HK, Kumar V, Pandev SP, Rao SB, Annapurna K (2006) Isolation and identification of natural endophytic rhizobia from rice (*Oryza sativa* L.) through rDNA PCR-RFLP and sequence analysis. Curr Microbiol 52:345–349

Singh MK, Singh SK, Singh DP (2018) Novel bacterium *Rhizobium undicola* isolated from the upland cultivated rice *Oryza sativa* L. promoting plant growth. J Pharmacogn Phytochem 7(3): 2437–2444

Smith LT, Allaith AM, Smith GM (1994a) Mechanism of osmotically-regulated N-acetylglutaminyl glutamine amide production in *Rhizobium meliloti*. Plant Soil 161:103–108

Smith LT, Smith GM, Desouza MR, Pocard JM, Le Rudulier D, Madkour MA (1994b) Osmoregulation in *Rhizobium meliloti*: mechanism and control by other environmental signals. J Exp Zool 268:162–292

Sofi P, Wani S (2007) Prospects of nitrogen fixation in rice. Asian J Plant Sci 6:203–213

Stoltzfus JR, So R, Malarvizhi PP, Ladha JK, de Bruijn FJ (1997) Isolation of endophytic bacteria from rice and assessment of their potential for supplying rice with biologically fixed nitrogen. Plant Soil 194:25–36

Tan Z, Hurek T, Vinuesa P, Müller P, Ladha JK, Reinhold-Hurek B (2001) Specific detection of *Bradyrhizobium* and *Rhizobium* strains colonizing rice (*Oryza sativa*) roots by 16S–23S ribosomal DNA intergenic spacer-targeted PCR. Appl Environ Microbiol 67:3655–3664

Tan KZ, Radziah O, Halimi MS, Khairuddin AR, Habib H, Shamsuddin ZH (2014) Isolation and characterization of rhizobia and plant growth-promoting rhizobacteria and their effects on growth of rice seedlings. Am J Agric Biol Sci 9(3):342–360

Tchan YT, Kennedy TR (1989) Possible N_2-fixing root nodule induced in non-legumes. Agric Sci 2:57–59

Teng Y, Wang X, Li L, Li Z, Luo Y (2015) Rhizobia and their bio-partners as novel drivers for functional remediation in contaminated soil. Front Plant Sci 6:32. https://doi.org/10.3389/fpls.2015.00032

Yanni YG, Dazzo FB (2010) Enhancement of rice production using endophytic strains of Rhizobium leguminosarum bv. trifolii in extensive field inoculation trials within the Egypt Nile delta. Plant Soil 336:129-142. https://doi.org/10.1007/s11104-010-0454-7

Yanni YG, Rizk RY, Corich V, Squartini A, Ninke K, Philip-Hollingsworth S, Orgambide G, Bruijn FD, Stoltzfus J, Buckley D, Schmidt TM (1997) Natural endophytic association between *Rhizobium leguminosarum* bv. *trifolii* and rice roots and assessment of its potential to promote rice growth. In: Ladha JK et al (eds) Opportunities for biological nitrogen fixation in rice and other non-legumes. Developments in plant and soil sciences, vol 75. Springer, Dordrecht. https://doi.org/10.1007/978-94-011-7113-7_10

Yanni YG, Rizk RY, Abd El-Fattah FK, Squartini A, Corich V, Giacomini A, De Bruijn F, Rademaker J, Maya-Flores J, Ostrom P, Vega-Hernandez M, Hollingsworth RI, Martinez-Molina E, Mateos P, Velazquez E, Wopereis J, Triplett E, Umali-Gracia M, Anarna JA, Rolfe BG, Ladha JK, Hill J, Mujoo R, Ng PK, Dazzo FB (2001) The beneficial plant growth promoting association of *Rhizobium leguminosarum* bv. *trifolii* with rice roots. Aust J Plant Physiol 28:845–870

Young JPW (1992) Phylogenetic classification of nitrogen-fixing organisms. In: Stacey G et al (eds) Biological nitrogen fixation. Chapman and Hall, New York, NY, pp 43–86

Zahran HH, Sprent JI (1986) Effects of sodium chloride and polyethylene glycol on root hair infection and nodulation of *Vicia faba* L. plants by *Rhizobium leguminosarum*. Planta 167:303–309

Zhang X, Sun L, Ma X, Sui XH, Jiang R (2011) *Rhizobium pseudoryzae* sp. nov. isolated from the rhizosphere of rice. Int J Syst Evol Microbiol 61(10):2425–2429

Zhang JJ, Liu TY, Chen WF, Wang ET, Sui XH, Zhang XX, Li Y, Li Y, Chen WX (2012) *Mesorhizobium muleiense* sp nov., nodulating with *Cicer arietinum* L. Int J Syst Evol Microbiol 62:2737–2742

Zhao JJ, Zhang J, Sun L, Zhang RJ, Zhang CW, Yin HQ, Zhang XX (2017) *Rhizobium oryziradicis* sp. nov. isolated from rice roots. Int J Syst Evol Microbiol 67(4):963–968

Zhao J, Zhao X, Wang J, Gong Q, Zhang X, Zhang G (2020) Isolation, identification and characterization of endophytic bacterium *Rhizobium oryzihabitans* sp. nov., from rice root with biotechnological potential in agriculture. Microorganisms 8(4):608. https://doi.org/10.3390/microorganisms8040608

Chapter 16
N-Fixation by Free-Living and Endophytic Bacteria and Their Impact on Field Crops with Emphasis on Rice

Motohiko Kondo, Rina Shinjo, and Takanori Okamoto

Abstract Effective use of N-fixation is important to establish crop production system with reduced dependency on chemical N fertilizer. This chapter reviews the past and recent findings on the ecophysiological and agronomic aspects of free-living and endophytic N-fixation in non-legume crops with an emphasis on rice and proposes the future research. Paddy fields are unique in that they have diverse N-fixation systems in soil, rhizosphere, and plant, due to waterlogging and the resulting various oxygen environment, and their effective utilization is valuable for sustainable food production. In particular, significant progress has been made in the microbial aspects by development of meta-DNA/RNA analysis, indicating that a variety of N-fixiation systems may be functioning in the soil and plant. Further elucidation of the metabolic aspects of these systems as microbial community with quantitative estimation on their contributions will help to promote their utilization. In addition, since C substrate is a driving factor in all N-fixation systems, a strategy to establish the C cycle to optimize N enrichment, C sequestration, and crop productivity in agricultural land will be useful from a micro- to macroscopic and long-term perspective with consideration on global warming.

Keywords Endophyte · N-fixation · Rhizosphere · Rice · Soil · Sustainable food production

16.1 Nitrogen (N) and Crops

N is an element that is essential for plant growth, yet the amount of N present in the soil is relatively small compared to the amount required by plants. For example, paddy soils contain only 0.24% as total N and 0.01% as mineralizable form on average (Oda et al. 1987), while approximately 2.0 kg of N is required to produce 100 kg of brown rice at the standard yield range in Japan (Toriyama 2002). For this

M. Kondo (✉) · R. Shinjo · T. Okamoto
Graduate School of Bioagricultural Sciences, Nagoya University, Nagoya, Aichi, Japan
e-mail: chokai@agr.nagoya-u.ac.jp

© The Author(s), under exclusive license to Springer Nature Singapore Pte Ltd. 2022

D. K. Maheshwari et al. (eds.), *Nitrogen Fixing Bacteria: Sustainable Growth of Non-legumes*, Microorganisms for Sustainability 36,
https://doi.org/10.1007/978-981-19-4906-7_16

reason, the supply of N is a key factor in determining crop production, and humans have devised various ways to supply N to crops. Organic N source has been used for crop production since ancient times till the time of the establishment of the Haber-Bosch process, industrial atmospheric N-fixation became possible, and chemical synthetic fertilizers began to replace the organic fertilizers used until then. In particular, semi-dwarf varieties in wheat and rice developed at the time of the Green Revolution responded well to N fertilization, and the amount of N applied worldwide increased tremendously. Cereals harvested N globally sums 1551 Tg of N, of which 48% was supplied from fertilizer-N from 1961 to 2010 (Ladha et al. 2016). It can be said that our current human activities are benefited greatly from N fertilizers. However, finite fossil fuels are used in the production of N fertilizer, which consumes a large amount of energy, equivalent to 1% of total energy consumption. Moreover, the use of excess N has serious ecological and other consequences. The outflow of nitrate from agricultural systems to water systems has a significant negative impact on ecological imbalance and even on human health. The concept of Planetary boundaries has been proposed to present nine environmental issues within which humanity can continue to develop and thrive for future generations, among which environmental pollution by N and P is the most urgent problem to be solved worldwide (Rockström et al. 2009). CO_2 produced in the production process of N fertilizers and N_2O generated from chemical fertilizers applied in farmland are also sources of greenhouse gases (IPCC 2014). Various fertilization technologies have been developed to improve N use efficiency. In particular, coated controlled-release fertilizers have greatly improved the recovery of applied N by crops and have become widely used. However, it is also regarded as a potential source of microplastic that threatens ecosystems. While excesses are a problem in some areas, there are also many areas where the use of N fertilizer is severely limited by economic and geographical circumstances. Therefore, in order to support the future growth of the world's population, there is a need for strategies to maintain food productivity while limiting dependence on chemical N fertilizers as much as possible. In this regard, it is becoming more important to utilize the N-fixing capacity of crops and farmland.

The legume–rhizobia symbiosis has been widely studied and understood. On the other hand, the utilization of N-fixation in the plant, soil, and rhizosphere of non-legume crops is also important for cereal production and sustainable food supply, and recent progress in research shows that new avenues may be found. In this chapter, we review ecophysiological and agronomic aspects of free-living and endophytic N-fixing bacteria associated with non-leguminous crops, with special emphasis on rice. Rice is a major cereal crop in the world, especially in Asia, and is becoming increasingly important as a staple food in other regions such as Africa. Paddy fields with submerged soils, where most rice plants with efficient oxygen conducting system from shoot to root are grown, create diverse oxygen conditions and thus provides a stage for various N-fixation systems.

16.2 N-Fixation in Paddy Fields

16.2.1 N-Fertility and N-Balance in Paddy Fields

In natural ecosystems, biological N-fixation plays a very important role in cycling nitrogen between the atmosphere and the biosphere as well as denitrification (Sullivan et al. 2014). N-fixing bacteria are widely present in ecosystems and contribute to the supply of N in various cropping soils. Paddy soils are known to have a high natural capacity to maintain fertility as compared to upland field conditions (Kawaguchi and Kyuma 1975; Kyuma 2004). This is partly due to the supply of nutrients contained in irrigation water and the slow decomposition of soil organic matter. In paddy fields under anaerobic conditions, organic matter is mainly decomposed by anaerobic bacteria, and the accumulation of organic matter is larger than in aerobic soil conditions. In addition to these factors, N-fixation capacity of paddy fields has been attracting attention to explain high ability to maintain soil N fertility since long time (Chakraborty and Gupta 1959; De 1939; Watanabe et al. 1951). Paddy field is a favorable environment for N-fixation activity because of its reductive conditions and abundance of labile organic matter (Ladha and Reddy 2003; Wada et al. 1978; Yoshida and Ancajas 1973). In rice paddies, there is an abundant supply of organic matter from roots, residues, aquatic plants, and micro-organisms, which provides an energy source for N-fixation. In paddy field, it has been shown that N absorption by rice can be maintained to some extent over long periods even when cultivated continuously without N-fertilization. Many estimates of positive N balance, i.e., enrichment, calculated from the long-term N-uptake by rice and the amount of residual N in the soil, have been reported from a wide range of regions. For example, in a cold region of Japan with gray lowland soil, N-uptake and relative N-uptake to N-fertilized plot in a no N-fertilized condition were almost maintained for more than 20 years even when the above-ground parts of the plants were removed out of the system after each harvest (Fig. 16.1). The positive balance between rice uptake and changes in soil N was estimated to be about 18 kg ha^{-1} per year, and the contribution from irrigation water was estimated to be less than 1 kg ha^{-1} per year. It is estimated that more than this amount of N was enriched in the soil by N-fixation since there was loss of N due to denitrification and leaching. In fact, the measured enrichment of N in the soil increased by 1–3 g m^{-2} per crop season (Yasuda et al. 2000), confirming the results of the N balance. The N enrichment is often estimated to be larger when organic matters and/or Ca are applied. The N balance under a long-term N omission in field trial in the tropics showed enrichment of 19–98 kg ha^{-1} (average 50 kg ha^{-1}) per crop season (Roger and Ladha 1992). In an early study, estimates by cumulative acetylene reduction activity (ARA) indicate that N-fixing activity is greater in submerged waterlogged conditions than in upland conditions, and that N-fixation increases with rice planting (Yoshida and Ancajas 1973). Global estimation indicated that rice profited more N from N-fixation at 22 kg ha^{-1} $year^{-1}$ larger than maize and wheat at 13 kg ha^{-1} $year^{-1}$ (Ladha et al. 2016). Although the use of paddy fields requires

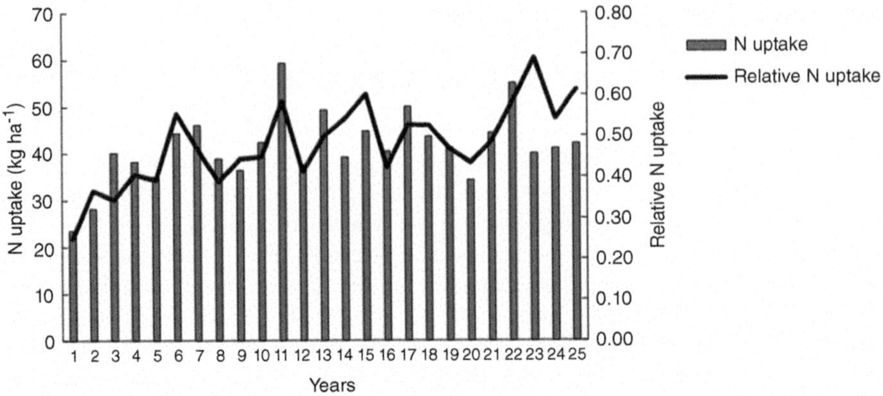

Fig. 16.1 Change in N-uptake and relative N-uptake to N-applied plot (80 kg N ha^{-1} $crop^{-1}$) in no-N plot in long-term experiment using rice in Tohoku region in Japan

irrigation and environmental conditions that guarantee water supply, the efficient use of paddy field systems for food production is considered to be an important option for global food security in the future.

Various N-fixation systems are known to function in paddy fields (Roger and Watanabe 1986; Ladha and Reddy 2003; Yoshida and Yoneyama 1980). The first is by free-living N-fixing bacteria in the soil, the second is by associative N-fixation in the rhizosphere, and the third is by endophytic N-fixing bacteria living inside the rice plant, such as in the roots, leaves, and stems. In general, the contribution to N enrichment is estimated to be largest from the soil. The fourth option is to use external inputs from N-fixing plants and green manure, such as aquatic plants symbiotic with the Rhizobia, such as *Sesbania* and *Aeschynomene* species, and cyanobacteria, such as *Azolla*, an aquatic fern symbiotic with *Anabaena azollae*. Although there are limitations in the use of these N-fixing green manure crops in terms of proper controls of the growth, labor, and cost, they have a large N-fixing potential in paddy fields under favorable conditions (Ladha and Reddy 2003).

16.2.2 N-Fixation in Soil and Water in Paddy Fields

In rice paddies, the contribution of N-fixation by phototrophs such as photosynthetic bacteria and cyanobacteria on the soil surface and in the surface-water is much larger than in the upland field because of the presence of standing water (Kobayashi and Haque 1971). In addition, N-fixation by heterotrophs functions in the soil below the surface layer. The reductive environment created by low oxygen conditions enhances the N-fixing activity of these free-living organisms in paddy soils. The presence of oxygen inhibits nitrogenase activity while providing energy through respiration. Therefore, the oxidation–reduction boundary is usually considered to be the most favorable condition, which is created in the soil surface, rhizosphere, and soil under intermittent irrigation. It is also known that the presence of consortium increases the activity of N-fixing bacteria in anaerobic microsites by cooperation with aerobic non-N-fixing bacteria, which also provide energy sources.

Accurate quantitative estimation of long-term N-fixation in surface soil and water is not easy due to the lack of reliable and feasible methods, but many semi-quantitative assessments have been made. Estimates of annual fixation by phototrophs from ARA in tropical rice paddies ranged from 10 to 80 kg ha^{-1} with an average of 30 kg ha^{-1} (Roger and Ladha 1992). On a 74-day experiment using a chamber with $^{15}N_2$, the contribution of N-fixation was generally higher in the light-exposed surface layer (Ma et al. 2019b). On the other hand, in the lower layers, C substrate supply generally limits N-fixing capacity, and it is inferred that external application of fresh organic matter increases the activity and contribution of heterotrophs in lower layers (Wada et al. 1978).

16.2.3 Effect of Organic Matter

The supply of organic matter as a substrate and energy source is a key environmental factor for N-fixation capacity in soil, and the application of organic matter to paddy fields promotes N-fixation in phototrophs and heterotrophs. In the tropics, straw application has been estimated to promote N-fixation of 2–4 kg N Mg^{-1} by the N balance method in pot trials (Santiago-Ventura et al. 1986). Rice stubble also maintains N-fixing activity when waterlogged (Yoneyama et al. 1977). Compost with low C/N is less effective, while rice straw and matured manure with high C/N ratio are more effective. Therefore, long-term continuous application of rice straw and manures not only supplies C to the soil but also contributes to the enhancement of N fertility through N enrichment capacity by increasing the potential for N-fixation (Kondo and Yasuda 2003a, b) (Table 16.1). Among the components of rice straw, sugars, hemicellulose, and cellulose contribute the most effective to stimulating N-fixation. According to the results of model experiments using straw containing various types of components indicated that hemicellulose had the highest contribution to N-fixation capacity under light conditions (48–53%), followed by cellulose (27–43%), glucose (0–9%), and lignin (<1%) (Yoo et al. 1991). One reason for these differences may reflect the accessibility for decomposition. High molecular weight C compounds are thought to be utilized by N-fixing bacteria mainly as low molecular weight compounds such as organic acids, sugars, and amino acids through decomposition. Cooperative N-fixation by cellulolytic bacteria and anaerobic N-fixing bacteria is thought to occur in rice straw and cellulosic soils (Adachi et al. 1989). C use efficiency (mg N fixed/g C consumed) was higher for cellulose (14.1) than glucose (3.4) in paddy soil (Kondo and Yasuda 2003b). The network of cooperative N-fixation between different groups of bacteria for such a less labile substrate is a topic that needs further understanding, which leads to a more efficient method of promoting N-fixation through the use of organic matter.

Table 16.1 Effect of long-term management of chemical fertilizer and organic matters on N enrichment and N-fixation activity of soils (modified from Kondo and Yasuda 2003a)

	Soil properties				N enrichment	ARA[a]
	T-C	T-N	Mineralizable N	Available P		
Soil management	(%)	(%)	(mg kg^{-1})	(mg P kg^{-1})	(mg N kg^{-1} season^{-1})	(nmol C_2H_4 g^{-1} h^{-1})
Chemical fertilizer (NPK)	2.08	0.17	24.4	131	46	6.8
Chemical fertilizer (NK)	1.89	0.15	14.2	23	−14	5.1
Compost: 10 Mg ha^{-1} year^{-1}	2.11	0.18	19.5	141	7	3.2
Straw: 10 Mg ha^{-1} year^{-1}	2.24	0.17	31.1	111	99	16.6
Manure: 36 Mg ha^{-1} year^{-1}	2.72	0.20	36.4	263	163	26.5

[a] Measured on 66 days after transplanting

The position of rice straw application influences the amount of N-fixation. When rice straw is applied to the surface, the amount of N fixed is greater than when it is plowed into the entire layer (Kanungo et al. 1997; Yasuda et al. 2000), mainly reflects the enhanced activity of cyanobacteria and photosynthetic bacteria. Rice straw application in anaerobic soil promotes an increase in methanogenic populations (Adachi et al. 1996) and substantially increases methane production (Oyediran et al. 1996). Methane is regarded as a greenhouse gas that needs to be reduced. The surface application of straw is also considered to be one of the effective options to mitigate methane emission.

16.2.4 Seasonal and Diurnal Change

Seasonal variation in N-fixation capacity of paddy soils has been observed (Quesada et al. 1998). In terms of ARA variation during the cropping season, there is a tendency to increase from the maximum tillering stage to the heading stage (Yoshida and Ancajas 1973; Wada et al. 1978). The N-fixing activity of soil, when organic matter is applied, varies depending on the degree of its decomposition and environmental factors such as light and temperature. When rice straw was applied, ARA peaked at the heading stage, but when cellulose, which is more easily decomposed, was applied, ARA peaked at an earlier stage. The rice straw applied immediately after harvest of the previous crop also increased N-fixing activity during the following crop-year, but to a lesser extent than that of the spring application (Kondo and Yasuda 2003a). Diurnal variation in *nifH* gene expression was observed (Mårtensson et al. 2009) and N-fixing activity by ARA showed a peak at mid-day (Balandreau et al. 1974).

16.2.5 Effect of N, P, and Trace Elements in Soil

In order to secure a certain level of yield of rice, it is important to combine the minimum required amount of fertilizer without suppressing the N-fixing activity of the soil. In general, high levels of inorganic N in the soil tends to suppress N-fixation activity (Rao 1976), while there was a case with no clear effect (Hsu and Buckley 2009). Comparison among soils under different fertilizer and organic matter applications showed that mineralizable N had a negative effect on ARA under dark, but it was positive in light, probably dominated by positive effect by labile C (Kondo and Yasuda 2003b). N source may also affect N-fixation activity. The lower colonization of *Azospirillum* sp. isolated from rice stems under NH_4^+ nutrition compared to NO_3^- nutrition was attributed to lower rhizosphere pH and lower organic acid secretion from roots (Naher et al. 2018). The effects of soil N on diazotroph community structure tended to be complex (Hsu and Buckley 2009; Tan et al. 2003). By elucidating these complex factors, clearer strategies can be developed to achieve both minimum fertilization and maximum soil N-fixing activity.

P also has a significant effect on phototrophic and heterotrophic N-fixation, especially in low available P in soil (Cholitkul et al. 1980; Kondo and Yasuda 2003a; Matsuguchi et al. 1975; Rao et al. 1986) (Table 16.1). In weathered soils, mainly found in the tropics, where the level of available P is low, the N-fixing activity is limited by P (Matsuguchi 1979). Application of Mo, a constituent element of nitrogenase, increased N-fixing activity, especially that in the cyanobacteria *Leptolyngbya* and *Microcoleus*, in low-Mo paddy soils (Ma et al. 2019a). Long-term manure application enhanced bioavailability of Mo by increasing reactivity of Fe oxide and N-fixation through an increased abundance of *nifH* genes and nitrogenase activity (Yu et al. 2021). The ecological significance of canonical (Mo-) nitrogenases (*nifD*) and alternative (Fe-only and V-) nitrogenases (*anfD* and *vnfD*) in N-fixation is still unclear in crop lands. Although alternative nitrogenase has lower specific activity and is often regarded as a backup to canonical nitrogenase (Eady 2003), its possible contribution in soil is suggested (Bellenger et al. 2014). The results of the analysis of canonical (*nifD*) and alternative (*anfD* and *vnfD*) nitrogenase amplicons in the coastal environment showed that both nitrogenase are present, and it is possible that alternative nitrogenases are also widely distributed (McRose et al. 2017). The higher aluminum oxide was found to enhance the N-fixation activity and *nifH* expression, which may be particularly important in strongly weathered soils such as Ultisol and Oxisol (Wang et al. 2019). Further elucidation of the relationship between soil constituents and N-fixation will have a significant impact on improving N-fixation in the tropics, where N fertility is often low and N fertilizers are not readily available.

16.2.6 Effects of Redox Potential and Other Soil Environments

The redox status of soil has a significant effect on N-fixation. At very low redox potentials, where methane is actively produced, N-fixing activity is lower than under higher redox potentials (Kanungo et al. 1997). Recently, it has been shown that Fe-reducing bacteria may have a significant contribution to N-fixation in paddy soil (Masuda et al. 2017). N-fixation was enhanced in the converted paddy field under rotation with upland crops (Kondo and Yasuda 2003b). The increase in Fe oxide due to oxidation under rotational use of field may enhance N-fixing activity after restoration to paddy fields although further evidence is needed. Field conversion may have a negative impact on soil fertility by depleting available associated with promoted decomposition of organic matter. Therefore, it is necessary to understand the balance between soil N depletion and enrichment for the rotational use of paddy fields that need further understanding on the relationships among redox status, oxides, organic matter decomposition.

Temperature may affect N-fixation efficiency through the decomposition process of organic matter and probably redox status. In a simple incubation experiment, while N-fixing rate decreased with low temperature, C use efficiency for N-fixation was similar or increase at low temperatures (Kondo and Yasuda 2003b). This may be partly attributed to the mechanism that the intermediate degradation products of polymeric organic matter increase under low-temperature conditions, making them more readily available for N-fixation although further examination on mechanisms should be conducted. A 2 °C increase in soil temperature was associated with changes in the bacterial flora in the subsoil (Okubo et al. 2014). A comparison of *nif* levels by climatic zones showed that *nif* levels were higher in the warm-temperate regions than in the tropics, suggesting that not only climatic conditions but also soil properties such as pH and C/P ratio affect N-fixing capacity and N-fixing bacterial flora (Wu et al. 2021).

16.2.7 Influence of Rice Genotypes

Several estimates showed a significant influence of rice varieties in N-fixation in soil and plant. Pot studies showed varietal differences in soil N balance (App et al. 1986), and a comparison of 69 rice lines, using ^{15}N dilution methods estimated that 1.5–21% of the variability in the contribution of N-fixation (Shrestha and Ladha 1996). The high contribution was mainly detected in local landrace varieties, but also found in improved varieties. Genetic analysis studies using early F2 populations suggested that multiple quantitative traits loci (QTLs) are involved in the variation in the contribution of N-fixation (Wu et al. 1995). The genetic control responsible for these genotypic variabilities seems to be complex and is still largely unknown. Variation in the compatibility of varieties with bacterial species and strains is thought to be present. The growth-promoting effect of *Bradyrhizobium* sp. isolated from rice fields in Thailand differed among rice varieties, with the effect being greater in Thai varieties than in Japanese varieties (Piromyou et al. 2015). N-fixing activity of hybrid indica (IIY) was higher than that of inbred japonica (W23), according to the cultivar comparison using $^{15}N_2$ chamber with higher abundance of cyanobacteria species, among heterocystous *Nostoc*, *Anabaena*, and *Cylindrospermum* based on *nifH* sequencing in hybrid indica (Ma et al. 2019b). Root exudates, root morphology, and growth rate are assumed to be the factors that cause such differences in bacterial flora and N-fixing activity among genotypes (Hirota et al. 1978), and their identification is expected.

16.3 Microflora Contributing to N-Fixation in Paddy Fields

16.3.1 Layered Dynamics of Bacterial Flora in Paddy Soil

Waterlogging soils in paddies develop anaerobic conditions overall, but heterogeneity exists in terms of soil layer and microsites. At the soil surface, there is diffusion of oxygen through the water and also oxygen generated from photosynthesis by cyanobacteria and other organisms. Oxygen released from roots creates oxidized condition also in rhizosphere. Metatranscriptomic analysis in Italian rice paddy soils showed that photosynthesis and methane oxidation were dominant in the oxidized layer, while methane production and aromatic compound degradation were characteristic in the reduced layer (Kim and Liesack 2015). These results reflect the succession of taxonomic compositions that are responsible for substrate dynamics in the oxidized layer and the reduced layer. The decomposition of organic matter and the production of energy, which are essential for N-fixation, involve a large number of bacterial flora, and their composition may vary depending on the stratum in which they are found. In the case of mixed culture of *Azotobacter* and photosynthetic bacterium *Rhodopseudomonas capsulatus*, slime substances are secreted and anaerobic microsites are formed to promote N-fixation with metabolic interaction between them (Okuda and Kobayashi 1961).

16.3.2 N-Fixing Microflora in Paddy Soil

In paddy soils, rhizospheric *Alphaproteobacteria*, *Betaproteobacteria*, and *Gammaproteobacteria* and phototrophic cyanobacteria are commonly found in N-fixing bacteria species (Mårtensson et al. 2009). Facultative anaerobes *Klebsiella* and *Enterobacter*, aerobes *Azospirillum* and *Pseudomonas* were found in rice rhizosphere in earlier studies (Ladha et al. 1983; Watanabe et al. 1979, 1987; Yoo et al. 1986). RT-PCR-DGGE analysis of the functional diversity and dynamics of *nifH* expressed in N-fixing bacterial populations revealed that many of them were clustered into *nifH* Cluster I and III, and were identified such as *Azoarcus* spp., *Methylococcus* spp., *Rhizobium* spp., *Methylocystis* spp., *Desulfovibrio* spp., *Geobacter* spp., and *Chlorobium* spp. (Mårtensson et al. 2009). Recently, the taxonomic composition of *nif* transcripts suggested that *Deltaproteobacteria* such as *Anaeromyxobacter* and *Geobacter* are also key groups for N-fixation (Masuda et al. 2017). The ability of these Fe-reducing bacteria to fix N in rice paddy soils was verified by the increase in N-fixing activity following the application of Fe oxide materials (Masuda et al. 2021). N-fixation by these groups of Fe-reducing bacteria may be of interest in paddy soils, which are generally rich in active Fe oxide. High-throughput sequencing together with network analyses showed that *Bradyrhizobium* and *Geobacter* were the major genera in paddy soil under double cropping in China (Dai et al. 2021).

The species and number of cyanobacteria in paddy fields are largely influenced by soil conditions, stratification, and climatic conditions (Quesada et al. 1998; Song et al. 2005). The direct analysis of ^{15}N-enriched *nifH* genes by ^{15}N-SIP has shown that heterocystous *Nostocales* and *Stigonematales* are the major cyanobacterial species (Wang et al. 2020). Soil comparison in a wide climatic zone in China showed a variation of N-fixation at 2.2 ± 0.5 to 20.1 ± 3.7 kg N ha^{-1} over 42 days using the $^{15}N_2$ labeling method, with the strongest effect of *Nostocales* species as the major species (Wang et al. 2019). On the other hand, in some cases, *Synechococcales* were also found in abundance (Song et al. 2005), suggesting that there may be environmental variation.

The contribution of cyanobacteria to N-fixation mainly occurred in soil surface. There was a case where 95% of fixed $^{15}N_2$ during feeding for 28 days was incorporated into the surface layer (0–0.5 cm), indicating a large contribution of cyanobacteria and photosynthetic bacteria, and this was higher under the rice crop 11.33 ± 1.90 kg N ha^{-1} and was larger than that in uncropped crop 3.55 ± 1.18 kg N ha^{-1} (Wang et al. 2020). Cyanobacteria inoculation has attracted attention in temperate and tropical area (Vaishampayan et al. 2001; Peoples et al. 1995). In an early study in India, sugarcane yields more than tripled in 3 years after waterlogging during the rainy season with cyanobacteria inoculation in Usar, an alkaline soil (Singh 1950). In Japan, a positive effect by inoculation with the cyanobacterium *Tolypothrix* on rice growth was found (Watanabe et al. 1951) and an increase in paddy rice yields by an average of 10.6% across several sites was reported in the 1950s and 1960s.

With recent progress in meta-DNA/RNA analysis together with metabolic analysis, it is expected that the N-fixing bacterial flora and their dynamics will be comprehensively and accurately clarified in relation to the substance dynamics associated with diverse redox conditions in soil layers.

16.4 N-Fixation by Endophytes in Rice

16.4.1 N-Fixing Endophytes in Non-leguminous Crops and Rice

N-fixing endophytes that reside inside plants are widely found in cereals and non-legume plants (Yoneyama et al. 2017). Since the isolation of *Beijerinckia Derx* from the rhizosphere of Brazilian sugarcane in the 1950s, which was inspired by the observation that soil N was not depleted by continuous cropping, indicating a possible contribution of N-fixation (Dobereiner 1961), we have made tremendous advancement in our understanding on plant-associated N-fixation, including endophytes (Baldani et al. 2002). In sugarcane, the quantitative contribution of N-fixation to N nutrition has been shown in ARA (Döbereiner et al. 1972) and also in N balance, with a maximum of 109–175 kg N ha^{-1} in hybrid varieties, confirming the significant magnitude of its contribution (Boddey et al. 1995; Lima et al. 1987). Estimation using $\delta^{15}N$ value indicated 0–72% dfa, 30% dfa on average in sugarcane grown in Brazil, Philippines, and Japan (Yoneyama et al. 1997). The significant contribution of endophytes to N-fixation has been also reported in various cereals and other crops other than sugarcane. 32 kg ha^{-1} or 26% dfa, a contribution of N fixation from atmosphere to total N accumulation, has been estimated by Avalue in Kallar grass (*Leptochloa fusca*) (Malik et al. 1988). In sweet potato (*Ipomoea batatas*), estimates by $\delta^{15}N$ have been up to 26–44% dfa (Yoneyama et al. 1998). These suggest the significance of the contribution of plant-associated N-fixation in non-legume crops, especially in high sugar crops.

It can be inferred that competition with other bacteria is lower in plant tissues than in soil, and that the lower oxygen concentration and higher C substrate, such as sugars, organic acids, and amino acids, contribute to offering the suitable habitat for activity of N-fixing bacteria (Baldani et al. 1997). One possible route of entry for endophytes into roots is at the base of the developing lateral roots and in the spaces between the cortical cells (James 2000). The bacteria entering from these sites may travel through the intercellular spaces and aeration tissues to the xylem, where it is transported by conduit flow to other parts of the root and to the aboveground. The endophytic bacteria are mainly found in the intercellular spaces, xylem, and dead tissues, and are rarely found in the vicinity of phloem or inside the cell (Bacon and Hinton 2007; Gyaneshwar et al. 2001; James et al. 2002). The extracellular location of N-fixing bacteria is different from that of rhizobium bacteroids.

The existence and contribution of endophytic N-fixing bacteria have been also well known in rice (Yoneyama et al. 2017), although their N-fixing capacities in rice are generally lower than high sugar plants. For example, the Most probable number (MPN) method counted 7.94×10^7 g dry wt^{-1} and 257×10^6 N-fixing bacteria in roots and stems, respectively, in a modern variety IR72, with a range of 10^3 to 10^7 in roots and 10^4 to 10^6 in stems among cultivars (Barraquio et al. 1997). Roots and stems are known to be the major habitat of N-fixing endophytes in rice. It has been known that among the above-ground parts, the lower part of the stem has high N-fixing activity (Ito et al. 1980). Recently, it was pointed out the possibility that high NSCs (non-structural carbohydrates), especially sugars, in the lower stems may be effective in enhancing N-fixing activity by endophytes (Okamoto et al. 2021). In rice, sugars and starch are the major storage NSCs with large genetic variation in their amounts (Arai-Sanoh et al. 2011). These variations possibly are related to diversity in degree of perenniality found in *Oryza sativa* and *O. glaberrima*, an African rice in part (Morishima et al. 1962). Recent progress in our understanding on genetic control of the accumulation of NSCs is expected to be utilized to lead to genotypic improvement in N-fixation.

16.4.2 Bacterial Flora of N-Fixing Endophytes

As an endophyte, *Gluconacetobacter diazotrophicus* was first isolated from sugarcane stalks in Brazil (Boddey 1995; Cavalcante and Döbereiner 1988) and has also been isolated from sugarcane in a wide area such as the Philippines (Asis et al. 2000) and India (Muthukumarasamy et al. 1999). The genus has also been isolated from other high-sugar crops such as sweet potato, pineapple (*Ananas comosus*), Napier grass (*Pennisetum purpureum*), as well as coffee (*Coffea arabica*) and rice (Jimenez-Salgado et al. 1997; Muthukumarasamy et al. 2005; Tapia-Hernandez et al. 2000). A variety of N-fixing endophytic bacteria have been found in various plants. N-fixing bacteria of the genera *Herbaspirillum*, *Burkholderia*, and *Azospirillum* have been isolated from sugarcane (Asis et al. 2000; Boddey et al. 2003; Reis Junior et al. 2000). *Klebsiella pneumoniae* have been isolated from *Zea mays* and *Zea luxurians* (Palus et al. 1996). *Azoarcus* spp. have been reported to be found in endophytically symbiotic rice, sugarcane, and Kallar grass (Hurek and Hurek 2003) and also been found in wild rice *Oryza longistaminata* (Demba Diallo et al. 2008). Rhizobium species such as *Bradyrhizobium* sp. and non-rhizobium species such as *Paraburkholderia* sp. have also been found in sweet potato (Terakado-Tonooka et al. 2008) and sugarcane stems (Fischer et al. 2012).

Proteobacteria are the major N-fixing species as endophytes in roots and stems at the Phylum level in rice (Bertani et al. 2016; Edwards et al. 2015; Ikeda et al. 2014; Okamoto et al. 2021). *Firmicute* and *Planctomycetes* were the next most common in root and shoot, respectively (Ikeda et al. 2014). *Proteobacteria* were also found to be major endophytes in wheat (Gdanetz and Trail 2017). Culturable endophytic microflora differed among plant parts and changed with growth stages (Mano et al. 2007). In both above-ground and root, *Proteobacteria* and *Alphaproteobacteria* were predominant, with rhizobium in above-ground and about 10% of *Alphaproteobacteria* in root were *Bradyrhizobium* (Ikeda et al. 2014). Isolation of rhizobial *Bradyrhizobium* has also been reported in Thailand under oligotrophic conditions in rice roots (Piromyou et al. 2015). In addition, *Burkholderia* and *Bradyrhizobium*, which have *nifH*, *iaaMH*, and *acdS* genes, were abundant in root under low-N conditions, suggesting that they may promote rice growth under oligotrophic conditions through N-fixation and also by others such as indole-3-acetic acid (IAA) biosynthesis and 1-aminocyclopropane-1-carboxylic acid (ACC) deaminase activity. *Bradyrhizobium* sp. has also been isolated from paddy soil as a bacterium with denitrification genes, and it is assumed to play a major role in N dynamics (Ikeda et al. 2014; Ishii et al. 2011). In the lower part of the stem, *nifH* analysis showed that *Alphaproteobacteria* such as *Bradyrhizobium* sp. were predominant followed by *Betaproteobacteria* such *Paraburkholderia* sp. with commonalities among rice varieties also associated with some specificities (Okamoto et al. 2021). *Herbaspirillum* sp. isolated from wild rice *Oryza officinalis* was observed to colonize the intercellular spaces of leaves (Elbeltagy et al. 2001).

Methane-oxidizing N-fixing bacteria have also been estimated to contribute significantly to N-fixation in rice roots. In roots, *Methylosinus* and *Methylocystis*, which belong to *Alphaproteobacteria* and Type II methanotrophs, were found in abundance, especially under low N conditions, and were highly active in methane oxidation (Ikeda et al. 2014). Metaproteomic analysis of root-associated bacteria from field-grown rice revealed that nitrogenase proteins were mainly derived from *Methylosinus* and *Methylocystis* (Bao et al. 2014a). *Methylosinus* sp. isolated from rice roots had $^{15}N_2$-fixing activity using methane as a substrate and the N-fixing activity was maintained even under 10% oxygen conditions (Shinoda et al. 2019). This suggests that Type II methanotrophs may perform N-fixation in the vicinity of vascular bundles with relatively high oxygen concentrations, which is interesting when considering the dual functions of methane oxidation and N-fixation. The *nifH* composition showed higher complexity in the rhizosphere than the phyllosphere with *Azorhizobium* and *Rhodopseudomonas*, as the most frequently found sequences, in phyllosphere while with *Rhizobium*, *Methylococcus*, *Dechloromonas*, *Anaeromyxobacter*, *Syntrophobacter*, and some methanogenic archaea in rhizosphere (Knief et al. 2012).

On the other hand, anaerobic *Clostridium* spp. has been isolated from shoot in rice and wild rice (Minamisawa et al. 2004). These anaerobic bacteria may form a consortium with aerobic bacteria. It seems that there is a diversity in the substrates and energy production systems involved in endophytic N-fixation in rice. *Gluconacetobacter diazotrophicus* ATTCC 49037 and *Herbaspirillum rubrisubalbicans* PAL8 isolated from sugarcane differed in their ability to utilize sucrose or aconitate as sole C source (Asis et al. 2003). An integrated understanding is expected for the relationship between the sites where endophytes express N-fixation activity, their oxygen environment and energy supply, as well as their N-fixation efficiency.

16.4.3 Growth-Promoting Effects by N-Fixing Endophyte

Growth promotion by inoculation of diazotrophic endophytes, *Herbaspirillum seropedicae* and *Burkholderia* spp., with significant N-fixation was demonstrated under controlled conditions (Baldani et al. 2000). Among species of diazotrophic endophyte found in rice plant, responsive species for N-fixation and their quantitative contribution to N acquisition are still not very clear in field-grown rice. A *nifH* gene expression profiling in rice root indicated that *Azoarcus*, *Burkholderia*, and *Azotobacter* are for active transcriptome but with large varietal differences among cultivated- and wild-rice genotypes (Knauth et al. 2005). The *nifH* RNA sequences in the stems and roots of sugarcane were similar to those of *Bradyrhizobium* sp. and *Azorhizobium caulinodans*, indicating possible contribution of rhizobia (Thaweenut et al. 2011).

In addition to N-fixation, many evidence suggested that number of other factors are involved in the promotion of plant growth by N-fixing bacteria living in the plant and rhizosphere. Inoculation of bacteria-free sugarcane with isolated N-fixing bacteria resulted in the uptake of $^{15}N_2$, and a mutant strain lacking N-fixing function also promoted growth, suggesting the role of plant hormones (Sevilla et al. 2001). Bashan et al. (1989) showed that growth promotion of tomato by *Azospirillum brasilense* was independent of N-fixation process using a mutant strain deficient in N-fixation capacity. A similar result was obtained with *Pseudomonas stutzeri* A15 (Pham et al. 2017). Furthermore, some reports have not shown a significant contribution of N-fixation by endophytes to N accumulation in plants (Boddey et al. 1986; Bremer et al. 1995). Several endophytes have been reported to synthesize and metabolize phytohormones and can stimulate or inhibit the plant's own hormone synthesis (Tsukanova et al. 2017). Inoculation of endophytes with such phytohormone-producing ability has been reported to increase lateral root growth and enhance the plant's ability to acquire N (Vacheron et al. 2013). Inoculation of *Serratia marcescens* to rice increased root length, but ARA was detected only when C sources (malate, succinate, or sucrose) were added in growth media (Gyaneshwar et al. 2001). The changes in root system structure may be due to the interference of endophytes and rhizosphere bacteria with auxin, cytokinin, ethylene, as well as gibberellins and abscisic acid (ABA), which are the major hormonal pathways controlling plant root development (Dodd et al. 2010; Moubayidin et al. 2009; Stepanova and Alonso 2009) Some bacterial species are capable of producing indole compounds such as indole-3-acetic acid (IAA), an auxin produced by plant growth-promoting bacteria (Spaepen et al. 2008). Lateral and adventitious roots are induced by high concentrations of IAA, whereas primary root growth is stimulated by relatively low concentrations of IAA (Meuwly and Pilet 1991; Pilet and Saugy 1987). However, the involvement of hormones produced by N-fixing bacteria in the regulation of hormone balance in plants has not been sufficiently understood. Some rhizosphere bacteria have been reported to promote nutrient absorption through increase root surface area (Lin et al. 1983; Mantelin et al. 2006); inoculation with *A. brasilense* promoted growth of wheat through increased inorganic N absorption (Saubidet et al. 2002). Co-inoculation of *A. brasilense* and *Pseudomonas fluorescens* also enhances N supply to rice through enhancement of N mineralization in the rhizosphere in addition to nitrogenase activity (Zhang et al. 2018). Shinjo et al. (2020) showed that *Burkholderia vietnamiensis*, an N-fixing endophyte, promotes growth and N accumulation in rice associated with increased root N absorption rate, which is at least partly explained by upregulation of nitrate transporters in roots (Table 16.2). This suggested that N-fixing endophytes also have the potential to increase the activity of transporters involved in nutrient absorption.

16.4.4 Utilization of N Fixed by Free-Living and Endophytic Diazotrophs by Plants

Most of the N fixed by N-fixing bacteria in soil is assumed to remain in the soil as biomass and is gradually mineralized for use by rice plants in the subsequent cropping seasons, but some of it is expected to be mineralized relatively quickly

Table 16.2 Effect of *B. vietnamiensis* RS1 on dry weight, N content, specific absorption rate (SAR), and N-fixation activity in rice (*O. sativa* cv. Nipponbare, 42 days after sowing) (modified from Shinjo et al. 2020)

Inoculation	Shoot dry weight (g $plant^{-1}$)		Root dry weight (g $plant^{-1}$)		N content (mg $plant^{-1}$)		SAR (specific absorption rate of N) (g N g^{-1} day^{-1})		ARA (nmol C_2H_4 h^{-1} g FW $root^{-1}$)	
Control	1.27 ± 0.07	*	0.15 ± 0.02	ns	67.4 ± 3.3	*	0.062 ± 0.003	*	1.00 ± 0.08	ns
B. vietnamiensis RS1	1.66 ± 0.05		0.20 ± 0.01		81.4 ± 2.4		0.076 ± 0.004		0.74 ± 0.12	

Values represent the mean ± standard error ($n = 4$). Asterisks indicate significant differences and "ns" indicates no significant differences between *B. vietnamiensis* RS1-inoculated and -uninoculated plants ($p < 0.05$, *t*-test)

(Inubushi and Watanabe 1987). These mineralized N is expected to show basically similar dynamics to that of fertilizer N. It was demonstrated that N fixed in rhizosphere could be translocated to shoot including panicle (Yoshida and Yoneyama 1980). In an experiment in which glucose was added to promote ^{15}N-fixation, most of the fixed N was converted to the amino form and then mineralized in which 34% of the fixed N was absorbed by the rice plants in 42 days (Ito and Watanabe 1981). In a 74-day experiment using a ^{15}N chamber, 1–2.5% of fixed N mainly in 0–1 cm layer was taken up by rice (Ma et al. 2019b). The fixed N is considered to also be transferred to other bacteria, probably in the community. N flow analysis indicated that fixed N was transferred more to ammonia-oxidizing bacteria K-strategists such as *Nitrosospira* sp., which grow slowly in nutrient-limited conditions (Bei et al. 2019).

Little is known about the dynamics of N fixed by endophytes in the intercellular spaces of the plant tissue and its assimilation by plants. As in the case of rhizosphere associative fixers, it is assumed that most of endophytically fixed N released by the decomposition of bacterial cells are mainly utilized by plants, but further verification is awaited. The fact that *Rhizobia* in free-living state export fixed N as NH_4^+ (O'Gara and Shanmugam 1976) indicates a possibility of excretion of NH_4^+ in apoplast from endophytic *Rhizobia*. Model experiments have shown that ^{15}N is transferred to yeast when *G. diazotrophicus* is cultured with yeast (Cojho et al. 1993). In sugarcane, the transfer and assimilation of fixed ^{15}N was observed to be relatively slow and remained as 80% ethanol-soluble fraction in fixed sites even after 3 days of feeding (Momose et al. 2009). The dynamics of fixed N may differ depending on the growth stage and site.

16.4.5 Ecology of N-Fixing Endophytes in Soil Ecosystem

It has been reported that high N fertilization suppresses bacterial numbers and activity in some cases, but there were also observations that early N fertilization may not depress N-fixation and proper fertilization may be useful to promote endophytic N-fixation in sugarcane. In sugarcane, high N levels (300 kg ha^{-1}) reduced *G. diazotrophicus* but had no effect on *Herbaspirillum* spp. (Reis Junior et al. 2000). Rhizoplane and endophytic colonization by *Serratia marcescens* and *Herbaspirillum seropedicae* in rice was suppressed by high NH_4^+, but not with NO_3^- (Gyaneshwar et al. 2000). *A. diazotrophicus* formed long pleomorphic, immobile cells with the high concentration of N, especially NH_4^+, which may be related to low detection of this species in sugarcane under high N (Muthukumarasamy et al. 2002). Type II methanotrophs, which have N-fixing activity in roots, increased under low N conditions (Ikeda et al. 2014; Shinoda et al. 2019). Attempts are also underway to reduce inhibitory effect by inorganic N and oxygen for nitrogenase by genetic modification of the bacteria (Ryu et al. 2020; Wen et al. 2021).

Soil properties and land use are also assumed to be critical to affecting flora and number of endophytes in plant. *Bradyrhizobium* sp., an oligotroph of rice roots and rhizosphere soil, is more abundant in fields with legumes (Guong et al. 2012; Piromyou et al. 2015). Photosynthetic *Bradyrhizobium* strains isolated from rice roots and rhizosphere are symbiotic with the aquatic legumes *Aeschynomene indica* and *Aeschynomene evenia*, and non-photosynthetic *Bradyrhizobium* strains were able to form nodules in symbiosis with *Aeschynomene americana*, *Vigna radiata*, and *Macroptilium atropurpureum*. Photosynthetic *Bradyrhizobium* strains, which form nodule on *Aeschynomene* legumes, were found as endophytes on wild rice *O. breviligulata* in West Africa and promoted growth by inoculation (Chaintreuil et al. 2000). These lead to speculation that rotation of rice with leguminous crops may be effective in maintaining endophytic *Bradyrhizobium* sp.

In agro-ecosystems including rice paddies, endophytes may be transmitted by propagules, seeds, plant residues left in the soil, and insect vectors, but their transfer and establishment processes seem not to be well understood. It is possible that rice stubbles can be used as an inoculum in the rice–legume crop rotation system (Piromyou et al. 2017). In some cases such as for *G. diazotrophicus*, bacteria isolated from inside the plant were not detected in the soil, suggesting a possibility that this N-fixing bacterium can be transmitted through vegetative organs. The genetic diversity of *G. diazotrophicus* in sugarcane from different regions was relatively limited (Caballero-Mellado and Martinez-Romero 1994).

16.4.6 Endophyte and Plant Interaction

Recent knowledge on plant–microbe interaction suggests that plants are able to regulate the bacterial community in the phytosphere (Ikeda et al. 2010). It is assumed that there is a selection of bacteria during the process of invasion and establishment from soil through roots. Most of the endophytes are assumed to invade non-specifically through roots and become established in the plant organs; i.e., *Azospirillum*, *Klebsiella*, *Rhizobium*, *Clostridium*, and *Bradyrhizobium* are detected both in the soil and in the plant. However, it is certain that there are some bacteria that preferentially colonize specific parts of the plants, and those specificities are related to physiological factors such as substrate requirement, mobility, oxygen requirement, quorum sensing, and biofilm formation (Sessitsch et al. 2012). The results of *nifH* amplicon analysis in sugarcane showed that *Bradyrhizobium*, *Methylocapsa*, *Burkholderia*, and *Azotobacter* were dominant in roots and stems, and *Azorhizobium* was more abundant in rhizosphere soil in addition to those found in plant (Gaby et al. 2018).

Differences in N-fixing activity and bacterial flora among rice genotypes also indicate endophyte–genotype specificity. *Klebsiella pasteurii* BDA134-6 and *Phytobacter diazotrophicus*, isolated from African rice *O. glaberrima* grown along the Niger River, also formed colonies and expressed N-fixing activity on *O. sativa*, but the establishment was higher on *O. glaberrima* (Bianco et al. 2021).

The presence of specific partnerships through the signal exchange, as in the case of rhizobia and host legumes, would be expected to be examined in no-leguminous crops. The common symbiosis pathway (CSP) involved in mycorrhizal symbiosis is known to exist in legume–rhizobial symbiosis, and orthologs of the genes constituting the CSP have been found in rice (Banba et al. 2008). It was suggested that *CCaMK*, a calcium/calmodulin-dependent protein kinase gene and a central factor of CSP, has roles in the promotion of the growth of methanotrophs, especially in rice under low-N conditions (Bao et al. 2014b; Ikeda et al. 2011; Minamisawa et al. 2016).

16.5 Research Issues on Sustainable Rice Production and N-Fixation

16.5.1 Metabolic Functions of Microbial Community

In recent years, meta-DNA/RNA analysis of the N-fixing bacteria has greatly expanded our knowledge on the microbial aspect of N-fixing bacteria in different N-fixing system. The great diversity in contribution of soil and plant-associated N-fixation among genotypes would offer an opportunity for improvement of N-fixing capacity in rice. On the other hand, there are large variations among reports in the quantitative contribution of soil and plant-associated N-fixation in rice and other crops, but the reasons for these variations are not necessarily clear in many cases. One reason for those is that the methods to quantify long-term N-fixation in the field remain a challenge: semi-quantitative estimations such as ARA and the $\delta^{15}N$ method have limitations, while the use of the $^{15}N_2$ method in the field remains costly and technical limitation. The other reason is due to the insufficient understanding of the physiological and ecological mechanisms of N-fixation by free-living and endophytic bacteria as microbial community under diverse and heterogenous environment both in soil and plant. A deeper understanding of energy supply to N-fixers and the metabolic pathway driving N-fixation in microbial community in soil and plant is expected to find a way to effectively enhance N-fixation in the soil–plant system.

16.5.2 Utilization of N-Fixation and C Flow in Rice Culture

N-fixation in agricultural fields is a process in which the supply of C sources, whether in the soil, rhizosphere, or plant, is a determining factor in activity. Therefore, optimizing C flow is considered to be important controlling factor for N-fixation driven by diverse microbial communities (Fig. 16.2). After the Green Revolution, high yielding hybrid and inbred varieties have been developed in

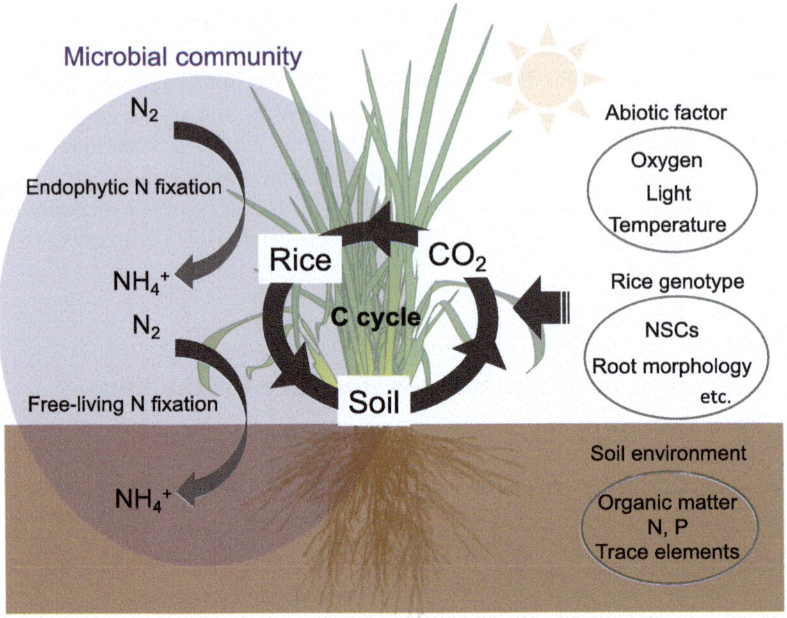

Fig. 16.2 Schematic relationships between C cycle and N-fixation in soil and rice plant driven by microbial community as affected by various biotic and abiotic factors

various regions until today. Tropical japonica genetic resources were utilized in the development of New Plant Type (NPT) at International Rice Research Institute (IRRI) started in the 1980s, strong culm and extremely heavy-panicle varieties. Indica and japonica-indica hybrid high-yielding varieties have also been developed in China, Korea, and Japan. These varieties have a yield potential of more than 15 and 30 t ha^{-1} as dry matter production under favorable conditions. Such high-yielding varieties also have large N requirements, for example, in some cases, 30 kg ha^{-1} of N was absorbed to produce 16 t ha^{-1} of grain (Okamura et al. 2022). It is obviously unsustainable to supply such a large N requirement with chemical fertilizers. For sustainable cultivation of such high-yielding varieties, enhancement of N-fixation capacity is crucial. Since high-yielding cultivars produce large amounts of photosynthate as NSC in stem, straw, root, and root exudate, the efficient use of these C sources for various N-fixation systems in the soil, rhizosphere, and plant is a desirable strategy. The use of by-products is also possible. Rice husk biochar has been reported to be a possible carrier for N-fixing bacteria and growth-promoting bacteria (Win et al. 2019). Since the N-fixing capacity of rice grown in paddy field is lower than that of legume–rhizobium, it would be effective to maximize the N-fixing capacity in an integrated manner in rice culture, by combination of N-fixation in the soil, endophytes, rhizosphere, and externally applied green manure, depending on the natural and cultivation environment. Although green manure was not the focus of this chapter, many studies have been conducted that showed its potentials in rice culture. For example, the high N-fixing ability of *Azolla*, which is unique to rice paddies, has been demonstrated in Asia and Africa as a source of readily available N to improve rice yield (Ito and Watanabe 1985; Kondo et al. 1989; Watanabe et al. 1989). Although practical and economic constraint should be solved, the integration of such green manure into the cropping system is an effective source of N and also C source to enhance N-fixation in soil and plant.

16.6 Conclusions

It is also necessary to consider how N-fixation can be utilized in accordance with global warming. N-fixing activity in soil is enhanced under elevated CO_2 (Cheng et al. 2001). This may be due to the fact that elevated CO_2 increases photosynthesis and growth, which in turn increases organic exudates from roots and enhances N-fixing activity. The increase in N-fixing capacity due to elevated CO_2 is also observed in legume–rhizobium. It will be interesting to see whether the future increase in CO_2 will promote N-fixation by enhancing C fixation, and whether this will be commensurate with the increased demand for N. There is also a possibility that land productivity can be further improved by intensified land use, such as double cropping, even in temperate regions, by taking advantage of the longer growing season due to elevated temperature; this may have a significant impact on the C cycle and microbial ecology in the soil–plant system. A model suggests that substitution of organic N in double cropping in China contributes to long-term improvement and stabilization of N-fixing capacity (Dai et al. 2021). The strategy to optimize C sequestration in soil and N-fixation of cropped soils with regards to microbial characteristics are to be examined from a long-term view with consideration of global warming.

Acknowledgment Not Applicable.

Conflict of Interest Author(s) declares no conflict of interest.

References

Adachi K, Watanabe I, Kobayashi M, Takahashi E (1989) Effect of application of glucose, cellulose, and rice straw on nitrogen fixation (acetylene reduction and soil-nitrogen components) in anaerobic soil. Soil Sci Plant Nutr 35:235–243. https://doi.org/10.1080/00380768.1989.10434756

Adachi K, Oyediran G, Senboku T (1996) Effect of application of rice straw and cellulose on methane emission and biological nitrogen fixation in a subtropical paddy field. Soil Sci Plant Nutr 42:713–723. https://doi.org/10.1080/00380768.1996.10416618

App A, Watanabe I, Ventura TS, Bravo M, Jurey CD (1986) The effect of cultivated and wild rice varieties on the nitrogen balance of flooded soil. Soil Sci 141:448–452

Arai-Sanoh Y, Ida M, Zhao R, Yoshinaga S, Takai T, Ishimaru T, Maeda H, Nishitani K, Terashima Y, Gau M, Kato N, Matsuoka M, Kondo M (2011) Genotypic variations in non-structural carbohydrate and cell-wall components of the stem in rice, sorghum, and sugar cane. Biosci Biotechnol Biochem 75:1104–1112. https://doi.org/10.1271/bbb.110009

Asis CA Jr, Kubota M, Chebotar VK, Ohta H, Arima Y, Nishiyama K, Tyschiya K, Akao S (2000) Endophytic bacterial population in Philippine sugarcane cultivars and isolation of nitrogen-fixing strains. Microbes Environ 15:209–216. https://doi.org/10.1264/jsme2.2000.209

Asis CA Jr, Shimizu T, Khan MK, Akao S (2003) Organic acid and sugar contents in sugarcane stem apoplast solution and their role as carbon source for endophytic diazotrophs. Soil Sci Plant Nutr 49:915–920. https://doi.org/10.1080/00380768.2003.10410356

Bacon CW, Hinton DM (2007) Bacterial endophytes: the endophytic niche, its occupants, and its utility. In: Gnanamanickam SS (ed) Plant-associated bacteria. Springer, Dordrecht, pp 155–194. https://doi.org/10.1007/978-1-4020-4538-7_5

Balandreau JP, Millier CR, Dommergues YR (1974) Diurnal variations of nitrogenase activity in the field. Appl Microbiol 27:662–665. https://doi.org/10.1128/am.27.4.662-665.1974

Baldani JI, Caruso L, Baldani VLD, Goi S, Döbereiner J (1997) Recent advances in BNF with non-legume plants. Soil Biol Biochem 29:911–922. https://doi.org/10.1016/S0038-0717(96)00218-0

Baldani D, Baldani J, Döbereiner J (2000) Inoculation of rice plants with the endophytic diazotrophs *Herbaspirillum seropedicae* and *Burkholderia* spp. Biol Fertil Soils 30:485–491. https://doi.org/10.1007/s003740050027

Baldani JI, Reis VM, Baldani VLD, Döbereiner J (2002) A brief story of nitrogen fixation in sugarcane - reasons for success in Brazil. Funct Plant Biol 29:417–423. https://doi.org/10.1071/PP01083

Banba M, Gutjahr C, Miyao A, Hirochika H, Paszkowski U, Kouchi H, Imaizumi-Anraku H (2008) Divergence of evolutionary ways among common sym genes: *CASTOR* and *CCaMK* show functional conservation between two symbiosis systems and constitute the root of a common signaling pathway. Plant Cell Physiol 49:1659–1671. https://doi.org/10.1093/pcp/pcn153

Bao Z, Okubo T, Kubota K, Kasahara Y, Tsurumaru H, Anda M, Ikeda S, Minamisawa K (2014a) Metaproteomic identification of diazotrophic methanotrophs and their localization in root tissues of field-grown rice plants. Appl Environ Microbiol 80:5043–5052. https://doi.org/10.1128/AEM.00969-14

Bao Z, Watanabe A, Sasaki K, Okubo T, Tokida T, Liu D, Ikeda S, Imaizumi-Anraku H, Asakawa S, Sato T, Mitsui H, Minamisawa K (2014b) A rice gene for microbial symbiosis, *Oryza sativa CCaMK*, reduces CH_4 flux in a paddy field with low nitrogen input. Appl Environ Microbiol 80:1995–2003. https://doi.org/10.1128/AEM.03646-13

Barraquio WL, Revilla L, Ladha JK (1997) Isolation of endophytic diazotrophic bacteria from wetland rice. Plant Soil 194:15–24. https://doi.org/10.1023/A:1004246904803

Bashan Y, Singh M, Levanony H (1989) Contribution of *Azospirillum brasilense* Cd to growth of tomato seedlings is not through nitrogen fixation. Can J Bot 67:2429–2434. https://doi.org/10.1139/b89-312

Bei Q, Xie Z, Cadisch G, Rasche F (2019) K-strategic ammonia-oxidizing bacteria capitalize on biological nitrogen fixation in a flooded, unfertilized rice soil. Biol Fertil Soils 55:713–722. https://doi.org/10.1007/s00374-019-01387-2

Bellenger JP, Xu Y, Zhang X, Morel FMM, Kraepiel AML (2014) Possible contribution of alternative nitrogenases to nitrogen fixation by asymbiotic N_2-fixing bacteria in soils. Soil Biol Biochem 69:413–420. https://doi.org/10.1016/j.soilbio.2013.11.015

Bertani I, Abbruscato P, Piffanelli P, Subramoni S, Venturi V (2016) Rice bacterial endophytes: isolation of a collection, identification of beneficial strains and microbiome analysis. Environ Microbiol Rep 8:388–398. https://doi.org/10.1111/1758-2229.12403

Bianco C, Andreozzi A, Romano S, Fagorzi C, Cangioli L, Prieto P, Cisse F et al (2021) Endophytes from African rice (*Oryza glaberrima* L.) efficiently colonize Asian rice (*Oryza sativa* L.) stimulating the activity of its antioxidant enzymes and increasing the content of nitrogen, carbon, and chlorophyll. Microorganisms 2021:1714. https://doi.org/10.3390/microorganisms9081714

Boddey R (1995) Biological nitrogen fixation in sugarcane: a key to energetically viable biofuel production. Crit Rev Plant Sci 14:263–279. https://doi.org/10.1080/07352689509701929

Boddey RM, Baldani VL, Baldani JI, Döbereiner J (1986) Effect of inoculation of *Azospirillum* spp. on nitrogen accumulation by field-grown wheat. Plant Soil 95:109–121. https://doi.org/10.1007/BF02378857

Boddey RM, de Oliveira OC, Urquiaga S, Reis VM, Olivares FL, Baldani VLD, Döbereiner J (1995) Biological nitrogen fixation associated with sugar cane and rice: contributions and prospects for improvement. Plant Soil 174:195–209. https://doi.org/10.1007/BF00032247

Boddey RM, Urquiaga S, Alves BJV, Reis V (2003) Endophytic nitrogen fixation in sugarcane: present knowledge and future applications. Plant Soil 252:139–149. https://doi.org/10.1023/A:1024152126541

Bremer E, Janzen H, Gilbertson C (1995) Evidence against associative N_2 fixation as a significant N source in long-term wheat plots. Plant Soil 175:13–19. https://doi.org/10.1007/BF02413006

Caballero-Mellado J, Martinez-Romero E (1994) Limited genetic diversity in the endophytic sugarcane bacterium *Acetobacter diazotrophicus*. Appl Environ Microbiol 60:1532–1537. https://doi.org/10.1128/aem.60.5.1532-1537.1994

Cavalcante VA, Döbereiner J (1988) A new acid-tolerant nitrogen-fixing bacterium associated with sugarcane. Plant Soil 108:23–31. https://doi.org/10.1007/BF02370096

Chaintreuil C, Giraud E, Prin Y, Lorquin J, Bâ A, Gillis M, de Lajudie P, Dreyfus B (2000) Photosynthetic bradyrhizobia are natural endophytes of the African wild rice *Oryza breviligulata*. Appl Environ Microbiol 66:5437–5447. https://doi.org/10.1128/AEM.66.12.5437-5447.2000

Chakraborty S, Gupta S (1959) Fixation of nitrogen by the rice plant. Nature 184:2033–2034. https://doi.org/10.1038/1842033a0

Cheng W, Inubushi K, Yagi K, Sakai H, Kobayashi K (2001) Effects of elevated carbon dioxide concentration on biological nitrogen fixation, nitrogen mineralization and carbon decomposition in submerged rice soil. Biol Fertil Soils 34:7–13. https://doi.org/10.1007/s003740100345

Cholitkul W, Tangcham B, Sangtong P (1980) Effect of phosphorus on N_2 fixation as measured by the field acetylene reduction technique in Thailand long-term fertility plots. Soil Sci Plant Nutr 26:291–299. https://doi.org/10.1080/00380768.1980.10431211

Cojho EH, Reis VM, Schenberg ACG, Dobereiner J (1993) Interactions of *Acetobacter diazotrophicus* with an amylolytic yeast in nitrogen-free batch culture. FEMS Microbiol Lett 106:341–346. https://doi.org/10.1111/j.1574-6968.1993.tb05986.x

Dai X, Song D, Guo Q, Zhou W, Liu G, Ma R, Liang G, Hea P, Sun G, Yuan F, Liu Z (2021) Predicting the influence of fertilization regimes on potential N fixation through their effect on free-living diazotrophic community structure in double rice cropping systems. Soil Biol Biochem 156:108220. https://doi.org/10.1016/j.soilbio.2021.108220

De PK (1939) The role of blue-green algae in nitrogen fixation in rice-fields. Proc R Soc B Biol Sci 127:121. https://doi.org/10.1098/rspb.1939.0014

Demba Diallo M, Reinhold-Hurek B, Hurek T (2008) Evaluation of PCR primers for universal *nifH* gene targeting and for assessment of transcribed *nifH* pools in roots of *Oryza longistaminata* with and without low nitrogen input. FEMS Microbiol Ecol 65:220–228. https://doi.org/10.1111/j.1574-6941.2008.00545.x

Dobereiner J (1961) Nitrogen-fixing bacteria of the genus *Beijerinckia Derx* in the rhizosphere of sugar cane. Plant Soil 15:211–216. https://doi.org/10.1007/BF01400455

Döbereiner J, Day JM, Dart PJ (1972) Nitrogenase activity in the rhizosphere of sugarcane and some other tropical grasses. Plant Soil 37:191–196. https://doi.org/10.1007/BF01578494

Dodd AN, Kudla J, Sanders D (2010) The language of calcium signaling. Annu Rev Plant Biol 61:593–620. https://doi.org/10.1146/annurev-arplant-070109-104628

Eady RR (2003) Current status of structure function relationships of vanadium nitrogenase. Coord Chem Rev 237:23–30. https://doi.org/10.1016/S0010-8545(02)00248-5

Edwards J, Johnson C, Santos-Medellin C, Lurie E, Podishetty NK, Bhatnagar S, Eisen JA, Sundaresan V (2015) Structure, variation, and assembly of the root associated microbiomes of rice. Proc Natl Acad Sci 112:E911–E920. https://doi.org/10.1073/pnas.1414592112

Elbeltagy A, Nishioka K, Sato T, Suzuki H, Ye B, Hamada T, Isawa T, Mitsui H, Minamisawa K (2001) Endophytic colonization and in planta nitrogen fixation by a *Herbaspirillum* sp. Isolated from wild rice species. Appl Environ Microbiol 67:5285–5293. https://doi.org/10.1128/AEM.67.11.5285-5293.2001

Fischer D, Pfitzner B, Schmid M, Simoes-Araujo JL, Reis VM, Pereira W et al (2012) Molecular characterisation of the diazotrophic bacterial community in uninoculated and inoculated field-grown sugarcane (*Saccharum* sp.). Plant Soil 356:83–99. https://doi.org/10.1007/s11104-011-0812-0

Gaby JC, Rishishwar L, Valderrama-Aguirre LC, Green SJ, Valderrama-Aguirre A, Jordan IK, Kostka JE (2018) Diazotroph community characterization via a high-throughput *nifH* amplicon sequencing and analysis pipeline. Appl Environ Microbiol 84:e01512–e01517. https://doi.org/10.1128/AEM.01512-17

Gdanetz K, Trail F (2017) The wheat microbiome under four management strategies, and potential for endophytes in disease protection. Phytobiomes J 1:158–168. https://doi.org/10.1094/PBIOMES-05-17-0023-R

Guong VT, Rosling A, Alström S, Chai B, Högberg N (2012) Different crop rotation systems as drivers of change in soil bacterial community structure and yield of rice, *Oryza sativa*. Biol Fertil Soils 48:217–225. https://doi.org/10.1007/s00374-011-0618-5

Gyaneshwar P, Reddy PM, Ladha JK (2000) Nutrient amendments influence endophytic colonization of rice by *Serratia marcescens* IRBG500 and *Herbaspirillum seropedicae* Z67. Microb Biotechnol 10:694–699

Gyaneshwar P, James EK, Mathan N, Reddy PM, Reinhold-Hurek B, Ladha JK (2001) Endophytic colonization of rice by a diazotrophic strain of *Serratia marcescens*. J Bacteriol 183:2634–2645. https://doi.org/10.1128/JB.183.8.2634-2645.2001

Hirota Y, Fujii T, Sano Y, Iyama S (1978) Nitrogen fixation in the rhizosphere of rice. Nature 276:416–417. https://doi.org/10.1038/276416a0

Hsu S-F, Buckley DH (2009) Evidence for the functional significance of diazotroph community structure in soil. ISME J 3:124–136. https://doi.org/10.1038/ismej.2008.82

Hurek T, Hurek BR (2003) *Azoarcus* sp. strain BH72 as a model for nitrogen-fixing grass endophytes. J Biotechnol 106:169–178. https://doi.org/10.1016/j.jbiotec.2003.07.010

Ikeda S, Okubo T, Anda M, Nakashita H, Yasuda M, Sato S, Kaneko T, Tabata S, Eda S, Momiyama A, Terasawa K, Mitsui H, Minamisawa K (2010) Community- and genome-based views of plant- associated bacteria: plant- bacterial interactions in soybean and rice. Plant Cell Physiol 51:1398–1410. https://doi.org/10.1093/pcp/pcq119

Ikeda S, Okubo T, Takeda N, Banba M, Sasaki K, Imaizumi, Anraku H, Fujihara S, Ohwaki Y, Ohshima K, Fukuta Y, Eda S, Mitsui H, Hattori M, Sato T, Shinano T, Minamisawa K (2011) The genotype of the calcium/calmodulin- dependent protein kinase gene (*CCaMK*) determines bacterial community diversity in rice roots under paddy and upland field conditions. Appl Environ Microbiol 77:4399–4405. https://doi.org/10.1128/AEM.00315-11

Ikeda S, Sasaki K, Okubo T, Yamashita A, Terasawa K, Bao Z, Liu D, Watanabe T, Murase J, Asakawa S (2014) Low nitrogen fertilization adapts rice root microbiome to low nutrient environment by changing biogeochemical functions. Microbes Environ 29:50–59. https://doi.org/10.1264/jsme2.ME13110

Inubushi K, Watanabe I (1987) Microbial biomass nitrogen in anaerobic soil as affected by N-immobilization and N_2-fixation. Soil Sci Plant Nutr 33:213–224. https://doi.org/10.1080/00380768.1987.10557567

IPCC (2014) AR5 synthesis report: climate change. IPCC, Geneva. https://www.ipcc.ch/report/ar5/syr/

Ishii S, Ashida N, Otsuka S, Senoo K (2011) Isolation of oligotrophic denitrifiers carrying previously uncharacterized functional gene sequences. Appl Environ Microbiol 77:338–342. https://doi.org/10.1128/AEM.02189-10

Ito O, Watanabe I (1981) Immobilization, mineralization and availability to rice plants of nitrogen derived from heterotrophic nitrogen fixation in flooded soil. Soil Sci Plant Nutr 27:169–176. https://doi.org/10.1080/00380768.1981.10431268

Ito O, Watanabe I (1985) Availability to rice plants of nitrogen fixed by Azolla. Soil Sci Plant Nutr 31:91–104. https://doi.org/10.1080/17470765.1985.10555220

Ito O, Cabrera D, Watanabe I (1980) Fixation of dinitrogen-15 associated with rice plants. Appl Environ Microbiol 39:554–558. https://doi.org/10.1128/aem.39.3.554-558.1980

James EK (2000) Nitrogen fixation in endophytic and associative symbiosis. Field Crop Res 65: 197–209. https://doi.org/10.1016/S0378-4290(99)00087-8

James KE, Gyaneshwar P, Mathan N, Barraquio WL, Reddy PM, Iannetta PPM, Olivares FL, Ladh JK (2002) Infection and colonization of rice seedlings by the plant growth-promoting bacterium *Herbaspirillum seropedicae* Z67. Mol Plant Microbe Interact 15:894–906. https://doi.org/10.1094/MPMI.2002.15.9.894

Jimenez-Salgado T, Fuentes-Ramirez LE, Tapia-Hernandez A, Mascarua-Esparza MA, Martinez-Romero E, Caballero-Mellado J (1997) *Coffea arabica* L., a new host plant for *Acetobacter diazotrophicus*, and isolation of other nitrogen-fixing acetobacteria. Environ Microbiol 63: 3676–3683. https://doi.org/10.1128/aem.63.9.3676-3683.1997

Kanungo PK, Ramakrishnan B, Rao RV (1997) Placement effects of organic sources on nitrogenase activity and nitrogen–fixing bacteria in flooded rice soils. Biol Fertil Soils 25:103-110. https://doi.org/10.1007/s003740050288

Kawaguchi K, Kyuma K (1975) Paddy soils in tropical Asia: Part 5. Soil fertility evaluation. S E Asian Stud 13:385–401. http://hdl.handle.net/2433/55817

Kim Y, Liesack W (2015) Differential assemblage of functional units in paddy soil microbiomes. PLoS One 10:e0122221. https://doi.org/10.1371/journal.pone.0122221

Knauth S, Hurek T, Brar D, Reinhold-Hurek B (2005) Influence of different Oryza cultivars on expression of *nifH* gene pools in roots of rice. Environ Microbiol 7:1725–1733. https://doi.org/10.1111/j.1462-2920.2005.00841.x

Knief C, Delmotte N, Chaffron S, Stark M, Innerebner G, Wassmann R, von Mering C, Vorholt JA (2012) Metaproteogenomic analysis of microbial communities in the phyllosphere and rhizosphere of rice. ISME J 6:1378–1390. https://doi.org/10.1038/ismej.2011.192

Kobayashi M, Haque MZ (1971) Contribution to nitrogen fixation and soil fertility by photosynthetic bacteria. Plant Soil 35:443–456. https://doi.org/10.1007/BF02661870

Kondo M, Yasuda M (2003a) Seasonal changes in N_2 fixation activity and N enrichment in paddy soils as affected by soil management in the northern area of Japan. J Agric Res Quart 37:105–111. https://doi.org/10.6090/jarq.37.105

Kondo M, Yasuda M (2003b) Effects of temperature, water regime, light, and soil properties on $^{15}N_2$ fixation associated with decomposition of organic matter in paddy soils. J Agric Res Quart 37:113–119. https://doi.org/10.6090/jarq.37.113

Kondo M, Kobayashi M, Takahashi E (1989) Effect of phosphorus on *Azolla* and its utilization in rice culture in Niger. Plant Soil 120:165–170. https://doi.org/10.1007/BF02377065

Kyuma K (2004) Paddy soil science. Kyoto University Press, Kyoto, p 280. ISBN 1920901000

Ladha JK, Reddy PM (2003) Nitrogen fixation in rice systems: state of knowledge and future prospects. Plant Soil 252:151–167. https://doi.org/10.1023/A:1024175307238

Ladha JK, Barraquio WL, Watanabe I (1983) Isolation and identification of nitrogen-fixing *Enterobacter cloacae* and *Klebsiella planticola* associated with rice plants. Can J Microbiol 29:1301–1308. https://doi.org/10.1139/m83-203

Ladha J, Tirol-Padre A, Reddy CK, Cassman KG, Verma S, Powlson DS et al (2016) Global nitrogen budgets in cereals: a 50-year assessment for maize, rice and wheat production systems. Sci Rep 6:19355. https://doi.org/10.1038/srep19355

Lima E, Boddey RM, Dobereiner J (1987) Quantification of biological nitrogen fixation associated with sugar cane using a ^{15}N aided nitrogen balance. Soil Biol Biochem 19:165–170. https://doi.org/10.1016/0038-0717(87)90077-0

Lin W, Okon Y, Hardy RW (1983) Enhanced mineral uptake by *Zea mays* and *Sorghum bicolor* roots inoculated with *Azospirillum brasilense*. Appl Environ Microbiol 45:1775–1779. https://doi.org/10.1128/aem.45.6.1775-1779.1983

Ma J, Bei Q, Wang X, Lan P, Liu G, Lin X, Liu Q, Lin Z, Liu B, Zhang Y, Jin H, Hu T, Zhu J, Xie Z (2019a) Impacts of Mo application on biological nitrogen fixation and diazotrophic communities in a flooded rice-soil system. Sci Total Environ 649:686–694. https://doi.org/10.1016/j.scitotenv.2018.08.318

Ma J, Bei Q, Wang X, Liu G, Cadisch G, Lin X, Zhu J, Sun X, Xie Z (2019b) Paddy system with a hybrid rice enhances cyanobacteria *Nostoc* and increases N_2 fixation. Pedosphere 29:374–387. https://doi.org/10.1016/S1002-0160(19)60809-X

Malik KA, Bilal R, Azam F, Sajjad MI (1988) Quantification of N_2-fixation and survival of inoculated diazotrophs associated with roots of Kallar grass. Plant Soil 108:43–51. https://doi.org/10.1007/BF02370098

Mano H, Tanaka F, Nakamura C, Kaga H, Morisaki H (2007) Culturable endophytic bacterial flora of the maturing leaves and roots of rice plants (*Oryza sativa*) cultivated in a paddy field. Microbes Environ 22:175–185. https://doi.org/10.1264/jsme2.22.175

Mantelin S, Desbrosses G, Larcher M, Tranbarger TJ, Cleyet-Marel J-C, Touraine B (2006) Nitrate-dependent control of root architecture and N nutrition are altered by a plant growth-promoting *Phyllobacterium* sp. Planta 223:591–603. https://doi.org/10.1007/s00425-005-0106-y

Mårtensson L, Díez B, Wartiainen I, Zheng W, El-Shehawy R, Rasmussen U (2009) Diazotrophic diversity, *nifH* gene expression and nitrogenase activity in a rice paddy field in Fujian, China. Plant Soil 325:207–218. https://doi.org/10.1007/s11104-009-9970-8

Masuda Y, Itoh H, Shiratori Y, Isobe K, Otsuka S, Senoo K (2017) Predominant but previously-overlooked prokaryotic drivers of reductive nitrogen transformation in paddy soils, revealed by metatranscriptomics. Microbes Environ 32:180–183. https://doi.org/10.1264/jsme2.ME16179

Masuda M, Shiratori Y, Ohba H, Ishida T, Takano R, Satoh S, Shen W, Gao N, Itoh H, Senoo K (2021) Enhancement of the nitrogen-fixing activity of paddy soils owing to iron application. Soil Sci Plant Nutr 67:243–247. https://doi.org/10.1080/00380768.2021.1888629

Matsuguchi T (1979) Factors affecting heterotrophic nitrogen fixation in submerged rice soils. In: Nitrogen and rice. International Rice Research Institute, Los Banos, pp 207–222

Matsuguchi T, Tangcham B, Patiyuth S (1975) Free-living nitrogen fixers and acetylene reduction in tropical rice field. J Agric Res Quart 8:253–256

McRose DL, Zhang X, Kraepiel AML, Morel FMM (2017) Diversity and activity of alternative nitrogenases in sequenced genomes and coastal environments. Front Microbiol 8:267. https://doi.org/10.3389/fmicb.2017.00267

Meuwly P, Pilet PE (1991) Local treatment with indole-3-acetic acid induces differential growth responses in *Zea mays* L. roots. Planta 185:58–64. https://doi.org/10.1007/BF00194515

Minamisawa K, Nishioka K, Miyaki T, Ye B, Miyamoto T, You M, Saito A, Saito M, Barraquio BL, Teaumroong N, Sein T (2004) Anaerobic nitrogen-fixing consortia consisting of clostridia isolated from gramineous plants. Appl Environ Microbiol 70:3096–3102. https://doi.org/10.1128/AEM.70.5.3096-3102.2004

Minamisawa K, Imaizumi-Anraku H, Bao Z, Shinoda R, Okubo T, Ikeda S (2016) Are symbiotic methanotrophs key microbes for N acquisition in paddy rice root? Microbes Environ 31:4–10. https://doi.org/10.1264/jsme2.ME15180

Momose A, Ohtake N, Sueyoshi K, Sato T, Nakanishi Y, Akao S, Ohyama T (2009) Nitrogen fixation and translocation in young sugarcane (*Saccharum officinarum* L.) plants associated with endophytic nitrogen-fixing bacteria. Microbes Environ 24:224–230. https://doi.org/10.1264/jsme2.ME09105

Morishima H, Hinata K, Oka HI (1962) Comparison between two cultivated rice species, Oryza sativa L. and *O. glaberrima* Steud. Japan J Breed 12:153–165. https://doi.org/10.1270/jsbbs1951.12.153

Moubayidin L, Di Mambro R, Sabatini S (2009) Cytokinin-auxin crosstalk. Trends Plant Sci 14: 557–562. https://doi.org/10.1016/j.tplants.2009.06.010

Muthukumarasamy R, Revathi G, Lakshminarasimhan C (1999) Diazotrophic associations in sugarcane cultivation in South India. Trop Agric 76:171–178

Muthukumarasamy R, Revathi G, Loganathan P (2002) Effect of inorganic N on the population, in vitro colonization and morphology of *Acetobacter diazotrophicus* (syn. *Gluconacetobacter diazotrophicus*). Plant Soil 243:91–102. https://doi.org/10.1023/A:1019963928947

Muthukumarasamy R, Cleenwerck I, Revathi G, Vadivelu M, Janssens D, Hoste B, Gum KU, Park KD, Cho Son CY, Sa T, Caballero-Mellado J (2005) Natural association of *Gluconacetobacter diazotrophicus* and diazotrophic Acetobacter peroxydans with wetland rice. Syst Appl Microbiol 28:277–286. https://doi.org/10.1016/j.syapm.2005.01.006

Naher K, Miwa H, Okazaki S, Yasuda M (2018) Effects of different sources of nitrogen on endophytic colonization of rice plants by *Azospirillum* sp. B510. Microbes Environ 33(3): 301–308. https://doi.org/10.1264/jsme2.ME17186

O'Gara F, Shanmugam KT (1976) Regulation of nitrogen fixation by *Rhizobia* export of fixed N_2 as NH_4^+. Biochim Biophys Acta 437:313–321. https://doi.org/10.1016/0304-4165(76)90001-5

Oda K, Miwa E, Iwamoto A (1987) Compact data base for soil analysis data in Japan. Jpn J Soil Sci Plant Nutr 58:112–131. (In Japanese). https://doi.org/10.20710/dojo.58.1_112

Okamoto T, Shinjo R, Nishihara A, Uesaka K, Tanaka A, Sugiura D, Kondo M (2021) Genotypic variation of endophytic nitrogen-fixing activity and bacterial flora in rice stem based on sugar content. Front Plant Sci 12:19259. https://doi.org/10.3389/fpls.2021.719259

Okamura M, Hosoi J, Nagata K, Koba K, Sugiura D, Arai-Sanoh Y, Kobayashi N, Kondo M (2022) Cross-locational experiments to reveal yield potential and yield-determining factors of the rice cultivar 'Hokuriku 193' and climatic factors to achieve high brown rice yield over 1.2 kg m^{-2} at Nagano in central inland of Japan. Plant Prod Sci 25:131. https://doi.org/10.1080/1343943X.2021.1981140

Okubo T, Tokida T, Ikeda S, Bao Z, Tago K, Hayatsu M, Nakamura H, Sakai H, Usui Y, Hayashi K, Hasegawa T, Minamisawa K (2014) Effects of elevated carbon dioxide, elevated temperature, and rice growth stage on the community structure of rice root- associated bacteria. Microbes Environ 29:184–190. https://doi.org/10.1264/jsme2.ME14011

Okuda A, Kobayashi M (1961) Production of slime substance in mixed cultures of *Rhodopseudomonas capsulatus* and *Azotobacter vinelandii*. Nature 192:1207–1208. https://doi.org/10.1038/1921207a0

Oyediran G, Adachi K, Senboku T (1996) Effect of application of rice straw and cellulose on methane emission and biological nitrogen fixation in a subtropical paddy field. Soil Sci Plant Nutr 42:701–711. https://doi.org/10.1080/00380768.1996.10416617

Palus JA, Borneman J, Ludden PW, Triplett EW (1996) A diazotrophic bacterial endophyte isolated from stems of *Zea mays* L. and *Zea luxurians* Iltis and Doebley. Plant Soil 186:135–142. https://doi.org/10.1007/BF00035067

Peoples MB, Herridge DF, Ladha JK (1995) Biological nitrogen fixation: an efficient source of nitrogen for sustainable agricultural production? Plant Soil 174:3–28. https://doi.org/10.1007/BF00032239

Pham VTK, Rediers H, Ghequire MGK, Nguyen HH, De Mot R, Vanderleyden J, Spaepen S (2017) The plant growth-promoting effect of the nitrogen-fixing endophyte *Pseudomonas stutzeri* A15. Arch Microbiol 199:513–517. https://doi.org/10.1007/s00203-016-1332-3

Pilet P-E, Saugy M (1987) Effect on root growth of endogenous and applied IAA and ABA: a critical reexamination. Plant Physiol 83:33–38. https://doi.org/10.1104/pp.83.1.33

Piromyou P, Greetatorn T, Teamtisong K, Okubo T, Shinoda R, Nuntakij A, Tittabutr P, Boonkerd N, Minamisawa K, Teaumroong N (2015) Preferential association of endophytic *Bradyrhizobia* with different rice cultivars and its implications for rice endophyte Evolution. Appl Environ Microbiol 81:3049–3061. https://doi.org/10.1128/AEM.04253-14

Piromyou P, Greetatorn T, Teamtisong K, Tittabutr P, Boonkerd N, Teaumroong N (2017) Potential of rice stubble as a reservoir of Bradyrhizobial inoculum in rice-legume crop rotation. Appl Environ Microbiol 83:e01488–e01417. https://doi.org/10.1128/AEM.01488-17

Quesada A, Nieva M, Leganes F, Ucha A, Martin M, Prosperi C, Fernandez-Valiente E (1998) Acclimation of cyanobacterial communities in rice fields and response of nitrogenase activity to light regime. Microb Ecol 35:147–155. https://doi.org/10.1007/s002489900069

Rao VR (1976) Nitrogen fixation as influenced by moisture content, ammonium sulphate and organic sources in a paddy soil. Soil Biol Biochem 8:445–448. https://doi.org/10.1016/0038-0717(76)90048-1

Rao VR, Rao JLN, Adhya TK (1986) Heterotrophic nitrogen fixation (C2H2 reduction) as influenced by phosphorus application in paddy soils. Plant Soil 92:125–132. https://doi.org/10.1007/BF02372273

Reis Junior FB, Reis VM, Urquiaga S, Döbereiner J (2000) Influence of nitrogen fertilisation on the population of diazotrophic bacteria *Herbaspirillum* spp. and *Acetobacter diazotrophicus* in sugar cane (*Saccharum* spp.). Plant Soil 219:153–159. https://doi.org/10.1023/A:1004732500983

Rockström J, Steffen W, Noone K, Persson A, Chapin FS, Lambin EF, Lenton TM, Scheffer M, Folke C, Schellnhuber HJ, Nykvist B, de Wit CA, Hughes T, van der Leeuw S, Rodhe H, Sörlin S, Snyder PK, Costanza R, Svedin U, Falkenmark M, Karlberg L, Corell RW, Fabry VJ, Hansen J, Walker B, Liverman D, Richardson K, Crutzen P, Foley JA (2009) A safe operating space for humanity. Nature 461:472–475. https://doi.org/10.1038/461472a

Roger P, Ladha JK (1992) Biological nitrogen fixation in wetland rice fields: estimation and contribution to nitrogen balance. Plant Soil 141:41–55. https://doi.org/10.1007/BF00011309

Roger PA, Watanabe I (1986) Technologies for utilizing biological nitrogen fixation in wetland rice: potentials, current usage, and limiting factors. Fertil Res 9:39–77

Ryu MH, Zhang J, Toth T, Khokhani D, Geddes BA, Mus F, Garcia-Costas A, Peters JW, Poole PS, Ané JM, Voigt CA (2020) Control of nitrogen fixation in bacteria that associate with cereals. Nat Microbiol 5:314–330. https://doi.org/10.1038/s41564-019-0631-2

Santiago-Ventura T, Bravo M, Daez C, Ventura V, Watanabe I, App AA (1986) Effects of N-fertilizers, straw, and dry fallow on the nitrogen balance of a flooded soil planted with rice. Plant Soil 93:405–411. https://doi.org/10.1007/BF02374291

Saubidet MI, Fatta N, Barneix AJ (2002) The effect of inoculation with *Azospirillum brasilense* on growth and nitrogen utilization by wheat plants. Plant Soil 245:215–222. https://doi.org/10.1023/A:1020469603941

Sessitsch A, Hardoim P, Döring J, Weilharter A, Krause A, Woyke T, Mitter B et al (2012) Functional characteristics of an endophyte community colonizing rice roots as revealed by metagenomic analysis. Mol Plant-Microbe Interact 25:28–36. https://doi.org/10.1094/MPMI-08-11-0204

Sevilla M, Burris RH, Gunapala N, Kennedy C (2001) Comparison of benefit to sugarcane plant growth and $^{15}N_2$ incorporation following inoculation of sterile plants with *Acetobacter diazotrophicus* Wild-Type and Nif− Mutant Strains. Mol Plant-Microbe Interact 14:358–366. https://doi.org/10.1094/MPMI.2001.14.3.358

Shinjo R, Tanaka A, Sugiura D, Suzuki T, Uesaka K, Takebayashi Y, Kojima M, Sakakibara H, Takemoto D, Kondo M (2020) Comprehensive analysis of the mechanisms underlying enhanced growth and root N acquisition in rice by the endophytic diazotroph, *Burkholderia vietnamiensis* RS1. Plant Soil 450:537–555. https://doi.org/10.1007/s11104-020-04506-3

Shinoda R, Bao Z, Minamisawa K (2019) CH_4 oxidation-dependent $^{15}N_2$ fixation in rice roots in a low-nitrogen paddy field and in *Methylosinus* sp. strain 3S-1 isolated from the roots. Soil Biol Biochem 132:40–46. https://doi.org/10.1016/j.soilbio.2019.01.021

Shrestha RK, Ladha JK (1996) Genotypic variation in promotion of rice dinitrogen fixation as determined by nitrogen-15 dilution. Soil Sci Soc Am J 60:1815–1821. https://doi.org/10.2136/sssaj1996.03615995006000060029x

Singh RN (1950) Reclamation of "Usar" lands in India through blue-green algae. Nature 165:325–326. https://doi.org/10.1038/165325b0

Song T, Mårtensson L, Eriksson T, Zheng W, Rasmussen U (2005) Biodiversity and seasonal variation of the cyanobacterial assemblage in a rice paddy field in Fujian, China. FEMS Microbiol Ecol 54:131–140. https://doi.org/10.1016/j.femsec.2005.03.008

Spaepen S, Dobbelaere S, Croonenborghs A, Vanderleyden J (2008) Effects of *Azospirillum brasilense* indole-3-acetic acid production on inoculated wheat plants. Plant Soil 312:15–23. https://doi.org/10.1007/s11104-008-9560-1

Stepanova AN, Alonso JM (2009) Ethylene signaling and response: where different regulatory modules meet. Curr Opin Plant Biol 12:548–555. https://doi.org/10.1016/j.pbi.2009.07.009

Sullivan BW, Smith WK, Townsend AR, Nasto MK, Reed SC, Chazdon RL, Cleveland CC (2014) Spatially robust estimates of biological nitrogen (N) fixation imply substantial human alteration of the tropical N cycle. Proc Natl Acad Sci 111:8101–8106. https://doi.org/10.1073/pnas.1320646111

Tan XY, Hurek T, Reinhold-Hurek B (2003) Effect of N-fertilization, plant genotype and environmental conditions on nifH gene pools in roots of rice. Environ Microbiol 5:1009–1015. https://doi.org/10.1046/j.1462-2920.2003.00491.x

Tapia-Hernandez A, Bustillos-Cristales MR, Jimenez-Salgado T, Caballero-Mellado J, Fuentes-Ramirez LE (2000) Natural endophytic occurrence of *Acetobacter diazotrophicus* in pineapple plants. Microb Ecol 39:49–55. https://doi.org/10.1007/s002489900190

Terakado-Tonooka J, Ohwaki Y, Yamakawa H, Tanaka F, Yoneyama T, Fujihara S (2008) Expressed nifH genes of endophytic bacteria detected in field-grown sweet potatoes (*Ipomoea batatas* L.). Microbes Environ 23:89–93. https://doi.org/10.1264/jsme2.23.89

Thaweenut N, Hachisuka Y, Ando S, Yanagisawa S, Yoneyama T (2011) Two seasons' study on nifH gene expression and nitrogen fixation by diazotrophic endophytes in sugarcane (Saccharum spp. hybrids): expression of *nifH* genes similar to those of rhizobia. Plant Soil 338:435–449. https://doi.org/10.1007/s11104-010-0557-1

Toriyama K (2002) Estimation of fertilizer nitrogen requirement for average rice yield in Japanese paddy fields. Soil Sci Plant Nutr 48:293–300. https://doi.org/10.1080/00380768.2002.10409204

Tsukanova K, Meyer J, Bibikova T (2017) Effect of plant growth-promoting Rhizobacteria on plant hormone homeostasis. S Afr J Bot 113:91–102. https://doi.org/10.1016/j.sajb.2017.07.007

Vacheron J, Desbrosses G, Bouffaud ML, Touraine B, Moënne-Loccoz Y, Muller D, Legendre L, Wisniewski-Dyé F, Prigent-Combaret C (2013) Plant growth-promoting rhizobacteria and root system functioning. Front Plant Sci 4:356. https://doi.org/10.3389/fpls.2013.00356

Vaishampayan A, Sinha RP, Ha der DP, Dey T, Gupta AK, Bhan U, Rao AL (2001) Cyanobacterial biofertilizers in rice agriculture. Bot Rev 67:453–516. https://doi.org/10.1007/BF02857893

Wada H, Panichsakpatana S, Kimura M, Takai Y (1978) Nitrogen fixation in paddy soils I. Factors affecting N_2 fixation. Soil Sci Plant Nutr 24(3):357–365. https://doi.org/10.1080/00380768.1978.10433115

Wang X, Liu B, Ma J, Zhang Y, Hu T, Zhang H, Feng Y, Pan H, Xu Z, Liu G, Lin X, Zhu J, Bei Q, Xie Z (2019) Soil aluminum oxides determine biological nitrogen fixation and diazotrophic communities across major types of paddy soils in China. Soil Biol Biochem 131:81–89. https://doi.org/10.1016/j.soilbio.2018.12.028

Wang X, Bei Q, Yang W, Zhang H, Hao J, Qian L, Feng Y, Xie Z (2020) Unveiling of active diazotrophs in a flooded rice soil by combination of NanoSIMS and $^{15}N_2$-DNA-stable isotope probing. Biol Fertil Soils 56:1189–1199. https://doi.org/10.1007/s00374-020-01497-2

Watanabe A, Nishigaki S, Konishi C (1951) Effect of nitrogen-fixing blue-green algae on the growth of rice plants. Nature 168:748–749. https://doi.org/10.1038/168748b0

Watanabe I, Barraquio WL, Guzman MR, Cabrera DA (1979) Nitrogen-fixing (acetylene reduction) activity and population of aerobic heterotrophic nitrogen-fixing bacteria associated with wetland rice. Appl Environ Microbiol 37:813–819. https://doi.org/10.1128/aem.37.5.813-819.1979

Watanabe I, So R, Ladha JK, Katayama-Fujimura Y, Kuraishi H (1987) A new nitrogen-fixing species of psedomonad: *Pseudomonas diazotrophicus* sp. nov. isolated from the root of wetland rice. Can J Microbiol 33:670–678. https://doi.org/10.1139/m87-117

Watanabe I, Ventura W, Mascariña G, Eskew DL (1989) Fate of *Azolla* spp. and urea nitrogen applied to wetland rice (*Oryza sativa* L.). Biol Fertil Soils 8:102–110. https://doi.org/10.1007/BF00257752

Wen A, Havens KL, Bloch SE, Shah N, Higgins DA, Davis-Richardson DA, Sharon J, Rezaei F, Mohiti-Asli M et al (2021) Enabling biological nitrogen fixation for cereal crops in fertilized fields. ACS Synth Biol 10:3264–3277. https://doi.org/10.1021/acssynbio.1c00049

Win KT, Okazaki K, Ookawa T, Yokoyama T, Ohwaki Y (2019) Influence of rice-husk biochar and *Bacillus pumilus* strain TUAT-1 on yield, biomass production, and nutrient uptake in two forage rice genotypes. PLoS One 14(7):e0220236. https://doi.org/10.1371/journal.pone.0220236

Wu P, Zhang G, Ladha JK, McCouch S, Huang N (1995) Molecular-marker-facilitated investigation on the ability to stimulate N_2 fixation in the rhizosphere by irrigated rice plants. Theor Appl Genet 91:117–1183. https://doi.org/10.1007/bf00220926

Wu C, Wei X, Hu Z, Liu Y, Hu Y, Qin H, Chen X, Wu J, Ge T, Zhran M, Su Y (2021) Diazotrophic community variation underlies differences in nitrogen fixation potential in paddy soils across a climatic gradient in China. Microb Ecol 81:425–436. https://doi.org/10.1007/s00248-020-01591-w

Yasuda M, Okada T, Nozoe T (2000) Characteristics of nitrogen enrichment by biological nitrogen fixation (BNF) on different management of paddy soils in the Tohoku District of Japan. Jpn J Soil Sci Plant Nutr 71:849–856. (In Japanese). https://doi.org/10.20710/dojo.71.6_849

Yoneyama T, Lee KK, Yoshida T (1977) Decomposition of rice residues in tropical soils. Soil Sci Plant Nutr 23:287–295. https://doi.org/10.1080/00380768.1977.10433048

Yoneyama T, Muraoka T, Kim TH, Dacanay EV, Nakanishi Y (1997) The natural ^{15}N abundance of sugarcane and neighbouring plants in Brazil, the Philippines and Miyako (Japan). Plant Soil 89:239–244. https://doi.org/10.1023/A:1004288008199

Yoneyama T, Terakado J, Masuda T (1998) Natural abundance of ^{15}N in sweet potato, pumpkin, sorghum and castor bean: possible input of N_2-derived nitrogen in sweet potato. Biol Fertil Soils 26:152–154. https://doi.org/10.1007/s003740050359

Yoneyama T, Terakado-Tonooka J, Minamisawa K (2017) Exploration of bacterial N_2-fixation systems in association with soil-grown sugarcane, sweet potato, and paddy rice: a review and synthesis. Soil Sci Plant Nutr 63:578–590. https://doi.org/10.1080/00380768.2017.1407625

Yoo ID, Fujii T, Sano Y, Komagata K, Yoneyama T, Iyama S, Hirota Y (1986) Dinitrogen fixation of rice-*Klebsiella* associations. Crop Sci 26:297–301. https://doi.org/10.2135/cropsci1986.0011183X002600020018x

Yoo ID, Kimura M, Wada H, Takai Y (1991) Organic constituents of rice straw contributed to the biological N_2 fixation in paddy field - model experiment. Jpn J Soil Sci Plant Nutr 62:219–225. https://doi.org/10.20710/dojo.62.3_219

Yoshida T, Ancajas RR (1973) Nitrogen-fixing activity in upland and flooded rice fields. Soil Sci Soc Am J 37:42–46. https://doi.org/10.2136/sssaj1973.03615995003700010017x

Yoshida T, Yoneyama T (1980) Atmospheric dinitrogen fixation in the flooded rice rhizosphere as determined by the N-15 isotope technique. Soil Sci Plant Nutr 26:551–559. https://doi.org/10.1080/00380768.1980.10431242

Yu GH, Kuzyakov Y, Luo Y, Goodman BA, Kappler A, Liu FF, Fu-Sheng Sun FS (2021) Molybdenum bioavailability and asymbiotic nitrogen fixation in soils are raised by iron (oxyhydr)oxide-mediated free radical production. Environ Sci Technol 55:14979–14989. https://doi.org/10.1021/acs.est.1c04240

Zhang J, Hussain S, Zhao F, Zhu L, Cao X, Yu S, Jin Q (2018) Effects of *Azospirillum brasilense* and *Pseudomonas fluorescens* on nitrogen transformation and enzyme activity in the rice rhizosphere. J Soils Sediments 18:1453–1465. https://doi.org/10.1007/s11368-017-1861-7

Part IV
Future and Significance

Chapter 17
Conclusions: The Rhizobial Eminence for Biological Nitrogen Fixation—Revisited and Refined

Piyush Pandey and Shrivardhan Dheeman

Biological nitrogen fixation (BNF) has been one of the most significant processes on Earth. Nitrogen being important for living and non-living forms, finds a gateway to enter in the domains of life processes through BNF. Rhizobia, a group of prokaryotes, has been researched and exploited for improving nitrogen status of agricultural fields, for their unique attribute of BNF in symbiotic association with plants. While there is no deficiency in literature, about the infinitesimal details related to cell-to-cell communications and molecular machinery that work during symbiotic nitrogen fixation process, yet this book has been a timely effort to revisit the concepts with extended perspectives. Up till now, the focus of the book on rhizobia-mediated BNF had been mainly focused over compatibility and efficiency of rhizobia with its respective host plants. However, the book entitled *Nitrogen Fixing Bacteria: Sustainable Growth of Non-legumes* has taken a broader outlook, to further elaborate the discussion on this symbiosis, which is beyond the rhizobia and respective host plant, and include a holistic approach with other possibilities. The book opens with an excellent epilogue, written by experienced researchers, followed by different sections.

The first section, 'Biological Nitrogen Fixation: Trends and Prospects' has four chapters that discuss the different aspects of BNF by prokaryotes, with emphasis on the diversity and processes that are not centred to legumes only. In fact, rhizobia have been demonstrated to function as a plant growth-promoting rhizobacteria (PGPR) when applied to non-leguminous crops. However, their interaction with non-legumes has not received the same attention as with host legumes. The molecular ecology viewpoint breaks the traditional approach to create new dimension for

P. Pandey (✉)
Department of Microbiology, Assam University, Silchar, India

S. Dheeman
Department of Microbiology, School of Allied Health Sciences, MVN University, Palwal, Haryana, India

© The Author(s), under exclusive license to Springer Nature Singapore Pte Ltd. 2022

D. K. Maheshwari et al. (eds.), *Nitrogen Fixing Bacteria: Sustainable Growth of Non-legumes*, Microorganisms for Sustainability 36,
https://doi.org/10.1007/978-981-19-4906-7_17

rhizobial application. This also creates a parallel to other well-known PGPR genera with established attributes of direct and indirect mechanisms for plant growth. In a similar line of discussion, another chapter explores the possibility of biotechnological approaches to improve the nitrogen nutrition in non-legume plants, with innovative suggestions such as applying nitrogenase to plant cells and/or introducing legume symbiosis for non-legumes. This may be challenging but such readings do provide opportunity to the open-end discussions and possibility of research in this area.

While humankind discovered the process of nitrogen fixation at the inception twentieth century, it was never too late to imagine the transfer of this physiological monopoly of prokaryotes to staple foods. Rice being one of the major crops was suggested to be the primary target for such genetic engineering efforts. Although earlier the complexity of molecular mechanisms behind BNF was realized and the task appeared to be unreal, but as the genome-level information for QTLs is available now, the possibility of nitrogen-fixing rice plant has been discussed. Not only rhizobia but also a different group of prokaryotes—Archaebacteria, which are otherwise different from eubacteria, but may fix nitrogen—should be explored for nitrogen enrichment in soil. These organisms are conferred with cellular ultrastructure, which make them survive in challenging microenvironments, and BNF in such conditions is itself an interesting phenomenon. These organisms have been described for their potential to be used as N-fixers in chemical-free agri-practices.

The second section, 'Plant Growth Promotion: Exploring Benefits', has six chapters. The role of rhizobia for their antifungal and nematocidal activities has been described, with special focus on genera such as *Rhizobium*, *Bradyrhizobium*, *Mesorhizobium* and *Ensifer* (*Sinorhizobium*) to non-legume plants. The cereal crops have been the major source of nutrition globally, and to achieve maximum productivity, often faces indiscriminate use of chemicals. Rhizobial nitrogen fixation in cereals has been less visited in research, yet the contributions of other direct and indirect methods including inducing systemic resistance hold promising outcomes for application in agricultural fields. This has been successfully elaborated.

Recently, the environmental genomics has gained popularity due to advances in sequencing strategies, with deeper understandings of microbial community structures and their functions. Soil microbiomes are being characterized to understand the holistic view of microbe–microbe and plant–microbe interactions, including the unculturable microbial populations too. Therefore, this approach is now extensively used to assess the presence of nitrifying, denitrifying, ammonia-oxidizing or nitrate/nitrite-reducing bacteria in the rhizosphere microbiome, as all these processes are detrimental for nitrogen mobilization in soil. This type of information, as I mentioned previously, makes this book to be highly meaningful in revisiting BNF with wider perspectives. In addition, the cash crops, such as coffee and cocoa, are also sought to have benefits of BNF, and hence a couple of strategies are applied, which include intercropping with N-fixing plant–microbe pair in agroforestry ecosystems and/or inoculation of diazotrophic rhizobacteria. It is interesting to learn the opportunities that is available to the biofertilizer industry to fill in the demand.

The third and last section, 'Application to Sustainable Agriculture', includes four chapters that provide opinions and information on the applications and challenges ahead for BNF as a sustainable approach. Associative and entophytic role of nitrogen fixation bacteria in various cereal and non-cereal crops has been discussed with prospects of improving soil fertility and crop production. In addition, the aspects of colonization with root hair conformity change in non-legumes under the influence of rhizobia have been described, with an emphasis on contribution and effects of plants. Such discussion also induces curiosity to a reader with scientific acumen, to understand the crosstalk, through signalling molecules involved in establishing symbiotic association between non-legume plants and N-fixing PGPR, which has been elaborated in another chapter. Finally, the ecophysiological and agronomic aspects of free-living and endophytic N-fixation in non-legume crops are required to be understood for the successful application of diazotrophs. In fact, as explained, rice being a crop that attracts much attention due to its high demand as well as distinctive growth conditions, including submerged anaerobic environment, its ecophysiological aspect has been explained with respect to N-fixation. In addition, the role of carbon sequestration as a driving factor in all N-fixation systems has been explained, as required, for optimal BNF.

The BNF in non-legumes has been an opportunity less visited than anticipated, and hence is an opportunity from agriculture, food security and environmental sustainability. Though environmental sustainability has not been a direct point of discussion here, it is relevant if I raise this issue with respect to BNF and climate change. In fact, according to one of the estimates, the C-storage in terrestrial ecosystems in 2100 will need an increase of 2.3–37.5 Pg of N, but the reactive N supply will only increase by 1.2–6.1 Pg of N (Hungate et al. 2003). Hence, the nitrogen supply would not be sufficient for the estimated amount of biological C sequestration. This clearly indicates the mandatory efforts to be made to achieve the desired levels of nitrogen through BNF. The challenges of using rhizobia with non-legumes are numerous, but the technological advantages have provided the opportunity to meet such challenges. Because of the complexity of nitrogenase synthesis and its sensitivity to atmospheric oxygen, engineering nitrogenase biosynthesis genes in non-legumes seems to be a difficult task (Rubio and Ludden 2008). An alternative to this could be to design non-leguminous hosts that release the nutrient requirements for target N-fixing rhizobia, resulting in a synthetic but effective symbiosis (Mus et al. 2016). Though the concerns of application and efficiency of this will require some experiments and optimization efforts, this seems to be a promising strategy. This book shall be able to induce thought-provoking discussions, and also help to reach some meaningful conclusions about BNF in non-leguminous plants. Given the expertise of authors, the insights presented here will be highly useful to the researchers and academicians, as this looks like an excellent effort to raise the issues at right time and at the right platform.

Acknowledgment Not applicable.

Conflict of Interest Author(s) declares no conflict of interest.

References

Hungate BA, Dukes JS, Shaw MR, Luo YQ, Field CB (2003) Nitrogen and climate change. Science 302:1512–1513

Mus F, Crook MB, Garcia K, Garcia Costas A, Geddes BA, Kouri ED, Paramasivan P, Ryu M-H, Oldroyd GED, Poole PS, Udvardi MK, Voigt CA, Ané J-M, Peters JW (2016) Symbiotic nitrogen fixation and the challenges to its extension to nonlegumes. Appl Environ Microbiol 82: 3698–3710. https://doi.org/10.1128/AEM.01055-16

Rubio LM, Ludden PW (2008) Biosynthesis of the iron-molybdenum cofactor of nitrogenase. Annu Rev Microbiol 62:93–111

Index

© The Author(s), under exclusive license to Springer Nature Singapore Pte Ltd. 2022

D. K. Maheshwari et al. (eds.), *Nitrogen Fixing Bacteria: Sustainable Growth of Non-legumes*, Microorganisms for Sustainability 36,
https://doi.org/10.1007/978-981-19-4906-7

Q

R

The manufacturer's authorised representative in the EU is Springer Nature Customer Service Centre GmbH, Europaplatz 3, 69115 Heidelberg, Germany. If you have any concerns regarding our products, please contact ProductSafety@springernature.com

Printed and bound by CPI Group (UK) Ltd, Croydon, CR0 4YY
30/09/2026
02225315-0011